Graph Theory, Combinatorics, and Applications

Volume 2

PROCEEDINGS OF THE SIXTH QUADRENNIAL INTERNATIONAL CONFERENCE ON THE THEORY AND APPLICATIONS OF GRAPHS

Western Michigan University

Edited by

Y. Alavi
G. Chartrand
O.R. Oellermann
A.J. Schwenk

A Wiley-Interscience Publication

JOHN WILEY & SONS, INC.

New York / Chichester / Brisbane / Toronto / Singapore

Library of Congress Cataloging in Publication Data:

International Conference on the Theory and Applications of Graphs (6th : 1988 : Western Michigan University)

Graph theory, combinatorics, and applications : proceedings of the Sixth Quadrennial International Conference on the Theory and Applications of Graphs / edited by Y. Alavi . . . [et al.].

p. cm.

"A Wiley-Interscience publication."

Includes bibliographical references.

ISBN 0-471-53245-2 (set). —ISBN 0-471-60917-X (v. 1)
—ISBN 0-471-53219 (v. 2)

1. Graph theory—Congresses. 2. Combinatorial analysis—Congresses. I. Alavi, Y. II. Title.

QA166.I55 1988

511′.5—dc20 90-28082

CIP

Printed in the United States of America

10 9 8 7 6 5 4 3 2 1

Contents

Volume 2

* Indicates the speaker at the Conference.

ON M-CONNECTED AND K-NEIGHBOUR-CONNECTED GRAPHS

G. Gunther

Sir Wilfred Grenfell College

B. L. Hartnell

Saint Mary's University

ABSTRACT

A graph is k-neighbour-connected (k-NC) if the removal of fewer than k vertices and all their neighbours does not disconnect the graph nor result in a complete graph. We shall examine briefly a number of problems dealing with k-neighbour-connected graphs and their relationship with the usual notion of connectivity. For instance, in an attempt to examine those graphs which, in some sense, are "between" k-NC and m-connected, we initiate an investigation into graphs which are still k-NC after the removal of any set of m vertices.

1. Introduction

The problems we look at in this paper originated out of a study of dominating sets. There, one wishes to look at properties of sets of vertices that dominate the entire graph. Our concern, however, has been to look for graphs that are in some sense difficult to dominate. With this end in mind, we have concentrated our efforts on studying properties of the set of vertices that remain undominated; that is, we have looked at the survivor set of vertices left remaining in a graph G after a set of closed neighbourhoods has been deleted from G.

The details of some of this work can be found in [5] - [14], as well as in [16]. Clearly, these types of questions are also relevant in the context of network reliability. See for example [1], [2], [3], [4] and [15].

One question we have pursued with some vigour in the last few years is concerned with determining those graphs which remain connected after the deletion of a number of closed neighbourhoods. We define a graph G to be *k-neighbour-connected*, or *k-NC*, if for any $X \subseteq V(G)$ with $|X| \leq k-1$, the graph $G - N[X]$ is a connected graph which is not complete. (**Note**:

$N[X]$ denotes the closed neighbourhood of X, and is defined by $N[X] = \cup N[x]$.) The possibility of a complete graph is excluded as

$$x \in X$$

the deletion of any further closed neighbourhood would completely destroy the graph.

It is not difficult to see that if G is k-NC, then G must be k-connected (in the standard sense of connectivity). See Corollary 1 of [8] for this result. It is also at once apparent that the converse of this is false - one needs only to consider any complete graph K_n, which is very highly connected, but which is not even 1-NC. Indeed, the property of being k-NC is very sensitive with respect to the number of edges present in the graph. As the example of the complete graph above shows, if there are too many edges present in the graph, the neighbour-connectivity actually decreases. It is unfortunately not at all clear what constitutes "too many" edges. The complicated relationship that holds between neighbour-connectivity and the number of edges present is manifested by the sequence of diagrams in Figure 1.

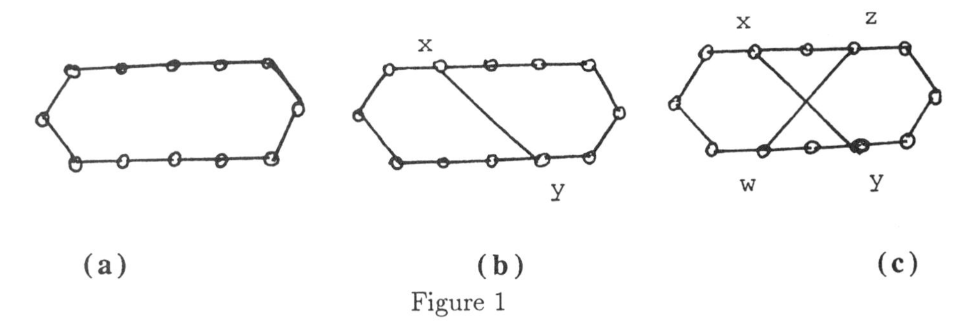

(a) (b) (c)

Figure 1

Diagram (a) shows C_{12}, which is 2-NC. In diagram (b), we have added edge xy to C_{12} - this graph is no longer 2-NC, for deleting $N[x]$ (i.e. x and its neighbours) disconnects the survivor graph. If, however, we add the additional edge zw (illustrated in (c)), then we have again re-established 2-neighbour-connectivity in the resultant graph.

2. Some Results and Open Problems

A number of families of k-NC graphs are described in [5], [6], [8] and[9]. All of these are k-regular in addition to being k-NC. It turns out that one parameter which governs the types of minimal k-regular, k-NC graphs G that are known is the size m of maximum cliques that one can find in G. In particular, in [9] one finds a classification of all minimal k-regular, k-NC graphs which contain k- cliques. These turn out to be of order $k(k+1)$. With these as seed graphs, the following construction yields new examples of k-NC graphs.

Lemma 1 Let G be a k-regular, k-NC graph which contains a k-clique $C = c_1, \cdots, c_k\}$. Form a new graph G' from G by adding an m-clique $D = \{d_1, d_2, \cdots, d_m\}$ where each d_i is joined to every $c_j \in C$ but to no other vertex in G. That is, C has been replaced by a $(k+m)$-clique. Then the new graph G' obtained from G in this way is k-NC.

Proof Suppose that we delete the $k-1$ closed neighbourhoods $N[x_1]$, $N[x_2], \cdots, N[x_{k-1}]$ with all $x_i \in G'$. If x_i is a vertex of C, then deletion of

$N[x_i]$ deletes all the new vertices in D, thus leaving the same survivor set in G' as would have been left in G. If all the vertices x_i belong to $G - C$, one must worry about the possibility that all of C might be conceivably be contained within $\cup N[x_i]$. If this were the case, then deleting all the $N[x_i]$ would disconnect D from the remainder of the survivor graph. However, Lemma 3.3 of [9] assures us that $|N[x] \cap V(C)| \leq 1$ and hence this particular situation cannot occur. A second potential problem might arise in the event where one of the vertices x_i belongs to D. In this case, deleting $N[x_i]$ will delete all of C, but will not affect any other vertex in the original vertex set of G. The only way we were able, in G, to delete all of C was by choosing some $c_j \in V(C)$ and deleting $N[c_j]$. Such a deletion, however, always removed exactly one vertex $z \in G - C$. The difference now is that we can delete C (from $d_i \in V(D)$) without touching vertex z. A possible concern is that there is some choice of the $k-1$ vertices $x_1, \cdots, x_{k-1}$ which results in the deletion of all k vertices in $N(z)$ without touching z itself. But for any $y \notin N[z]$, we have $|N(y) \cap N(z)| \leq 1$, for otherwise we could find some $s, t \in N(y) \cap N(z)$ so that $G[\{z, s, y, t\}]$ would be either a 4-cycle or a 4-cycle with one diagonal, both of which are impossible configurations in a k-regular, k-NC graph (Lemma 3.2 in [9]). This completes the proof.

We describe without proof, a second construction which yields new examples of k-NC graphs.

Lemma 2 Let G be a k-NC graph. Choose a vertex $v \in G$, and adjoin to G a new vertex v' which is joined to exactly all vertices in $N[v]$. Then the new graph G' for which $V(G') = V(G) \cup \{v'\}$ is also k-NC.

These constructions allow us to conclude the following result.

Proposition 1 For all positive integers k and all integers $p \geq k(k+1)$, there exist k-NC graphs on p vertices.

Proof By [9], there exists a k-regular, k-NC graph on $k(k+1)$ vertices which has a k-clique. By Lemma 1, the result follows.

In an attempt to encourage investigation that would lead to a better understanding of the relation that holds between the concepts of k-connectivity and k-neighbour-connectivity, we pose three problems.

First, observe that with traditional k-connectivity, the allowable survivor set of vertices after the deletion of $k-1$ or fewer vertices is a connected graph which is not a single point (considered trivial since removing one more point destroys the graph), whereas with k-neighbour-connectivity, the allowable survivor set is a connected graph which is not a clique (again considered trivial since removing any further neighbourhood completely annihilates the graph).

Problem 1. Call a graph G, a *weak-k-neighbour-connected* graph if the deletion of fewer than k closed neighbourhoods neither disconnects G nor leaves only a single vertex. For each k, determine the minimum number, p, of vertices required to have a weak-k-neighbour- connected graph on p vertices.

We observe that the minimum weak-2-NC graph is the 5-cycle (not the 6-cycle as it was for 2-NC). In general the numbers achieved by k-NC graphs give an upper bound for weak-k-NC graphs. By Proposition 1 this shows k^2+k to be achievable. The construction given in the following proposition improves this bound.

Proposition 2 For all positive integers k there exists a weak-k-NC graph on k^2+1 vertices.

Proof For any positive integer k consider a k-regular graph G on k^2+1 vertices where G consists of k cliques, say $C_1, C_2, \cdots, C_k$, each of size k and one additional point, say x. There is exactly one edge between each pair of cliques C_i and C_j and between each clique and x. As G is k-regular each vertex in a clique C_i has precisely one neighbour not in that clique.

To see that G is a weak-k-NC graph we first observe that since G is k-regular the deletion of fewer than k closed neighbourhoods leaves more than a single point.

Furthermore we note that if u and v are any two vertices which remain after such a deletion they must still be connected. This follows since it is impossible to destroy the link between any two cliques without entirely destroying one of the cliques. This is also true for the edge between x and any of the cliques (x or the clique itself must be removed else the edge remains).

Figure 2 illustrates the preceding construction for $k = 2$ (the 5-cycle) and $k = 3$. In these cases it is straightforward to verify that these are in fact the minimum order graphs for weak-k-NC. We conjecture that the bound $k^2 + 1$ is sharp for all k.

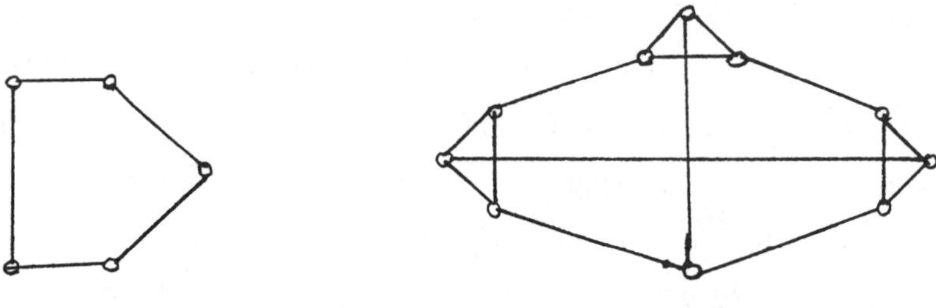

Figure 2

We would like to comment briefly on another subtlety that distinguishes neighbour-connectivity from regular connectivity. In dealing with k-connectivity, one worries about the effects that the deletion of a number of vertices might have. It is clearly immaterial in which order these vertices are deleted. When one deletes entire neighbourhoods, however, this question of order becomes relevant. Indeed, in the definition of k-NC, it implicitly follows that all the closed neighbourhoods are to be deleted *simultaneously*. This means, in particular, that two neighbourhoods $N[x]$ and $N[y]$ may be deleted even when x and y are themselves adjacent to each other.

If we weaken this definition, but in a manner different from that of Problem 1, we might wonder about graphs that remain connected after the successive deletions of a number of neighbourhoods. This forces the deleted vertices to be independent. In particular, let us call a graph G *k-I-neighbour- connected* (k-INC) if for any independent set $X \subseteq V(G)$ with $|X| \leq k-1$, the graph

$G - N[X]$ is connected but not complete.

Problem 2. Determine the minimum number of vertices required in a k-I-neighbour- connected graph. Note that now, of course, we will no longer have the possibility of deleting N[x] and N[y] for adjacent vertices.

Comment Clearly, this distinction between deleting simultaneously or successively is meaningless in the case of 2-INC graphs, since there we only allow the deletion of one closed neighbourhood. The minimum graph here is still C_6, the cycle on six vertices. If we look for 3- INC graphs, i.e., where the neighbourhoods are to be deleted in succession, we quickly realize that our results change. A simple argument shows that the smallest 3-INC graph is the familiar Peterson graph of order 10, which we show in Figure 3.

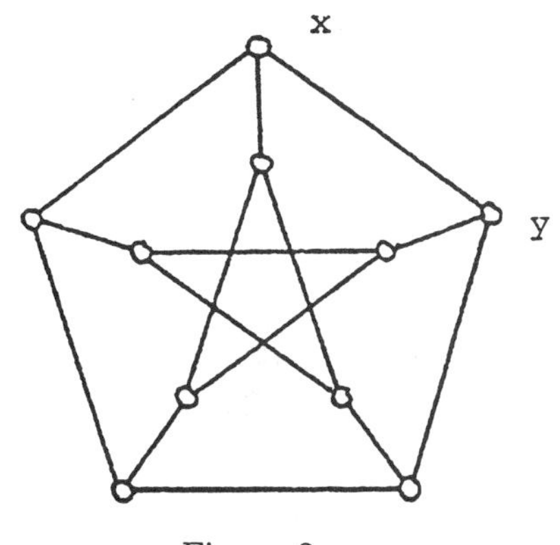

Figure 3

This graph is 1-transitive on its vertices, so we may pick any vertex to begin our deletion. Note that deleting N[x] leaves as a survivor set the 6-cycle, which we know is 2-NC. If, however, we delete N[x] and N[y] simultaneously, we end up with a disconnected survivor set. Thus, this graph provides us with an example of a graph which is not 3-NC (i.e., under simultaneous deletions), but which is 3-INC (i.e., under successive deletions).

Let us try to extend this pattern by searching for a smallest 4-INC graph H which has the property that for any vertex $x \in V(H)$, the survivor graph

H-N[x] is isomorphic to the Petersen graph. That is, we wish to adjoin a new vertex z with $t+3$ neighbours each of which is joined to $t+2$ vertices in the Petersen graph. Each vertex in the Petersen graph has t new edges to the neighbours of z. Hence

$$10t = (t+3)(t+2)$$

$$10t = t^2 + 5t + 6$$

$$t^2 - 5t + 6 = 0$$

$$\therefore t = 2 \text{ or } t = 3$$

For the smallest case we consider $t = 2$.

How should the new edges connect $N(z)$ to the Petersen graph? We illustrate the situation in Figure 4, numbering the vertices from 1 to 16 for convenience.

We ask first whether vertex 2 may be joined to two neighbouring vertices x, y in the Petersen graph?

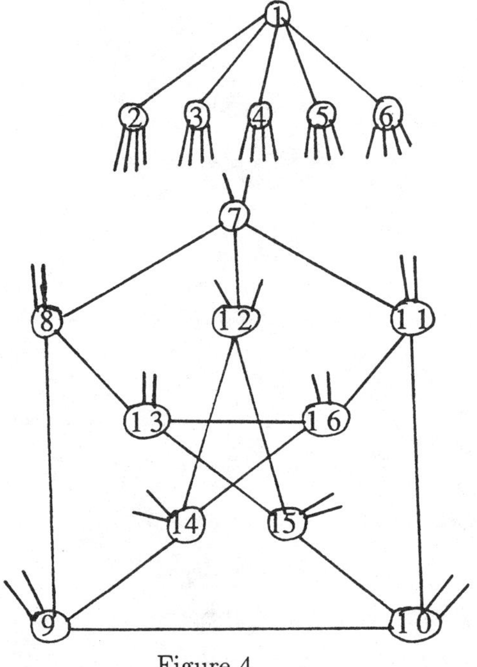

Figure 4

Assume that this happens. Then a simple count shows that there must exist at least two vertices u, v in the Petersen graph, neither of which is connected to 2 or x or y. But then if we delete N[u] and look at the survivor graph H-N[u], we note that this graph would contain the 3-cycle $2 - x - y$, which cannot happen as the Petersen graph does not contain any 3-cycles.

Consequently, the four edges out of vertex 2 must go to four vertices in the Petersen graph which are pairwise not connected. But there exist precisely five such sets of four vertices in the Petersen graph which are pairwise not connected. These are, using the numbering of Figure 4, the five quadruples 7-10-13-14, 7-9-15-16, 9-11-12-13, 8-11-14-15 and 8-10-12-16. Since the numbering of vertices 2,3,4,5,6 is arbitrary, we may without any loss of generality connect these five vertices to the five quadruples listed, so that 2 is connected to each of 7,10,13,14 and so on. It is easy to check that the resultant graph H on 16 vertices is 5-regular and has the property that the deletion of any closed neighbourhood leaves a Petersen graph as the survivor graph. Consequently, H is the smallest 4-INC graph.

We show a symmetric representation of H in Figure 5.

Finally, we observe that, in general, k-NC graphs, which we know exist for $k(k+1)$ vertices, give us a bound for k-INC graphs.

We conclude by describing a family of graphs in which the deletion of m vertices and k closed neighbourhoods still leaves an allowable survivor set. Again, there are several possibilities for "allowable". Here we will demand that the survivor graph be connected and not complete.

Lemma 3 Let G be k-NC. Let s, t be two natural numbers with $s + t < k$. If we delete s closed neighbourhoods and t additional vertices in G, then the survivor graph remains connected and is not a complete graph.

Proof Deleting s neighbourhoods results in a graph which is still (k - s)-NC, and hence (k - s)-connected. But, $k - s > t$, and thus we can still delete t vertices without jeopardizing connectivity. Also, since the survivor set we would leave after the deletion of $s+1$ *neighbourhoods* is not a clique, it follows that the survivor set now cannot be a clique either.

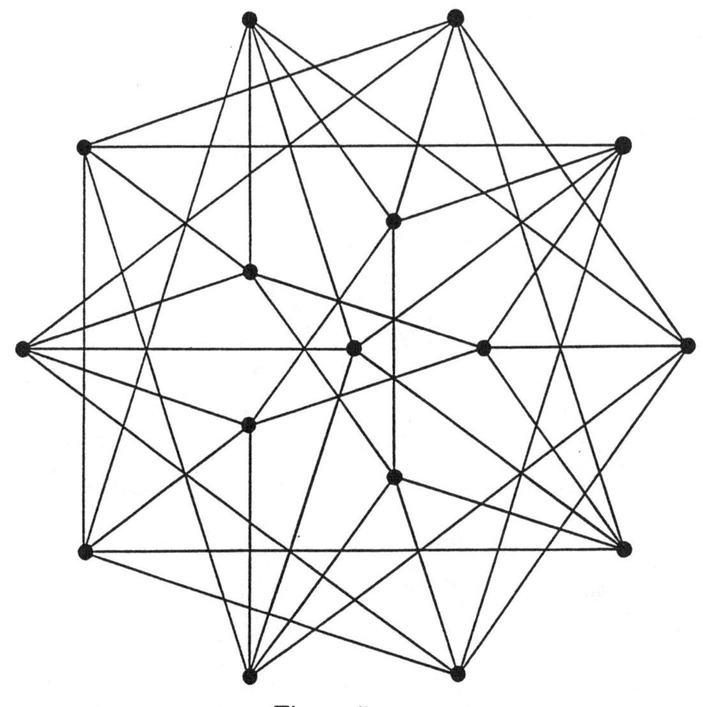

Figure 5

It turns out that we can find smaller graphs which remain connected in the allowable way after the deletion of a mixture of neighbourhoods and vertices. For example, the smallest graph which remains connected (but is not complete) after the deletion of one neighbourhood and one further vertex, is the graph on nine vertices shown in Figure 6.

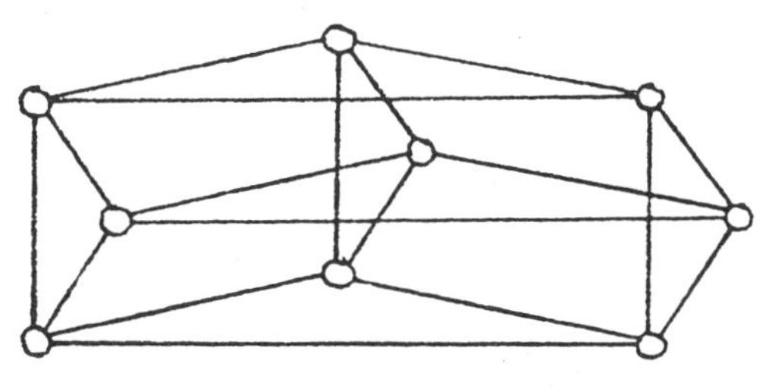

Figure 6

We recognize in this graph the product $K_3 \times K_3$.

This example generalizes nicely, as we note in the following lemma.

Lemma 4 Let $G = K_k \times K_m$. Then the following hold for G.

a) G is a $(k + m - 2)$-regular graph of order km

b) $G$is $(k + m + 2)$-connected

c) G is p-NC, where $p = \min\{m - 1, k - 1\}$

d) If in G we delete s neighbourhoods, where $s < p$, then the survivor graph is still at least $(k + m - 2 - 2s)$-connected.

Proof Part (a) is immediate and part (b) follows from Menger's Theorem. Part (c) is a consequence of the fact that if $x \in G$, then $G - N[x] = K_{k-1} \times K_{m-1}$. This result, in combination with (b), is sufficient to force (d) as well. We conclude with,

Problem 3 Determine the minimum number, p, of vertices such that there exists a graph on p vertices such that the deletion of no more than k closed neighbourhoods and m additional vertices leaves a connected graph which is not a clique.

REFERENCES

[1] K.S. Bagga, L. W. Beineke, M. J. Lipman and R. E. Pippert, "On the edge-integrity of graphs," *Congressus Numerantium* 60 (1987), 141-144.

[2] F. T. Boesch and A. P. Felzer, "A general class of invulnerable graphs," *Networks* 2 (1972) 261-283.

[3] F. T. Boesch and C. L. Suffel, "An overview of discrete methods in the synthesis of reliable networks," *Large Scale Systems* 7 (1984), no. 2-3, 191-196.

[4] S. Goodman and D. Shier, "On designing a reliable hierarchical structure," *SIAM J. Appl Math.* 32(2) (1977) 418-430.

[5] G. Gunther, "Minimum k-neighbour-connected graphs and affine flags," *Ars Combinatoria* 26(1988) 83-90.

[6] G. Gunther and B. L. Hartnell, "Flags and neighbour-connectivity," *Ars Combinatoria* 24 (1987) 31-38.

[7] G. Gunther, "On the existence of neighbour-connected graphs", *Congressus Numerantium* 54 (1986) 105-110.

[8] G. Gunther, B. L. Hartnell and R. Nowakowski, "Neighbour-connected graphs and projective planes," *Networks* 17 (1987) 241-247.

[9] G. Gunther and B. L. Hartnell, "Neighbour-connectivity in regular graphs," *Disc. Appl. Math.* 77 (1985) 233-243.

[10] G. Gunther and B. L. Hartnell, "Optimal k-secure graphs," *Disc. Appl. Math* 2 (1980) 225-231.

[11] G. Gunther and B. L. Hartnell, "On minimizing the effects of betrayals in a resistance movement," *Proc. Eighth Manitoba Conference on Numerical Mathematics and Computing* (1978) 285-306.

[12] B. L. Hartnell, "The Optimum defence against random subversions in a network," *Proc. Tenth Southeastern Conference on Combinatorics, Graph Theory and Computing*, Boca Raton (1979) 293-299.

[13] B. L. Hartnell, "Some problems on minimum dominating sets," *Proc. Eighth Southeastern Conference on Combinatorics, Graph Theory and Computing*, Baton Rouge (1977) 317- 320.

[14] B. L. Hartnell and R. Nowakowski, "Indivisible graphs," *Congressus Numerantium* 33 (1981) 55-62.

[15] H. Heffes and A. Kumar, "Incorporating dependent node damage in deterministic connectivity analysis and synthesis of networks," *Networks* 16 (1986) 51-65.

[16] A. Meir and J. Moon, "Survival under random coverings of trees," *Graphs and Combinatorics* 4 (1988) 49-65.

GRAPHS AND THE STRONG LAW OF SMALL NUMBERS

Richard K. Guy

ABSTRACT

Graphs furnish many examples of the fickle nature of The Strong Law of Small Numbers [12,14]: "There aren't enough small numbers to carry out the many tasks demanded of them."

It is the enemy of mathematical discovery. On the one hand, a few early exceptions can hide an underlying pattern. On the other hand, spurious similarities can cause careless conjectures. The Second Strong Law of Small Numbers [15] states that: "If two numbers look equal, they aren't necessarily so."

A common form of puzzle, or intelligence test, is to give a few members of a sequence and to ask for the next few members. A mathematician knows that it is possible to make a rule whereby any given finite sequence can be continued with any other given finite sequence of additional terms, so such exercises are not always fair. However, in particular cases, some continuation rules are very much more plausible than others. As an example, give the next three terms of the sequence

(1), 2, 3, 5, 7, 11, 13, 17, 19, 23, 29, 31, 37, 41, 43, 47, 53, ...

But in this paper we give you the rules. An earlier draft had the format of [14]: a long list of examples, followed by a long list of answers. Here, at the referee's suggestion, we give six sets of examples of the appearance of six different sequences in six different sections. The reader is invited to decide, within each set, whether the coincidences are genuine or not. Are there, in fact, many more sequences? At the end of each section we give the answers, insofar as we know them.

1. Introduction

The first sequence

$$1, \quad 1, \quad 2, \quad 5, \quad 14, \quad \ldots$$

is exemplified by

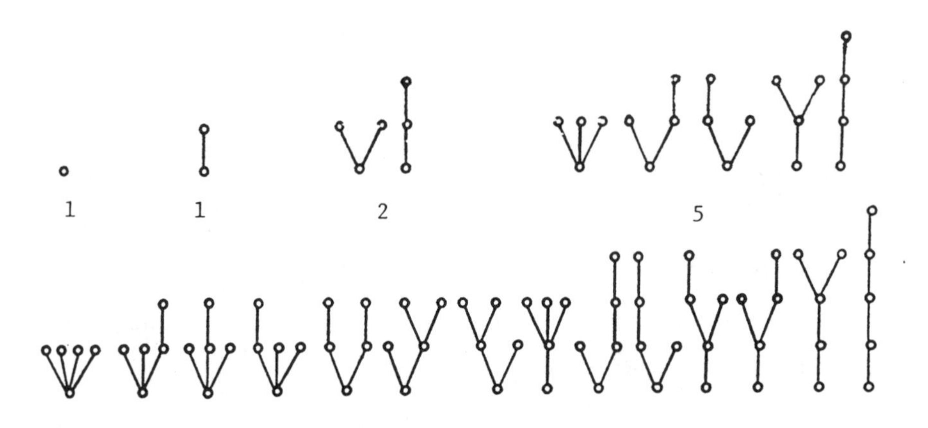

Example 1A The numbers of rooted plane trees with n edges and $n+1$ vertices $(n = 0, 1, 2, \ldots)$, where by ***plane*** we mean to distinguish between an embedding of a tree in the plane and its reflextion in a vertical axis by:

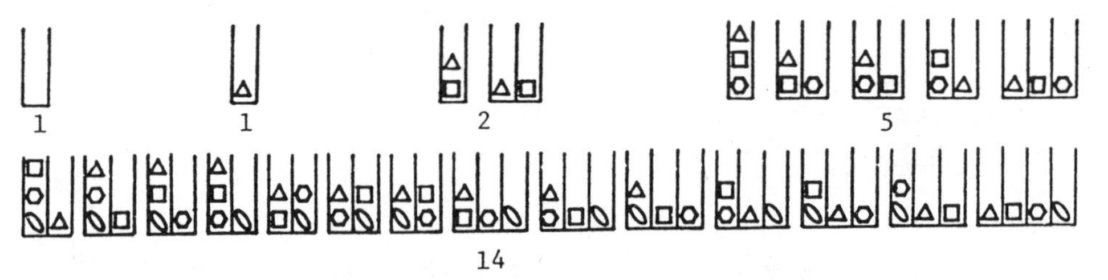

Example 1B The numbers of distributions of n distinguishable objects in indistinguishable boxes, with at most 3 objects in a box by:

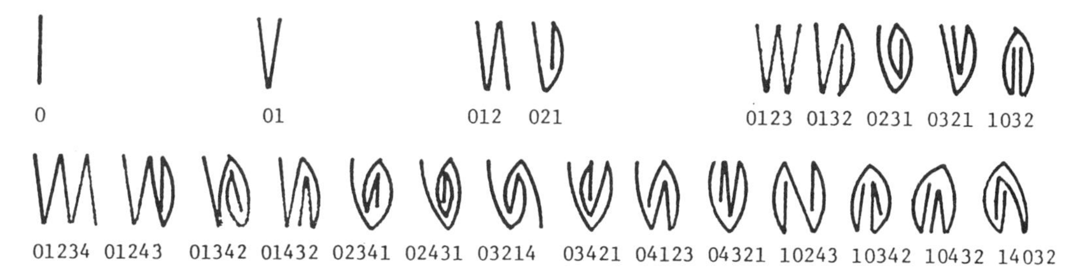

Example 1C The numbers of ways of making n folds in a strip of $n+1$ postage stamps, where we don't distinguish between the back and front, or the top and bottom, or the left and right of a stamp by:

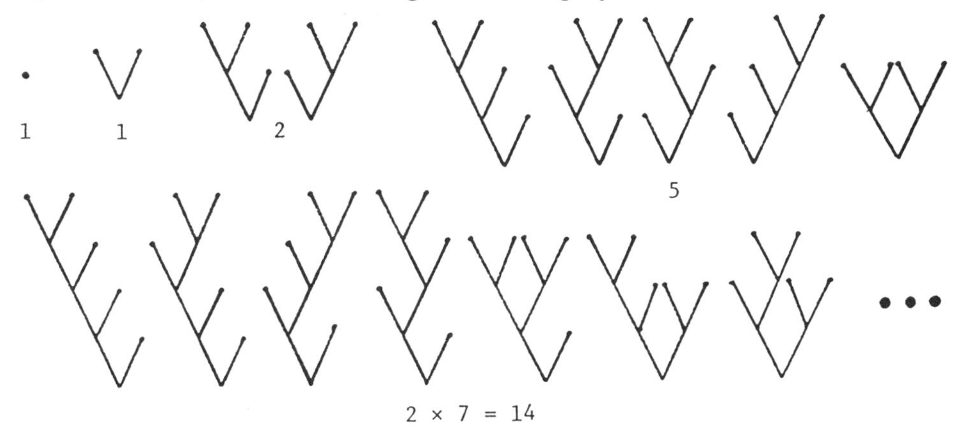

Example 1D The numbers of plane binary trees with n decisions and $n+1$ outcomes; again the word *plane* means that we distinguish between left and right, and the dots at the end of the figure indicate that there are seven more trees, the reflextions of the first seven; notice that when n is even and positive, the number of trees will be even: and, in addition by:

Example 1E The numbers of different groups, up to isomorphism, of order 2^n.

Although the one-one correspondence between them is not quite immediate, the objects in Examples 1A and 1D, the rooted plane trees and the plane binary trees, are both manifestations of the ubiquitous Catalan numbers [2,

13, 37, #577], $\binom{2n}{n}/(n+1), n = 0, 1, 2, ...$

$$1, 1, 2, 5, 14, 42, 132, 429, 1430, 4862, 16796, 58786, 208012, 742900, ...$$

Kutchiniski [23] relates them by putting each in correspondence with certain sequences of zeros and ones, while Conway and Guy [7] do so via the various ways of parenthesizing a repeated exponentiation.

On the other hand, the sequence defined by Example 1B [27, 37 #579]

$$1, 1, 2, 5, 14, 46, 166, 652, 2780, 12644, 61136, 312676, 1680592, 9467680, ...$$

and that in example 1C [22, 37 #576], see also [38],

$$1, 1, 2, 5, 14, 39, 120, 358, 1176, 3527, 11622, 36627, 121622, 389560, ...$$

(the sixth member is given erroneously as 38 in [11]) are not connected with the Catalan numbers, nor with one another. Also, the number of groups of order $2^n, n = 0, 1, ...$ [37 #581]

$$1, 1, 2, 5, 14, 51, 267, ...$$

is again different, and perhaps not known beyond $n = 6$.

The nth term of the sequence in Example 1B is given by $n!\Sigma 1/a.!b!c!2^b6^c$, where the sum is taken over all triples of positive integers satisfying $a+2b+3c = n$.

2. The Second Sequence,

$$1, \ 1, \ 1, \ 2, \ 3, \ 6, \ 11, \ \cdots$$

manifests itself as

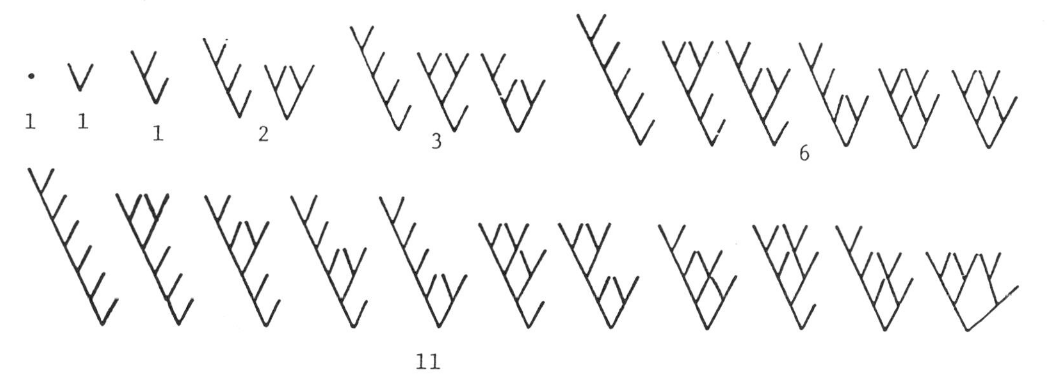

Example 2A The numbers of binary trees with $0, 1, 2, 3, 4, 5, 6$ branchings, where we no longer distinguish between left and right;

Example (and Figure) 2B The number of ways of arranging a knock-out tournament, among $1, 2, 3, 4, 5, 6, 7$ players, by numbers of matches in each round: at first sight, this appears to be isomorphic to 2A (read the numbers of branchings at each level, from the top downwards, the winner being the root of the tree):

$$\phi, \quad 1, \quad 11, \quad 111 \text{ or } 21, \quad 1111 \text{ or } 211 \text{ or } 121, \cdots$$

but with 6 players, i.e. 5 matches, we have 11111 or 2111 or 1211 or 1121 or the first three arrangements of Figure 2B, where the first two are the same binary tree, but different arrangements of matches (1 winner in the first round of the first arrangement gets a bye in the second round, while 2 players get byes in the first round of the second arrangement) and the last two are counted as the same arrangement, 221, although they are different binary trees: similarly for the second and third sets of three arrangements in Figure 2B, involving 7 players; the eleven different arrangements for this number of players being 111111, 21111,12111, 11211, 11121, 2211, 2121, 1221, 321, 1311 and 3111;

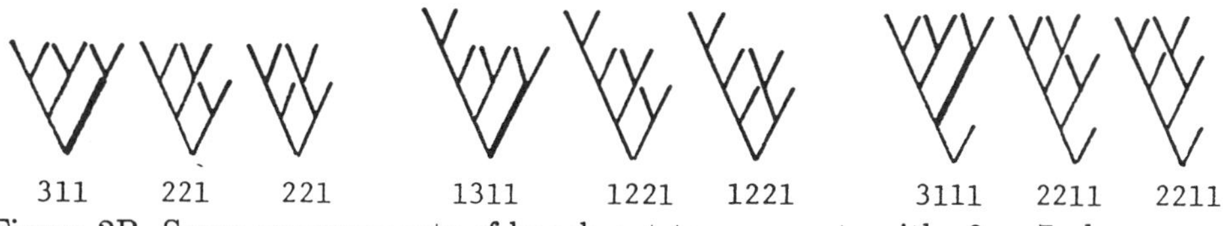

Figure 2B. Some arrangements of knock-out tournaments with 6 or 7 players

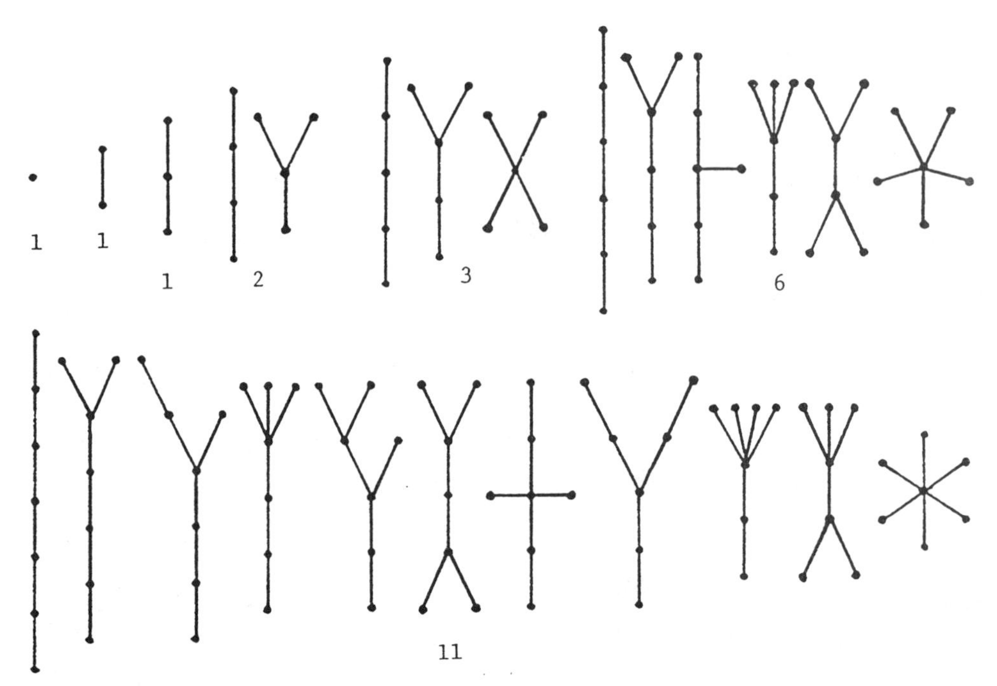

Example 2C The numbers of trees with 1,2,3,4,5,6,7 vertices, where the trees are now neither rooted nor labelled in any way; and as

Example 2D The coefficients of $x^n (0 \le n \le 6)$ in the expansion of the generating function

$$\frac{1}{2}\{\,1+\frac{1-3x^2}{(1-2x)(1-x^2)}\,\},$$

which can also be described recursively, after the initial values, by the rule: alternately double, and double and subtract one.

No two of the sequences in section two are connected. The binary trees in Example 2A are a manifestation of the Wedderburn-Etherinton numbers [9,40, 37 #298, 6, p. 54]. They are

$$1, 1, 1, 2, 3, 6, 11, 23, 46, 98, 207, 451, 983, 2179, 4850, 10905, 24631, 56011, \ldots$$

The arrangements, by numbers of matches in the various rounds of a knock-out tournament, could be called Capell-Narayana numbers [3, 37 #297]:

$$1,\ 1,\ 1,\ 2,\ 3,\ 6,\ 11,\ 22,\ 42,\ 84,\ 165, \ldots$$

and don't appear to have been calculated any further; perhaps because the arrangements are mostly of interest to the organizers of the tournament. What the general public wants to know, in addition, is the order in which the players appear in the arrangement, who is "seeded", who gets a bye, and so on. The numbers of *labelled* knock-out tournaments for 1,2,3,... players is

$$1, 1, 3, 15, 105, 945, 10395, 135135, 2027025, 34459425, 654729075, \ldots$$

These are known as double factorials [37 #1217]: if $m(= n - 1)$ is the number of matches, they may be written as $(2m)!/2^m m!$ Even more interesting, of course, is: who is the winner? The number of ways in which he emerges is

$$1, 2, 12, 120, 1680, 30240, 665280, 17297280, 518918400, 17643225600, \ldots,$$

i.e. $(2m!)/m!$ [37, #808]. These numbers appear as coefficients when representing powers in terms of Hermite polynomials [36]. Is there some inversion formula whereby we can recapture the Capell-Narayana numbers from these more familiar sequences?

The number of trees with n vertices, $n = 1, 2, \cdots$ (Example 2C)

$$1, 1, 2, 3, 6, 11, 23, 47, 106, 235, 551, 1301, 3159, 7741, 19320, 48629, 123867, \ldots$$

was first obtained by Otter [30], see also [37 #299; 33 pp. 135-138; 16, p. 232; 26].

The generating function given as Example 2D was suggested by Jonathan Schaer. It is a red herring, swimming in the narrow straits between the Wedderburn-Etherington and Capell-Narayana numbers. The coefficients of the power series are

$$1, 1, 1, 2, 3, 6, 11, 22, 43, 86, 171, 342, 683, 1366, 2731, 5462, 10923, 21846, \ldots$$

and can be simply described after examining them briefly.

3. The Third Sequence

$$1, \quad 1, \quad 2, \quad 4, \quad 9, \cdots$$

appears in

Example 3A The number of possible score sequences in a round-robin tournament between 1,2,3,4,5 players, there being no ties:

1	0	1
2	10	1
3	210,111	2
4	3210,3111,2220,2211	4
5	43210,43111,42220,42211,33310,33220,33211,32221,22222	9

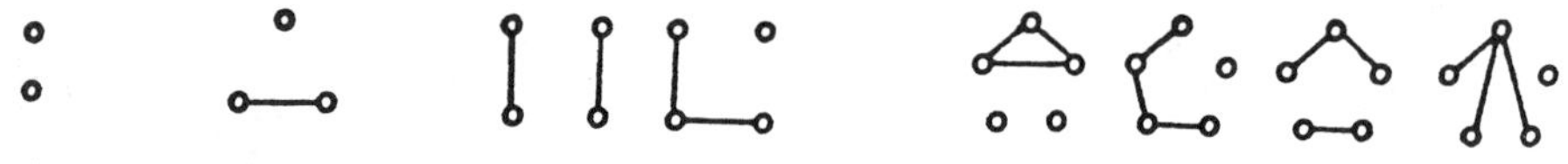

1 1 2 4

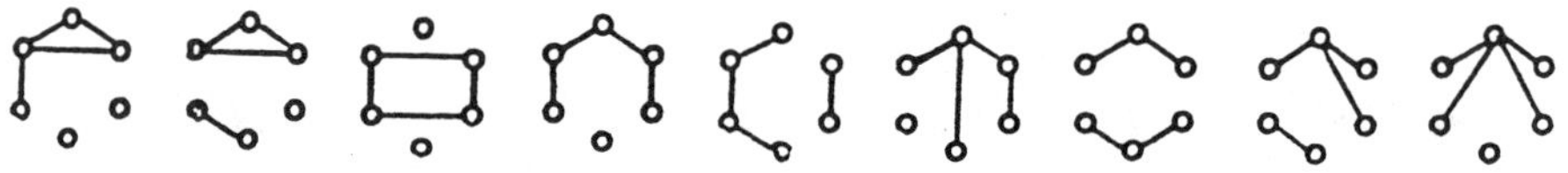

9

in Example 3B, graphs with $n+1$ vertices and $n-1$ edges, $n = 1, 2, 3, 4, 5$:

in Example 3C, the values of k for which $(15 \times 2^k) + 1$ is prime in:

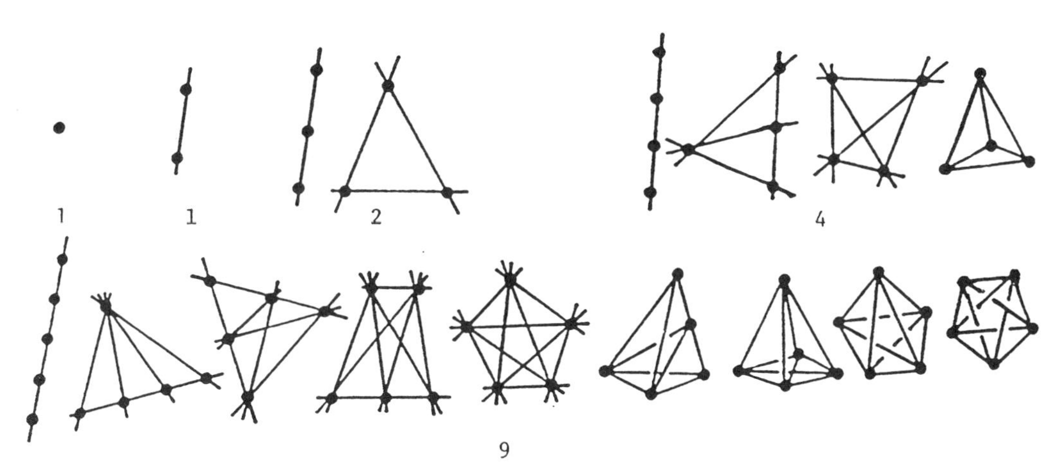

Example 3D Combinatorial geometries with 1,2,3,4,5 points; notice that the last two diagrams with four points differ in that they have dimensions 2 and 3; the nine diagrams with five points are one of dimension 1, four of dimension 2, three of dimension 3, and one of dimension 4: and finally in

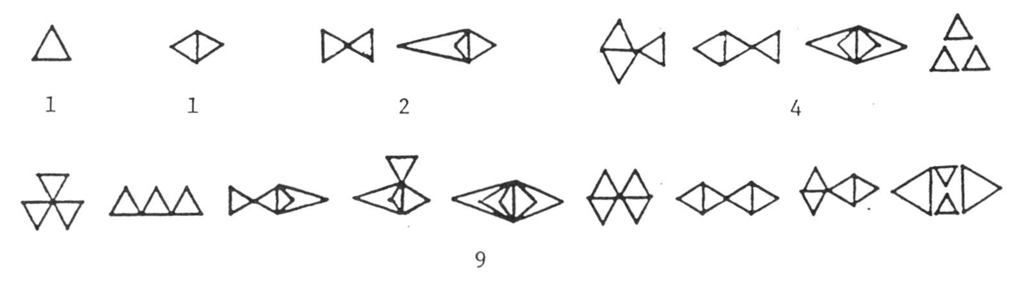

Example 3E Minimal triangle graphs with $n+2$ vertices, $n = 1,2,3,4,5$; notice that the last examples with 6 and 7 vertices have only 3 and 4 triangles respectively, not 4 and 6. It may be easier to visualize these as minimal (but not necessarily minimum) sets of triples which cover a set of $n+2$ elements (vertices).

All five of the examples beginning 1,1,2,4,9 are different sequences! The numbers of score sequences in Example 3A are

$$1, 1, 2, 4, 9, 22, 59, 167, 490, 1486, 4639, 14805, 48107, 158808, 531469, 1799659, \ldots$$

and might be called Narayana-Bent numbers [27]; see also [28, p. 66; 37 #459]. The numbers of graphs with $n+1$ vertices and $n-1$ edges (necessarily disconnected; Example 3B) are

$$1, 1, 2, 4, 9, 21, 56, 148, 428, 1305, 4191, 14140, 50159, 185987, 720298, 2905512, \ldots$$

and are listed in [33, p. 146; 37 #458]. The expression of $15 \times 2^k + 1$ is prime for $k =$

$$1, 2, 4, 9, 10, 12, 27, 37, 38, 44, 48, 78, 112, 168, 229, 297, 339, \ldots$$

[37 #445]. The numbers of combinatorial geometries with $1, 2, \ldots$ points are

$$1, 2, 4, 9, 26, 101, 950, \ldots$$

[8, 37 #462] and may not have been calculated for more than 7 points. The numbers of minimal triangle graphs with $3, 4, 5, \ldots$ vertices,

$$1, 1, 2, 4, 9, 48, 117, 307, 821, 2277, \ldots$$

[1, 37 #450] also do not appear to be known for more than 12 vertices.

4. The Fourth Sequence

$$1, \quad 1, \quad 2, \quad 3, \quad 5, \quad 7, \quad 11, \ \ldots$$

features in

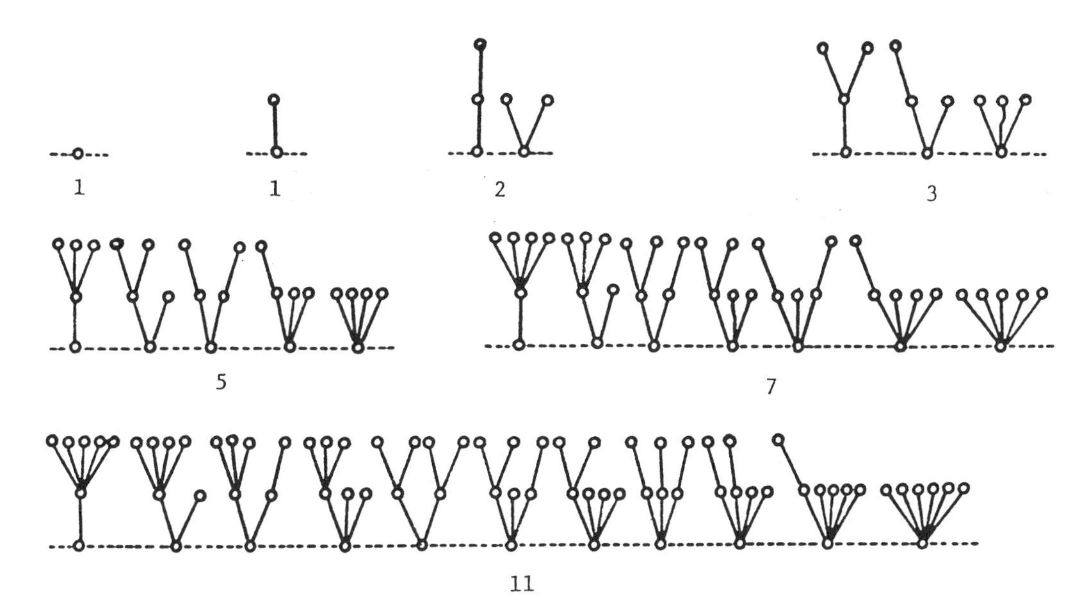

Example 4A Trees of height at most two, with 0, 1,2,3,4,5,6 edges:

Example 4B The number of partitions of n (it is conventional to define $p(0) = 1$):

1	1
2,11	2
3,21,111	3
4,31,22,211,1111	5
5,41,32,311,221,2111,11111	7
6,51,42,411,33,321,3111,222,2211,21111,111111	11

Example 4C (Apart from the initial ones), the primes: and

Example 4D The integer part of $(3/2)^n$

$n =$	0	1	2	3	4	5	6
$(3/2)^n =$	1	1.5	2.25	3.375	5.0625	7.59375	11.390625

There is a neat correspondence between the trees of height at most two with n edges and the partitions of n (Examples 4A and 4B): after defining

$p(0) = 1$, we simply ask what are the valences of the vertices at height one in each tree:

$$1 \; ; \; 2, 11 \; ; \; 3, 21, 111 \; ; \; 4, 31, 22, 211, 1111 \; ; \; 5, 41, \ldots$$

Conversely, to construct a tree corresponding to a given partition of n, take as many vertices at height one as there are parts, connect them to a root vertex, and add enough vertices at height two to make the total $n+1$, and connect them to the height one vertices to make the valences correspond to the sizes of the parts. Of course, the values of the partition function [37 #244]

$$1, 1, 2, 3, 5, 7, 11, 15, 22, 30, 42, 56, 77, 101, 135, 176, 231, 297, 385, 490, \ldots$$

are rarely prime, after the first few; nor does the sequence of integer parts of powers 1.5 coincide any longer [37 #245]:

$$1, 1, 2, 3, 5, 7, 11, 17, 25, 38, 57, 86, 129, 194, 291, 437, 656, 989, 1477, 2216, 3325, \ldots$$

5. The Members of the Fifth Sequence

$$1, \quad 2, \quad 4, \quad 6, \quad 10, \quad 14, \; \ldots$$

are just one less than those in the fourth. They appear in Example 4A $(n = 0, 1, 2, 3, 4)$ as numbers of trees of height two with $n+2$ edges; for each value of n, omit the trees of height one; and in Figure 5A $(n = 5)$:

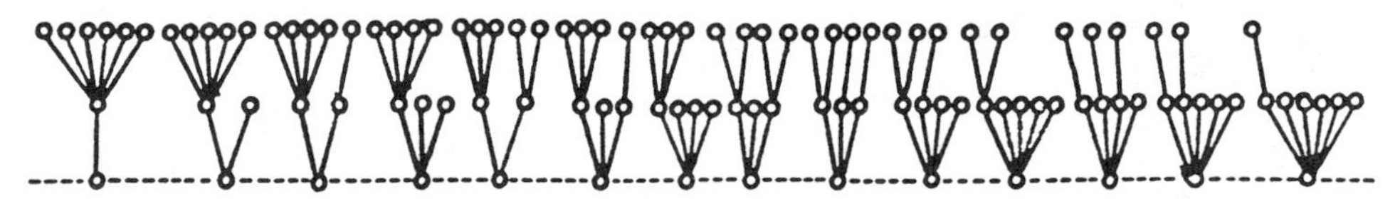

Figure 5A. Fourteen trees of height two, with 7 edges

in Example 5B As numbers of partitions of $2n$ into parts which are powers of two (define this as 1 for $n = 0$): exponents indicate repetitions of part size:

n = 1	$2, 1^2$	2
2	$4, 2^2, 21^2, 1^4$	4
3	$42, 41^2, 2^3, 2^2 1^2, 21^4, 1^6$	6
4	$8, 4^2, 42^2, 421^2, 41^4, 2^4, 2^3 1^2, 2^2 1^4, 21^6, 1^8$	10
5	$82, 81^2, 4^2 2, 4^2 1^2, 42^3, 42^2 1^2, 421^4, 41^6, 2^5, 2^4 1^2, 2^3 1^4, 2^2 1^6, 21^8, 1^{10}$	14

Example 5C Here, the values of k which make k^2+1 prime, $(0^2+1), 1^2+1=2,\ 2^2+1=5,\ 4^2+1=17,\ 6^2+1=37, 10^2+1=101,$

$14^2+1=197, \ldots$; and

1

2

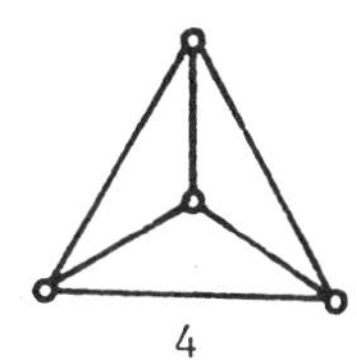

4

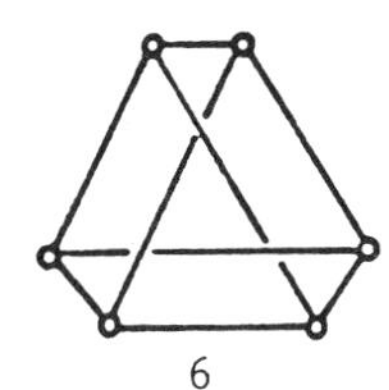

6

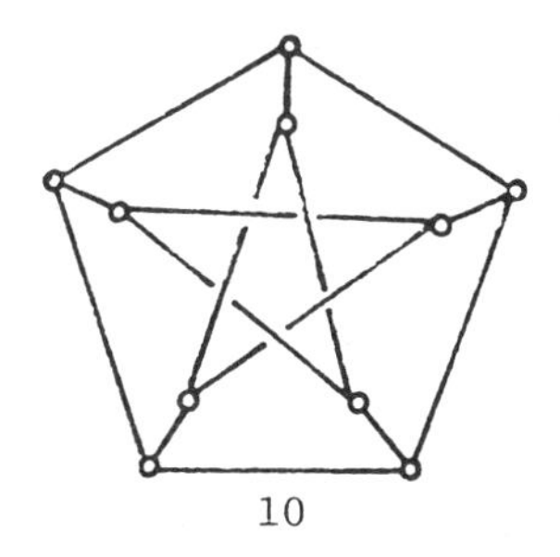

10

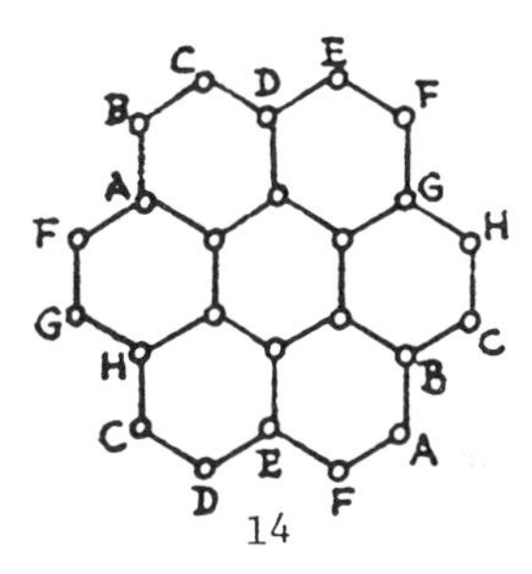

14

Example 5D The least number of vertices in a regular trivalent graph of girth $n+1$. For $n=0$ and 1 we have to use non-Michigan graphs, and count a loop as contributing only 1 to the valence. But for $n=2,3,4,5$ we have those standbys without which no paper on graph theory is complete: $K_4, k_{3,3}$, the Petersen graph and the Heawood map on the torus. In this last, identify vertices bearing the same label.

The trees of height two (Figure 5A) are just one less numerous than the trees of height at most two, so the number with n edges is $p(n)-1$. However, the numbers of partitions of $2n$ into parts which are powers of 2,

$(1), 2, 4, 6, 10, 14, 20, 26, 36, 46, 60, 74, 94, 114, 140, 166, 202, 238, 284, 330, 390, 450, \ldots$

are not the same thing [4, 5, 37 #378]. Nor, of course, are the values of k making k^2+1 prime:

$$1,2,4,6,10,14,16,20,24,26,36,40,54,66,74,84,90,94,110,116,120,124,126,...$$

nor the numbers of vertices in the minimal regular trivalent graphs of girth $n+1$.

$$1,2,4,6,10,14,24,30,...$$

which have only been determined up to girth 8[39,34,35,25,37 #380].

6. The Sixth and Final Sequence

1, 2, 5, 15, ...

occurs as

Example 6A The number of arrangements of 1,2,3,4 different objects in indistinguishable boxes, at most four in a box; as in Figure 1B, but add a box containing four objects to the set of fourteen; as

Example 6B The first members of a sequence of "convergent sets" for a ternary continued fraction, the ratio of alternate members being good approximations to the real root of $x^3-7x^2+3x-1=0$; and as

Example 6C The number of ways of answering the questions in this paper; i.e. the number of equivalence relations on 1,2,3,4 objects, the number of rhyme schemes for verses with 1,2,3,4 lines:

1		1
12;	1,2	2
123;	1,23; 2,13; 3,12; 1,2,3	5
1234;	1,234; 2,134; 3,124; 4,123; 12,34; 13,24	
	14,23; 1,2,34; 1,4,23; 2,3,14; 2,4,13; 3,4,12; 1,2,3,4	15

The final sequence, in the forms of Example 6C, is the sequence of Bell numbers [13,37 #585]

$$1, 2, 5, 15, 52, 203, 877, 4140, 21147, 115975, 678570, 4213597, 276544437, \ldots$$

One way of describing them is the number of ways of arranging n distinguishable objects in indistinguishable boxes, with no restriction on the number of objects in a box. So the members of Example 6A fall short by increasing numbers:

$$1, 2, 5, 15, 51 = 52 - 1, 196 = 203 - 7, 827, 3795, 18755, 99146, 556711, 3305017,$$

[27, 37 #584]. Example 6B is

$$1, 2, 5, 15, 32, 99, 210, 650, 1379, 4268, 9055, 28025, 59458, 184021, 390420, \ldots$$

arising in an investigation [24,37 #582] into a problem of Jacobi.

7. Conclusion

There are still many unsolved problems in enumerative combinatorics, in spite of the recent advances that have been made [17,32,41]. Take a look at three problems which have a superficial look of similarity. First the now classical one of enumeration of polyominoes: you can adjoin four squares to a single quare in the obvious way; then three more to one of these, but not to all four of them, because of overlap. Second, the problem of the number of different deltahedra we can make from regular tetrahedra: we can again adjoin four tetrahedra to the faces of a single one; then twelve more, one to each of the three exposed faces on these four; but at the third stage we run into restrictions, since at most five tetrahedra will fit round an edge. Third, the problem of the number of isomers of the saturated hydrocarbons, C_nH_{2n+2} [18-21,31] has been tackled as a mathematical, but not as a chemical problem. A tetravalent carbon atom can be associated with four other such: can each of these four be associated with another three, and each of these twelve with...? Clearly there is a physical bound.

But even as mathematical problems, many enumerations have not always been clearly defined: the literature contains inconsistencies, eg. [10]: exact formulas remain beyond our grasp: and even the best known asymptotic formulas are often weak.

[Note: The next three terms of the sequence in the Introduction are, of course, 59, 60, 61: the orders of the simple groups!]

REFERENCES

[1] Robert Bowen "The generation of minimal triangle graphs," *Math. Comput.* 21 (1967) 248-250.

[2] W. G. Brown, "Historical note on a recurrent combinatorial problem," *Amer. Math. Monthly* 72 (1965) 973-977.

[3] P. Capell and T. V. Narayana, "On knock-out tournaments," *Canad. Math. Bull.* 13 (1970) 105-109.

[4] R. F. Churchhouse, "Binary partitions," in Atkin, Birch (eds) Computers in Number Theory, Atlas Symposium, Oxford, 1969, Academic Press, London, 1971, 397-400.

[5] R. F. Churchhouse, "Congruence properties of the binary partition function," *Proc. Cambridge Philos. Soc.*, 66 (1969)371-376.

[6] L. Comtet, "Advanced Combinatorics," D. Reidel, Drodrecht, Holland, 1974.

[7] John Conway and Richard Gury, "The Book of Numbers," Scientific American Library (in preparation).

[8] Henry H. Crapo and Gian-Carlo Rota, "On the foundations of combinatorial theory II."Combinatorial geometries, *Stud. Appl. Math.* 49 (1970) 109-133.

[9] I. M. H. Etherington, "Non-associate powers and a functional equation," *Math. Gaz.* 21 (1937) 36-39.

[10] R. A. Fisher, "Some combinatorial theorems and enumerations connected with the numbers of diagonal types of a latin square," *Ann. Eugenics*, 11 (1942) 395-401; see review by Coxeter, MR 4, 183e.

[11] Martin Gardner, "Mathematical games: permutations and paradoxes in combinatorial mathematics," *Sci. Amer.* 209 #2 (Aug. 1963) 112-119; esp. p. 114; see also #3 (Sept.) p. 262.

[12] Martin Gardner, "Mathematical games: patterns in primes are a clue to the strong law of small numbers," *Sci. Amer.* 243 #6 (Dec. 1980) 18-28.

[13] Henry W. Gould, "Bell and Catalan Numbers: Research Bibliography of Two Special Number Sequences," Morgantown, WV, 6th edition, 1985.

[14] Richard K. Guy, "The strong law of small numbers," *Amer. Math. Monthly*, 95 (1988).

[15] Richard K. Guy, "The second strong law of small numbers" (in preparation)

[16] F. Harary, "Graph Theory," Addison-Wesley, 1969.

[17] Frank Harary and Edgar M. Palmer, "Graphical Enumeration," Academic Press, New York, 1973.

[18] Henry R. Henze and Charles M. Blair, "The number of structurally isomeric alcohols of the methanol series," *J. Amer. Chem. Soc.* 53 (1931) 3042-3046.

[19] Henry R. Henze and Charles M. Blair, "The number of isomeric hydrocarbons of the methane series," *J. Amer. Chem. Soc.* 53 (1931) 3077-3085.

[20] Henry R. Henze and Charles M. Blair, "The number of structurally isomeric hydrocarbons of the ethylene series," *J. Amer. Chem. Soc.* 55 (1933) 680-686.

[21] Henry R. Henze and Charles M. Blair, "The number of structural isomers of the more important types of aliphatic compounds, *J. Amer. Chem. Soc.* 56 (1934) 157.

[22] John E. Koehler, "Folding a strip of stamps," *J. Combin. Theory*, 5 (1968) 135-152.

[23] Mike Kuchinski, "Catalan Structures and Correspondences," M.Sc. thesis, W. Virginia University, 1977.

[24] D. N. Lehmer, "On ternary continued fraction," *Tôhoku Math. J.* 37 (1933) 436-445.

[25] W. F. McGee, "A minimal cubic graph of girth seven," *Canad. Math. Bull.* 3 (1960) 149-152; MR 23 #A1550.

[26] Z. Melzak, "A note on homogeneous dendrites," *Canad. Math. Bull.* 11 (1968) 85-93; MR 37 #3954.

[27] F. L. Miksa, L. Moser and M. Wyman, "Restricted partitions of finite sets," *Canad. Math. Bull.* 1 (1958) 87-96; MR 20 #1636.

[28] J. W. Moon, "Topics on Tournaments," Holt, New York, 1968.

[29] T. V. Narayana and D. H. Bent, "Computation of the number of score sequences in round-robin tournaments," *Canad. Math. Bull.* 7 (1964) 133-136.

[30] Richard Otter, "The number of trees," *Annals of Math.* (2) 49 (1948) 583-599; MR 10, 53.

[31] Douglas Perry, "The number of structural isomers of certain homologs of methane and methanol," *J. Amer. Chem. Soc.* 54 (1932) 2918-2920.

[32] George Pólya and R. C. Read, "Combinatorial Enumeration of Groups, Graphs and Chemical Compounds," Springer-Verlag, New York, 1987.

[33] John Riordan, "An Introduction to Combinatorial Analysis," Wiley, New York, 1958.

[34] H. Sachs, "On regular graphs with given girth," in M. Fiedler (ed) Theory of Graphs and its Applications, Smolenice, 1963, 91-97, MR 30 #3467.

[35] H. Sachs, "Regular graphs with given girth and restricted circuits," *J. London Math. Soc.* 38 (1963) 423-429; MR 28 #1613.

[36] Herbert E. Salzer, "Coefficients for expressing the first thirty powers in terms of the Hermite polynomials, *Math. Tables Aids Comput.* 3 (1948) 167-169.

[37] Neil J. A. Sloane, "A Handbook of Integer Sequences," Academic Press, New York, 1973.

[38] Jacque Touchard "Contributions á l'étude du problème des timbres-poste," *Canad. J. Math.* 2 (1950) 385-398.

[39] W. T. Tutte, "A family of cubical graphs," *Proc. Cambridge Philos. Soc.* 43 (1947) 459-474; MR 9, 97.

[40] J. H. M. Wedderburn, "The functional equation $g(z^2) = 2ax + [g(x)]^2$," *Ann. of Math.* 24 (1923) 121-140.

[41] I. P. Goulden and David M. Jackson, "Combinatorial Enumeration," John Wiley, Chichester, 1983.

Combinatorial Additive Number Theory and Cayley Graphs

Yahya Ould Hamidoune
Université Pierre et Marie Curie

ABSTRACT

We obtain the following generalization of the Cauchy–Davenport Theorem.

Let G be an Abelian group and let S be a finite subset of G. There exists $s \in S$ such that the subgroup H generated by s has the following property. For every $A \subset G$ such that $A \cap H \neq \emptyset$, either $|A \cup (A + S)| \geq |A| + |S|$ or $H \subset A \cup (A + S)$.

0. Introduction

Addition theorems have many applications in Analysis, Number Theory and Combinatorics.

These theorems give some relations between the cardinality of the sum of two subsets of an Abelian group and the cardinalities of these sets. The book of H.B. Mann [12] is an appropriate reference for these questions

In this paper we mention a relation between addition theorems and the strong–connectivity of Cayley directed graphs.

By a *directed graph* we mean an ordered pair (V, E), where V is a set and E is a subset of V x V. We consider only finite directed graphs.

Let X be a directed graph. The vertex (resp. edge)–set of X will be denoted by V(X) (resp. E(X)).

Let A be a subset of V(X). The set of vertices incident from A will be denoted by $N^+_X(A)$, or simply $N^+(A)$. More precisely

$$N^+_X(A) = \{y \in V(X)\setminus A : \exists\, x \in A \text{ such that } (x, y) \in E(X)\}.$$

Let X be a directed graph. The *strong–connectivity* of X, denoted by $\kappa(X)$, is defined as follows.

$$\kappa(X) = \min\{|N^+(A)| : |A| = 1 \text{ or } A \cup N^+(A) \neq V(X)\}.$$

It is easy to see that $\kappa(X)$ is the minimum number of vertices whose deletion reduces X to a single vertex or to a nonstrongly connected directed graph.

In [6, 7, 8, 9], we have only considered directed graphs without loops. We allow loops in this paper in connection with additive number theory. But the results of [6, 7, 8, 9] apply in this situation since the strong–connectivity is not affected if we delete the loops.

More details about the strong–connectivity of directed graphs and its relation with groups can be found in [6, 7, 8, 9].

1. Theorems of Lagrange and Cauchy

The following theorem is due to Lagrange, cf. [3]. It was a piece in the proof of his famous theorem that any integer is the sum of 4 squares.

Theorem 1.1 (Lagrange [11]) *Given a prime number* p, $x, y \not\equiv 0 \bmod p$. *There are* $t, u \in \mathbf{Z}$ *such that* $y \equiv t^2 + xu^2 \bmod p$.

Cauchy found a combinatorial proof to the following more general result.

Let $X_1, X_2, \ldots, X_r$ be distinct variables and $f \in \mathbf{Z}[X_1, \ldots, X_r]$. Let $A(f)$ be the number of distinct residues mod p among all the values of f.

Theorem 1.2 (Cauchy [3]) *Let* $X_1, X_2, \ldots, X_r, \ldots, X_{r+s}$ *be distinct variables. Let* $f \in \mathbf{Z}[X_1, \ldots, X_r]$, $g \in \mathbf{Z}[X_{1+r}, \ldots, X_{r+s}]$. *Then either* $|A(f+g)| = p$ *or* $|A(f+g)| \geq |A(f)| + |A(g)| - 1$.

Let $L : \mathbf{Z}_p \times \mathbf{Z}_p \to \mathbf{Z}_p$, be the mapping $L(t, u) = t^2 + xu^2$. Lagrange's Theorem states that L is surjective. Cauchy's Theorem easily implies this fact.

2. The Cauchy–Davenport Theorem

Let G be an Abelian group and let $A, B \subset G$. We write $A + B = \{x + y \mid x \in A$ and $y \in B\}$.

Theorem 2.1 (The Cauchy–Davenport Theorem) *Let* p *be a prime number, and let* A *and* B *be two proper subsets of* $\mathbf{Z}_p$ *such that* $B + A \neq \mathbf{Z}_p$. *Then* $|A + B| \geq |A| + |B| - 1$.

This result is proved by Davenport [4]. Davenport mentioned in his historical note [5] that this theorem is contained in [3]. Theorem 2.1 is known as the Cauchy–Davenport Theorem.

The equivalence between this theorem and Cauchy's Theorem is due to the well known lemma saying that every subset of $\mathbf{Z}_p$ can be represented as the set of values of a polynomial.

We shall show now how to reduce this problem to a connectivity problem on Cayley directed graphs.

Let G be a finite group and S be a subset of G. We shall use the additive notation. The *Cayley directed graph* of G with respect to S is the graph Cay(G, S) = (G, E), where $E = \{(x, y) \mid -x + y \in S\}$.

Remark 1 Let X = Cay(G, S). Then $N^+(A) = (A + S)\backslash A$.

Theorem 2.2 (Hamidoune [6]) *Let* $S \subset \mathbf{Z}_p$. *Then* $\kappa(Cay(\mathbf{Z}_p, S)) = |S\backslash 0|$.

Proof of the Cauchy–Davenport Theorem using Theorem 2.2. We may assume without loss of generality that $0 \in B$. Let $X = Cay(\mathbf{Z}_p, B)$. Since $0 \in B$, we have $A \subset A + B$. By Remark 1, we have

$$A + B = A \cup N^+(A). \qquad (1)$$

It follows that $A \cup N^+(A) \neq V(X)$. By the definition of $\kappa(X)$ and Theorem 2.2, we have

$$|N^+(A)| \geq \kappa(X) = |B\backslash 0| = |B| - 1. \qquad (2)$$

By (1) and (2), we have $|A + B| = |A \cup N^+(A)| \geq |A| + |B| - 1$. ❑

3. Theorems of Shepherdson and Alon

Using the Davenport transfer argument, Shepherdson proved the following.

Theorem 3.1 (Shepherdson [14]) *Let G be a finite Abelian group and let A, S be two subsets of G such that* $S \subset A$. *Then either* $|A \cup (A + S)| \geq |A| + |S|$ *or there is* $s \in S$ *such that* $A \cup (A + S)$ *contains the subgroup generated by s.*

As an application of Theorem 3.1, Shepherdson obtained the following.

Theorem 3.2 (Shepherdson [13]) *Let G be a finite Abelian group and let A and let S be a subset of G. Let* $k \in N$ *be such that* $k|S| \geq |G|$. *Then there is a sequence of natural numbers* $(n_s; s \in S)$ *such that* $\sum n_s s = 0$ *and* $1 \leq \sum n_s \leq k$.

This result is only stated by Shepherdson for $\mathbf{Z}_n$ and $\mathbf{Z}_n^*$ (the group of units of $\mathbf{Z}_n$). But Shepherdson's method works for all finite Abelian groups.

We proved in [8] the following.

Theorem 1.D (Hamidoune [8]) *Let G be a finite group and let $S \subset G$. Then there is a nonvoid sequence of elements of S with product $= 1$ and length $\leq \lceil |G| / |S| \rceil$.*

We did not realize in [8] that Theorem 1.D extends Shepherdson's Theorem (3.2) to the case where the group is not necessarily Abelian. Indeed we were unaware of Shepherdson's Theorem.

We obtained in [8] a more general lower bound for the length of a directed cycle in a directed graph with a transitive group of automorphisms.

Recently, Theorem 3.2 has been rediscovered by Alon [1]. The proof of Alon uses a result of Scherk [13].

Let $n, m \in \mathbf{N}$. We put $[n] = \{1, \dots, n\}$. Define $f(n, m)$ as the smallest integer k such that $\forall A \subset [n]$ $(|A| > k \Rightarrow (\exists B \subset A ; B \neq \emptyset$ and $\sum_{x \in B} x = m)$.

It was conjectured by Erdös and Graham that $f(n, 2n) = (\frac{1}{3} + o(1))n$. This conjecture was proved by Alon [1], as a corollary of the following.

Theorem 3.3 (Alon [1]) *For every $\varepsilon > 0$ and $k > 1$, there is an n_0 such that for every $n > n_0$, and $A \subset \mathbf{Z}_n$ with $|A| > (\frac{1}{k} + \varepsilon)n$, there is a nonvoid subset of A with sum $= 0$ and cardinality $\leq k$.*

Alon proved a similar statement for Abelian groups of odd order.

Theorem 3.4 (Alon [1]) *For every $\varepsilon > 0$ and $k > 1$, there is an n_0 such that for every Abelian group G of odd order $n > n_0$, and $A \subset G$ with $|A| > (\frac{1}{k} + \varepsilon)n$, there is a nonvoid subset of A with sum $= 0$ and cardinality $\leq k$.*

4. A New Addition Theorem

The proofs of the results announced in this section can be found in [10]. These proofs use the theory of atoms of Cayley directed graphs [6, 7, 8, 9]. Our main result is the following.

Theorem 4.1 [10] *Let G be an Abelian group and let S be a finite subset of G. There exists an $s \in S$ such that the subgroup H generated by s has the following property. For every $A \subset G$ such that $A \cap H \neq \emptyset$, either $|A \cup (A + S)| \geq |A| + |S|$ or $H \subset A \cup (A + S)$.*

Theorem 4.1 generalizes the Cauchy–Davenport Theorem and Shepherdson's Theorem (3.1). As applications we obtain the following generalizations to Alon's Theorem (3.3).

Theorem 4.2 *For every* $\varepsilon > 0$ *and* $k > 1$, *there is an* n_0 *such that for every* $n > n_0$, *and* $A \subset \mathbf{Z}_n$ *with* $|A| > (\frac{1}{k} + \varepsilon)n$, *there exists an* $a \in A \backslash \{0\}$ *such that any element of the subgroup generated by* a *is a sum of a nonvoid subset of* A *cardinality* $\leq k$.

Theorem 4.3 *For every* $\varepsilon > 0$ *and* $k > 1$, *there is an* n_0 *such that for every group* G *of odd order* $n > n_0$, *and for every* $A \subset G$ *with* $|A| > (\frac{1}{k} + \varepsilon)n$, *there exists an* $a \in A \backslash \{0\}$ *such that any element of the subgroup generated by* a *is a sum of a nonvoid subset of* A *with cardinality* $\leq k$.

Theorem 4.2 implies the following result proved by Alon [2].

$\forall \varepsilon > 0$, $\exists$ *an* n_0, *such that* $\forall$ *prime* $p > n_0$, $x \in \mathbf{Z}_p$ *and* $A \subset \mathbf{Z}_p$ *with* $|A| > (\frac{1}{k} + \varepsilon)p$, *there is* $A \subset \mathbf{Z}_p$ *such that* $1 \leq |A| \leq k$ *and* $\sum_{y \in A} y = x$.

REFERENCES

[1]. N. Alon, Subset sums, *J. Number Theory* **27** (1987), 196 – 205.

[2]. N. Alon, private communication.

[3]. A. Cauchy, Recherches sur les nombres, *J. Ecole Polytechnique* **9** (1813), 99 – 116.

[4] H. Davenport, On the addition of residue classes, *J. London Math. Soc.* **10** (1935), 30 – 32.

[5] H. Davenport, A historical note, *J. London Math. Soc.* **22** (1947), 100 – 101.

[6] Y.O. Hamidoune, Sur les atomes d'un graphe orienté, *C.R. Acad. Sc. Paris A* **284** (1977), 1253 – 1256.

[7] Y.O. Hamidoune, Sur la séparation dans les graphes de Cayley abeliens, *Discrete Math.* **55** (1985), 323 – 326.

[8] Y.O. Hamidoune, An application of connectivity theory in graphs to factorization of elements in groups, *Europ. J. of Combinatorics* **2** (1981), 108 – 112.

[9] Y.O. Hamidoune, On the connectivity of Cayley digraphs, *Europ. J. of Combinatorics* **5** (1984), 309 – 312.

[10] Y.O. Hamidoune, On subset products in finite groups, preprint.

[11] J.L. Lagrange, *Nouv. Mémoires* Acad. Berlin (1770), 123 – 133, Ouvres 3, 189 – 201.

[12] H.B. Mann, Addition theorems, *Interscience tracts* **18** John Wiley and Sons, (1965)

[13] P. Scherk, *Amer. Math. Monthly* **62** (1955), 46 – 47.

[14] J.C. Shepherdson, On the addition of elements of a sequence, *J. London Math. Soc.* **22** (1947), 85 – 88.

Recent Results and Unsolved Problems on Hypercube Theory

Frank Harary
New Mexico State University

ABSTRACT

The study of graph theoretic properties of hypercubes is relatively uninvestigated and offers a rich field for original research. After reviewing seven presentations of a hypercube Q_n, we look into various properties. First we characterize the edge sums of a hypercube, i. e., the numbers which can arise as the sum of the numerical labels on two adjacent nodes. Then we study cubical graphs which are the subgraphs of some hypercube. The cubical dimension of a cubical graph G is the smallest n such that G is contained in Q_n. The topological cubical dimension of an arbitrary graph G is the smallest n such that some subdivision of G is a subgraph of Q_n. There are several unsolved problems involving these two dimensions. For practical computer-reasons, it would be helpful to have a characterization of the structure of those trees and other graphs which span a hypercube. Partial progress toward this difficult problem is reported. Finally several new elementary problems on changing and unchanging the diameter of a hypercube are given.

000. Introduction

The utilization of hypercube architectures in massively parallel computers has motivated considerable research activity on hypercube theory. While much is known about the structure of hypercubes, there is very much more which has not yet been investigated. This is the inspiration for the survey of hypercube theory in [11], and for my present research.

My current doctoral student, Niall Graham, is also fascinated by hypercube theory. I am grateful for his kind assistance in the preparation of this manuscript.

Table of Contents

001. Presentations

There are many logically equivalent ways to present the information that a given graph G is a hypercube, or more specifically that G and Q_n are isomorphic. These are summarized in [9] where the following seven presentations, P1 to P7, of a hypercube are given in detail, and others.

P1. Binary Sequences

The n–dimensional *hypercube* $Q_n = (V_n, E_n)$ is the graph whose node set V_n consists of all the 2^n binary n–strings. A pair of nodes $x = (x_1, \ldots, x_n)$ and $y = (y_1, \ldots, y_n)$ are adjacent in Q_n if and only if $\Sigma|x_i - y_i| = 1$, i. e., the binary words x and y differ in exactly one place. Figure 1.1 shows the first four hypercubes Q_1 to Q_4 with this binary labeling of their nodes.

Thus Q_n is regular of degree n, is bipartite, has $p = 2^n$ nodes, and hence has $q = n2^{n-1}$ edges, a quantity which has been jokingly described as "the derivative of 2^n with respect to 2."

P2. Cartesian Products

Referring to [7, p. 21] for the definition of the cartesian product $G \times H$ of two graphs, we have the following recursive definition of Q_n:

$$Q_1 = K_2 ; \quad Q_n = Q_{n-1} \times Q_1 \tag{1.1}$$

Thus in particular $Q_2 = C_4$, the 4–cycle seen in Figure 1.1, and $Q_4 = C_4 \times C_4$ which naturally embeds on the surface of a torus. Figure 1.2 brings out the toroidal nature of Q_4 as the cartesian product of two cycles.

A company which manufactures hypercube computers has distributed a poster which vividly displays an extension of the drawing in Figure 1.2 to the next hypercube Q_5.

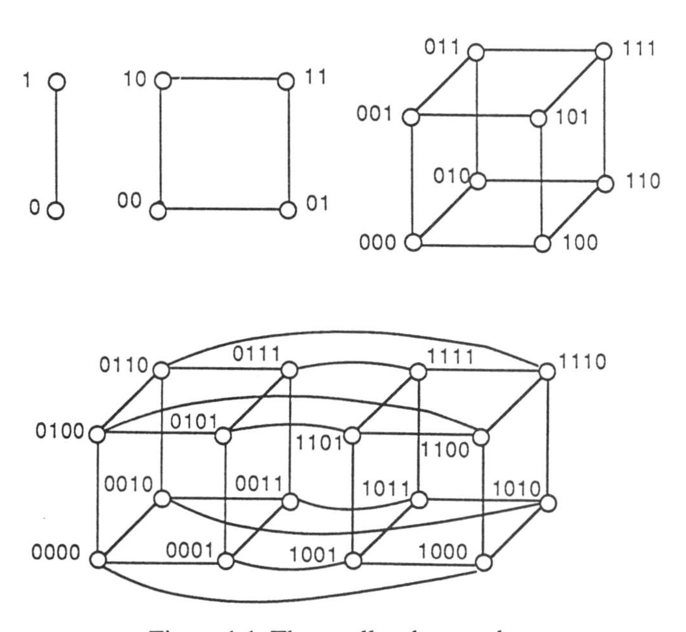

Figure 1.1 The smallest hypercubes

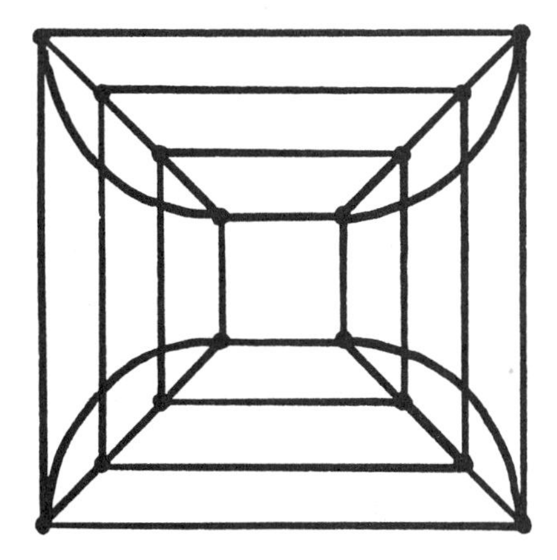

Figure 1.2 A drawing of Q_4 with crossing number 8

P3. Unit Cube in R^n

In n–dimensional euclidean space R^n, the hypercube Q_n may be regarded as the expected extension of the unit interval $[0,1] = Q_1$, the unit square Q_2, the Platonic solid Q_3, and so forth. Thus Q_n is the unit hypercube in R^n with coordinates which are the binary sequences stated in presentation P1 and shown in Figure 1.1. Hence it is meaningful to consider hyperplanes in R^n which "cut through" Q_n without passing through any vertex of Q_n. The set of all nodes on either side of such a cutting hyperplane has been called a *linearly separable boolean function.* Such functions have been studied extensively. The natural extension of the boolean functions determined by two or more cutting planes has not yet been investigated.

P4. The Power Set of an n–Set

Writing $S = N_n = \{1, 2, \ldots, n\}$, the *power set* 2^S is the collection of all the 2^n subsets of S. This gives the node set $V_n = 2^S$ of Q_n. The obvious 1–1 correspondence between a subset $X \subset S$ and a binary sequence is given by the characteristic function of X. The edge set E_n is then defined as in P1.

P5. Lattices

The collection of all subsets of a finite set N_n gives the complete lattice L_n whose covering relation, the Hasse diagram $H(L_n)$, is precisely Q_n. It is this presentation of a hypercube which proved useful in determining (in our last section) the maximum number of new edges which can be added to Q_n without decreasing the diameter.

P6. Finite Boolean Algebras

Whether defined by any one of literally dozens of axiom systems or in any other correct manner, a finite boolean algebra with n atoms is precisely realized by Q_n. Its n atoms correspond to the nodes $(\delta_{i1}, \delta_{i2}, \ldots, \delta_{in})$, for $i = 1$ to n, where δ_{ij} is the familiar Kronecker delta. Each binary sequence label on a node stands for the subset of N_n as in presentation P4.

In the light of this presentation, every theorem about finite boolean algebras is a theorem about hypercubes, *and conversely!* Thus each new result about hypercubes is a disguised form of a new theorem about finite boolean algebras, a subject which logicians and algebraists have long regarded as closed.

Let $u = u_1, u_2, \ldots, u_n$ be a binary sequence, i. e., a node of Q_n. Then u can be regarded as one term in the disjunctive normal form of a boolean function on n variables x_i as in the following illustration for $n = 4$:

$$u = 1101 \rightarrow \text{term } x_1\, x_2\, x'_3\, x_4$$

Thus a subset S of $V(Q_n)$ determines a boolean function in a $1-1$ manner. The automorphism group of Q_n provides the equivalence relation which defines the type of a boolean function. Further, each boolean function written $S \subset V(Q_n)$ yields the induced subgraph $\langle S \rangle$, called "the graph of a boolean function."

P7. Adjacency matrices

It is well known that an adjacency matrix A of a graph G specifies G up to isomorphism. When G has p nodes, A is a $p \times p$ matrix with zero diagonal. Referring to the labeling in Figure 1, we have the illustration:

$$A(Q_2) = \begin{array}{c} \\ 00 \\ 01 \\ 10 \\ 11 \end{array} \begin{array}{c} \begin{array}{cccc} 00 & 01 & 10 & 11 \end{array} \\ \begin{bmatrix} 0 & 1 & 1 & 0 \\ 1 & 0 & 0 & 1 \\ 1 & 0 & 0 & 1 \\ 0 & 1 & 1 & 0 \end{bmatrix} \end{array}$$

When G is bipartite, a more concise matrix $B = B(G)$ contains all the information in $A(G)$. Following [1, 6], the *bipartite adjacency matrix* $B = [bij]$ of graph G with node color sets U and W has $b_{ij} = 1$ when nodes u_i and w_j are adjacent; and $b_{ij} = 0$ otherwise. Thus for Q_2, we have

$$B(Q_2) = \begin{array}{c} \\ 00 \\ 11 \end{array} \begin{array}{c} \begin{array}{cc} 01 & 10 \end{array} \\ \begin{bmatrix} 1 & 1 \\ 1 & 1 \end{bmatrix} \end{array}$$

We observed in [3] that writing $B_n = B(Q_n)$, we have the recursive formulation:

$$B_1 = [1] \qquad B_{n+1} = \begin{bmatrix} B_n & I \\ I & B_n \end{bmatrix} \tag{1.2}$$

We then used this recursion to evaluate the number of perfect matchings in Q_n, written $f_1(Q_n)$, with the symbol f_1 serving as a mnemonic for 1–factor, by applying the straightforward permanent formula,

$$f_1(Q_n) = \mathrm{per}(B_n).$$

Other Presentations

This is a dynamic field of research and various other presentations and characterizations of a hypercube are currently being developed.

010. Edge Sums

The usual binary sequences associated as in Figure 1.1 with the nodes of Q_n can be regarded either as binary numbers or as the numbers from 0 through $2^n - 1$. It is natural to ask which numbers are obtained when one writes on each edge the sum of the labels of its two incident nodes. A positive integer k is an *edge sum of a hypercube* if it appears in this way on an edge of some Q_n. We show in Figure 2.1 the edge sums of Q_2 and Q_3.

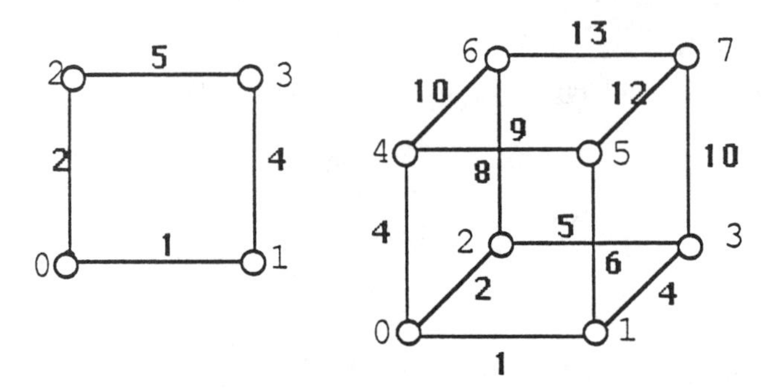

Figure 2.1 The edge sums of two hypercubes

We determined in [4] exactly which numbers are edge sums:

Theorem 2A *A natural number x is an edge sum if and only if $x \not\equiv 3 (mod\ 4)$, i.e., the binary expression for x does not end in 11.*

To prove this sufficiency, note that for arbitrary binary substrings, α and β, the sum $\alpha 00\beta + \alpha 10\beta = \alpha 010\beta 0$. Thus any positive even number may be an edge sum. The sum $\alpha 0 + \alpha 1 = \alpha 01$ generates any binary number ending in 01. Hence every binary number not ending in 11 is indeed an edge sum.

For the necessity we observe that to achieve a sum that ends in 11 which is odd, we must vary the last digit. But we have just seen that such a sum is of the form $\alpha 01$, affirming that only strings of the form $\alpha 11$ cannot arise as edge sums.

This modest exercise illustrates the fact that there exist new, elementary, interesting results about hypercubes to be found.

011. Cubical graphs

An *n–cubical graph* G is isomorphic to a subgraph of Q_n. A *cubical graph* is n–cubical for some n. Garey and Graham [2] proved that the problem of deciding whether a given graph G is n–cubical is NP–complete. They accomplished this by the study of minimal noncubical graphs G, that is, graphs G which are not cubical but the removal of any one edge results in a cubical graph.

As Q_n is bipartite, we know that every cubical graph is bipartite. However the converse does not hold: the smallest counterexample is provided by the complete bipartite graph $K_{2,3}$ as mentioned in [8].

It is well known that every tree is cubical and so is every unicyclic graph whose unique cycle is even. Every square–cell animal is cubical and so is every polyhex. The only cubical graphs among the complete bipartite graphs $K_{r,s}$ with $r \leq s$ are the stars $K_{1,s}$ and the 4–cycle $K_{2,2} = C_4$.

It is easy to see that every cubical graph G can be embedded in some hypercube Q_n as an induced subgraph.

100. Cubical dimensions

The *cubical dimension* cd(G) of a cubical graph G is the minimum n such that $G \subset Q_n$ (subgraph). This concept and the above notation were the subject of [8]. Various easily verified observations which give exact values of cd(G) for several graphs G include the following formulas for stars, paths, even cycles, meshes, full binary trees T_h of height h, and matchings rK_2.

(4.1) $cd(K_{1,s}) = s$

(4.2) $cd(P_r) = \lceil \log_2 r \rceil$

(4.3) $cd(C_k) = \lceil \log_2 k \rceil$ for k even

(4.4) $cd(M_{r,s}) = \lceil \log_2 r \rceil + \lceil \log_2 s \rceil$

(4.5) $cd(T_h) = h + 2, \ h \geq 2$

(4.6) $cd(rK_2) = 1 + \lceil \log_2 r \rceil$

Unsolved problems

1. Among all trees T with p nodes, the star has the largest cubical dimension and the path has the smallest. What is the distribution of cubical dimensions among all other trees? What are the mean, median and mode among all these values of cd(T)?

2. What is the exact value of cd(T) for the following two special classes of trees T:
   (a) starlike trees $S(a_1, a_2, \ldots, a_n)$ in which branch i is a path of length a_i
   (b) caterpillars $C(a_1, a_2, \ldots, a_k)$ where the spine is the path $P_k = v_1, v_2, \ldots, v_k$ and a_i is the number of endnodes adjacent to spinal node v_i

When a graph G is cubical, say cd(G) = n, then G is necessarily isomorphic to an induced subgraph of some hypercube Q_m, with $m > n$ unless G is already an induced subgraph of Q_n. This fact immediately suggests defining the *induced cubical dimension* of G, written icd(G), as the smallest such n. The exact determination of icd(G) for various cubical graphs G, including paths and cycles, appears to offer an intrinsically intractable problem area, as we shall see in the last section.

101. Embedding

For practical computing purposes, it is necessary to embed graphs G into hypercubes even when G is not cubical, as reported in Livingston and Stout [13]. This is done by embedding not G itself but a subdivision of G.

Folklore Theorem *Every graph has a cubical subdivison.*

Folklore Proof It is sufficient to verify that this holds for every complete graph, K_p, as every graph G is a spanning subgraph of the complete graph $G \cup \overline{G}$. Recall that the *weight* of a node $u = (u_1, \ldots, u_n)$ of Q_n is the sum Σu_i. Thus the origin $(0, \ldots, 0)$ is the only node of weight 0 and there are just n nodes of weight 1.

To construct a subdivision S of the complete graph K_{n+1} which is a subgraph of Q_n, take the n + 1 nodes of weight at most 1 as the node set of K_{n+1} and define S to be the subgraph of Q_n induced by all nodes of weight 0, 1 or 2.

The graph S has the n edges incident with the origin as unsubdivided edges of K_{n+1}, and each of the remaining $\binom{n}{2}$ edges of K_n is subdivided by one node of degree 2. These degree-2 nodes of S are just the nodes of weight 2 in Q_n, as illustrated in Figure 5.1 for n = 3.

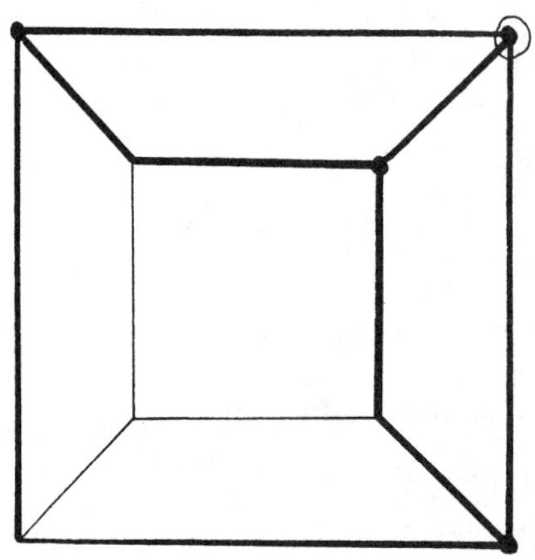

Figure 5.1 Topological embedding of K_4 in Q_3

It follows at once that for every graph G with p nodes, some subdivision SG is $(p-1)$-cubical.

The *topological cubical dimension* $tcd(G)$ of an arbitrary graph G is the smallest n such that some subdivision SG satisfies $SG \subset Q_n$. When G is cubical, it is obvious that $tcd(G) = cd(G)$. The invariant $tcd(G)$ was studied in [10].

110. Spanning subgraphs

The investigation of various families of spanning subgraphs of a hypercube is the subject of a series of papers with M. Lewinter. Our progress to date is summarized in the survey article [12] in this volume. Of course, a tree T which spans Q_n has 2^n nodes. As Q_n has 2^{n-1} nodes of even weight and 2^{n-1} of odd weight, it is a bipartite graph with the same number of nodes in each of its two color classes. A bipartite graph with this property is called *equitable.* Obviously, every connected spanning subgraph of Q_n is an equitable bipartite graph. A *starlike tree* is a subdivision of a star $K_{1,k}$ with $k \geq 3$.Just these two necessary conditions prove to be sufficient for starlike trees:

A starlike tree T spans Q_n if and only if T has 2^n nodes and is equitable.

The general problem of characterizing the structure of trees which span Q_n appears quite intractable. There is an adage which recommends that if one cannot solve a problem for graphs, he should attempt to settle it for trees. This maxim may be modified to suggest that a difficult problem for trees should first be tried for caterpillars. Even this restricted problem has only yielded partial progress.

However, we did succeed in characterizing those meshes which span Q_n. Recall that a *mesh* is the cartesian product of a finite number d of nontrivial paths. As usual, P_h denotes the path with h nodes. We call P_h a *binary path* if $h = 2^k$ for some k. Then a *binary mesh* M is the cartesian product of binary paths, as illustrated in Figure 6.1. Our result is:

A mesh M spans Q_n if and only if M is a binary mesh with 2^n nodes.

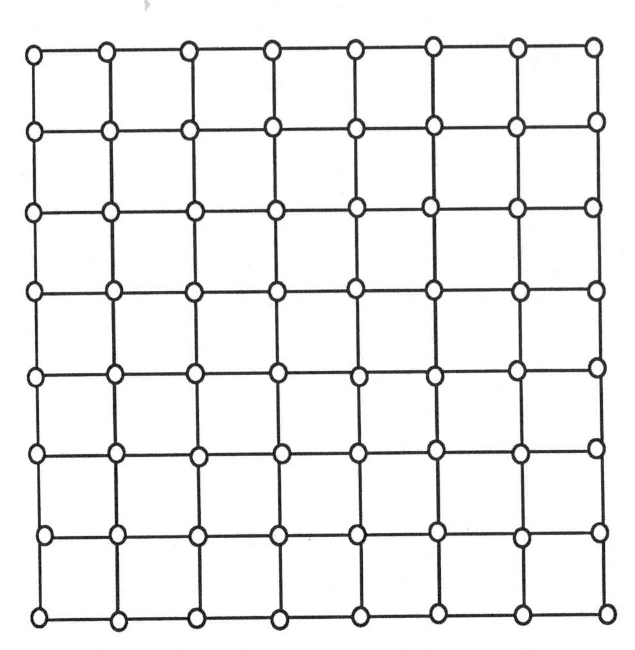

Figure 6.1 A binary mesh which spans Q_6

111. Diameter

In a parallel processor the diameter of the underlying graph architecture imposes a bound on the communication time between an arbitrary pair of processors. Motivated by an investigation of the diameter in the presence of edge faults, the following extremal invariants of a hypercube suggest themselves. These changing and unchanging invariants were named and studied in [5].

Let $ch^+(Q_n)$ be the least number of edges whose addition to Q_n decreases the diameter, and let $ch^-(Q_n)$ be the least number of edges whose deletion increases the diameter, ch standing for changing. Similarly, let $un^+(Q_n)$ be the maximum number of edges which can be added to Q_n without changing (decreasing) the diameter, and let $un^-(Q_n)$ be the greatest number of edges whose removal from Q_n does not affect (increase) the diameter, un standing for unchanging. The following results were found in [5] concerning these invariants.

$$ch^-(Q_n) = n - 1 \tag{7.1}$$

In words, the removal of $n - 1$ edges and no fewer is sufficient to increase the diameter. This is only achieved by removing all but one edge incident to the same node.

$$ch^+(Q_n) = 2 \tag{7.2}$$

Two additional edges are required to decrease the diameter. There are several solutions. The simplest is to augment any subcube Q_2 in Q_n into a copy of K_4 by adding two diagonal edges.

$$(n-3)2^{n-1} + 2 \le un^-(Q_n) < (n-2)2^{n-1} \tag{7.3}$$

The lower bound was achieved by constructing a family of graphs defined in terms of spanning trees of Q_n. More recently, a lower bound asymptotic to $(n - 20/7)2^{n-1}$ was found, offering a slight improvement. The upper bound is a consequence of the above mentioned fact that the minimum degree $\delta \ge 2$. A consequence of the lower bound is that

$$\lim_{n\to\infty} un^-(Q_n)/n2^{n-1} = 1 \tag{7.4}$$

In words "almost all" the edges of a hypercube may be removed without altering the diameter!

$$un^+(Q_n) = \binom{2n}{n-1} + \frac{1}{2}\binom{2n}{n} - (n+1)2^{n-1} \tag{7.5}$$

The derivation of this quantity is routine as exactly one pair of nodes, namely, $00\ldots0$ and $11\ldots1$, remain at distance n apart after the addition of all the new edges which join nodes whose weights differ by 1.

The minimum diameter d of a spanning tree T of Q_n is given by:

$$\min(d : T \text{ spans } Q_n) = 2n - 1$$

As every node in Q_n has eccentricity n, the center of a spanning tree must lie on a path joining two nodes whose distances from the center are respectively at least n and $n-1$. A family of spanning trees with this property are easily constructed.

REFERENCES

[1] R. Brualdi, F. Harary and Z. Miller, Bigraphs versus digraphs via matrices. *J. Graph Theory* **4** (1980) 51–57.

[2] M. R. Garey and R. L. Graham, On cubical graphs. *J. Combin. Theory* **B18** (1975) 84–95.

[3] N. Graham and F. Harary, The number of perfect matchings in a hypercube. *Applied Math. Letters* **1** (1988) 41–44.

[4] N. Graham and F. Harary, Edge sums in hypercubes. *Irish Math. Soc. Bull.* **2** (1989) 8–12.

[5] N. Graham and F. Harary, Changing and unchanging the diameter of a hypercube, to appear.

[6] F. Harary, Permanents, determinants and bipartite graphs. *Math. Mag.* **42** (1969) 146–148.

[7] F. Harary, *Graph Theory.* Addison–Wesley, Reading (1969).

[8] F. Harary, Cubical graphs and cubical dimensions. *Comput. Math. Appl.* **15** (1988) 271–275.

[9] F. Harary, Presentations of a hypercube and their roles. *Proc. Intl. Conf. on Computer Science, Hong Kong* (1988).

[10] F. Harary,The topological cubical dimension of a graph, to appear.

[11] F. Harary, J. P. Hayes and H. J. Wu, A survey of the theory of hypercube graphs. *Comput. Math. Appl.* **15** (1988) 277–289.

[12] F. Harary and M. Lewinter, Spanning subgraphs of a hypercube VI: Survey and unsolved problems. This volume.

[13] M. Livingston and Q. F. Stout, Embeddings in hypercubes. *Math. Comput. Modeling* **11** (1988) 222–227.

Spanning Subgraphs of a Hypercube VI: Survey and Unsolved Problems

Frank Harary
New Mexico State University

Martin Lewinter
SUNY at Purchase

ABSTRACT

The hypercube Q_n *is defined recursively by* $Q_1 = K_2$, *while* $Q_n = Q_{n-1} \times K_2$. *A characterization of the spanning trees of* Q_n *is at present unknown. We present a survey of results in this area and propose various unsolved or partially solved problems. The n–dimensional mesh* $M(a_1, ..., a_n)$ *is the cartesian product of the paths* P_{a_i}, $i = 1, ..., n$. *While we have shown that this mesh spans a hypercube if and only if each* a_i *is a power of 2, there are several open questions concerning embedding meshes in hypercubes.*

1. Introduction

Hypercubes are of considerable current interest in computer science, particularly in computer architecture, as they are being successfully utilized in parallel processing.

The *hypercube* Q_n is defined recursively by $Q_1 = K_2$ while $Q_n = Q_{n-1} \times K_2$. We employ the graph–theoretic terminology of [1] except that we use *nodes* and *edges* rather than points and lines. The node set $V_n = V(Q_n)$ is the set of all binary n–tuples, so that $|V_n| = 2^n$. Let nodes $x, y \in V(Q_n)$ have binary labels $(x_1, ..., x_n)$ and $(y_1, ..., y_n)$. Then $xy \in E_n = E(Q_n)$ if and only if $\Sigma |x_i - y_i| = 1$. More generally, the distance $d(x, y) = \Sigma |x_i - y_i|$. Note that $|E_n| = n2^{n-1}$.

We require several definitions. A tree is *equitable* if in the 2–coloring of V(T), both colors sets have the same cardinality. If the cardinalities differ by one, it is *nearly equitable;* otherwise it is called *skewed.* A *caterpillar* is a tree which becomes a path when its endnodes are removed. This path is called the *spine* of the caterpillar.

The path P_k has k nodes. Denoting, as usual, the maximum degree of a graph G by $\Delta(G)$, a *k–legged* caterpillar T satisfies $\Delta(T) = k + 2$. When the degree set of a caterpillar T is $\{1, 2, k+2\}$, $k \geq 2$, it is called *strictly* k–legged. A *starlike* tree T is homeomorphic to a star. If $\Delta(T) = k$, the starlike tree is called *k–branched.* A *double star* is a tree with exactly two nodes which are not endnodes, i.e., it is a caterpillar with spine K_2. A *double starlike* tree is a subdivision of a double star S in which the edge joining the two central nodes of S is not subdivided.

The *n–dimensional mesh* denoted by $M = M(a_1, ..., a_n)$ has node set $V(M)$ consisting of all the ordered n–tuples $x = (x_1, ..., x_n)$ such that $1 \leq x_i \leq a_i$ for each i between 1 and n. As for hypercubes, two nodes $x, y \in V(M)$ are adjacent if and only if $\Sigma |x_i - y_i| = 1$. If each $a_i = 2$, then $M = Q_n$. Obviously $M(a_1, ..., a_n)$ is the cartesian product of the paths $P_{a_1}, ..., P_{a_n}$. If each a_i is a power of 2, then M is a *binary mesh.*

An integer–valued function f on a set S of trees *interpolates* if given trees T, $T' \in S$ and an integer m such that $f(T) < m < f(T')$, then there exists a tree $T'' \in S$ with $f(T'') = m$. If m must be an even (odd) number, then f has the *even (odd) interpolation property.* In Section 2, we summarize various results on spanning trees of hypercubes and present several open questions. Section 3 deals with embedding meshes in hypercubes.

2. Spanning Trees of a Hypercube

An open question is to characterize the spanning trees of Q_n. In [3] we showed that equitable k–branched starlike trees on 2^n nodes with $3 \leq k \leq n$ spans Q_n, thereby extending a result of Nebesky [10], who proved this for $k = 3$. We proved in [4] that if an equitable double starlike tree T on 2^n nodes satisfies $\Delta(T) = 3$, then T spans Q_n. This result has not yet been settled for $\Delta(T) > 3$, although it is conjectured to hold.

Havel and Liebel in [9] showed that equitable 1–legged caterpillars on 2^n nodes span Q_n. It is not known exactly which k–legged caterpillars span Q_n. In [8], we presented several classes of strictly 2–legged caterpillars which span Q_n and gave a tight upper bound for the number of nodes of degree 4 in the next result.

Theorem A *There exist strictly 2–legged caterpillars T which span Q_n having the following properties:*

(1) $n \geq 5$ is odd and T has $(2^n - 2)/3$ nodes of degree 4.

(2) $n \geq 6$ is even and T has $(2^n - 4)/3$ nodes of degree 4.

We also proved in [7] that the number of nodes of degree 4 in hypercube–spanning strictly 2–legged caterpillars is even and has the even interpolation property.

We ask the following question: Given $n \geq 5$, what is the shortest possible spine among all the caterpillars which span Q_n? The 3–legged caterpillar of Figure 1 spans Q_5 and its spine is P_{10}. Note that the sequence which lists the number of endnodes (consecutively) of the spinal nodes, i.e., the *caterpillar code*, is the palindrome (3, 3, 3, 2, 0, 0, 2, 3, 3, 3). The construction involves considering Q_5 as four copies of Q_3 and using the spine to span the two "opposite" copies of Q_2 in the upper left and lower right copies of Q_3. This requires two nodes (with code 0) to connect these spinal segments.

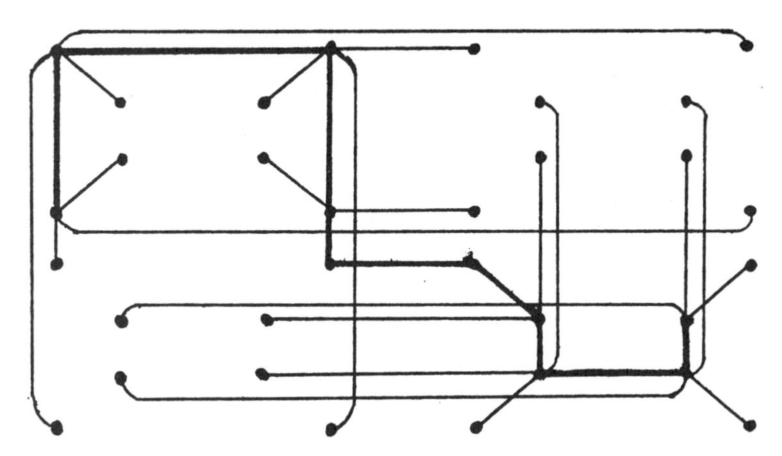

Figure 1. A 3–legged caterpillar which spans Q_5 in which Q_5 is represented as $Q_3 \times Q_2$.

For $n \geq 5$, a similar construction yields a 3–legged caterpillar with a spine of order $2^{n-2} + 2$, whose code is (3, ..., 3, 2, 0, 0, 2, 3, ..., 3). We conjecture that this is the shortest such spine.

Observe that the number of endnodes of this shortest caterpillar is $2^n - (2^{n-1} + 2) = 3 \cdot 2^{n-2} - 2$ which asymptotically equals 75% of $|V(Q_n)|$. This raises the open question: What is the maximum possible number of endnodes of a spanning tree of Q_n? If the above observation always gives the maximum, then the answer for caterpillars is $3 \cdot 2^{n-2} - 2$, since the number of endnodes plus the number of spinal nodes equals 2^n.

In [7], we call a subgraph H of Q_n a *subcube of codimension* k if $H = Q_{n-k}$. Given an embedded spanning tree T of Q_n and a subcube H where $T \cap H$ is connected and is, in fact, a spanning tree of H, then H is called a *spanned subcube* of Q_n relative to T. We exhibited in [7], using techniques developed in [6], a method of determining whether a copy of Q_n with an embedded spanning tree T contains a spanned subcube of codimension 1. Each spanning tree of Q_3 determines at least one spanned subcube of codimension 1, while for $n \geq 4$, there exists spanning trees which do not. The analogous situation for spanned subcubes of codimension greater than 1 remains to be examined.

3. Embedding Meshes in Q_n

We characterized in [5] the meshes which span a hypercube:

Theorem B *The m–mesh $M(a_1, ..., a_m)$ spans Q_n if and only if $a_1 a_2 ... a_m = 2^n$. In words, a mesh M spans some hypercube if and only if M is a binary mesh.*

As a consequence, if a tree spans a binary mesh, then it spans a hypercube. In particular, if a tree T spans a binary ladder $M(2, 2^{n-1})$, then T spans Q_n.

For the mesh $M(k, k)$ to be contained in Q_n, we must have $k^2 \leq 2^n$, from which we obtain $\lceil 2 \log_2 k \rceil \leq n$. Now since Q_m contains a hamiltonian path P_{2^m}, it follows that the mesh $P_{2^m} \times P_{2^m}$ spans $Q_m \times Q_m = Q_{2m}$. As $M(k, k) = P_k \times P_k$, we observe that $M(k, k)$ embeds in $Q_s \times Q_s = Q_{2s}$, where $s = \lceil \log_2 k \rceil$. Thus the minimum n for which $M(k, k) \subset Q_n$ satisfies

$$\lceil 2 \log_2 k \rceil \leq n \leq 2 \lceil \log_2 k \rceil. \tag{1}$$

Now for any positive number x, we have $2\lceil x \rceil - \lceil 2x \rceil \leq 1$, hence the left and right expressions in (1) differ by at most 1. It is noted in the survey article [2] that the upper bound in (1) is exactly the smallest n.

By a similar argument, the minimum n such that the d–dimensional mesh $M(k,k,...,k)$ embeds in Q_n satisfies

$$\lceil d \log_2 k \rceil \leq n \leq d \lceil \log_2 k \rceil. \tag{2}$$

Again, the upper bound gives the exact value.

REFERENCES

[1] F. Harary, *Graph Theory.* Addison–Wesley, Reading (1969).

[2] F. Harary, J.P. Hayes and H.J. Wu, A survey of the theory of hypercube graphs. *Comput. Math. Appl.* 15 (1988) 277 – 289.

[3] F. Harary and M. Lewinter, The starlike trees which span a hypercube. *Comput. Math. Appl.* 15 (1988) 299 – 302.

[4] F. Harary and M. Lewinter, Spanning subgraphs of a hypercube II: Double starlike trees, *Math Comput. Modelling* 11 (1988) 216 – 217.

[5] F. Harary and M. Lewinter, Spanning subgraphs of a hypercube III: Meshes, *International Journal of Computer Mathematics,* 25 (1988) 1 – 4.

[6] F. Harary and M. Lewinter, Spanning subgraphs of a hypercube IV: Rooted trees, *Comput. Math. Appl.,* to appear.

[7] F. Harary and M. Lewinter, Spanning subgraphs of a hypercube V: Spanned subcubes, *Proc. First China–USA Conf. on Graph Theory*, to appear.

[8] F. Harary, M. Lewinter and W. Widulski, On two–legged caterpillars which span hypercubes. *Congress Numerantium*, to appear.

[9] I. Havel and P. Liebl, One–legged caterpillars span hypercubes. *J. Graph Theory* 10 (1986) 69 – 77.

[10] L. Nebesky, On quasistars in n–cubes, *Casopis Pest. Mat.* 109 (1984) 153 – 156.

The Ramsey Number of $K_{3,3}$

Heiko Harborth

Ingrid Mengersen

Technische Universität Braunschweig, West Germany

ABSTRACT

The Ramsey numbers $r(K_{2,3},K_{3,3}) = 13$ and $r(K_{3,3}) = 18$ are proved.

The Ramsey number $r = r(G,H)$ is the smallest number r such that in every two–coloring of the edges of the complete graph K_r there is a copy of G with all edges of the first color (green) or a copy of H with all edges of the second color (red). If $G = H$ then $r(G,G) = r(G)$ denotes the so–called diagonal Ramsey number.

Burr [1] has listed the diagonal Ramsey numbers $r(G)$ for all graphs G with up to six edges and Hendry [5] has extended this list to all graphs with up to seven edges. Except for the complete graph of order five, $r(G)$ is known for every graph G with at most five vertices (see [1], [2], [3], [6], [7]). The proof of $r(K_5 - e) = 22$ was found recently [3]. At present it is known that $43 \leq r(K_5) \leq 55$.

So the Ramsey number is unknown for one of the two Kuratowski graphs, K_5. What about the other one, $K_{3,3}$?

The case where G is a complete bipartite graph is of fundamental interest. In only a few of the cases is $r(G)$ known exactly. It is known that

$$r(K_{1,n}) = \begin{cases} 2n & \text{if } n \text{ is odd,} \\ 2n-1 & \text{if } n \text{ is even.} \end{cases}$$

Also, $r(K_{2,2}) = 6$ and $r(K_{2,3}) = 10$ are found in Burr's table [1]. Some general estimations are given in [2]. It is the purpose of this note to prove $r(K_{3,3}) = 18$.

Let $ge(v)$ or $re(v)$ denote the numbers of green or red edges, respectively, incident to a vertex v in a two–coloring of G. The sets of neighbors of v incident to v in green or red will be denoted by G_v or R_v, respectively. The numbers of green or red edges joining two vertex sets S and T are denoted ge(S,T) or re(S,T), respectively. If S consists of a single vertex x, then ge(x,T) or re(x,T) is used.

Theorem 1 $r(K_{1,3},K_{m,n}) = m + n + 2$.

Proof A green cycle graph C_{m+n+1} proves $r(K_{1,3},K_{m,n}) > m + n + 1$. In K_{m+n+2} without a green $K_{1,3}$ at most two green edges are incident to any vertex. Then m vertices of K_{m+n+2} can be chosen such that at most two of the remaining vertices are adjacent each by one green edge to one of the m vertices. The remaining n vertices together with these m vertices determine a red $K_{m,n}$, and $r(K_{1,3},K_{m,n}) \leq m + n + 2$ is proved. ❑

Theorem 2 $r(K_{2,3},K_{3,3}) = 13$.

Proof Figure 1 shows the green edges of a two–colored K_{12} without a green $K_{2,3}$ and without a red $K_{3,3}$. This proves $r(K_{2,3},K_{3,3}) > 12$.

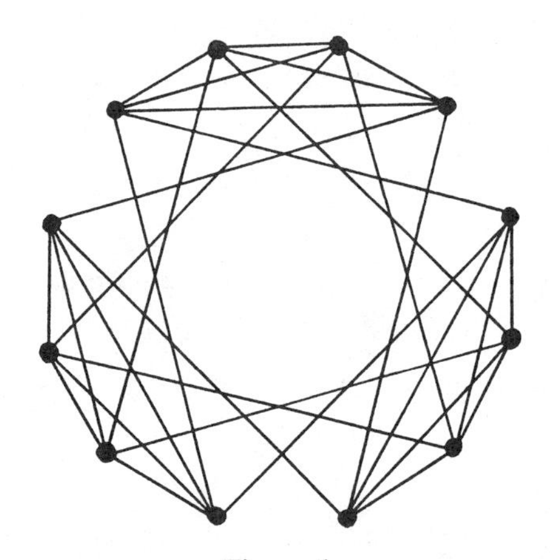

Figure 1

The following sequence of conclusions guarantees a green $K_{2,3}$ or a red $K_{3,3}$ in every two–coloring of the edges of K_{13}.

If $re(v) \leq 4$ in a two–coloring of K_{13}, then $ge(v) \geq 8$, and together with $r(K_{1,3},K_{3,3}) = 8$ a green $K_{2,3}$ or a red $K_{3,3}$ occurs. If $re(v) \geq 10$, then $r(K_{2,3},K_{2,3}) = 10$ guarantees a green $K_{2,3}$ or a red $K_{3,3}$.

Assume $5 \leq re(v) \leq 6$ for all vertices v in a two–coloring of K_{13}. Then a green $K_{2,3}$ exists, since otherwise $re(R_v,G_v) \geq re(v)(ge(v) - 2) \geq 24$ for a fixed vertex v, $re(w,G_v) \geq ge(v) - 3 \geq 3$ for every vertex $w \in G_v$, and thus $re(w_0) = re(w_0,R_v) + re(w_0,G_v) \geq 7$ for at least one vertex $w_0 \in G_v$, which contradicts the assumption.

In the remaining case a vertex v exists with $7 \leq re(v) \leq 9$.

Assume first $ge(R_v,G_v) > \min\{2re(v), re(v) + \binom{ge(v)}{2}\}$, that is, $ge(R_v,G_v) \geq 15$, 15, or 13 for $re(v) = 12 - ge(v) = 7, 8$, or 9, respectively. Then either one vertex of R_v is joined by three green edges to G_v, or two vertices of R_v together with two vertices of G_v determine a green $K_{2,2}$. In both cases together with vertex v a green $K_{2,3}$ occurs.

Assume second $re(R_v,G_v) > 4ge(v)$, that is $re(R_v,G_v) \geq 21$, 17, or 13 for $re(v) = 12 - ge(v) = 7, 8$, or 9, respectively. Then $re(w,R_v) \geq 5$ for at least one vertex $w \in G_v$. If w is adjacent by six red edges to a set S of six vertices of R_v, then either a red $K_{3,3}$ occurs, or two vertices of G_v are joined by at least eight green edges to S. These two vertices together with v and two vertices of S determine a green $K_{2,3}$. If w is adjacent by exactly five red edges to a set T of five vertices of R_v, then either v or w are joined by three green edges to a set, say C, of three vertices. If no red $K_{3,3}$ occurs, then $ge(T,C) \geq 9$, and v together with one vertex of T and all vertices of C, or with two vertices of T and two vertices of C determine a green $K_{2,3}$.

It remains

$$re(R_v,G_v) + ge(R_v,G_v) = re(v)ge(v) \leq 4ge(v) + \min\{2re(v), re(v) + \binom{ge(v)}{2}\},$$

which, however, is impossible for $re(v) = 7$, 8, and 9. Thus the proof of $r(K_{2,3},K_{3,3}) \leq 13$ is complete. ❑

Theorem 3 *$r(K_{3,3}) = 18$.*

Proof For the proof of $r(K_{3,3}) > 17$ it can be checked that the well–known unique two–coloring of K_{17} without a monochromatic K_4 does not contain a monochromatic $K_{3,3}$.

The proof of $r(K_{3,3}) \leq 18$ is partitioned into the following Lemmas.

Lemma 1 *If a two–colored complete graph contains a $K_{2,4}$ of one and a $K_{2,5}$ of the other color such that at most the sets of two vertices intersect one another, then there is a monochromatic $K_{3,3}$.*

Proof To avoid a monochromatic $K_{3,3}$ the four vertices of $K_{2,4}$ are joined by at least 12 edges of one color to the five vertices of $K_{2,5}$, and these are joined by at least 10 edges of the other color to the four vertices of $K_{2,4}$, however, there exist only 20 edges. ❑

Lemma 2 *If for some vertex v in a two–colored complete graph and for some $w \in G_v$, one has $ge(v) \geq 8$ and $re(w,R_v) \geq 5$ or else $ge(v) \geq 6$ and $re(w,R_v) \geq 6$, then there is a monochromatic $K_{3,3}$. — The same conclusion holds if colors green and red are interchanged.*

Proof Assume $ge(v) \geq 8$, and let v be joined by green edges to w and to $a_1, a_2, ..., a_7$, and v and w both be joined by red edges to $b_1, b_2, ..., b_5$. Either a red $K_{3,3}$ occurs or $ge(\{a_1, ..., a_7\},\{b_1, ..., b_5\}) \geq 21$. Then one vertex of $\{b_1, ..., b_5\}$, say b_5, is adjacent by green edges to five vertices of $\{a_1, ..., a_7\}$. Together with v there exists a green $K_{2,5}$. Moreover, v, w and $\{b_1, ..., b_4\}$ determine a red $K_{2,4}$, and Lemma 1 can be used. — The corresponding proofs of the second part of Lemma 2 and with colors interchanged are left to the reader. ❑

In the following, $s_i(v)$ denotes the number of vertices $w \in G_v$ with $re(w,R_v) = i$, and $t_i(v)$ denotes the number of vertices $w \in R_v$ with $ge(w,G_v) = i$.

Lemma 3 *If*

$$\sum_{i \geq 3} \binom{i}{3} s_i(v) > \binom{re(v)}{3},$$

or

$$\sum_{i \geq 3} \binom{i}{3} s_i(v) = \binom{re(v)}{3}, \text{ and } re(w,R_v) \geq 3 \text{ for one } v \in R_v,$$

or the corresponding expressions with colors green and red interchanged are valid, then there is a monochromatic $K_{3,3}$.

Proof This is an easy consequence of the definitions of $s_i(v)$ and $t_i(v)$. ❑

Lemma 4 *If for some vertex v in a two-colored K_{18} one has $ge(v) = 9$ or $re(v) = 9$, then there is a monochromatic $K_{3,3}$.*

Proof Assume $ge(v) = 9$, and thus $re(v) = 8$. Either $a \in G_v$ exists with $re(a,R_v) \geq 5$, or $b \in R_v$ with $ge(b,G_v) \geq 5$, and then Lemma 2 can be used. — For $re(v) = 9$ interchange colors green and red. ❑

Lemma 5 *If for some vertex v in a two-colored K_{18} one has $ge(v) = 11$ or $re(v) = 11$, then there is a monochromatic $K_{3,3}$.*

Proof Assume $ge(v) = 11$, and therefore $re(v) = 6$. Furthermore assume that no monochromatic $K_{3,3}$ occurs.

Lemma 2 implies $t_i(v) = 0$ for $i \geq 6$, and thus $re(R_v,G_v) \geq 36$. Also Lemma 2 implies $s_i(v) = 0$ for $i \geq 5$. By Lemma 3 then $s_3(v) + 4s_4(v) \leq 20$ can be assumed. This together with $re(R_v,G_v) \geq 36$ leaves the unique possibility $s_4(v) = 3$, $s_3(v) = 8$, $s_3(v) + 4s_4(v) = 20$ and $re(R_v,G_v) = 36$. It follows $ge(R_v,G_v) = 30$, $ge(w,G_v) = 5$ and $re(w,G_v) = 6$ for all $w \in R_v$, and Lemma 3 yields $re(w,G_v) \leq 2$ for all $w \in R_v$.

For all $w \in R_v$ one has $re(w) = 1 + re(w,G_v) + re(w,R_v) = 7, 8,$ or 9. Lemma 4, however, leaves only $re(w) = 7$, and thus $re(w,R_v) = 0$ for all $w \in R_v$. This means, R_v is a green K_6 which contains a green $K_{3,3}$, in contradiction to the assumption. — For $re(v) = 11$ interchange colors green and red.

Lemma 6 *If for some vertex v in a two–colored K_{18} one has $ge(v) = 10$ or $re(v) = 10$, then there exists a monochromatic $K_{3,3}$.*

Proof Assume $ge(v) = 10$, and therefore $re(v) = 7$. Furthermore assume that no monochromatic $K_{3,3}$ occurs.

Lemma 2 implies $s_i(v) = 0$ for $i \geq 5$. By Lemma 3 it follows $s_3(v) + 4s_4(v) \leq 35$, and thus $s_4(v) \leq 8$, $re(G_v,R_v) \leq 8 \cdot 4 + 2 \cdot 3 = 38$, $ge(R_v,G_v) = 7 \cdot 10 - re(G_v,R_v) \geq 32$, and together with Lemma 2 then $t_i(v) = 0$ for $i \geq 6$ and $t_5(v) \geq 4$.

Let w_i for $1 \leq i \leq 7$ denote the vertices of R_v, and let $re(w_i,G_v) = 5$, that is $ge(w_i,G_v) = 5$, for $1 \leq i \leq t_5(v)$. If $re(w_i) \geq 10$ for some $i \leq t_5(v)$, then v and w_i are the sets of two vertices of a green $K_{2,5}$ and a red $K_{2,4}$, and Lemma 1 can be used. If $re(w_i) = 9$ or $re(w_i) = 8$ for some $i \leq t_5(v)$, then Lemma 4 works. If $re(w_i) = 6$ for some $i \leq t_5(v)$, then Lemma 5 can be applied. Since $re(w_i) \geq re(w_i,G_v) + 1 = 6$ for $i \leq t_5(v)$, it remains $re(w_i) = 7$, and thus $re(w_i,R_v) = 1$ for all $1 \leq i \leq t_5(v)$.

Now in R_v two vertices w_i and w_j with $i, j \leq t_5(v)$, which are joined in red, together with w_k, $k \leq t_5(v)$, determine a green $K_{3,3}$. Thus all edges (w_i,w_j), $i, j \leq t_5(v)$, can be assumed green.

If $t_5(v) \geq 5$, then w_6 or w_7, say w_7, is connected by at least three red edges to vertices w_i, $i \leq t_5(v)$, say to w_1, w_2, w_3. Then w_1, w_2, w_3 together with w_4, w_5, w_6 determine a green $K_{3,3}$.

It remains $t_5(v) = 4$. Then $ge(G_v,R_v) \geq 32$ implies $4t_4(v) + 3t_3(v) + ... \geq 12$, and this is possible only for $t_4(v) = 3$ and $ge(G_v,R_v) = 32$. It follows $re(w_i,G_v) = 6$ for $i = 5, 6, 7$. Three red edges from w_5, w_6 or w_7 to $\{w_1,w_2,w_3,w_4\}$ force a green $K_{3,3}$ as before. Thus for example w_5 is joined by exactly two red edges, say to w_1 and w_2. Then w_1, w_2 both are joined by green edges to w_3, w_4, w_6, and w_7. Any green edge from w_5 to w_6 or to w_7 forces a green $K_{3,3}$. Two red edges (w_5,w_6) and (w_5,w_7), however, imply $re(w_5) = 11$, which is covered by Lemma 5. — For $re(v) = 10$ interchange colors green and red. ❑

Lemma 7 *If in a two–colored K_{18} for all vertices v either $ge(v) = 12$ or $re(v) = 12$, then there exists a monochromatic $K_{3,3}$.*

Proof Assume that $re(v) = 12$ for at least 9 vertices v of K_{18}. If every pair of them is connected by a red edge, then a red K_9 occurs which contains a red $K_{3,3}$. If at least one pair is connected by a green edge, then both vertices are joined by red edges to at least eight other vertices. Then, however, $r(K_{3,3},K_{1,3}) = 8$ (Theorem 1) guarantees a monochromatic $K_{3,3}$. ❑

Now the proof of $r(K_{3,3}) \leq 18$ can be finished. By Lemmas 4 to 7 one vertex exists in a two–coloring of K_{18} with $ge(v) \leq 4$ or $ge(v) \geq 13$. Then $r(K_{3,3},K_{2,3}) = r(K_{2,3},K_{3,3}) = 13$ (Theorem 2) forces a monochromatic $K_{3,3}$, and Theorem 3 is proved. ❑

Thus $K_{3,3}$ is another graph which has 18 as its diagonal Ramsey number just as the path P_{13}, the star $K_{1,9}$, the book B_4, the complete graph K_4, and some other graphs. How large is the set of all graphs G with $r(G) = 18$?

A final remark: If the edges of a complete bipartite graph $K_{s,t}$ instead of K_r are two–colored, then all pairs (s, t) are known [4] such that any two–coloring of $K_{s,t}$ contains a monochromatic $K_{3,3}$, but $K_{s-1,t}$ and $K_{s,t-1}$ do not $(s \leq t)$, and these pairs are (5,41), (7,29), (9,23) and (13,17).

REFERENCES

[1] S.A. Burr, Diagonal Ramsey numbers for graphs, *J.Graph Theory* **7** (1983), 57 – 69.

[2] F.R.K. Chung and R.L. Graham, On multicolor Ramsey numbers for complete bipartite graphs, *J. Combinatorial Theory* **18** (B) (1975), 164 – 169.

[3] C. Clapham, G. Exoo, H. Harborth, I. Mengersen and J. Sheehan, The Ramsey number of $K_5 - e$, *J. Graph Theory* **13** (1989), 7 – 15.

[4] F. Harary, H. Harborth and I. Mengersen. Generalized Ramsey theory for graphs XII: Bipartite Ramsey sets, *Glasgow Math. J.* **22** (1981), 31 – 41.

[5] G.R.T. Hendry, Diagonal Ramsey numbers for graphs with seven edges, *Utilitas Mathematica* **32** (1987), 11 – 34.

[6] G.R.T. Hendry, Ramsey numbers for graphs with five vertices, *J. Graph Theory* **13** (1989), 245 – 248.

[7] H. Harborth and I. Mengersen, All Ramsey numbers for five vertices and seven or eight edges, *Discrete Math.* **73** (1988/89), 91–98.

ON GRAPHS WITH (I,n)-REGULAR INDUCED SUBGRAPHS

M. A. Henning

Henda C. Swart

P. A. Winter

University of Natal, Durban

ABSTRACT

For $n \geq 2$, a graph G is said to be (I,n)-regular of degree k if each vertex of G is an end vertex of exactly k non-trivial paths of length vertex of G is an end vertex of exactly k non-trivial paths of length at most $n-1$. For $n \geq 3$, a complete characterization of connected graphs of girth at least n for which every vertex-induced, connected, non-trivial, proper subgraph is (I,n)-regular is obtained.

1. Introduction

Th terminology and notation of [1] will be used throughout. In particular G will denote a connected graph with vertex set V, edge set E, order p, size q and girth g. If $v \in V$, the degree of v in G is written as deg v and $\delta =$ min $\{degv : v \in V\}, \Delta = \max\{degv : v \in V\}$. The degree set of G is denoted by D_G.

If n is an integer, $n \geq 2$ and $v \in V$, we denote by $p_n(v)$ the number of paths of length $n-1$ in G which have v as an end vertex and by $deg(n,v)$ the number of non- trivial paths of length $\leq n-1$ with v as an end vertex; hence $deg(n,v) = \sum_{m=2}^{n} p_m(v)$. The graph G is said to be *(I,n)-regular of degree* k if $deg(n,v) = k$ for all v in V. The graph G is said to be *highly (I,n)-regular* if G and all its non-trivial, vertex-induced, connected subgraphs are (I, n)-regular.

It is our purpose to characterize connected graphs of girth $\geq n$ for which every vertex-induced, connected, non-trivial, proper subgraph is (I, n)-regular for all integers $n \geq 2$, and also to characterize highly (I, n)-regular graphs of girth $\geq n$.

These concepts find application in the fields of Sociology, Epidemiology, etc., as follows: Given that an idea (infection, etc.) may be transferred from the originator and transmitted sequentially to $n-1$ others, without losing its integrity, design experiments to test the potency of ideas such that all participants in the experiment have equal potential for disseminating their ideas. We associate the participants in the experiment with the vertices of a graph G, two vertices of G being adjacent if and only if ideas may be transferred directly between the corresponding participants. The demands of the experiment are met if G if (I, n)-regular, where G is to be of girth at least n to avoid feed-back. However, since such experiments may be conducted over a long period, in which some participants are expected to withdraw and the results processed only for a proper subset of participants (those remaining to the end), it is preferable that G should have the property that each of its proper, connected, non- trivial, vertex-induced subgraphs be (I, n)-regular for all integers $n \geq 3$. Highly (I, n)-regular graphs would be preferable, but their use is not always feasible.

Obviously a graph is (I, n)-regular if and only if it is regular. The authors [2] obtained a complete characterization of (I, n)-regular graphs of girth at least n.

Theorem 1 For $n \geq 3$, a connected graph G of girth $\geq n$ is (I, n)-regular if and only if it satisfies one of the following conditions:

(a) G is r-regular.

(b) G is a tree of diameter $\leq n-1$.

(c) G is a bipartite graph with bipartition (X, Y), such that $degx = \Delta$ and $degy = \delta$ for all vertices $x \in X$ and $y \in Y$ and n is odd.

(d) If j is a proper factor of $n-1, G$ is obtained from a graph H satisfying condition (a) or (c) above by the replacement of each edge of H by a path of length j.

(e) If $j = n-1, G$ is obtained from a Δ-regular multigraph H by the replacement of each edge of H by a path of length j.

In particular, a connected graph G is $(I,3)$-regular if and only if it satisfies (a) (and $r^2 = k$) or (c) (and $k = \delta\Delta$).

We note that, if G satisfies condition (d) or (e), then j is the shortest distance between a vertex of degree Δ and another vertex of degree > 2 in G.

Theorem 2 A connected graph G of order $p \geq 3$ has the property that every vertex-induced, connected, non-trivial, proper subgraph is $(I,3)$-regular if and only if it satisfies one of the following conditions:

(i) G is a connected graph of order 3 of 4;

(ii) $p \geq 5$ and G is Kp of $K(p-t,t)$ for $1 \leq t \leq p-1$.

Proof If G is one of the graphs listed in (i) or (ii) above, then, clearly, every vertex-induced, connected, non-trivial, proper subgraph of G is $(I,3)$-regular.

Let G be a graph which satisfies the hypotheses of the theorem. Since there is nothing to show if $p \leq 4$, we may assume that $p \geq 5$.

If G is a tree, let v be an end-vertex of G. From the comment following Theorem 1 and the observation that G contains no path of length 3 as a vertex-induced subgraph, we obtain $G-v = K(p-2,1)$ and $G = K(p-1,1)$.

For the remainder of the proof we shall assume that G is not a tree. Let C be a cycle in G of minimum order. Necessarily C is a vertex-induced subgraph of G. If C is a cycle of order ≥ 5, then G would contain a vertex-induced proper subgraph which is a path of length > 2 and is not $(I,3)$-regular. Thus $C = C_3$ or $C = C_4$.

We next show that G is 2-connected. If this is not the case, then there is a cut-vertex v of G. Since G contains the cycle C, at least one component of $G - v$ is of order ≥ 2; let G_1 denote such a component and G_2 any other component of $G - v$. Then there exist vertices $u_1 \in V(G_1)$ and $u_2 \in V(G_2)$, each of which is adjacent to v in G. Let w be any neighbour of u_1 in G_1. Necessarily the vertex-induced subgraph on $\{u_1, u_2, v, w\}$ is either P_4 or $K_1 + (K_1 \cup K_2)$, neither of which is $(I, 3)$-regular, which contradicts our choice of G. Hence G is 2-connected.

We proceed by induction on p.

Case $C = C_3$: Suppose that $p = 5$ and $V(G) = V(C) \cup \{u_1, u_2\}$. Since G is 2-connected, each of u_1 and u_2 is adjacent to some vertex of C. Further, since the vertex-induced subgraph on $V(C) \cup \{u_1\}$ contains a triangle, the comment following Theorem 1 implies that this subgraph is K_4. Similarly, the vertex-induced subgraphs on $V(C) \cup \{u_2\}$ and $V(G) - \{v\}$, where $v \in V(C)$, are K_4. Consequently, G is K_5.

Suppose $p > 5$ and assume validity of the theorem for all connected graphs of order $p - 1$. This assumption, together with the fact that G contains a triangle and the observation that $G - v$ is connected for all vertices $v \in V$, implies that the vertex-induced subgraph on $V - \{v\}$ is K_{p-1} for every vertex $v \in V$. Therefore, obviously, $G = K_p$.

Case $C = C_4$: Suppose that $p = 5$ and that $V(G) = V(C) \cup \{u\}$. In view of the fact that G is 2-connected and triangle free, u must be adjacent to exactly two non-adjacent vertices of C. Therefore $G = K(2, 3)$.

Suppose $p > 5$ and assume validity of the theorem for all connected graphs of order $p - 1$. This assumption, together with the fact that G is triangle free and the observation that $G - v$ is connected for all vertices $v \in V$, implies that the vertex-induced subgraph on $V - \{v\}$ is $K(p - 1 - t_1, t_1)$ or $K(p - 1 - t_2, t_2)$ where $1 < t_1 < t_2 < p - 2$ and $t_2 - t_1 = 1$. Thus the complement of G is disconnected and G is reconstructible. Therefore G is $K(p - t_2, t_2)$, as required.

Theorem 3 For $n \geq 4$, a connected graph G of girth $\geq n$ has the property that every vertex-induced, connected, non-trivial, proper subgraph of G is (I,n)- regular if and only if it satisfies one of the following conditions:

(i) $G = K(2,3)$ and $n = 4$;

(ii) G is a tree of diameter $\leq n-1$;

(iii) $G = P_{n+1}$;

(iv) $G = C_n$;

(v) $G = C_{n+1}$;

(vi) $G = C_n \cup < u > +uw$, where $w \in V(C_n)$.

Proof If G is one of the graphs listed in (i) to (vi) above, then, clearly, every vertex-induced, connected, non-trivial, proper subgraph of G is (I,n)-regular.

Let G be a graph which satisfies the hypotheses of the theorem and which does not satisfy conditions (ii), (iii), (iv) or (v). Since G does not satisfy (ii) or (iii), Theorem 1 implies that G is not a tree. Let C be a cycle of minimal order in G. Note that since C is minimal it is a vertex-induced subgraph of G. Any vertex-induced subgraph of C is a vertex-induced subgraph of G and cannot be P_{n+1}, whence $|V(C)| \leq n+1$. Since the girth of $G \geq n$, it follows that $|V(C)| \in \{n, n+1\}$. Because G does not satisfy (iv) or (v), $C \neq G$.

Let u be a vertex in $V(G) - V(C)$ which is adjacent to a vertex of C. If u is adjacent to more than one vertex of C, then $C = C_4$ and the vertex-induced subgraph G_1 on $V(C) \cup \{u\}$ is $K(2,3)$ with $n = 4$. If $G_1 \neq G$, then G_1 must be (I,n)-regular, which is false by Theorem 1. Therefore $G = G_1 = K(2,3)$.

Assume u is adjacent to only one vertex of C. Let G_1 be the vertex-induced subgraph of G on $V(C) \cup \{u\}$. Then $C \neq C_{n+1}$, since otherwise G_1 would contain a vertex-induced path of length n which is not (I,n)-regular. Thus $G_1 = C_n \cup < u > +uw$ for some $w \in V(C)$. If $G_1 \neq G$, then G_1 must be (I,n)-regular which is false by Theorem 1. Therefore $G = G_1$.

The above theorems now yield a complete characterization of highly (I,n)-regular graphs of girth $\geq n$:

Corollary 1 For $n \geq 3$, a connected graph G of girth $\geq n$ is highly (I, n)-regular if and only if it satisfies one of the following conditions:

(a) $G = K_p$ and $n = 3$;

(b) $G = K(p - t, t)$ where $1 \leq t \leq p - 1$ and $n = 3$;

(c) G is a tree of diameter $\leq n - 1$;

(d) $G = C_n$;

(e) $G = C_{n+1}$.

The authors thank the referee for improvements to the paper.

REFERENCES

[1] G. Chartrand and L. Lesniak, *Graphs and Digraphs* (Second Edition), Wadsworth, Monterey, 1986.

[2] M. A. Henning, Henda C. Swart and P. A. Winter, "On (I,n)-regular graphs," *Ars Combinatoria* 23A (1987), 141-152.

MATCHING EXTENSION AND CONNECTIVITY IN GRAPHS ii

Derek Holton*

University of Otago - New Zealand

M. D. Plummer**

Vanderbilt University

1. Introduction

All graphs considered are finite, undirected, connected and simple (i.e., they have no loops or parallel lines). Let n and p be positive integers with $n \leq (p-2)/2$ and let G be a graph with p points having a perfect matching. Graph G is said to be *n- extendable* if every matching of size n in G extends to a perfect matching. (We shall call graph G *0-extendable* if it contains a perfect matching.) For a discussion of the role of the concept of n-extendability within the general framework of matching theory in graphs and for an historical résumé of the development of n-extendability, the reader is referred to the book [3], to [6] or to [7].

*Work supported by Grant UGC 32-635

** Work supported by ONR Contract #N00014-85-K-0488, NSF New Zealand-U.S.A. Cooperative Research Grant INT-8521818, and the University of Otago

In this paper we continue work begun in [6] and [7]. In Section 2 of this paper, we improve a general connectivity result as follows. In [7] it was shown that if one has a point cutset S of size $n+1$ in an n-extendable graph G where $n \geq 2$, then the graph $G-S$ has at most $n+1$ components with equality holding if and only if $G = K_{n+1,n_1}$. We sharpen this result to show that for $n \geq 2$, if G is n-extendable and $\neq K_{n+1,n+1}$, and if S is a point cutset of size $n+1$, then $G-S$ must have *exactly two* components.

In Section 3, we address a problem on matching extension in bipartite graphs. In [5], it was shown that no planar graph can be 3-extendable. In light of that result, it seemed natural to investigate 2-extendable planar graphs and this study was begun in [1]. In that paper it was shown, in particular, than any cubic 3-connected planar graph which is cyclically 5-connected must be 2-extendable. The planarity property played a strong role in the proof and it occurred to the authors to ask whether in a *non*-planar $n+1$-regular $n+1$-connected graph there was *any* lower bound on cyclic connectivity which would guarantee that the graph be n-extendable. We now show that the answer is yes, at least in the case when the graph is bipartite. Moreover, the bound we obtain is sharp by a recent result of Lou and Holton [2].

2. A Point-Connectivity Result

There are two basic theorems about n-connected graphs which we will need repeatedly in this paper, and we list them for the reader's convenience. Proofs of each may be found in [4].

1980A. Theorem If $n \geq 2$ and G is n-extendable, then G is $(n-1)$-extendable.

1980 B. Theorem If $n \geq 1$ and G is n-extendable, then G is $(n+1)$-connected.

We now present the main result of this section.

Theorem 2.1 Suppose n is a positive integer. Suppose further that G is an n-extendable graph and S is a point cutset of G with $|S| = n+1$. then:

(a) $G = K_{n+1,n+1}$, or

(b) if $n = 1$, then $G - S$ has at least 2 even components, but no odd components, or exactly 2 odd components, but no even components;

(c) if $n \geq 3$ and n is odd, then $G-S$ has exactly 2 odd components and no even components, or exactly 2 even components and no odd components; or

(d) if $n \geq 2$ and n is even, then $G - S$ has exactly 2 components, one of which is odd and the other, even.

Proof In this proof, in addition to Theorems 1980A and 1980B stated above, we will also make use of the simple observation that if $n \geq 1$ if G is *any* n-extendable graph and if $e = ab$ is any line in G, then the graph $G - a - b$ is $(n-1)$-extendable. The proof will proceed by induction on n. First we deal with the "anomalous" case when $n = 1$.

Suppose $G - S$ has an even component C. Let e_1 be a line joining C to S. Let M be a perfect matching containing e_1. Then by parity, M must match S into C. Thus $G - S$ has no odd components and at least 2 even components, as claimed.

Now suppose $G - S$ has no even components and hence at lest 2 odd components. But then since G is 1-extendable, $G - S$ must have exactly 2 odd components.

Now suppose $n = 2$. First suppose $G-S$ has at least 2 odd components (and hence by parity at least 3 odd components). Furthermore, let us suppose that $G - S$ has an even component C_e. Let e_1 be a line joining S to C_e. Then e_1 does not extend to a perfect matching, a contradiction. Hence $G-S$ has only odd components and at least of 3 of them.

Suppose one of these odd components, C_0, has at least 3 points. Let e_1 and e_2 be a matching from C_0 to S. Then $\{e_1, e_2\}$ does not extend to a perfect matching and again we have a contradiction. Thus all components of $G-S$ are singletons. Since G has a perfect matching, $G-S$ must consist of exactly 3 singletons. But since G is 3- connected, all of these singletons must have degree 3. Also recall from Theorem 2.2 of [7] that set S is independent. Thus $G = K_{3,3}$.

Suppose now that $G - S$ has exactly 1 odd component and hence at least 1 even component.

First we shall suppose that $G - S$ has at least 2 even components, say C_1 and C_2. Let $\{e_1, e_2\}$ be a matching of one point of C_1 and one of C_2 into S. Then the matching $\{e_1, e_2\}$ does not extend, a contradiction. Thus $G - S$ has exactly one even component as claimed.

Now let us suppose that $n = 3$. Suppose $G - S$ has at least 2 even components. Let C_3, C_4 be two such. Let $\{e_3, e_4\}$ be a matching of size 2 where e_i matches a point of C_i into S. Then $\{e_3, e_4\}$ extends to a perfect matching which must match a point in each of C_3 and C_4 to the remaining two points of S. Hence $G - S$ has no odd components.

Suppose now that $G - S$ has at least 3 even components. Let C_5, C_6 and C_7 be 3 such even components and let $\{e_5, e_6, e_7\}$ be a matching of size 3 of C_5, C_6 and C_7 respectively into S. Then $\{e_5, e_6, e_7\}$ does not extend to a perfect matching, a contradiction.

Thus G must have exactly 2 even components and hence no odd components. Note further here that it is impossible in this case that both even components are K_2's for if both were K_2's, graph G would not be 2-extendable. Hence by Theorem 1980A, G would not be 3-extendable either and this is a contradiction.

Suppose next that $G - S$ has no even components. Thus $G - S$ has at least 2 odd components.

Suppose one of these odd components has at least 3 points. Call it C_8. Let e_8, e_9 and e_{10} be a matching from components C_8 into S. Since this matching extends to a perfect matching, it follows that $G - S$ has precisely 2 odd components.

Now suppose that all the odd components of $G - S$ are singletons. Then again, since S is independent, it follows that $G = K_{4,4}$.

Finally, suppose that $G-S$ has exactly 1 even component C_e. (Hence G has at least 2 odd components.) Suppose that $|V(C_e)| \geq 4$. Then let e_{11}, e_{12} and $\{e_{13}\}$ be a matching of size 3 from C_e into S. Extending $\{e_{11}, e_{12}, e_{13}\}$ to a perfect matching is then impossible, contrary to the assumption that G is 3-extendable.

So $|V(C_e)| = 2$ and hence $C_e = K_2$.

Now suppose $G - S$ has at least 3 odd components. Let C_9, C_{10} and C_{11} be 3 such. Let $\{e_{14}, e_{15}, e_{16}\}$ be a matching where e_{14} joins C_9 to S, e_{15} joins C_{10} to S and e_{16} joins C_e to S. Then $\{e_{14}, e_{15}, e_{16}\}$ does not extend and again we have a contradiction.

Thus $G-S$ has exactly 2 odd components. Call these 2 components C_{12} and C_{13}. (Let us denote the even components C_e which is a single line by f.) Let $\{e_{17}, e_{18}\}$ be a matching of C_{12} (respectively C_{13}) into S. Then $\{e_{17}, e_{18}, f\}$ does not extend to a perfect matching, a contradiction.

This completes the proof for $n = 3$.

1. First suppose that n is even, $n \geq 4$ and that G is n-extendable.

1.1 Suppose $G - S$ has an even component C_e. Let $e = ab$ be a line joining S and C_e, where $a \in S$ and $b \in V(C_e)$. Let $H = G - a - b$. Then H is $(n-1)$-extendable (by Theorem 1980A) and $n-1$ is odd. Also $S' = S - a$ is a (minimum) cutset of H (by Theorem 1980B). Thus by the induction hypothesis, there are three possibilities: either $H = K_{n,n}, H - S'$ consists of 2 odd components and no even components, or $H - S'$ consists of 2 even components and no odd components.

1.1.1 Suppose $H = K_{n,n}$. Let $T = V(G) - S - b$. Now we already know that $S' = S - a$ is independent has cardinality n. Hence S' must be one of the two color classes of $H = K_{n,n}$. But then T must be the other color class of $H = K_{n,n}$ and hence T is independent in H. It now follows that all odd components of $G - S$ are singletons and that C_e is the only even component of $G - S$. Furthermore, since C_e is connected and $|V(C_e)| \geq 2$, component C_e must contain a line which must be of the form bc, where

$c \in V(C_e)$. Both line bc does not extend to a perfect matching and so G is not 1-extendable. Hence, yet again by Theorem 1980A, graph G is not n-extendable, a contradiction.

1.1.2. Next suppose that $H-S'$ consists of precisely two odd and no even components. Call the odd components A and B. Without loss of generality, suppose that $B \subseteq C_e$. Now $b \in V(C_e)$ and since S is a cutset of G, b is not adjacent to any point of A. Thus $G-S$ consists of precisely 2 components, one odd, the other even, as claimed.

1.1.3. Now suppose that $H-S'$ has precisely 2 even and no odd components. Call the even components D and E. Again, since S is a cutset in G, point b is adjacent to at most one of D and E and hence to exactly one of D and E, say E. But then $C_e = G[V(E) \cup \{b\}]$ is an odd component of $G-S$, a contradiction.

1.2. Let us now suppose that $G-S$ has no even components.

1.2.1. First suppose that all odd components of $G-S$ are singletons. Then since G is $(n+1)$-connected, we must have $G = K_{n+1,n+1}$.

1.2.2. Next suppose that at least one of the odd components of $G-S$ has at least 3 points, say component C_0. Let $ab = e$ be a line joining C_0 to S. Let $H = G-a-b$. Then H is $(n-1)$-extendable), and $S' = S-a$ is a minimum cutset of H, as before. So by the induction hypothesis, either $H = K_{n,n}, H-S'$ has precisely 2 odd components and no even components, or $H-S'$ has precisely 2 even components and no odd components.

1.2.2.1. Suppose $H = K_{n,n}$. It then follows that $G = K_{n+1,n+1}$.

1.2.2.2. Suppose $H-S'$ has 2 odd components and no even ones. Let the odd components be called D and E. Again, as before, point b is adjacent to exactly one of D and E, say E. This $G-E$ has an even component, namely $G[V(E) \cup \{b\}]$, contradiction.

1.2.2.3. Now suppose that $H-S'$ has 2 even components and no odd components. Again, point b is adjacent to exactly one of D and E, say E. Thus $G-S$ has an even component, namely D, and once more we have a contradiction.

This completes the proof in Case 1.

2. Now let us suppose that n is odd, $n \geq 5$ and G is n-extendable.

2.1 Suppose $G-S$ has an even component, C_e. Let H, a, b and S' be as in Case 1.1. Then by the induction hypothesis, either $H = K_{n,n}$ or $H - S'$ has exactly two components, one odd and the other, even.

2.1.1. Suppose $H = K_{n,n}$. As before, S' is an independent set of size n in H and as such, must be one of the two color classes of $H = K_{n,n}$. But then T must be the other color class of $H = K_{n,n}$ of size n and thus again $G-S$ has exactly one even component C_e (and $C_e = K_2$) and $n-1$ singleton (odd) components. But since S is independent, the line constituting C_e cannot extend to a perfect matching. Hence G is not 1-extendable, and hence by Theorem 1980A, G is not n-extendable, a contradiction.

2.1.2. Now suppose $H - S'$ has one odd component-call it A-and the other even - call it B. As before, point b is adjacent only to points of one of A and B, and since $b \in C_e$, b is adjacent to points of the odd component A. So $G[V(A) \cup \{b\}] = C_e$ and $G - S$ has exactly 2 even components.

2.2 So, finally, suppose $G - S$ has no even components, and hence at least 2 odd components.

2.2.1. Suppose one of these odd components, C_0, has at least 3 points. Choose a line $e = ab$ joining a point a in S to a point b in C_0. Let $H = G - a - b$ as before. By the induction hypothesis, graph H has precisely one odd component and one even. Let the odd component be A and the even, B. But then $G - S$ has exactly 2 odd components; namely A and $G[V(B) \cup \{b\}] = C_0$.

2.2.2. So we may suppose that all the odd components of $G - S$ are singletons. But then $G = K_{n+1,n+1}$ and the proof of the theorem is complete.

In closing this section, we remark that each of the classes (b), (c) and (d) of n-extendable graphs mentioned in the conclusion of the above theorem is non-empty. To see this, let us introduce some notation. Let G and H

denote any 2 graphs. Then by $G+H$ we mean the *join* of G and H. The join is defined to be the graph obtained from graphs G an H by joining each point of G to each point of H. Now let F, G and H be any three graphs. Let us denote by $F+G+H$ the join $G+(F \cup H)$. Finally, let K_2 denote the graph consisting of r disjoint liens. With this notation, we note that class (b) contains the infinite family $\overline{K_2}+rK_2$, for $r \geq 2, K_{2_r}+\overline{K_2}+K_{2s}$, for $r \geq 2$ and $s \geq 2$ and $K_{2r+1}+\overline{K_2}+K_{2s+1}$, for $r, s \geq 2$.

In class (c) for $n \geq 3, n$ odd, we have $K_{r+2}+\overline{K_{n+1}}+K_s, r, s \geq n$ and both odd, as well as $K_{r+3}+\overline{K_{n+1}}+K_{s+1}, r, s \geq n$, and both odd.

Finally, in class (d) for $n \geq 2, n$ even, we have $K_{r+1}+\overline{K_{n+1}}+K_{s+2}, r, s, \geq n$ and both even.

In closing our discussion of Theorem 2.1, it is worth observing that none of the conditions (b), (c) or (d) implies that G is n-extendable, so that a converse of Theorem 2.1 of this type fails to hold.

3. A Cyclic Connectivity Result for Regular Bipartite Graphs

In this section we will be concerned with extending matchings in regular bipartite graphs. Let us begin by recalling the concept of cyclic connectivity. The *cyclic connectivity* of a connected graph G is the cardinality of a smallest cutset of lines in G the deletion of which leaves 2 components each containing a cycle. We denote this number by $c\lambda(G)$.

Before presenting our main result, we state the next Lemma, the proof of which is trivial and is left to the reader.

Lemma 3.1 Let F be a forest with no isolated points. Suppose the bipartition of $V(F)$ is $A \cup B$, where $|A| = a, |B| = b$ and $a > b$. Then F contains a tree with at least 2 endpoints in A.

Now we are prepared for the main result of this section.

Theorem 3.2 Let n be a positive integer and let G be an $(n+1)$-regular, $(n+1)$-connected bipartite graph. Then if $c\lambda(G) \geq n^2$, G is n-extendable.

Proof If $n = 1$, the conclusion is immediate since G is just an even cycle.

So let us assume that $n \geq 2$. Suppose graph G satisfies the hypotheses of this theorem and suppose further that the bipartition of its points is $V(G) = A \cup B$. Now suppose that G is not n-extendable. Then there are n independent lines $e_1 = a_1b_1, \cdots, e_n = a_nb_n$ where each point $ai \in A$ and $b_i \in B$ and such that $G' = G - a_1 - \cdots - a_n - b_1 - \cdots - b_n$ has no perfect matching. Thus by the well-known theorem of Philip Hall on bipartite matching, we may assume, without loss of generality, that there exists a point set $A_1 \subseteq A$ with $|\Gamma_{G'}(A_1)| < |A_1|$. (Here $\Gamma_{G'}(A_1)$ denotes the set of neighbors of set A_1 in graph G'.) Moreover, since graph G is $(n+1)$-connected, $\{b_1, ..., b_n\}$ is not a cutset of G and hence A_1 contains no isolates in G'. Let us denote $\Gamma_{G'}(A_1)$ by B_1. Let $G_1 = G[A_1 \cup B_1]$ and let $G_0 = G[A_0 \cup B_0]$ where $A_0 = A - (A_1 \cup \{a_1, ..., a_n\})$ and $B_0 = B - (B_1 \cup \{b_1, ..., b_n\})$.

Note that here at the outset that the pair (A_1, B_1) where $\Gamma_{G'}(A_1) = B_1$, point set A_1 contains no isolates and $|\Gamma_{G'}(A_1)| = |B_1| < |A_1|$ is certainly not necessarily unique. Let us call such a pair (A_1, B_1) a *Hall barrier* in G'. (For more on the barrier concept in general, see [3].)

Note that $B_1 \neq \phi$ for if $B_1 = \phi$, then $\{b_1, ..., b_n\}$ separates A_1 from $G - A_1$, thus contradicting the assumption that G is $(n+1)$-connected. Thus $|B_1| \geq 1$ and hence $|A_1| \geq 2$. (Similarly, $B_0 \neq \phi$ and since $\{a_1, ..., a_n\}$ is not a cutset, we have $|A_0| \geq 1$ and hence $|B_0| \geq 2$.)

For the rest of this proof, let us adopt the following terminology:

$m =$ the number of lines from A_1 to $\{b_1, ..., b_n\}$,

$n_0 =$ the number of lines from B_0 to $\{a_1, ..., a_n\}$,

$n_1 =$ the number of lines from B_1 to $\{a_1, ..., a_n\}$,

$n_2 =$ the number of lines from B_1 to $\{a_1, ..., a_n\}$,

$n_3 =$ the number of lines from $\{b_1, ..., b_n\}$to A_0, and

$n_4 =$ the number of lines in $G[S]$,

where $S = \{a_1, ..., a_n\} \cup \{b_1, ..., b_n\}$. (See Figure 3.1 below. Line cut L shown in that figure will be discussed later in this proof.)

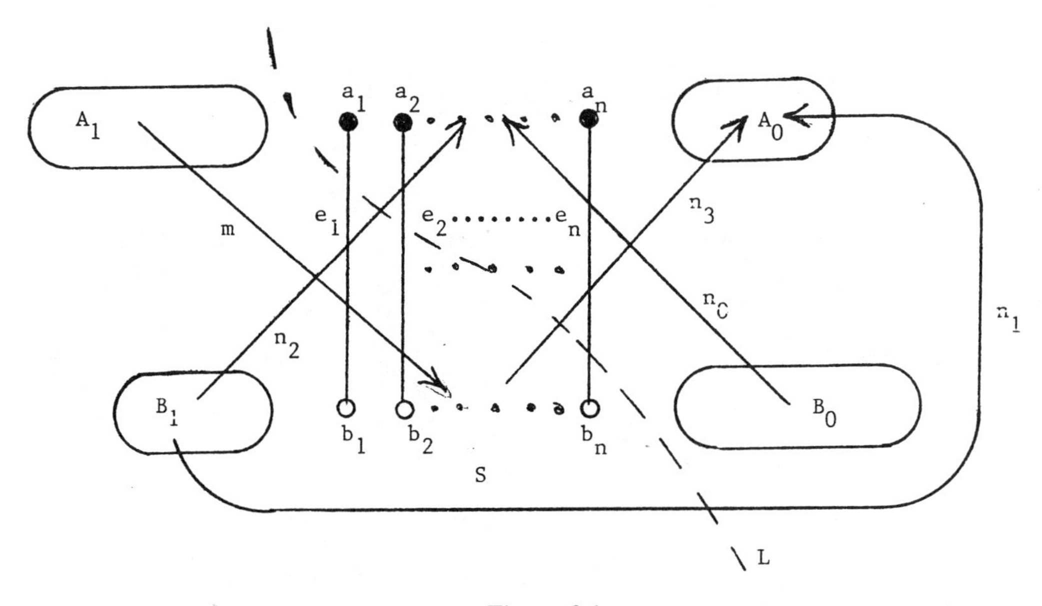

Figure 3.1

Now counting lines in G_1, we have $|A_1|(n+1) - m = |B_1|(n+1) - n_1 - n_2$, or

$$|A_1|(n+1) = |B_1|(n+1) - n_1 - n_2 + m. \tag{1}$$

Also $|A_1| \geq |B_1| + 1$, so

$$|A_1|(n+1) \geq (|B_1|+1)(n+1) = |B_1|(n+1)+n+1. \qquad (2)$$

Combining (1) and (2), we get $|B_1|(n+1)+n+1 \leq |B_1|(n+1)-n_1-n_2+m$, or

$$n_1 + n_2 \leq m - n - 1. \qquad (3)$$

Claim 1 If G_0 is a forest, then $|A_0|+|B_0| \leq 2n$. Since G_0 is a forest, $|V(G_0)| \geq |E(G_0)|+1$, and so

$$\begin{aligned} 2(|V(G_0)|) &= 2(|A_0|+|B_0|) \\ &\geq 2(|E(G_0)|+1) \\ &= \sum_{v\in V(G_0)} degv + 2 \\ &= |A_0|(n+1) - n_1 - n_3 + |B_0|(n+1) - n_0 + 2 \end{aligned}$$

or

$$(|A_0|+|B_0|)(n-1) \leq n_0 + n_1 + n_3 - 2. \qquad (4)$$

Now we also have

$$n_0 \leq n^2 - n_2, \qquad (5)$$

and

$$n_3 \leq n^2 - m. \qquad (6)$$

Thus substituting (3), (5) and (6) into (4), we obtain

$$\begin{aligned}(|A_0| + |B_0|)(n-1) &\le n^2 - n_2 + n_1 + n_3 - 2 \\ &\le n^2 - n_2 + m - n - 1 - n_2 + n_3 - 2 \\ &\le n^2 - 2n_2 + m - n - 3 + n^2 - m \\ &= 2n^2 - n - 3 - 2n_2 \\ &\le 2n^2 - n - 3 \\ &= n^2 - n + n^2 - 3\end{aligned}$$

So

$$\begin{aligned}|A_0| + |B_0| &\le n + \frac{n^2 - 3}{n-1} \\ &< n + \frac{n^2 - 1}{n-1} \\ &= 2n + 1.\end{aligned}$$

So $|A_0| + |B_0| \le 2n$, as claimed.

Claim 3 $n_1 + n_2 + n_3 + n_4 \le n^2 - 1$.

For we have

$$\begin{aligned}n_1 + n_2 + n_3 + n_4 &= n_1 + n_2 + n_3 + n(n+1) - m - n_3 \\ &\le m - n - 1 + n(n+1) - m \\ &= n^2 - 1.\end{aligned}$$

Now among all Hall barriers in G', let (A_1, B_1) be one with the smallest number of points. Without loss of generality, we may assume that $A_1 \subseteq A, B_1 \subseteq B$ and $|A_1| \ge |B_1| + 1 \ge 2$. Let $G_1 = G[A_1 \cup B_1]$. We now define $H_0 = G[A_0 \cup B_0 \cup \{a_1, ..., a_n\}]$ and $H_1 = G[A_1 \cup B_1 \cup \{b_1, ..., b_n\}]$.

Claim 3 Subgraph H_1 contains a cycle.

Suppose not. Then H_1 is a forest and hence so is G_1. Since G_1 is a Hall barrier, it contains no isolates. Since $|A_1| > |B_1|$, we may apply Lemma 3.1 to conclude that G_1 contains a tree T_1 with at least 2 endpoints in A_1. But since G is $(n+1)$- regular, each of these endpoints is adjacent to all of

$\{b_1, ..., b_n\}$ and hence, since $n \geq 2$, subgraph H_1 must contain a 4-cycle, a contradiction.

Claim 4 $|A_0| = |B_0| - 1$.

We will show that $|A_1| = |B_1| + 1$, and the claim will follow. Suppose $|A_1| \geq |B_1| + 2$.

Suppose $\alpha \in A_1$. Furthermore, suppose that every neighbor in B_1 of α is adjacent to some point in $A_1 - \alpha$. Then $(A_1 - \alpha, B_1)$ is a smaller Hall barrier than $A_1, B_1)$, a contradiction.

So for every $\alpha \in A_1$ there must exist a $\beta \in B_1$ such that α is adjacent to β, but β is adjacent to no other point in A_1. But then we must have a matching of A_1 into B_1 and hence $|A_1| \leq |B_1|$ which is again a contradiction.

Claim 5 H_0 contains a cycle.

Let $\ell = |B_0|$. Then $|A_0| = \ell - 1$ by Claim 4. Now suppose H_0 does not contain a cycle. Then H_0 is a forest (and hence so is G_0). Hence

$$\begin{aligned}
|V(H_0)| &= 2\ell - 1 + n \\
&\leq |E(H_0)| + 1 \\
&= (n + \ell - 1)(n + 1) - (n_1 + n_2 + n_3 + n_4) + 1 \\
&\leq (n + \ell - 1)(n + 1) - (n_2 - 1) + 1 \\
&= \ell n + \ell + 1.
\end{aligned}$$

So, in particular,

$$\begin{aligned}
\ell n &\leq 2\ell + n - \ell - 2 \\
&= \ell + n - 2 \\
&< 2n - 2
\end{aligned}$$

where we have used Claims 2 and 1 respectively.

But $\ell \geq 2$ and we have $2n < 2n - 2$, a contradiction.

Thus by Claims 3 and 5, if L is the set of lines counted by $n_1+n_2+n_3+n_4$ (see Figure 3.1), then L is a *cyclic* line cut and $c\lambda(G) \leq |L| \leq n^2 - 1$, contradicting the hypothesis and completing the proof of the theorem.

Remark 1 Figure 3.2 below shows a 3-regular 3-connected bipartite graph G with $c\lambda(G) = 3$ which is not 2-extendable, thus showing that our lower bound in Theorem 3.2 is sharp, at least for the value $n = 2$. Very recently, Lou and Holton [2] have shown that the lower bound in Theorem 3.2 is sharp for *all* $n \geq 2$ by constructing, for all such n, $(n+1)$-regular $(n+1)$-connected bipartite graphs which have $c\lambda(G) = n^2 - 1$, but which are not n-extendable.

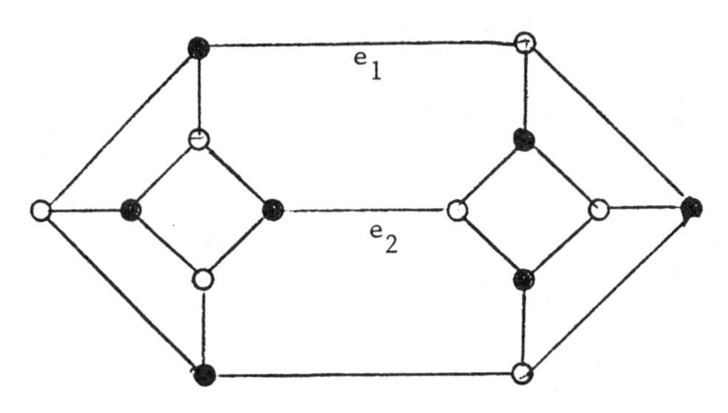

Figure 3.2

Remark 2 It is also interesting to compare the case $n = 2$ in Theorem 3.2 above with Theorem 3.1 of [1] which says that if G is a 3-regular 3-connected planar (but not necessarily bipartite) graph with $c\lambda(G) \geq 4$ and if G has no faces of size 4, then G is 2-extendable.

REFERENCES

[1] D. A. Holton and M. D. Plummer, "2-extendability in 3-polytopes," *Proceedings of the Seventh Hungarian Symposium on Combinatorics (Eger, 1987)* 52, Colloq. Math. Soc. János Bolyai, Akadémiai Kiadó, Budapest, 281-300.

[2] Dingjun Lou and D. A. Holton, "Lower bound of cyclic edge connectivity for n- extendability of regular graphs, preprint, 1988.

[3] L. Lovász and M. D. Plummer, *Matching Theory*, Ann. Discrete Math. 29, North- Holland, Amsterdam, 1986.

[4] M. D. Plummer, "On n-extendable graphs," Discrete Math. 31, 1980, 201-210.

[5] ——————, "A theorem on matchings in the plane," *Graph Theory in Memory of G. A. Dirac*, Eds.: Andersen, et al., Ann. Discrete Math. 41, North-Holland, Amsterdam, 1989, 347-354.

[6] ——————, "Matching extension in bipartite graphs," *Proceedings of the Seventeenth Southeastern Conference on Combinatorics, Graph Theory and Computing*, Congress. Numer. 54, Utilitas Math., Winnipeg, 1986, 245-258.

[7] ——————, "Matching extension and connectivity in graphs," *Proceedings of the 250th Anniversary Conference on Graph Theory,*" Eds.: K. S. Bagga, et al., Congress. Numer. 63, Utilitas Math., Winnipeg, 1987, 147-160.

ORTHOMORPHISMS AND NEAR ORTHOMORPHISMS

D. Frank Hsu

Fordham University

ABSTRACT

We briefly survey the development of orthomorphism, near orthomorphism, orthomorphism graph and their applications. Using a classification scheme similar to that used in the study of neofields, we exhibit orthomorphisms and near orthomorphisms of cyclic group of small order.

1. Introduction

Let $|G|$ be the order of a finite group $(G,.)$. We define a *K-Complete Mapping* where $K = \{k_1, ..., k_n\}$ to be an arrangement of the non-identity elements of G into s cyclic sequences of lengths $k_1 ..., k_s : (g_{11} \cdots g_{1k_1}), (g_{21} \cdots g_{2k_2}), \cdots, (g_{s1} \cdots g_{sk_s})$, so that the elements $g_{i,j}^{-1} g_{i,j+1})$, together with the elements range over all the non-identity elements of G when i ranges from 1 to the positive integer s. It is obvious that $\sum_{i=1}^{s} k_i = |G| - 1$. Similarly, a *K-Near Complete Mapping* where $K = \{k; k_1, ..., k_s\}$ is an arrangement of the elements of $|G|$ into a sequence with length h and s cyclic sequences of length $k_1, \cdots, k_s : [g'_1 \cdots g'_h](g_{11} \cdots g_{1k_1}) \cdots (g_{s_1} \cdots g_{sk_s})$ such that the elements $(g'_i)^{-1} g'_{1+1}$ and $g_{i,j}^{-1} g_{i,j+1}$ together with $g_{i,k_i}^{-1} g_{i,1}$ consist of each of the non-identity elements of G exactly once. Here we have $h + \sum_{i=1}^{s} k_1 = |G|$. A *Generalized Complete Mapping* is either a K-complete Mapping or a K-Near Complete Mapping.

Generalized Complete Mapping was defined and studied in Hsu and Keedwell [14] and [15] in a more general context. The concept of a generalized complete mapping subsumes those of a complete mapping, a starter, an even starter, a sequencing, a near-sequencing (or R-sequencing) and a left (or right) neofield. Generalized complete mappings have been used to characterize left neofields (see Hsu and Keedwell [14], to provide new constructions of block designs of Mendelsohn type (see Hsu and Keedwell [15] and to give graceful labelings of directed graphs (see Bloom and Hsu [1]). Reprints have been collected in Hsu [12] and a survey on the aspect of neofields and combinatorial designs is given in the collection. For other works of generalized complete mapping and its applications, the reader is referred to Hsu [13]. A K- Complete Mapping has been also studied under the notion of an orthomorphism in the construction of mutually orthogonal latin squares and finite projective planes. Since most of the constructions are in abelian group (especially in cyclic group), *orthomorphism* is defined addictively as follows: Let G be a group of order n written addictively whether abelian or not, an orthomorphism of G is a bijection $f : G \to G$ such that the mapping defined by $g(x) = f(x) - x$ is also a bijection. Without loss of generality, we can assume $g(0) = f(0) = 0$, when 0 is the identity of G.

In the following section, we briefly survey the development of orthomorphism, its structure and applications. We also define the notion of a near orthomorphism and show its relation to K-near complete mappings. In section 3, we introduce a classification scheme for the class of orthomorphisms and near orthomorphisms on cyclic groups. This scheme is analogous to that used in Hsu [11] for neofield with cyclic multiplicative group. Using this classification, we exhibit orthomorphisms and near orthomorphisms of cyclic groups of order up to $n = 9$.

2. Orthomorphism and Near Orthomorphism

The notion of an orthomorphism was first defined and studied in Johnson, Dulmage and Mendelsohn [16]. It is equivalent to complete mapping defined and studied earlier by Hall and Paige [10]. In fact, Mann [18] was the first to define and study complete mappings. More specifically, if $x \to f(x)$ is an orthomorphism, then $x \to -f(x)$ is a complete mapping. Hall and Paige [10] showed that a group G with a nontrivial cyclic Sylow 2-subgroup does not admit orthomorphisms. Although they did not prove the converse, they have shown that that converse is true for certain classes of groups, such as non-cycle 2-groups and the symmetric group $S_n, n > 3$.

In [11], Hsu studied neofields which have cyclic multiplication group and combinatorial designs using orthomorphism. Algorithm was given and a FORTRAN program was run on the Michigan Terminal System which lists all orthomorphisms of cyclic group of odd order. Orthomorphism does not exist in cyclic group of even order. However, certain mapping very similar to an orthomorphism does exist in cyclic group of even order. In general, we have the following definition.

Definition 1 Let G be a group written additively with 0 as identity. Let f be a bijection from G to G. Then f is called a *near orthomorphism* if $\{f(x) - x | x \in G\} = G - \{h\}$ where h is some element in G.

Just as an orthomorphism may be thought of as a K-complete mapping, so a near orthomorphism may be regarded as a K-near complete mapping. Moreover, given a K-near complete mapping, one can construct a near orthomorphism. Let f be a K-near complete mapping written addictively as follows:

$$[g'_1 g'_2 \cdots g'_h](g_{11} g_{12} \cdots g_{1k_1})(g_{21} g_{22} \cdots g_{2k_2}) \cdots (g_{s1} g_{s2} \cdots g_{sk_s})$$

where the elements $g'_{i+1} - g'_i$ and $g_{i,j+1} - g_{ij}$ together with the elements $g_{i,1} - g_{i,k_l}$ comprise the non-identity elements of G. In other words, the mappings $f(g'_i) = g'_{i+1}$ and $f(g_{i,j}) = g_i, j+1$ together with $f(g_i, k_i) = g_{i,1}$ map $G - \{g_h\}$ one-to-one onto $G - \{0\}$ and $\{f(x) - x / x \in G - \{g'_h\}\} = G - \{0\}$. Without loss of generality, we assume $g'_1 = 0$. Let the mapping g be defined as follows:

$$\begin{cases} g(x) = f(x) \quad + g'_h \cdot \forall x \neq g'_h \\ g(g'_h) = g g'_h \end{cases}$$

Hence the mapping g from G to G is one-to-one. For if $g(x) = g(y)$, we have either $f(x) = f(y)$ or $f(x) = 0$. Since $f(x) \neq 0$, it follows that $f(x) = f(y)$. This is impossible since f is one-to-one. Since $\{f(x) - x | x \in G - \{g'_h\}\} = G - \{0\}$, we have $\{g(x) - x | x \in G - \{g'_h\}\} = G - \{g'_h\}$. Therefore, $g(x) - x | x \in G\} = G - \{g'_h\}$ since $g(g'_h) - g'_h = g'_h - g'_h = 0$. Hence g is a near orthomorphism.

Using Theorem 3.5 of Hsu and Keedwell [14], we have

Theorem 1 For abelian groups, the existences of an orthomorphism and a near orthomorphism are mutually exclusive. An abelian group $(G, +)$ has an orthomorphism if and only f the sum of all its elements is 0. It has a near orthomorphism if and only if it has a unique element of order 2 and then the sum of all its elements is this unique element of order 2.

The proof of Theorem 3.5 in Hsu and Keedwell [14] is based on the results of L.J. Paige [21] and B. Gordon [9].

If an abelian group $(G, +)$ has a near orthomorphism f and equivalently K-near complete mapping with $g'_1 = 0$, then the element g'_h is the unique element of order 2 in G. Here we give two examples of near orthomorphism.

Example (1) The following mapping g is a near orthomorphism of Z_6 with its corresponding K-near complete mapping f where $K = \{3 : 3\}$:

x	0	1	2	3	4	5
g(x)	1	5	2	3	0	4
g(x)-x	1	4	0	0	2	5

x	0	1	2	3	4	5
f(x)	4	2	5	*	3	1
$f(x) - x$	4	1	3	*	5	2

Example (2) A K-near complete mapping f where $K = \{2, 3, 3, 3, 3\}$ is given in the dihedral group $D_7 = gp\{a, b : a^7 = b^2 = \in, ab = ba^{-1}\}$ together with its corresponding near orthomorphism g. For convenience, the group is wither multiplicatively.

x	e	a	a^2	a^3	a^4	a^5	a^6	b	ba	ba^2	ba^3	ba^4	ba^5	ba^6
$f(x)$	ba^3	a^4	a	ba	a^2	ba^2	ba^6	a^6	ba^5	ba^4	*	a^5	a^3	b
$x^{-1} \times f(x)$	ba^3	a^3	a^6	ba^4	a^5	b	ba^5	ba^6	a^4	a^2	*	ba^2	ba	a

x	a	a	a^2	a^3	a^4	a^5	a^6	b	ba	ba^2	ba^3	ba^4	ba^5	ba^6
g(x)	e	ba^6	ba^2	a^2	ba	a	a^4	ba^4	a^5	a^6	ba^3	ba^5	b	a^3
$x^{-1} \times g(x)$	e	b	ba^4	a^6	ba^5	a^3	a^5	a^4	ba^6	ba	e	a	a^2	ba^2

The *orthomorphism graph Orth(G) of a group* G is the graph with vertex set consisting of all orthomorphisms of G, two orthomorphisms α, β of G being adjacent if and only if the mapping $\delta : G \to G$ defined by $\delta(x) = \alpha(x) - \beta(x)$ is a bijection. The notion of an orthomorphism graph was first defined by Evans ([4], [5] and [6]) and used to study mutually orthogonal latin square and to construct affine planes (see also Johnson, Dulmage and Mendelsohn [16]. More specifically, an n-clique of the orthomorphism graph can be used to construct a set of $n+1$ mutually orthogonal latin squares from the Cayley table of G by permutating its columns while an affine plane of order n can be constructed from an $(n-2)$-clique of the orthomorphism graph of the group G. Recently, orthomorphism graphs have been studied extensively. However, the proofs have tended to use computer searches rather than cohesive theoretical approaches. A theory of orthomorphism graphs of groups needs to be developed.

Among the best examples of orthomorphism graphs studied are two particular finite groups Z_{11} and $Z_2 \times Z_6$. The clique number of Orth(Z_{11}) is 9 since the mappings $x \to ax. 2 \le a \le 10$ form a 9-clique. In 1961 Johnson, Dulmage and Mendelsohn [16] showed that Orth(Z_{11}) has only one 9-clique and found that the number of vertices of Orth(Z_{11}) is 3441 by using a computer. This was confirmed by Hsu in 1980 and Evans and McFarland in 1984. In his study of neofields, Hsu [11] showed that the number of cyclic neofields (i.e. neofields with a cyclic multiplicative group $G = Z_{11}$) is 3441. While in 1984, Evans and McFarland [8] found Orth (Z_{11}) to have 660 vertices of degree 3, 135 vertices of degree 8 and 6 vertices of degree 162, the remaining 2640 vertices are isolated.

The clique number of Orth($Z_2 \times Z_6$) is 4. The constructions of the 4-cliques were done independently by Bose. Chakravarti and Knuth [3] in 1960 using a Hadamard Matrix of order 12 and a computer and by Johnson, Dulmage and Mendelsohn [16] in 1961 using a search algorithm for computation. These three authors in 1961 also mentioned that Parker and Van Duren proved that Orth ($Z_2 \times Z_6$) contained no 5-cliques. This was reconfirmed in 1973 by Baumert and Hall [2] using computers. Construction of a 4-clique in Orth ($Z_2 \times Z_6$) was given by Mills [20] in 1977. However, no proof of the non-existence of a 5-clique has ever been given. A survey of the work done on Orth(Z_p), p an odd prime was given by Evans [6]. Other works on orthomorphisms, orthomorphism graphs and relation to difference matrices and generalized Hadamard Matrices can be found in Evans ([4] and [5]) and Jungnickel ([17] and [18]). Recently Evans and Hsu [7] used orthomorphisms to construct various types of triple systems such as directed triple systems and steiner triple systems.

3. Existence and Enumeration

All orthomorphisms and near orthomorphisms of $Z_n, 2 \leq n \leq 9$ are exhibited in the Appendix. As stated in Theorem 1, orthomorphisms are mutually exclusive for abelian groups. For cyclic groups, it is known that orthomorphism exists in every Z_n, n odd and near orthomorphism exists in every Z_n, n even. Johnson, Dulmage and Mendelsohn [15] gave an analysis of some special cases for orthomorphisms when the order n of the group is small. We briefly discuss orthomorphisms and near orthomorphisms for non-cyclic groups.

For $n = 4$, the group $Z_2 \times Z_2$ has 3 orthomorphisms. For $n = 8$, the group $Z_2 \times Z_4$ has no orthomorphisms which are automorphisms but has 49 orthomorphisms. The group $Z_2 \times Z_2 \times Z_2$ does not have orthomorphisms which are not automorphisms. However, the automorphisms group of $Z_2 \times Z_2 \times Z_2$ is the simple group of order 168. Following Sylow's theorem there exist 8 subgraphs of order 7 and each of these subgroups consists of elements which are orthomorphisms. For $n = 9$, other than the 225 orthomorphisms of Z_9 listed in the Appendix, the group $Z_3 \times Z_3$ has 28 orthomorphisms among the 48 automorphisms it has. There are plenty of other orthomorphisms of $Z_3 \times Z_3$ which are not part of a clique in Orth($Z_3 \times Z_3$). For the case $n = 12$, the group $Z_2 \times Z_6$ has only two orthomorphic automorphisms other than the identity. Other calculations using transversals shows that there exist 4-cliques in Orth($Z_2 \times Z_6$) as stated before. Apart from the identity, it was shown that there are exactly 16.512 orthomorphisms of $Z_2 \times Z_6$.

In general, the quest for the construction of orthomorphisms and near orthomorphisms is a challenging job. A theory together with applications has to be developed. It is known that the only finite abelian groups which have no orthomorphism are those with a unique element of order 2. It is also known that if the finite group G has an orthomorphism then G does not contain a cyclic Sylow 2-subgroup. It is commonly conjectured that the converse is true. In the case of near orthomorphism, by Theorem 1 and the discussion above, an abelian group G can not have a near orthomorphism unless it has a unique element of order 2. The converse of this is true since this kind of abelian group has a sequencing which is a near orthomorphism. One natural question to ask is: which finite non-abelian groups have orthomorphisms, which have near orthomorphism? Most research has been centered on dihedral groups. The reader is referred to Hsu [12] and Hsu [13] and references listed for further reading. The following orthomorphism is given in Hsu and Keedwell [14], where G is the non-abelian group of order 21 gp$\{a, b : a^7 = b^3 = \epsilon, ab = ba^2\}$.

$$(a\ a^2\ a^4)\ (a^3\ a^6\ a^5)\ (b\ b^2)\ (ba\ b^2\ a^3)\ (ba^2\ b^2\ a^6)\ (ba^3\ b^2\ a^2)$$
$$(ba^4\ b^2\ a^5)\ (ba^5\ b^2\ a)\ (ba^6\ b^2\ a^4).$$

This orthomorphism was used to construct a neofield of order 22 when the night distributive law holds.

The updated information for the number $\sharp(n)$ of orthomorphisms or near orthomorphisms other than the identity in cyclic group Z_n is:

n	3	4	5	6	7	8	9	10	11	12	13	14	15
$\sharp(n)$	1	2	3	8	19	64	225	928	3441	*	79.259	*	2.424.195

Here we introduce a classification scheme which is different from the study of orthomorphism graphs. Rather, this classification is analogous to the study of cyclic neofields. This method is more algebraic than combinatorial. Consequently, it may help in the quest for a solution to the enumeration problem, although we are not clear about its significance in the construction of mutually orthogonal latin squares.

Let G_n be the set of all orthomorphisms or near orthomorphisms of cyclic group Z_n of order n (we restrict to cyclic group for the purpose of explaining the Appendix). We define three operations dr, dc and tp on the elements of G_n as follows:

For every $f \in G_n : f(k) = t$ implies

$$\begin{aligned} f(k-t+h) &= -t \quad &\text{in} \quad & dr(f) \\ f(-k) &= t-k \quad &\text{in} \quad & dc(f) \\ f(t+h) &= k+h \quad &\text{in} \quad & tp(f), \end{aligned}$$

where

$$h = \begin{cases} 0 & \text{if n is odd} \\ n/2 & \text{if n is even} \end{cases}$$

Define $\Delta = tp * dc$. Then it is easily shown that $\Delta^3 = (dr)^2 = (dc)^2 = (tp)^2 = i$, the identity mapping. Since dc, dr, tp and Δ are well-defined mapping from G_n to G_n we consider G_n as a graph with elements of G_n as vertices. Two elements f and g in G_n are adjacent to each other if f is related to g by one of the four mappings dc, dr, tp and Δ. Moreover, $\{i.dc, dr, tp, \Delta, \Delta^2\}$ has subgroups $\{i\}, \{i, di\}, \{i, dr\}, \{i, tp\}$ and $\{i, \Delta, \Delta^2\}$. In 1980, Hsu [10] showed that the graph G_n consists of isolated cliques of K_1, K_2, K_3, and K_6 in the study of cyclic neofields.

In the case of K_3, three vertices of G_n are connected by dc.tp or dr. For K_2, two vertices of G_n are adjacent through dc, dr and tp where $dc(f) = dr(f) = tp(f)$. For the case of K_1, the vertex f is fixed by all mappings. In other words, $dr(f) = dc(f) = tp(f) = \Delta(f) = i(f)$. Therefore, the total number of orthomorphisms and near orthomorphisms, i.e. the number of vertices in G_n, is equal to $g(n) = 1 \times a_1 + 2 \times a_2 + 3 \times a_3 + 6 \times a_6$ where a_i is the total number of complete subgraphs K_i in G_n. In the appendix, AB, BC, CD, DE, EF, FA, AC, and AD represent the adjacency dr, dc, dr, dc, dr, dc, Δ and dt respectively.

Appendix

All orthomorphisms and near orthomorphisms of $Z_n, 3 \leq n \leq 9$ are listed here in the form of a generalized complete mapping. Following Theorem 1, all listed mappings are orthomorphisms if n is odd, and are near orthomorphisms if n is even. In the appendix, AB, BC, CD, DE, EF, FA, AC and AD represent the adjacency dr, dc, dr, dc, dr, dc, dr, dc, Δ and dt respectively.

	A	B	C	D	E	F
Z_3	(12)					
Z_4	[013]					
	[031]					
Z_5	(14)(23)	(13 42)		(12 43)		
Z_6	[021453]					
	[045213]					
	[015243]	[043](125)		[013](254)		
	[051423]	[023](154)		[053](124)		
Z_7	(13)(26)(45)					
	(15)(46)(23)					
	(13 26 45)	(154623)				
	(16)(25)(43)	(142)(356)		(124)(365)		
	(145 326)	(156243)	(134265)	(125463)	(136452)	(162354)
	(154236)	(132564)	(146523)	(126435)	(153462)	(163245)
Z_8	[04](127)(365)					
	[04](176)(235)					
	[04](1276)(35)	[04](2365)(17)				
	[04](167)(253)	[04](13)(27)(56)		[04](12)(36)(57)		
	[04](172)(356)	[04](16)(23)(57)		[04](13)(25)(67)		
	[05762314]	[06735214]		[05617324]		
	[03126574]	[02153674]		[03271564]		
	[05326714]	[027314](56)		[057364](12)		
	[03562174]	[061574](32)		[031524](76)		
	[01362754]	[06371254]		[01653724]		
	[07526134]	[02517634]		[07235164]		
	[01726354]	[023754](16)		[013764](25)		
	[07162534]	[065134](72)		[075124](63)		
	[04](17)(2563)	[04](132756)	[04](136752)	[04](125763)	[04](165723)	[04](1672)(35)
	[0254](1673)	[054](13)(267)	[014](263)(57)	[0214](3756)	[0564](1372)	[0164](2573)
	[0634](1572)	[034](162)(57)	[074](13)(256)	[0674](1325)	[0324](7516)	[0724](6315)
	[06127354]	[054](17)(236)	[014](276)(35)	[06523714]	[05376124]	[01736524]
	[02761534]	[034](17)(265)	[074](126)(35)	[02365174]	[03512764]	[07152364]

	A	B	C	D	E	F
Z_9	(12)(36)(48)(57)	(187245)(36)		(154278)(36)		
	(12)(37)(46)(58)	(187346)(25)		(164378)(25)		
	(18)(27)(36)(45)	(157842)(36)		(124875)(36)		
	(14)(26)(35)(78)	(126538)(47)		(183562)(47)		
	(15)(24)(36)(78)	(127548)(36)		(184572)(36)		
	(13)(25)(48)(67)	(137645)(28)		(154673)(28)		
	(16)(28)(34)(57)	(167243)(58)		(134276)(58)		
	(17)(24)(38)(56)	(14)(275683)		(14)(238657)		
	(15)(23)(47)(68)	(17)(235486)		(17)(268453)		
	(15278)(364)	(16378)(254)	(18736)(245)	(12)(375846)	(12)(364857)	(18725)(346)
	(164)(25783)	(17256438)	(18346527)	(142)(37)(568)	(124)(37)(586)	(146)(23875)
	(143)(26)(578)	(16534728)	(18274356)	(16742)(358)	(12476)(385)	(134)(26)(587)
	(15463)(278)	(15428)(367)	(18245)(376)	(136752)(48)	(125763)(48)	(13645)(287)
	(14652783)	(1428)(37)(56)	(1824)(37)(56)	(146752)(38)	(125764)(38)	(13872564)
	(146)(27853)	(17342658)	(18562437)	(16834752)	(12574386)	(164)(23587)
	(178325)(46)	(17325648)	(18465237)	(1452)(37)(68)	(1254)(37)(68)	(152387)(46)
	(18)(265)(347)	(1462)(3785)	(1264)(3587)	(12538746)	(16478352)	(18)(256)(374)
	(15647328)	(1783)(2546)	(1387)(2645)	(12537684)	(14867352)	(18237465)
	(17326458)	(17846)(253)	(16487)(235)	(125)(374)(68)	(152)(347)(68)	(18546237)
	(12748)(365)	(142635)(78)	(153624)(78)	(12638)(475)	(18362)(457)	(18472)(356)
	(1658)(2374)	(14352786)	(16872534)	(12643857)	(17583462)	(1856)(2473)
	(126538)(47)	(17826435)	(15346287)	(12685473)	(13745862)	(18476)(235)
	(1438)(2675)	(16537824)	(14287356)	(12746583)	(13856472)	(1834)(2576)
	(17548)(236)	(15)(247863)	(15)(236874)	(127)(35486)	(172)(36845)	(18457)(263)
	(1758)(26)(34)	(16)(247853)	(16)(235874)	(12743586)	(16853472)	(1857)(26)(34)
	(14375628)	(16524783)	(13874256)	(12764)(358)	(14672)(385)	(18265734)
	(14537628)	(13)(265)(478)	(13)(256)(487)	(12835)(476)	(15382)(467)	(18267354)
	(158)(27634)	(136)(27854)	(163)(24587)	(128436)(57)	(163482)(57)	(185)(24367)
	(15238)(476)	(13786254)	(14526873)	(12846537)	(17356482)	(18325)(467)
	(15347628)	(13)(26)(4785)	(13)(26)(4587)	(128476)(35)	(167482)(35)	(18267435)
	(13826754)	(153824)(67)	(142835)(67)	(13)(2748)(56)	(13)(2847)(56)	(14576283)
	(173)(268)(45)	(15328467)	(17648235)	(1325)(4867)	(1523)(4768)	(137)(286)(45)
	(1682)(3745)	(18435267)	(17625348)	(13264785)	(15874623)	(1286)(3547)
	(16823754)	(15243867)	(17683425)	(1327)(4856)	(1723)(4658)	(14573286)
	(175)(268)(34)	(167)(24853)	(176)(23584)	(13274586)	(16854723)	(157)(286)(34)
	(1534)(2768)	(13854267)	(17624583)	(13284756)	(16574823)	(1435)(2867)
	(1756)(2438)	(16758324)	(14238576)	(134)(27)(586)	(143)(27)(568)	(1657)(2834)
	(176)(24538)	(13267584)	(14857623)	(13472865)	(15682743)	(167)(28354)
	(15764)(238)	(13867254)	(14527683)	(137)(284)(56)	(173)(248)(56)	(14675)(283)
	(16457)(283)	(17256843)	(13486527)	(14237685)	(15867324)	(17546)(238)
	(178365)(24)	(148)(27563)	(184)(23657)	(14572)(386)	(12754)(368)	(156387)(24)
	(175846)(23)	(17324586)	(16854237)	(1527)(34)(68)	(1725)(34)(68)	(164857)(23)

REFERENCES

[1] G. S. Bloom and D. F. Hsu: "On graceful directed graphs. *SIAM J. of Algebraic and Discrete Methods.* v.6 (1985) p. 519-536.

[2] L. Baumert and M. Hall, Jr. "Nonexistence of Certain Planes of Order 10 and 12," *J. of Combinatorial Theory* A14 (1973) p. 273-280.

[3] R. C. Bose, I. M. Chakravarti and D. E. Knuth, "On Methods of Constructing Sets of Mutually Orthogonal Latin Squares using a computer I," *Technometries* 2 (1960) p. 507-516.

[4] A. B. Evans, "Generating Orthomorphisms of $GF(q)_+$, *Discrete Math.* 63 (1987) p. 21-26.

[5] A. B. Evans, "Difference Matrices, Generalized Hadamard Matrices and Orthomorphism Graphs of Groups," *JCMCC, V I* (1987), p. 97-105.

[6] A. B. Evans, "Orthomorphism Graphs of Z_p," *ARS Combinatoria* 25B (1988) p. 141-152.

[7] A. B. Evans and D. F. Hsu, "Triple Systems and Orthomorphisms," to appear.

[8] A. B. Evans and R. L. McFarland, "Planes of Prime Order with Translations," *Congressus Numerantium* 44 (1984) p. 41-46.

[9] B. Gordon, "Sequences in Groups with Distinct Partial Products," *Pacific J. Math.* 11 (1961) p. 1309-1313.

[10] M. Hall, Jr. and L. J. Paige, "Complete Mappings of Finite Groups," *Pacific J. of Math.* 5(1955), p. 541-549.

[11] D. F. Hsu, "Cyclic Neofields and Combinatorial Designs," LNM #824, Springer-Verlag, Berlin, 1980.

[12] D. F. Hsu, "Advances in Discrete Mathematics," Volume 1, *Neofields and Combinatorial Designs (edited)*, Hadronie Press, Nonantum, MA, 1985.

[13] D. F. Hsu, "Advances in Discrete Mathematics, Volume 2," *Generalized Complete Mappings (edited)*, Hadronie Press, Nonantum, MA, 1987.

[14] D. F. Hsu and A. D. Keedwell, "Generalized Complete Mappings, Neofields, Sequenceable Groups and Block Designs I," *Pacific J. Math.* 111 (1984), p. 317-332.

[15] D. F. Hsu and A. D. Keedwell, "Generalized Complete Mappings, Neofields, Sequenceable Groups and Block Designs II," *Pacific J. Math.* 117 (1985), p. 291-312.

[16] D. M. Johnson, A. L. Dulmage and N. S. Mendelsohn, "Orthomorphisms of Groups and Orthogonal Latin Squares. I., " *Canadian J. Math.* 13 (1961), p. 356-372.

[17] D. Jungnickel, "On Difference Matrices, Resolvable Transversal Designs and Generalized Hadamard Matrices," *Math. Z.*, 167 (1979), p. 49-60.

[18] D. Jungnickel, "On Difference Matrices and Regular Latin Squares," *Abh. Math. Sem. Univ. Hamburg* 50 (1980), p. 219-231.

[19] H. B. Mann, "The Construction of Orthogonal Latin Squares," *Annals of Math. Statistics* 12 (1942), p. 418-423.

[20] W. H. Mills, "Some Mutually Orthogonal Latin Squares," Proc. 8th SE Conf. on Combinatorics, Graph Theory and Computing," (1977), p. 473-487.

[21] L. J. Paige, "A Note on Finite Abelian Groups," *Bull. Amer. Math. Soc.* (A) 53 (1947), p. 590-593.

Cartan Matrices and Strong Perfect Graph Conjecture

Siming Huang
The University of Iowa

ABSTRACT

In this paper, we introduce the Cartan matrices of graphs. We prove that if G is a critical imperfect graph and its Cartan matrix is affine type, then $G = C_{2m+1}$, that is, an odd circuit. Furthermore, there exists a polynomial algorithm to check whether a Cartan matrix is affine type.

1. Introduction

In this paper, we only deal with simple graphs, that is, no multiple edges and no loops. A graph G is said to be *perfect* if G and each of its induced subgraphs have the property that the chromatic number χ equals the size of a maximum clique ω. A graph is *critically imperfect* if it is not perfect but each of its proper induced subgraphs is perfect. The Strong Perfect Graph Conjecture, posed by Berge in 1960, is that a graph G is perfect if and only if neither G nor its complement $\bar{G}$ contains as an induced subgraph an odd circuit of length greater than 3. An equivalent version of the Strong Perfect Graph Conjecture can be formulated as follows: The only critically imperfect graphs are the odd circuits and their complements.

In this paper, we introduce the Cartan matrices of graphs. We will state some of its properties and classifications in section 2, and prove that there exists a polynomial algorithm to determine whether the Cartan matrix of a graph is affine type. In section 3, we will prove the main theorem of this paper: If G is a critical imperfect graph and its Cartan matrix is affine type, then $G = C_{2m+1}$. The converse is also true.

2. Cartan Matrices of Graphs

Now we introduce the Cartan matrix of a graph G. More generally, we have the following definition.

Definition The $n \times n$ matrix A is called a generalized *Cartan matrix* if it satisfies the following conditions:

(C1): $a_{ii} = 2$ for $i = 1, 2, \ldots, n$;

(C2): $a_{ij} \leq 0$ i, j integers, $i \neq j$;

(C3): $a_{ij} = 0$ implies $a_{ji} = 0$.

A matrix A is called *decomposable* if, after reordering the indices (i.e., a permutation of its rows and the same permutation of the columns), A decomposes into the form

$$\begin{bmatrix} A_1 & 0 \\ 0 & A_2 \end{bmatrix}$$

Let G be any simple graph, and A be its adjacent matrix, we define $C(G) = 2I - A$ the Cartan matrix associated to G. For convenience, we denote $C(G) = C$.

The following theorem is obvious.

Theorem 2.1 *C is indecomposable if and only if G is a connected graph.*

We will give a complete classification of Cartan matrices. More generally, we consider the following matrices:

Let $A = (a_{ij})$ be a real $n \times n$ matrix which satisfies the following three properties:

(P1): A is indecomposable;

(P2): $a_{ij} \leq 0$ for $i \neq j$;

(P3): $a_{ij} = 0$ implies $a_{ji} = 0$.

The following theorem [1] gives a complete classification of all the matrices satisfying (P1) – (P3), hence of all Cartan matrices of connected graphs.

Theorem 2.2 *Let A be a real $n \times n$ matrix satisfying (P1) – (P3). Then one and only one of the following three possibilities holds for A:*

(Fin.) $\det A \neq 0$; there exists $u > 0$ such that $Au > 0$; $Av \geq 0$ implies $v > 0$ or $v = 0$;

(Aff.) $\operatorname{rank}(A) = n - 1$; there exists $u > 0$ such that $Au = 0$; $Av \geq 0$ implies $Av = 0$;

(Ind.) there exists $u > 0$ such that $Au < 0$; $Av \geq 0$, $v \geq 0$ implies $v = 0$.

Recall that a matrix of the form $(a_{ij})_{i,j \in S}$, where $S \subset \{1, 2, \ldots, n\}$, is called a *principal submatrix* of $A = (aij)$; we will denote it by A_S. The determinant of a principal submatrix is called a *principal minor*.

The following two lemmas are essentially the Lemma 4.4 and Lemma 4.5 of [1].

Lemma 2.3 *If A is of finite or affine type, then any proper principal submatrix of A decomposes into a direct sum of matrices of finite type.*

Lemma 2.4 *A symmetric matrix $A = (a_{ij})$ satisfying (P1) – (P3) is of finite (resp. affine) type if and only if A is positive definite (resp. positive semidefinite of rank $n - 1$).*

Theorem 2.5 *There exists a polynomial algorithm to determine whether a symmetric matrix $A = (a_{ij})$ satisfying (P1) – (P3) is of finite, affine, or indefinite type.*

Proof Let A_k denote the principal submatrix of A, where

$$A_k = \begin{bmatrix} a_{11} & \cdots & a_{1k} \\ & & \\ a_{k1} & \cdots & a_{kk} \end{bmatrix}, \quad k = 1, 2, \ldots, n.$$

Then by Lemma 2.4, A is of finite (resp. affine) type if and only if A is positive definite (resp. positive semidefinite of rank $n - 1$). By the well–known results of linear algebra, A is positive definite (resp. positive semidefinite of rank $n - 1$) if and only if $\det(A_k) > 0$ for $k = 1, 2, \ldots, n$ (resp. $\det(A_k) > 0, k = 1, 2, \ldots, n - 1$, $\det(A) = 0$). The complexity for computing $\det(A_k)$ is k^3, so the total complexity for determining whether A is positive definite (resp. positive semidefinite of rank $n - 1$) is $\Sigma k^3 = O(n^4)$. By Theorem 2.2, these three types are exclusive, hence we have proved the theorem. ❑

Corollary 2.6 *There exists a polynomial algorithm to determine whether the Cartan matrix of a connected graph G is of finite, affine, or indefinite type.*

Proof Since the Cartan matrix of a connected graph G is symmetric and satisfies (P1) – (P3), Theorem 2.5 applies. ❑

The following theorem will play a crucial rule in proving our main theorem in section 3.

Theorem 2.7 *Let G be a simple connected graph which contains a circuit of length greater than or equal to 3 and let C be its Cartan matrix. If C is of finite or affine type, then G is a circuit.*

Proof Suppose G contains a circuit with length ≥ 3, then there exists a principal submatrix B of C of the form

$$\begin{bmatrix} 2 & -1 & 0 \dots & 0 & -1 \\ -1 & 2 & -1 \dots & & 0 \\ \dots & \dots & \dots & \dots & \dots \\ 0 & 0 & \dots & 2 & -1 \\ -1 & 0 & \dots & -1 & 2 \end{bmatrix}$$

By Lemma 2.3, B is of finite or affine type. As in this case $\det B = 0$, Lemmas 2.3 and 2.4 imply that $B = C$, hence G is a circuit. ❑

3. The Main Theorem

A graph G of order $\alpha\omega + 1$ $(\alpha,\omega \geq 2)$ is called an (α,ω)–graph if, for every vertex v, the vertices of $G - v$ can be partitioned into α cliques of order ω and into ω independent sets of order α.

Theorem 3.1 [3] *Every critically imperfect graph G is an (α,ω)–graph with $\alpha = \alpha(G)$ and $\omega = \omega(G)$; where $\alpha(G)$ denotes the independence number of G and $\omega(G)$ denotes the clique number of G.*

The properties given in the next theorem were first established for critically imperfect graphs by Padberg [4], and then shown to apply to all (α,ω)–graphs by Bland, Huang, and Trotter [5].

Theorem 3.2 [4], [5] *Let G be an (α,ω)–graph of order n. Then*

(1) G has exactly n ω–cliques;
(2) G has exactly n independent α–sets;
(3) every vertex is in exactly ω ω–cliques;
(4) every vertex is in exactly α independent α–sets;
(5) every ω–clique is disjoint from exactly one independent α–set;
(6) every independent α–set is disjoint from exactly one ω–clique.

Now we are ready to prove our main theorem.

Theorem 3.3 *G is critical imperfect with C(G) affine if and only if $G = C_{2m+1}$, $m \geq 1$.*

Proof If $G = C_{2m+1}$, then G is critically imperfect and it is easy to check that C(G) is affine type.

If G is critically imperfect and C(G) is affine, then G is an (α,ω)–graph by

Theorem 3.1. We consider two cases:

1) If $\omega \geq 3$, then G has a cycle of length ≥ 3 by Theorem 3.2 (1). So G is a cycle by Theorem 2.7 since C(G) is affine. Furthermore, G is an odd cycle since G is imperfect.
2) If $\omega = 2$, then by properties of an (α,ω)–graph in Theorem 3.2, it is easy to prove that G is an odd cycle, hence $G = C_{2m+1}$. ❑

From Theorem 3.3, we can state an equivalent version of the Strong Perfect Graph Conjecture.

Strong Perfect Graph Conjecture *If G is critically imperfect, then either the Cartan matrix of G or its complement $\bar{G}$ is affine type.*

The following theorem states that in order to prove the above conjecture we may try to consider both the Cartan matrices C(G) and $C(\bar{G})$.

Theorem 3.4 *Let G be a graph, and $C = C(G)$, $\bar{C} = C(\bar{G})$ be the Cartan matrices of G and $\bar{G}$ respectively, then we have*

a) C and $\bar{C}$ cannot be both finite type if $n > 4$;

b) C and $\bar{C}$ cannot be both affine type if $n > 5$; C and $\bar{C}$ both are affine type for critically imperfect graphs if and only if $G = C_5$;

c) It cannot be the case that C is finite and $\bar{C}$ is affine if $n > 4$ or vice versa.

Proof a) Suppose C and $\bar{C}$ are finite type, and let $e = (1, \ldots, 1)^T$ be a vector with all components one. Then by Lemma 2.4, we have

$$e^T C e = 2n - \Sigma d_G(v_i) > 0$$

$$e^T \bar{C} e = 2n - \Sigma d_{\bar{G}}(v_i) > 0$$

which implies that

$$0 < e^T C e + e^T \bar{C} e = 4n - \Sigma\{d_G(v_i) + d_{\bar{G}}(v_i)\}$$

$$= 4n - \Sigma(n-1)$$

$$= n(5-n).$$

Hence we get $n < 5$, a contradiction.

The proofs of b) and c) are similar to the proof of a), so we will not state them here. ❑

From Theorem 3.4, we can conclude that if we are able to prove that C and $\bar{C}$ cannot be both indefinite or one is indefinite and the other is finite for critically imperfect graphs, then we have proved the above conjecture, i.e., the Strong Perfect Graph Conjecture. It needs further research and work.

REFERENCES

[1] Victor G. Kac, *Infinite Dimensional Lie–Algebras*, Second Edition, Cambridge University Press 1985.

[2] C. Berge and V. Chvatal, Topics on perfect graphs. *Annals of Discrete Mathematics* **21**.

[3] L. Lovasz, Perfect graphs, in *Selected Topics in Graph Theory 2* (eds. Lowell W. Beineke and Robin J. Wilson).

[4] M.W. Padberg, Perfect zero–one matrices. *Math. Programming* **6** (1974), 180 – 196.

[5] R.G. Bland, H.C. Huang and L.E. Trotter, Jr., Graphical properties related to minimal imperfection. *Disc. Math* **27** (1979), 11 – 22.

SHORTNESS PARAMETERS OF R-REGULAR R-CONNECTED GRAPHS

Bradley W. Jackson

San Jose State University

ABSTRACT

A graph G is said to be r-regular if every vertex of G has degree r. We also say that G is r-connected if it cannot be reduced to a disconnected graph (or a single vertex) by the removal of fewer than r vertices. In addition, we say that G is s-cyclically-edge- connected if it cannot be disconnected into two components, both containing a cycle, by the removal of fewer than s edges.

1. Introduction

In this paper we discuss techniques for measuring longest cycles in r-regular r-connected graphs. Known results for various families of r-regular r-connected s-cyclically-edge-connected are presented and several open problems are discussed.

2. Hamiltonian Cycles in Classes of 3-regular 3-connected Graphs

A Hamiltonian cycle in a graph G is a cycle that contains every vertex of G. If G has a hamiltonian cycle, we say that G is a hamiltonian graph has and otherwise we say that G is nonhamiltonian.

We denote by F_1 the family of 3-regular 3-connected graphs. It is well known that the family F_1 has nonhamiltonian members. The smallest nonhamiltonian 3-regular 3-connected graph has 10 vertices. It is known as the Petersen graph and is pictured in figure 1.

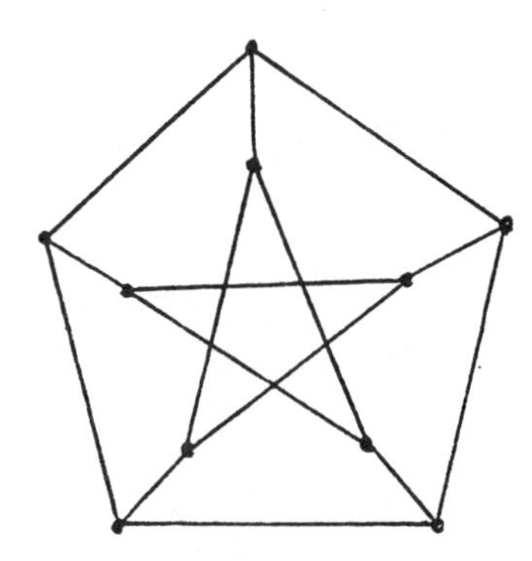

Figure 1. The Petersen graph

We denote by F_2 the family of 3-regular 3-connected planar graphs. The problem of finding hamiltonian cycles in 3-regular 3-connected planar graphs was first studied because of its relationship to the four color problem [18]. In 1946, Tutte presented a nonhamiltonian member of the family F_2 containing 46 vertices [19]. The smallest known nonhamiltonian 3-regular 3-connected planar graph has 38 vertices. It was discovered by Lederberg [14] in 1967. It is also known that any 3-regular 3- connected planar graph with 34 vertices or fewer has a hamiltonian cycle [1].

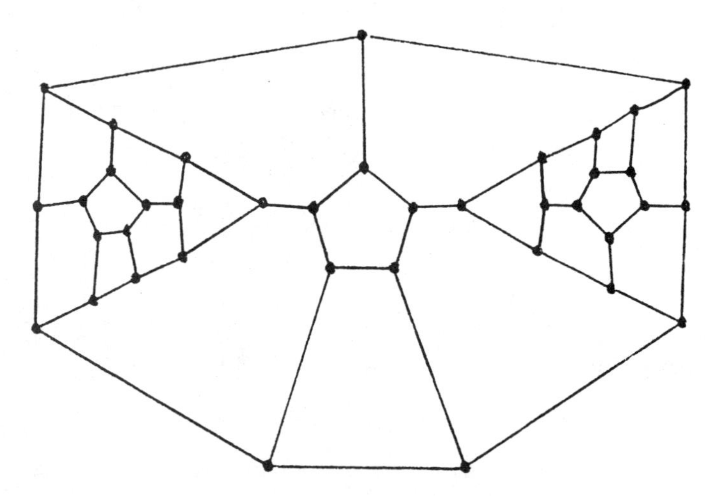

Figure 2. The Lederberg graph

We denote by F_3 the family of 3-regular 3-connected bipartite graphs. It was conjectured by Tutte [21] that every member of F_3 had a hamiltonian cycle, but Horton discovered a member of F_3 had a hamiltonian cycle, but Horton discovered a nonhamiltonian member of this family containing 96 vertices [2]. The smallest known nonhamiltonian 3-regular 3-connected bipartite graph has 54 vertices. It was discovered by Ellingham and Horton [5] and is pictured in figure 3.

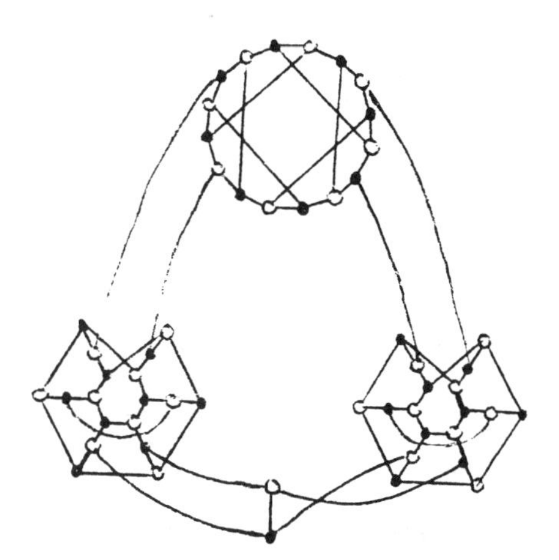

The Ellingham-Horton Graph

Finally we consider F_4, the family of 3-regular 3-connected bipartite planar graphs. Barnette has made the following conjecture.

Problem 1 Every 3-regular 3-connected bipartite planar graph has a hamiltonian cycle.

Figure 4. A counterexample to Barnette's conjecture
(to be supplied by the reader)

Although Barnette's conjecture remains open, it has been shown [9] that any counterexample must have at least 66 vertices.

2. Shortness Parameters of Families of Graphs

Let G be any graph. We denote by $c(G)$, the number of vertices in a longest cycle of G, and we let $n(G)$ represent the total number of vertices in G. We say that $a(G) = \frac{c(G)}{n(G)}$, is the shortness coefficient of G. For a given family of graphs F, we define the shortness coefficient $a(F)$ by $a(F) = \text{limin } f\{a(G)|G \in F\}$. If $S\underline{C}\ F$ is a subfamily of F and liminf $\{a(G) \mid GCS\} = c_1$, then we way that $a(F) \leq c_1$. On the other hand, if $a(G) \geq c_2$, for all G in F, we say that $a(F) \geq c_2$. We say that F has a nontrivial shortness coefficient if $a > c_1 \geq a(F) \geq c_2 > 0$.

In a similar way we define the shortness exponent of a graph G to be $b(G) = \frac{\log c(G)}{\log n(G)}$. As above, we can also discuss the shortness exponent of a family of graphs. Note that we have the following relationships between shortness exponents and shortness coefficients, $a(F) > 0$ implies that $b(F) = 1$ and $b(F) < 1$ implies that $a(F) = 0$.

3. Constructing Families of Graphs with Small Shortness Parameters

One operation that has been successfully exploited in several constructions of families of r-regular r- connected graphs with small shortness parameters is the removal and replacement of vertices. Suppose that H is an r-regular r-connected graph and let v be any vertex of H. We denote by $H\backslash v$ the topological set obtained by removing v from H but leaving the r free edges that were incident to v.

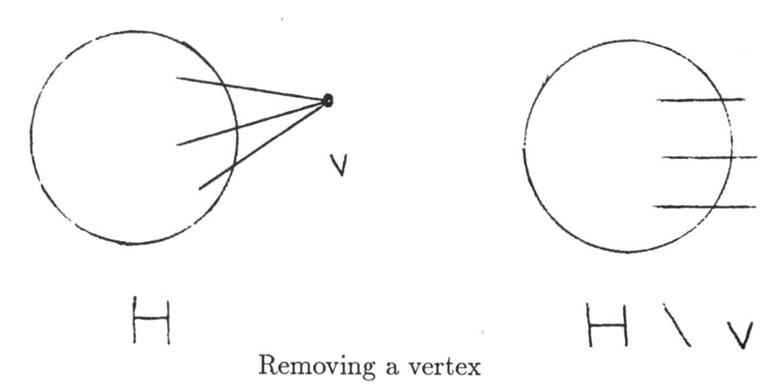

Removing a vertex

If G is an r-regular r-connected graph then $H\backslash v$ can be used to replace any vertex u in G by removing that vertex and identifying the r free edges of $H\backslash v$ with the r free edges of G/u in some order. It is well known [12] that the resulting graph is also r-regular and r-connected. We denote by $G[H]$ any graph obtained from G by replacing each vertex of G with $H\backslash v$ in this manner. Note that the graphs obtained in this manner have cyclic connectivity r but no higher.

In 1980, Bondy and Simonvits [3] described the following recursive construction. Let $G_0 = P$ be the Petersen graph and for $i \geq 1$, let $G_i = G_{i-1}$ [P].

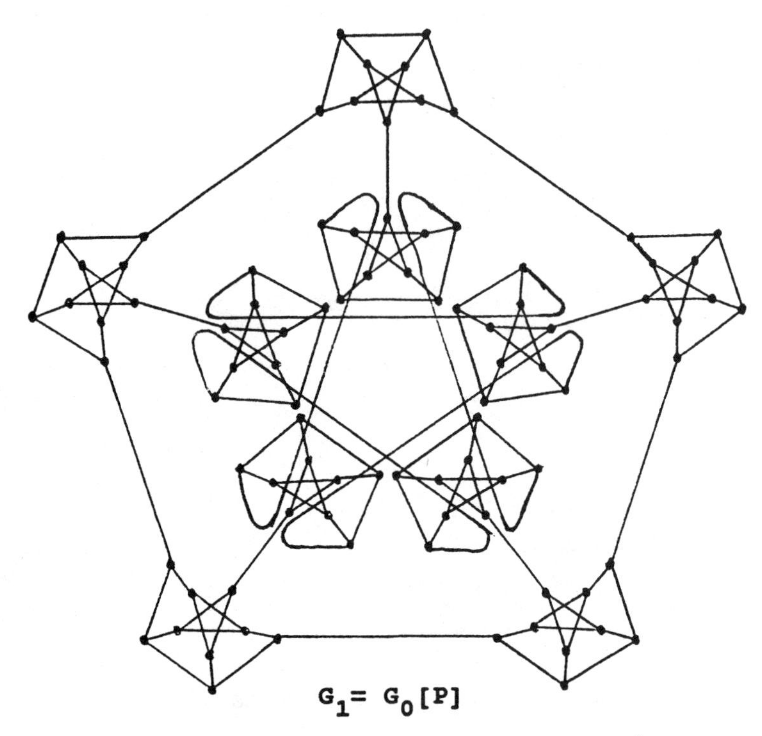

$G_1 = G_0[P]$

It is easy to see that $g(G_i) = 10(9^i)$ and $c(G_i) = 9(8^i)$. Thus $b(F_1) \leq \frac{\log 8}{\log 9}$ for the family of 3-regular 3-connected graph.

For the family F_2 a similar construction starting with the Lederberg graph (38 vertices) shows that $b(F_2) \leq \frac{\log 36}{\log 37}$. For the family of 3-regular 3-connected bipartite graphs a construction starting with the Ellingham-Horton graph (54 vertices) shows that $b(F_3) \leq \frac{\log 52}{\log 53}$. Likewise any counterexample to Barnette's conjecture (see figure 4) could be used to construct a family of 3-regular 3-connected bipartite planar graphs with shortness exponent $b < 1$.

In a recent paper, Bill Jackson [10] has shown that $b(F_1) \geq \log_2(1 + \sqrt{5}) - 1$. Thus showing that F_1 has a nontrivial shortness exponent. His result also gives a lower bound for the shortness exponents of the other families, but no further improvements in his bound for F_2, F_3, or F_4 have been computed. To construct families of 3-regular 3-connected graphs with cyclic-edge-connectivity greater than 3, different constructions are needed. One such construction involves the removal of an edge. Suppose that H is a 3-regular 3-connected graph and let e be any edge of H. We denote by $H\backslash e$ the topological set obtained from H by removing e and its endpoints. If x and y are the endpoints of e in H, then we label by x the two free edges of $H\backslash e$ that were incident to x and by y the two free edges of $H\backslash e$ that were incident to y.

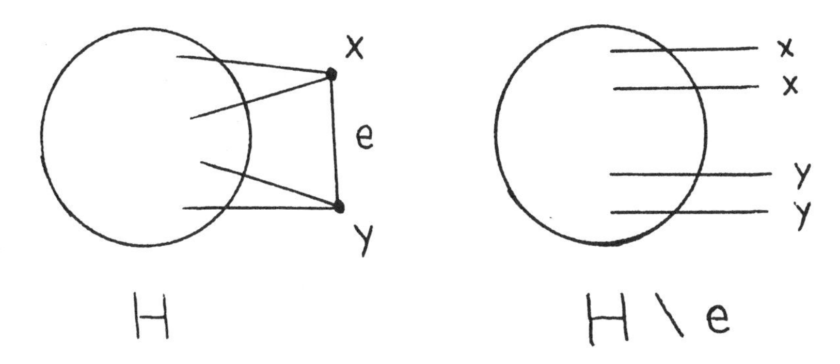

Figure 7. Removing an edge

We denote by $m[H]$ the graph obtained from m copies of $H\backslash e$ by identifying both x-edges of the ith copy H_i with both y-edges of the i+1st copy H_{i+1} mod m. Note that if H is 4-cyclically-edge-connected, then so is $m[H]$.

Consider the graph $m[P]$ in figure 8 which is constructed from m copies of the Petersen graph. Since the Petersen graph is 5-cyclically-edge-connected than $m[P]$ is 4-cyclically-edge-connected and since each copy of $P\backslash e$ has 8 vertices then $n(m[P]) = 8m$. If C is a cycle in $m[P]$ that contains all of the vertices of the ith copy P_i then either C enters P_i once using both x-edges and a second time using both y-edges, or it enters P_i once using both x-edges and a second time using both y-edges. From this it is easy to check that C contains all of the vertices of the ith copy P_i then either C enters P_i once using both x-edges or both y-edges, or it enters P_i once using both x-edges and a second time using both y-edges. From this it is easy to check that C contains all of the vertices in at most two copies of P and that $c(m[P]) = 7m + 2$. Thus the shortness coefficient of the family of 3-regular 3-connected 4-cyclically-edge-connected graphs is at most $7/8$.

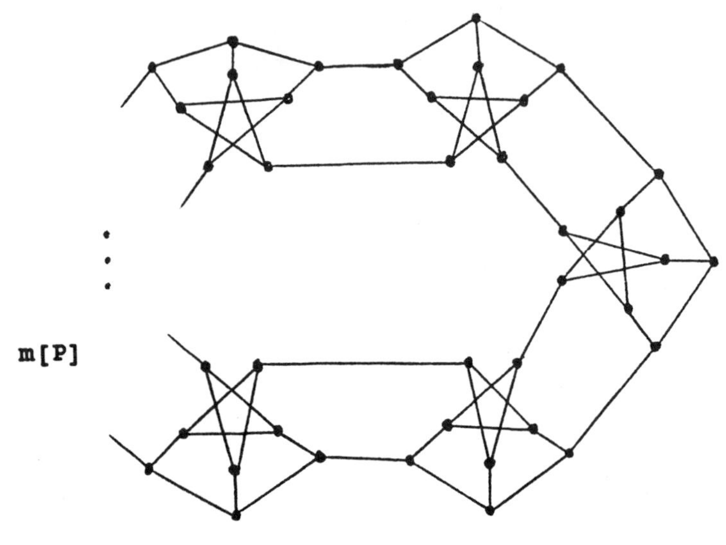

Figure 8

Bondy has made the following conjecture.

Problem 2 The shortness coefficient for the class of 3-regular 3-connected 4- cyclically-edge-connected graphs has a nontrivial lower bound $k > 0$.

For planar graphs the shortness coefficient has been well studied. In 1974, Grunbaum and Malkevitch [7] showed that the shortness coefficient of any 3-regular 3-connected 4-cyclically-edge-connected planar graph is at least $3/4$. The smallest known nonhamiltonian 3-regular 3-connected 4-cyclically-edge-connected planar graph has 42 vertices. It was discovered by Honsberger in 1973 and is pictured in figure 9. By removing an edge from this graph and joining m copies as above one can construct a family of 3-regular 3-connected 4-cyclically-edge-connected planar graphs with shortness coefficient $a = 39/40$. Thus this family of graphs has a nontrivial shortness coefficient.

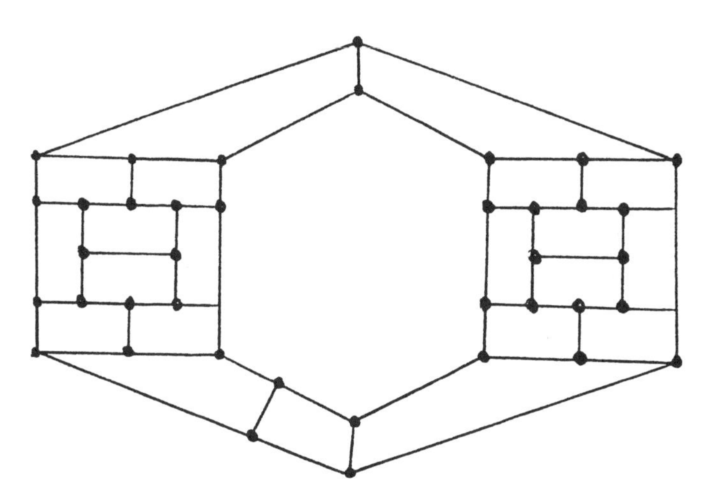

Figure 9. The Honsberger graphs

The Ellingham-Horton graph in figure 3 is a 3-regular 3-connected 4-cyclically-edge-connected bipartite graph with 54 vertices. By removing an edge from this graph and joining m copies one can construct a family of 3-regular 3-connected 4-cyclically-edge connected bipartite graphs with shortness coefficient $a = 51/52$.

There is also a 3-regular 3-connected 5-cyclically-edge connected planar graph with 44 vertices. It was discovered by Tutte in 1972 and is pictured in figure 10.

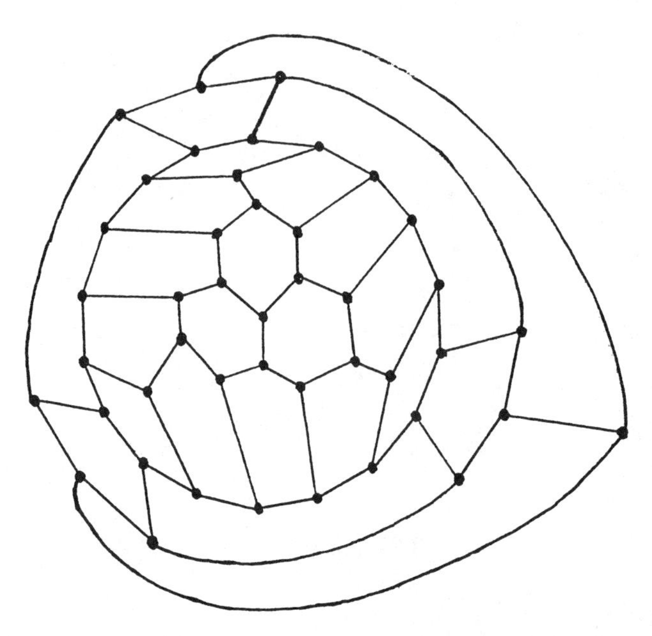

Figure 10. A Tutte graph with cyclic connectivity 5

In 1982, Zaks [22] constructed a family of 3-regular 3-connected 5-cyclically-edge connected planar graphs, containing only faces with 5,14,20 or 35 sides, with shortness coefficient equal to $a = 84/85$. His construction uses the subgraph Q_1 in figure 11 with 85 vertices. When Q_1 is a subgraph of a graph G, connected to the rest of G by the edges e_1, e_2, e_3, e_4, e_5, then no cycle in G contains all the vertices of Q_1.

A similar construction for arbitrary 3-regular 3-connected 5-cyclically-edge-connected graphs can b used to construct a family of graphs with shortness coefficient $a = 34/35$. It uses the subgraph Q_2 in figure 12 which has 35 vertices. As above when Q_2 is a subgraph of a graph G, connected to the rest of G by the edges e_1, e_2, e_3, e_4, e_5, then no cycle in G contains all the vertices of Q_2. There are no known 3-regular 3-connected bipartite graphs with cyclic connectivity equal to five.

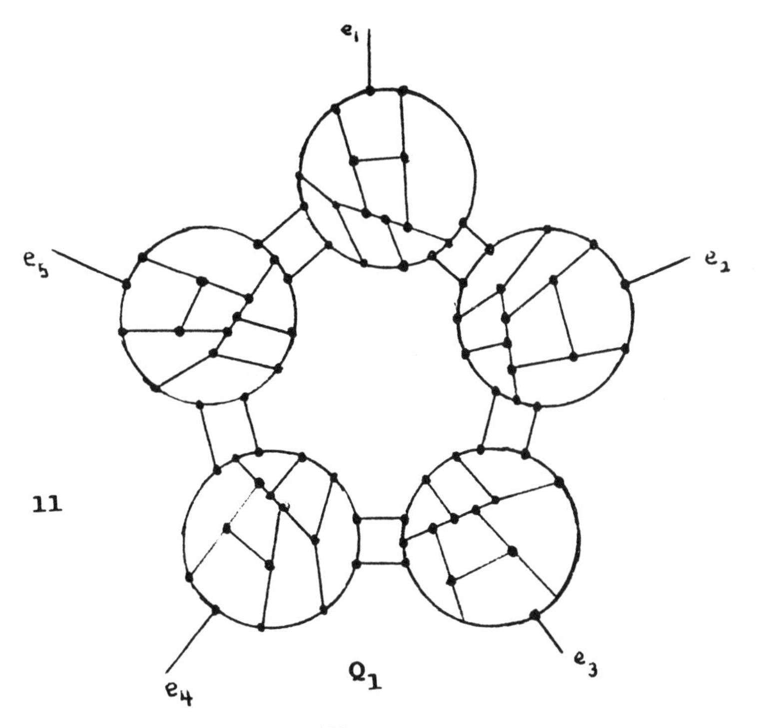

Figure 11

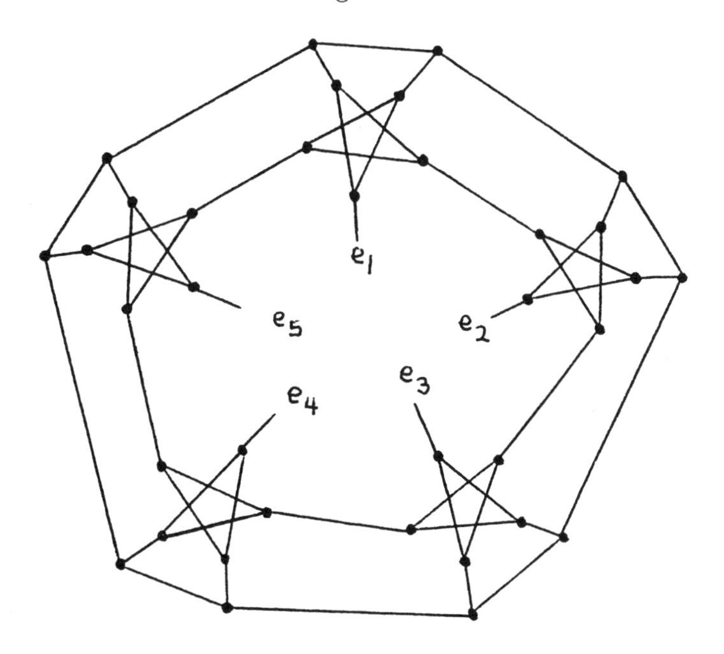

Figure 12

Little is known about nonhamiltonian 3-regular 3-connected graphs with cyclic connectivity greater than five. Note that there are no 3-regular 3-connected planar graphs with cyclic connectivity greater than five. However, for the other two classes of 3-regular 3-connected graphs that we have encountered we have the following problems.

Problems 3-4 For the class of 3-regular 3-connected graphs F_1 (bipartite graphs F_3) is there an s_i sufficiently large, so that all s_i-cyclically-edge-connected graphs in F_i have a hamiltonian cycle.

Note that there is a 3-regular 3-connected 7-cyclically-edge-connected non-hamiltonian graph with 28 vertices. It is known as the Coxeter graph and is pictured in figure 13.

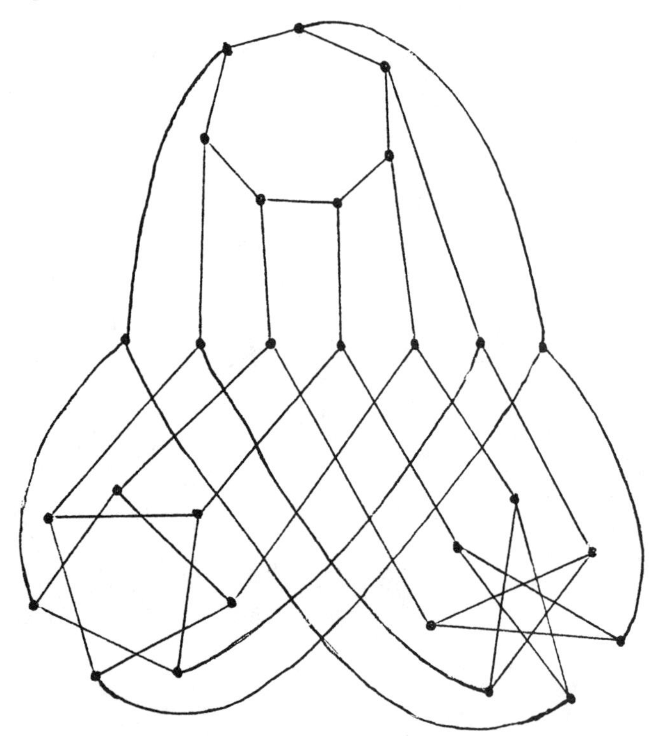

Figure 13. The Coxeter graph

4. Hamiltonian Cycles in Classes of 4-regular 4-connected Graphs In 1971 C. Nash- Williams [17] conjectured that every 4-regular 4-connected graph was hamiltonian. However in 1973, Meredith [16] constructed nonhamiltonian r-regular r-connected graphs, for every $r > 3$. The construction starts with a nonhamiltonian r-regular r-edge-connected multigraph obtained by adding edges to the Petersen graph. Each vertex is then replaced by a copy of $K_{r,r}$ with one vertex removed. The resulting graph is an r-regular r-connected nonhamiltonian graph with $10(2r-1)$ vertices. A 4- regular 4-connected nonhamiltonian graph with 70 vertices is pictured in figure 14.

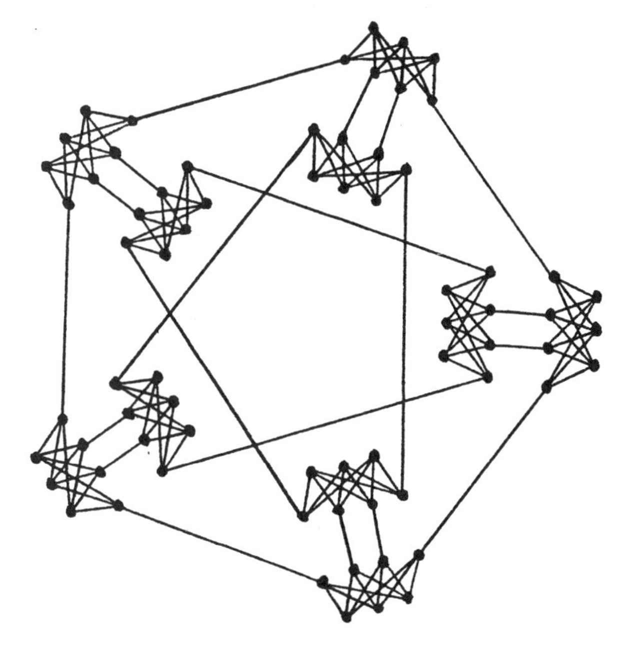

Figure 14. The Meredith graph for $r = 4$

The smallest known 4-regular 4-connected nonhamiltonian graph has 52 vertices. It was discovered in 1987 by Brad Jackson [11] and is pictured in figure 15.

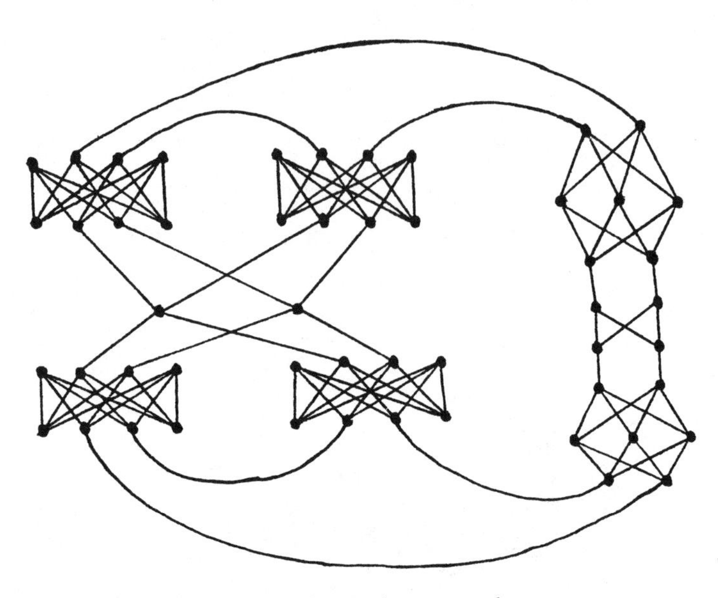

Figure 15. The Jackson graph

As for other classes of 4-regular 4-connected graphs, Tutte [19] proved in 1956 that every 4-connected planar graph has a hamiltonian cycle. By starting with a nonhamiltonian 3-regular 3-connected bipartite graph and adding edges to obtain a nonhamiltonian r-regular r-edge-connected bipartite multigraph, Meredith's construction can also be used to obtain nonhamiltonian r-regular r-connected bipartite graphs, for every $r > 3$. The smallest known 4-regular 4-connected nonhamiltonian bipartite graph was constructed by McCuaig and Rosenfeld [15] in 1984. It has 84 vertices and is pictured in figure 16.

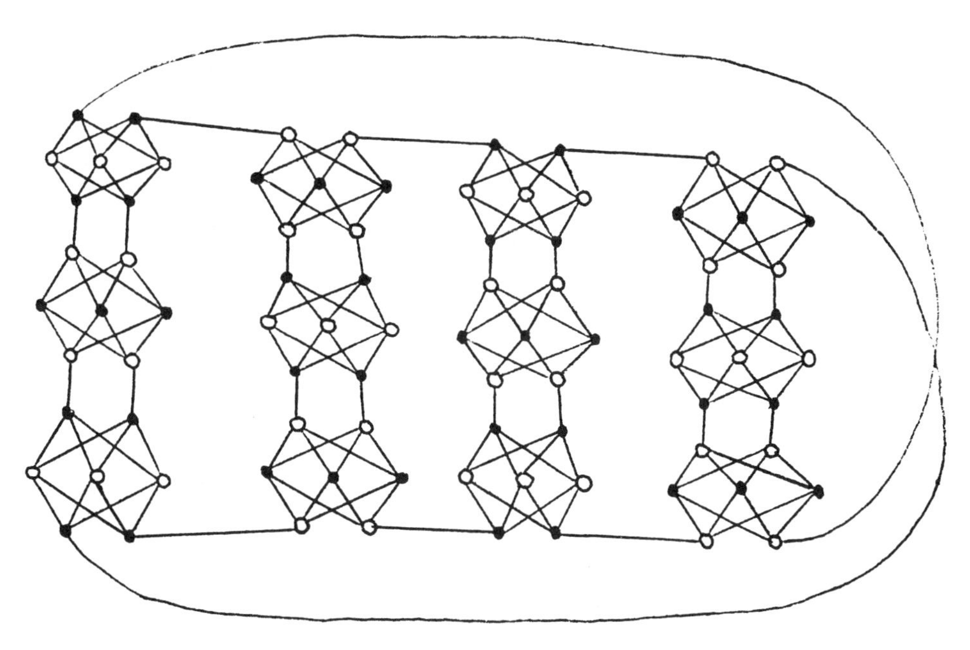

Figure 16. The McCuaig-Rosenfeld graph

5. Shortness Parameters of 4-regular r-connected Graphs

In 1981, Jackson and Parsons [13] found a family of r-regular r-connected graphs with a nontrivial shortness exponent. Starting with a r-regular r-edge-connected multigraph with 10 vertices their construction would show that the shortness exponent of the family of r-regular r-connected graphs is at most $\frac{\log(10(r-1)-1)}{\log(10(r-1))}$. Note that these graphs have cyclic connectivity r but no greater since the construction involves vertex replacement. In a similar way it is possible to find an upper bound for the shortness exponent r-regular r-connected bipartite graphs. This leads to the following problems.

Problem 5-6 Find a lower bound for the shortness exponent of r-regular r- connected (bipartite) graphs.

Problem 7 Show that the class of r-regular r-connected $r+1$-cyclically-edge- connected graphs has a nontrivial shortness coefficient. For s even and $s < \frac{4}{3}r$ there are r-regular r-connected s-cyclically-edge-connected graphs with no hamiltonian cycle. A 5-regular 5-connected 6-cyclically-edge-connected nonhamiltonian graphs is pictured in figure 17.

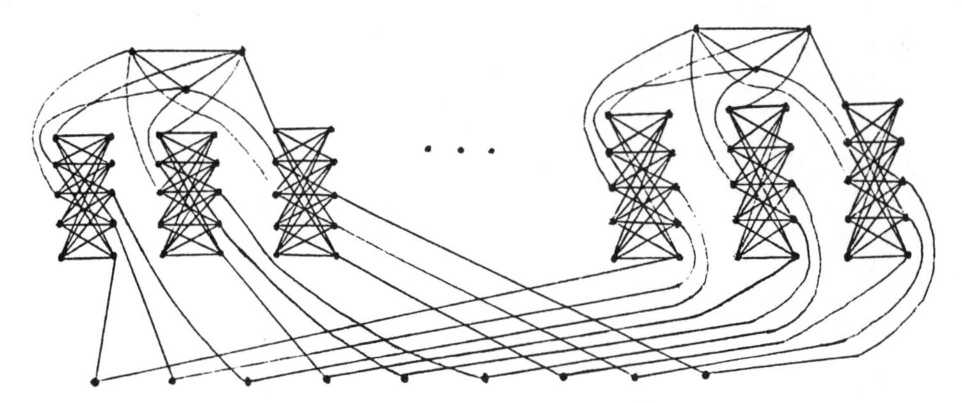

Figure 17. A 5-regular 5-connected nonhamiltonian graph

We conclude with the following set of problems.

Problems 8 - ∞ For a given $r > 3$ and s sufficiently large, show that all r-regular r-connected s-cyclically-edge-connected graphs are hamiltonian.

REFERENCES

[1] D. Barnette, "Every Simple 3-Polytope with 34 vertices is Hamiltonian," *Discrete Math.* 62(1986) 1-20.

[2] J. Bondy and U. Murty, "Graph Theory with Applications," Macmillan and Co., London, 1976.

[3] J. Bondy and M. Simonovits, "Longest cycles in 3-connected 3-regular graphs," *Canad. J. Math.*, 32(1980) 987-992.

[4] M. Capobianco and J. Molluzo, "Examples and Counterexamples in Graph Theory," Chapter 9 *(Traversability)*, North-Holland, New York, 1978.

[5] M. Ellingham and J. Horton, "Honhamiltonian 3-connected cubic bipartite graphs," *J. Combinatorial Th. Ser. B4* (1983) 350-353.

[6] G. Faulkner and D. Younger, "Nonhamiltonian cubic planar graphs," *Discrete Math.*, (1974)67-74.

[7] B. Grunbaum and J. Malkevitch, "Pairs of edge-disjoint hamiltonian circuits," *Aequationes Math.* 14(1976) 191-196.

[8] B. Grunbaum and H. Walther, "Shortness exponents of families of graphs," *J. Combinatorial Theory Ser. A*, 14 (1973), 364-385.

[9] D. Holton, B. Manvel, B. McKay, "Hamiltonian cycles in Cubic 3-connected bipartite planar graphs," *J. Combinatorial Th. Ser. B,* 38 (1985) 279-297.

[10] Bill Jackson, "Longest Cycles in 3-connected cubic graphs," *J. Combinatorial Theory Ser. B*, 41 (1986), 17-26.

[11] Brad Jackson, "Small r-regular r-connected nonhamiltonian graphs," *Ars Combinatoria* 24(1987) 77-83.

[12] Brad Jackson and T. Parsons, "On r-regular r-connected nonhamiltonian graphs," *Bull. Australian Math. Soc.* 24 (1981) 205-220.

[13] Brad Jackson and T. Parsons, "A shortness exponent for r-regular r-connected graphs, *J. Graph Theory*, 6(1982) 169-176.

[14] J. Lederberg, "Hamiltonian circuits of convex trivalent polyhedra (up to 18 vertices)," *American Math. Monthly* 74 (1967) 522-527.

[15] W. McCuaig and M. Rosenfeld, "Cyclability of r-regular r-connected graphs," *Bull. Australian Math. Soc.*, 29 (1984) 1-11.

[16] G. Meredith, "Regular n-valent n-connected nonhamiltonian non-n-edge-colorable graphs," *J. Combinatorial Theory Ser. B*, 14 (1973) 55-60.

[17] C. Nash-Williams, "Possible directions in graph theory," *Combinatorial Mathematics and its Applications*, D. Welsh, editor, Academic Press, New York, 1971.

[18] P. Tait, "Remarks on the coloring of maps," *Proc. Royal Society Edinburgh Ser. A.*, 10(1980) 729.

[19] W. Tutte, "On hamiltonian circuits," *J. London Math. Soc.* 21 (1946) 98-101.

[20] W. Tutte, "A theorem on planar graphs," *Trans. American Math. Soc.* 82 (1956) 99-116.

[21] W. Tutte, "On two-factors of bipartite regular graphs," *Discrete Math.* 41(1982) 35-41.

[22] J. Zaks, "Shortness Coefficient of Cyclically 5-connected cubic planar graphs," *Aequationes Math.* 24 (1982) 97-102.

AN INTRODUCTION TO TOLERANCE INTERSECTION GRAPHS

M. S. Jacobson
F. R. McMorris
University of Louisville

H. M. Mulder
Erasmus Universiteit

ABSTRACT

A masterplan of tolerance intersection graphs is proposed. Any class of intersection graphs can be generalized by introducing tolerances. As an example, nested families of sets with tolerances are studied, by which new characterizations of threshold graphs, interval graphs and coTT graphs are obtained. Some questions and problems are posed.

1. Introduction

The Masterplan of Tolerance Intersection Graphs An intriguing theme in graph theory is that of the intersection graph of a family of subsets of a set: the vertices of the graph are represented by the subsets of the family and adjacency is defined by a non-empty intersection of the corresponding subsets. By specifying conditions on the family of subsets one gets interesting classes of intersection graphs. Two notable examples are interval graphs and chordal graphs. An interval graph is the intersection graph of a family of closed intervals on the real line. These graphs have many nice characterizations, such as by forbidden subgraphs [8]. Chordal graphs are defined as graphs having no induced cycle C_n of length $n \geq 4$. They were proven to be the intersection graphs of subtrees of a tree [2].

Golumbic and Monma in [5] introduced a generalization of interval graphs using tolerances (see also [6]). In this case, we have a family of intervals $\{I_v\}_{v \in V}$ and each interval I_v is assigned a positive real number t_v, its tolerance. Now we construct a graph with vertex set V by joining two vertices u an v whenever $|I_u \cap I_v| \geq \min\{t_u, t_v\}$, where $|I|$ denotes the length of interval I. A full characterization of these graphs is still open. Besides this problem some other natural questions present themselves. What happens when all tolerances are equal or when $t_v = |I_v|$, for each vertex v? In [5] it was shown that with constant tolerance c we get precisely the interval graphs and with tolerances $t_v = |I_v|$ we get precisely the permutation graphs.

This idea of tolerances on intervals suggests a *Masterplan.*

Let Z be a set and let μ be a measure on Z assigning to each non-empty subset S of Z a positive real number $\mu(S)$. Let $S = \{S_v\}_{v \in V}$ be a (finite) family of non-empty subsets of Z, and let $t : S \rightarrow R+$ be a mapping, which assigns to each subset S_v a *tolerance* t_v. Finally, let $\phi : R^+ \times R^+ \rightarrow R^+$ be a function assigning a positive real number to each pair of positive reals. Then the *tolerance intersection graph* of (S, μ) and (t, ϕ) is the graph $G = (V, E)$ with vertex-set V and

$$uv \in E \text{ if and only if } \mu(S_u \cap S_v) \geq \phi(t_u, t_v).$$

By specifying conditions on S, μ and ϕ we get special classes of tolerance graphs, which could be termed *ϕ-tolerance (S, μ)-graphs.* For each such class we can try to answer the canonical natural questions of finding a characterization of the class, and establishing the prototype theorems (I) on constant tolerances and (II) on tolerances $t_v = \mu(S_v)$, for v in V. It is important to note that being a tolerance intersection graph is hereditary; that is, every induced subgraph is also a tolerance intersection graph of the same type. Note also that this implies that a forbidden subgraph list exists for each such class.

A characteristic line of approach in the spirit of the Masterplan is to start with an intersection representation of a nice class P of graphs. Generalize this class to that of (ϕ-)*tolerance P graphs* by introducing tolerances, and then establish prototype theorems I and II and try to find characterizations of this class. From this point of view it is more appropriate to name the above graphs introduced in [5] by Golumbic and Monma, (min-) tolerance interval graphs instead of "tolerance graphs".

The aim of this paper is to show some of the possibilities of the Masterplan, to open up new vistas, and to present some questions and problems. In Section 2 we consider intersection representations for split graphs and chordal graphs. Apparently these representations are not the right ones to yield interesting generalizations, because in all cases, by introducing tolerances we obtain all graphs. In Section 3 we choose a different line of approach. The family of subsets is in all cases a family of nested sets, and we vary the function on the tolerances. Although prototype theorems I and II reduce to trivia, we obtain some nice new characterizations of known classes of graphs. Finally, in Section 4, we state some problems and suggest several promising generalizations.

In the following we will restrict ourselves to families consisting of intervals on the real line, finite sets, or subgraphs. As a measure of the respective families we use length of intervals, the cardinality of the finite sets, or the number of vertices of the subgraphs.

2. Subtrees of a Tree with Tolerances Chordal graphs, split graphs, and threshold graphs all have intersection representations involving subtrees of a tree. In this section we study the effects of introducing tolerances, where the measure is the number of vertices in the subtree and ϕ is the minimum of the tolerances. First we recall some relevant definitions and results that are needed here and in the next section. They can all be found in [3].

A graph $G = (V, E)$ is a *threshold graph* if we can associate a *threshold* Θ with G and assign weights w_v to the vertices v in V such that

$$uv \in E \text{ if and only if } w_u + w_v > \Theta.$$

The typical structure of a threshold graph is depicted in Figure 1. The vertex-set of G is partitioned into $D_0, D_i, ..., D_m$, where D_0 may be empty and $D_{\lceil m/2 \rceil}$ only exists if m is odd. The sets D_i are indicated by cells in Figure 1, and a line between cells D_i and D_j means that each vertex in D_i is adjacent to each vertex in D_j. Moreover, the cells on the left hand side consist of independent sets and the cells on the right hand side are cliques. So the left hand side is an independent set and the neighborhoods of its vertices are ordered by inclusion downwards. And the right hand side is a clique and the closed neighborhoods of its vertices are ordered by inclusion upwards.

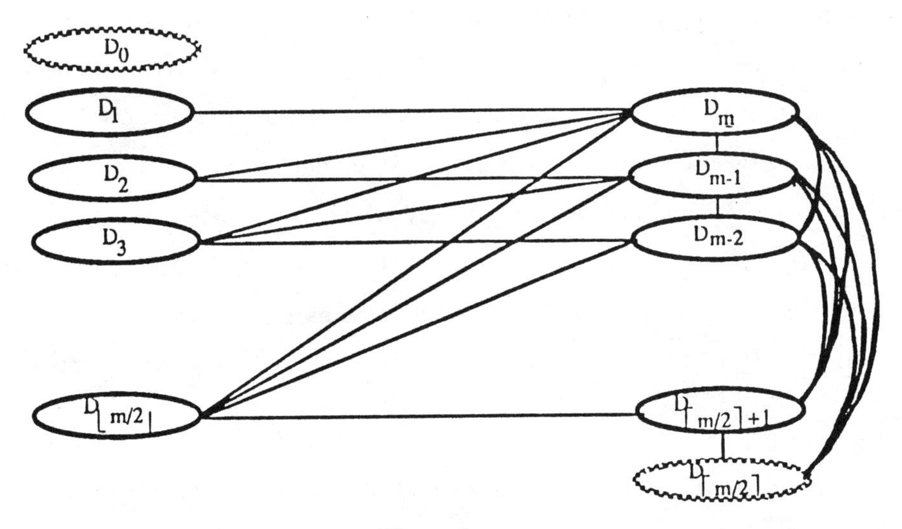

Figure 1

A *split graph* $G = (V, E)$ is a graph whose vertex-set V can be partitioned into an independent set I and a clique K. So, in particular, a threshold graph is split. Clearly, threshold graphs and split graphs are special instances of chordal graphs.

In [9] it was observed that a graph is split if and only if it is the intersection graph of a family of substars of a star $K_{1,n}$ (for some n). The substars in the family consisting of only a pendant vertex of the large star constitute the independent set of the split graph. All other substars in the family contain the center of the large star and hence constitute a clique of the split graph, yielding the desired partition. However, when we introduce tolerances in this case we do not get a very interesting generalization of split graphs. This is an immediate consequence of the following lemma which is known [4], but we include a proof for completeness.

Lemma 1 Let $G = (V, E)$ be a graph. Then G is the edge-intersection graph of substars of a star.

Proof Set $m = |E|$ and label the pendant vertices of the star $K_{1,m}$ with the edges of G. Let c be the center of $K_{1,m}$. For each vertex v of G, let S_v be the substar of $K_{1,m}$ consisting of the center of $K_{1,m}$ and the pendant vertices labeled with the edges incident with v. Then uv is an edge in G if and only if S_u and S_v have an edge in common, viz. the edge between the center c and the pendant vertex labeled with uv in E.

We proceed by considering prototype theorems I and II for the substars of a star case.

Theorem 2.1 Every graph G is a min-tolerance intersection graph of substars of a star for some constant tolerance.

Proof From Lemma 1, using constant tolerance 2, we get any graph from substars of a star.

Recall that a comparability graph is obtained from a partially ordered set $(V, \leq)$, with u and v adjacent if and only if $u < v$ or $v < u$.

Theorem 2.2 A graph G is a min-tolerance intersection graph of substars of a star with tolerances equal to the number of vertices in substars if and only if it is a comparability graph.

Proof If the tolerances are equal to the number of vertices of the substars, then adjacency is equivalent to substar containment. Thus we have a containment graph of substars of a star, which is known to be equivalent to being a comparability graph [7].

Chordal graphs are the intersection graphs of subtrees of a tree. Theorems 2.1 and 2.2 imply that in this case we do not get a nice generalization of chordal graphs. We simply state prototype theorems 1 and 2.

Theorem 3.1 Any graph is the min-tolerance intersection graph of subtrees of a tree for some constant tolerance.

Theorem 3.2 A graph is the min-tolerance intersection graph of subtrees with tolerances equal to the number of vertices in the subtree if and only if it is a comparability graph.

Finally, a threshold graph is an intersection graph of substars of a star, where the substars containing the center of the large star are nested when restricted to pendant vertices that are substars. In this case we get the same results. All graphs in prototype theorem 1 and the comparability graphs in prototype theorem 2.

We also have considered the continuous versions of the above. That is, each edge is replaced by a line segment. Then the size of an intersection is the sum of the lengths of the line segments in the intersection.

Even in this case, however, we get precisely the same classes of graphs as previously obtained in prototype theorems 1 and 2.

3. Nested Families with Tolerances

In this section we restrict ourselves to nested families of sets (i.e. linearly ordered by inclusion), and we vary the function ϕ on the tolerances. For ϕ we consider the minimum, the maximum, and the sum of the two tolerances. Other such functions will be considered in future work. Although prototype theorems 1 and 2 reduce to trivia, we attain new characterizations of some important classes of chordal graphs when arbitrary tolerances are considered. For the nested families we consider intervals on the real line, with their lengths as measure, and finite sets, with their cardinalities as measure. We will refer to the measure of a set as the *set-size*. First we observe that without loss of generality we may choose as nested intervals $N = \{[0, r_i] : i = 1, 2, ..., \}$

with $0 < r_1 \leq r_2 \leq ... \leq r_n$, and in the case of finite sets we may choose them to be of the form $\{1, ..., k_i\}$ with positive integers k_i satisfying $1 \leq k_1 \leq k_2 \leq ... \leq k_n$. Furthermore, in the case of finite sets $\{1, ..k_i\}$ we may convert them into intervals $[0, k_i]$, and vice versa, if we have intervals $[0, r_i]$ with integer r_i, then we may convert them into finite sets $\{11, ..., r_i\}$ without changing the tolerance-intersection structure. Let us call the associated tolerance intersection graphs ϕ *tolerance chain graphs*, depending on the choice of ϕ.

A graph is said to admit a *P-elimination scheme* if its vertices can be ordered $v_1, v_2, \cdots, v_n$ such that each v_i is of type P in the subgraph induced by $v_i, v_{i+1}, \cdots, v_n$. The ordering $v_1, v_2, \cdots, v_n$ is a *P-elimination ordering* of the vertices. A *simplicial vertex* is a vertex having a clique (a set inducing a complete graph) as its neighborhood. A nice and well-known characterization of chordal graphs (see [3]) is that they are the graphs having a simplicial elimination scheme, where any simplicial vertex may start a simplicial elimination ordering. An equivalent formulation of this property is that every induced subgraph of a chordal graph contains a simplicial vertex. A *central vertex* is a vertex adjacent to all other vertices of the graph. The following fact follows immediately from the structure of threshold graphs as given in Figure 1.

Lemma 4 A graph G is a threshold graph if and only if G admits an (isolated or central)-elimination ordering. Here any isolated or central vertex may start an elimination ordering.

First we consider the min-tolerance chain graphs. As we noted earlier, the prototype theorems reduce to trivia.

Theorem 5.1 A graph G is a min-tolerance chain graph with constant tolerances if and only if G is the disjoint union of a complete graph and an edge-less graph.

Theorem 5.2 A graph G is a min-tolerance chain graph with tolerances equal to the set-sizes if and only if G is a complete graph.

With arbitrary tolerances we get a new characterization of threshold graphs.

Theorem 6 A graph is a threshold graph if and only if it is a min-tolerance chain graph.

Proof First let $N = \{[0, r_i] : i = 1, 2, ..., n\}$ be a nested family of intervals with $0 < r_1 \leq r_2 \leq ... \leq r_n$, where interval $[0, r_i]$ has tolerance t_i, for $i = 1, ..., n$. Let G be the associated min-tolerance chain graph with v_i represented by $[0, r_i]$, for $i = 1, ...n$. In view of the fact observed above, it suffices to show that G has a central vertex or an isolated vertex. If $t_1 \leq r_1$, then v_1 is adjacent to all other vertices. If $t_1 > r_1$ and v_1 is not isolated, then let v_j be a neighbor of v_1. This implies that $t_j \leq r_1$, whence v_j is a central vertex. For the converse we have indicated in Figure 2 a min-tolerance chain representation for the typical threshold graph. For every cell the representing set and tolerance for all vertices in the cell are given.

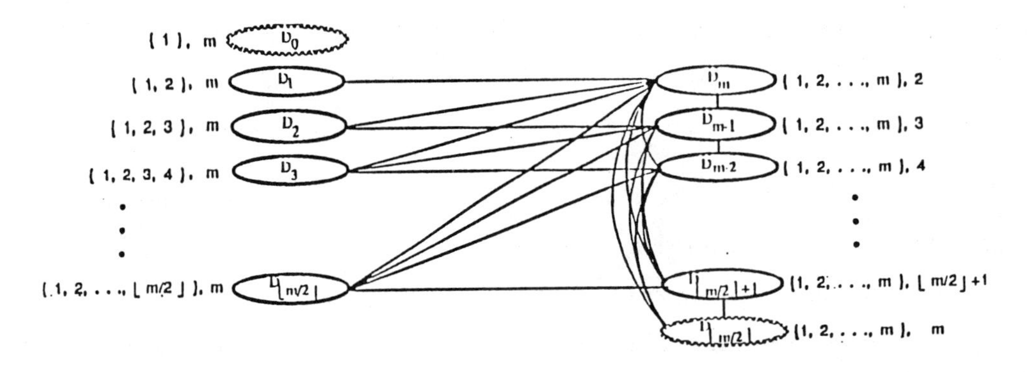

Figure 2

Next we consider the max-tolerance chain graphs.

Theorem 7.1 A graph G is a max-tolerance chain graph with constant tolerances if and only if G is the disjoint union of a complete graph and an edge-less graph.

Theorem 7.2 A graph G is a max-tolerance chain graphs with tolerances equal to set-sizes if and only if G is a disjoint union of complete graphs.

With arbitrary tolerances we get a new characterization of interval graphs. This can be proved by a transformation of the nested intervals with tolerances into arbitrary intervals with intersections, and vice versa. We also include a second proof using the tolerances. It is somewhat longer, but it is more illuminating with respect to the effects of the tolerances.

Theorem 8 A graph is a max-tolerance chain graph if and only if it is an interval graph.

First Proof Let $N = \{[0, r_i] : i = 1, 2, ..., n\}$ be a nested family of intervals with $0 < r_1 \leq r_2 \leq ... \leq r_n$ and with respective tolerances t_i. Let $G = (V, E)$ be the associated max-tolerance chain graph, where v_i is represented by $[0, r_i]$, for $i = 1, ..., n$.

We may assume that $t_i \leq r_i$, for $i = 1, ..., n$, for if $t_i > r_i$, then v_i is isolated and we can disregard v_i. consequently, v_i and v_j are adjacent if and only if

$$\min\{r_i, r_j\} \geq \max\{t_i, t_j\}.$$

This is true if and only if the two intervals $[t_i, r_i]$ and $[t_j, r_j]$ have a non empty intersection.

Thus, the max-tolerance chain graph of N is the interval graph of the intervals $\{[t_i, r_i] : i = 1, 2, ..., n\}$, and vice versa.

Second Proof Let N, t_i and G be as in the first proof. We may assume that G is connected, so that in particular $t_i \leq r_i$, for $i = 1, ..., n$. We show that v_1 is simplicial. Let v_1 and v_j be distinct neighbors of v_1. Then we have $t_i, t_j \leq r_1 \leq r_i, r_j$, so that v_i and v_j are also adjacent. Clearly, any neighbor of v_1, except v_2, is adjacent to v_2, since $r_1 \leq r_2$. So we have a simplicial elimination ordering $v_1, v_2, ..., v_n$ for G such that, for each i, all neighbors of v_i among $v_{i+2}, ..., v_n$ are adjacent to v_{i+1}. It is easily checked that all forbidden subgraphs for interval graphs (see [8]) fail to have such an elimination ordering. Hence G is an interval graph.

Conversely, an interval graph H has the property that its maximal cliques can be linearly ordered as $C_1, ..., C_m$ such that any vertex of H occurs in consecutive maximal cliques. Now, if vertex v is in cliques $C_i, C_{i+1}, ..., C_j$, then we associate with v the set $\{1, 2, ..., j\}$ and tolerance i. It is easily checked that the chain of sets with tolerances thus obtained is a max-tolerance chain representation of H.

In the second proof we have shown that a graph is an interval graph if and only if it admits a simplicial elimination ordering $v_1, v_2, ..., v_n$ with the additional property that for each i, the neighbors of v_i among $v_{1+2}, ..., v_n$ are all adjacent to v_{i+1}. As far as we know this seems to be another new characterization of interval graphs.

Now we turn to the sum-tolerance case. Again we state prototype theorems I and II.

Theorem 9.1 A graph G is a sum-tolerance chain graph with constance tolerances if and only if G is the disjoint union of a complete graph and an edge-less graph.

Theorem 9.2 A graph G is a sum-tolerance chain graph with tolerances equal to set-sizes if and only if G is an edge-less graph.

In a recent paper Monma, Reed and Trotter [11] have introduced what they call threshold tolerance graphs, which generalize threshold graphs. With each vertex v of a graph we associate a weight t_v and a tolerance r_v. Then in u and v are adjacent if and only if

$$t_u + t_v > \min \{r_u, r_v\}.$$

The main results in [11] concern complements of threshold tolerance graphs, called *coTT- graphs*. Here adjacency is defined by

$$t_u + t_v \leq \min\{r_u, r_v\}.$$

Hence, coTT graphs are equivalent to the sum-tolerance chain graphs with nested intervals $N = \{[0,_i] : i = 1,2,...,n\}$ and tolerances t_i. We now analyze coTT from our perspective. Because our proofs are considerably different, we will reprove some results in [11] as Lemmas 12 and 13.

Lemma 10 Let G b the sum-tolerance chain graph of the nested intervals N with tolerances t_i. Then any vertex with maximum tolerance is simplicial in G.

Proof Let v_j be a vertex with maximum tolerance, and let v_i and v_k be distinct neighbors of v_j. Then we have

$$t_i + t_k \leq t_i + t_j \leq \min\{r_i, r_j\} \leq r_i$$

and

$$t_i + t_k \leq t_j + t_k \leq \min\{r_j, r_k\} \leq r_k,$$

so that

$$t_i t_k \leq \min\{r_i, r_k\}.$$

Hence v_i and v_k are adjacent.

Thus, sum-tolerance chain graphs are chordal.

Lemma 11 lt $u_1, u_2, ..., u_p$ be an induced path in a sum-tolerance chain graph, where s_i is the tolerance of u_i, for $i = 1, .., p$. Then either

$$s_1 = s_2 < s_3 < ... < s_p,$$

or

$$s_1 > s_2 > ... > s_{p-1} = s_p,$$

or

$$s_1 > s_2 > ... > s_j < s_{j+1} < ... < s_p,$$

for some j with $1 \leq j \leq p$.

Proof Let $[0, x_i]$ be the interval associated with u_i, for $i = 1, ..., p$. Assume that $s_i \leq x_{i+1}$, for some i. Since u_{i+1} and u_{i+2} are adjacent, we have

$$x_{i+2}, x_{i+1} \geq s_{i+2} + s_{i+1} \geq s_{i+2} + s_i$$

Since u_{i+2} and u_i are not adjacent, it follows that

$$s_{i+2} + s_i > x_i.$$

Now u_{i+1} and u_i are adjacent, so that $x_i \geq s_{i+1} + s_i$. This implies that

$$s_{i+2} > s_i.$$

Now suppose that $s_i = s_{i+1}$, for some i with $1 < i < p - 1$. Then, by the previous argument, we have $s_{i-1} > s_i = s_{i+1} < s_{i+2}$, with, say, $s_{i+2} \leq s_{i-1}$.

Then we have

$$x_{i+2}, x_{i+1} \geq s_{i+2} + s_{i+1} = s_{i+2} + s_i > x_i \geq s_{i-1} + s_i,$$

which is impossible. These observations immediately imply the assertion in the lemma.

An m-*trampoline*, or m-*sun*, with $m \geq 3$ consists of a complete graph with vertex set $\{1, 2, ..., m\}$ together with vertices $m+1, m+2, ..., 2m$, where $m + 1$ is adjacent to i and $i + 1$ (mod m). The *strongly chordal graphs* are the chordal graphs without induced m-suns, for $m \geq 3$ (see [1]).

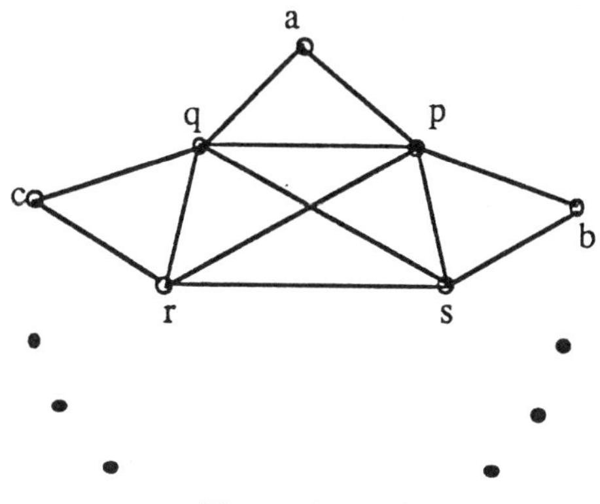

Figure 3

Lemma 12 A sum-tolerance chain graph is strongly chordal.

Proof It suffices to show that m-suns do not have a sum-tolerance chain representation. Assume the contrary. In Figure 3 an m-sun is depicted, where $r = s$ in the case $m = 3$. The vertices are labeled with their tolerances, and the intervals of vertices b, p, q, and c are, respectively, [0,B], [0P], [0,Q], and [0,C]. Without loss of generality, we may assume that in Figure 3 we have $a \geq b \geq c$. Then we have

$$P \geq p + b \geq p + c > C \geq q + c,$$

so $p > q$. On the other hand, we have

$$Q \geq a + q \geq b + q > B \geq b + p,$$

whence $q > p$. This contradiction completes the proof. In Figure 4 we see two other graphs that are forbidden in sum-tolerance chain graphs.

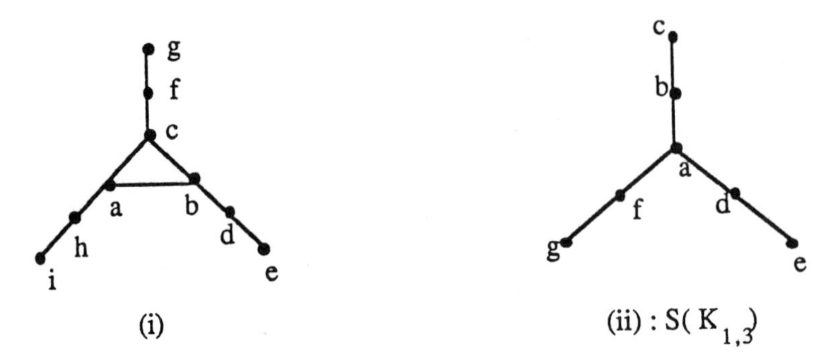

Figure 4

Lemma 13 The graphs in Figure 4 do not occur as induced subgraphs in a sum- tolerance chain graph.

Proof Assume that the graphs of Figure 4 each have a sum-tolerance chain representation. Again the vertices are labeled with their tolerances and the associated intervals are of the form [0,X], where X is associated with vertex and tolerance x. First consider the graph of Figure 4(i). Since the edges of the triangle are each internal edges of some induced path, the tolerances on the triangle are all distinct, say, $a < b < c$ by Lemma 11. Again by Lemma 11, we have $b < d < e$ and $c < f < g$. Now, if $d \leq c$, then we get

$$A, C \geq c + a \geq d + a > D \geq d + b > d + a,$$

which is impossible. Hence $d > c$ and we have

$$D \geq d + e > d + c. \qquad (*)$$

On the other hand, we have

$$F, C \geq f + c > f + b > B \geq d + b,$$

so $f > d$. Hence we have

$$C \geq f + c > d + c. \qquad (**)$$

Now (*) and (**) force d and c to be adjacent. This impossibility shows that Figure 4(i) depicts a forbidden subgraph for sum-tolerance chain graphs.

Second consider the graph of Figure 4(ii). By Lemma 10, one of the pendant vertices must have maximal tolerance, say c is maximal. There are three candidates for the next maximal tolerance, viz. b, e, and g. Assume that $b \leq e$. Then we have

$$D \geq d + e \geq d + b > B \geq c + b \geq d + b,$$

which is impossible. So we have $b > e$, and similarly $b > g$. Now assume that $a \leq d$. Then we get

$$E \geq d + e \geq a + e > A \geq a + b > a + e,$$

which is impossible. So we have $a > d$. Similarly, we infer that $a > f$. But now we are in conflict with Lemma 11, which completes the proof. The graphs of Figure 4 are minimal forbidden subgraphs as is shown by the graphs in Figures 5 and 6. Note that the graph of Figure 4(ii) minus a pendant vertex is induced in the graph of Figure 6. This last graph shows how we can realize arbitrary long paths and caterpillars (with extra backbone) using Fibonacci's sequence. This will be used for Theorem 15.

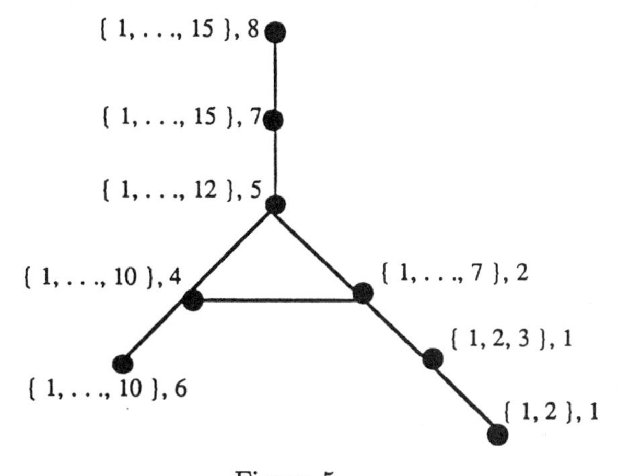

Figure 5

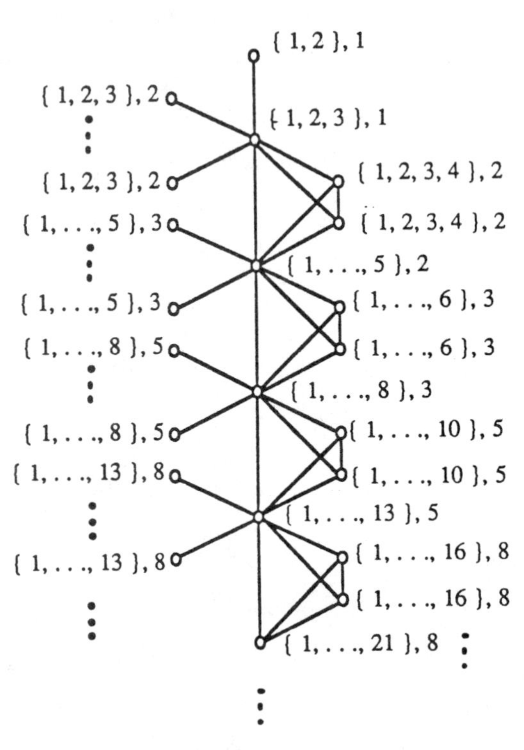

Figure 6

Theorem 14 A sum-tolerance chain graph is a strongly chordal graph without any of the graphs of figure 4 as induced subgraphs.

Proof Use Lemmas 12 and 13.

We conclude this section showing that trees which are sum-tolerance chain graphs are precisely those with no induced $S(K_{1,3})$.

Theorem 15 A tree T is a sum-tolerance chain tree if and only if T contains no induced subgraph isomorphic to $S(K_{1,3})$.

Proof We only need show that if T is a tree with no induced $S(K_{1,3})$'s then T is a sum-tolerance chain tree. We proceed by induction. Assume the result is true for all such trees with less than n vertices (it is easy to derive a representation for "small" trees) and let T be a tree with no induced $S(K_{1,3})$ of order n. It T were a path, as previously noted, T would be a sum-tolerance graph. Thus, we may assume that T contains a vertex x of degree at least 3. Furthermore, since T contains no induced $S(K_{1,3})$ we may assume that x is adjacent to at least one pendant vertex v. By the induction hypothesis $T - v$ has a sum-tolerance representation.

Choose z and y neighbors of x in $T - v$, so that t_y and t_z are the smallest two values over all neighbors of x. If both t_y and t_z were at most t_x, then

$$t_y + t_z \leq t_x + t_z \leq \min\{r_x, r_z\} \leq r_z,$$

$$t_y + t_z \leq t_y + t_x \leq \min\{r_y, r_x\} \leq r_y,$$

which implies $t_y + t_z \leq \min\{r_y, r_z\}$, hence $yz \in E(T - v)$ which is a contradiction. Thus at least one of t_z or t_y is greater than t_x. Without loss of generality, say $t_y > t_x$. Since $t_y > t_x$. Since $t_x + t_y \leq r_x$, it follows that $t_y \leq r_x - t_x$. To complete the sum-tolerance representation for T, let $t_v = r_x - t_x$ and $r_v = r_x$. Clearly, if t_z were also greater than t_x, then it would follow that v would only be adjacent to x, and T would result. The only possibility remaining is to suppose $t_y > t_x \geq t_z$. Since $t_v = r_x - t_x \geq t_y > t_x$, it follows that all vertices non-adjacent to x would also be non-adjacent to v. Hence we need only show that z and v are not adjacent.

First, suppose $r_z \geq r_z$. This implies $t_z + t_y \leq t_x + t_y \leq \min\{r_x, r_y\} \leq \min\{r_2, r_y\}$ which would imply that zy is an edge. Since this cannot be the case, we can conclude that $r_z < r_x$.

Now suppose $t_v + t_z \leq \min\{r_v, r_z\}$. If $r_z \leq r_y$, then

$$r_z = \min\{r_z, r_y\} < t_z + t_y \leq t_z + t_v = t_z + r_x - t_x \leq r_z$$

Thus $r_y < r_z$, which implies $t_x + t_y \leq \min\{r_x, r_y\} = r_y = \min\{r_z, r_y\} < t_z + t_y$, which yields the contradiction $t_z > t_x$.

Consequently, $t_v + t_z > r_z$ which implies that $t_v + t_z > \min\{r_v, r_z\}$ and hence v and z are not adjacent so that the resulting representation gives precisely T.

Note that in [6], the trees of Theorem 15 were shown to be bounded min-tolerance interval trees.

4. Open Problems and Questions

We have seen that in the case of interval graphs by introducing tolerances one gets a very interesting class of graphs ([5], [6]). In section 2 we tried to obtain similar classes for threshold split and chordal graphs with somewhat unsatisfactory results. We pose the problem of searching for intersection representations for these classes so that one gets meaningful and interesting classes of tolerance threshold graphs, tolerance slit graphs, and tolerance chordal graphs.

A question that presents itself after section 3 is whether there are functions ϕ on the tolerances such that the ϕ-tolerance chain graphs are not chordal.

Along the lines of the previous section we may consider more general families than chains of intervals or finite sets. A natural candidate is a family of sets called a "tree of subsets" (with respect to set inclusion, the sets form a tree ordered set.) Every chain in a tree of subsets is characterized by the results in the previous section. In particular, it is easy to show that with min-tolerance and max-tolerance the smallest sets correspond to simplicial or isolated vertices. Hence in these cases we have again classes of chordal graphs. In section 3 the prototype theorems I and II reduced to trivia, but in this situation we expect non-trivial results for prototype theorems I and II.

Another promising tolerance intersection problem is the following. In [10] McMorris and Scheinerman observed that the chordal graph are precisely the intersection graphs of families of leaf-generated subtrees of some full binary tree. Here a full binary tree is a rooted tree with root of degree 2, all leaves (pendant vertices) at equal distance from the root and all other vertices have degree 3. A leaf-generated subtree is obtained by choosing some leaves of the binary tree and then taking the subtree having precisely the chosen leaves as pendant vertices. Surprisingly, it turns out that different constant tolerances yield different classes of graphs, and this will be reported elsewhere. However, much work needs to be done on this representation.

The authors hope that the above problems and questions will stimulate much work in the near future.

REFERENCES

[1] M. Farber, "Characterizations of strongly chordal graphs," *Discrete Math.* 43 (1983), 173-189.

[2] F. Gavril, "The intersection graph of subtrees in trees are exactly the chordal graphs," *J. Combin. Theory B* 16 (1974), 47-56.

[3] M. C. Golumbic, Algorithmic Graph Theory and Perfect Graphs, Academic Press, New York, 1980.

[4] M. C. Golumbic and R. E. Jamison, "The edge intersection graphs of paths in a tree, *J. Combin. Theory* B 38 (1985), 8-22.

[5] M. C. Golumbic and C. L. Monma, "A generalization of interval graphs with tolerances," *Congressus Numerantium* 35 (1982), 321-331.

[6] M. C. Golumbic, C. L. Monma and W. T. Trotter, "Tolerance graphs," *Discrete Appl. Math.* 9(1984), 157-170.

[7] M. C. Golumbic and E. R. Scheinerman, "Containment graphs, posets and related classes of graphs," to appear in *Proc. 3rd International Conf. on Combinatorial Math.,* New York Academy of Sciences.

[8] C. G. Lekkerkerker and J. C. Boland, "Representation of a finite graph by a set of intervals on the real line," *Fund. Math.* 51(1962), 45-64.

[9] F. R. McMorris and D. Sheir, "Representing chordal graphs on $K_{1,n}$," *Comm. Math. Univ. Carolinae* 24(1983), 489-494.

[10] F. R. McMorris and E. R. Scheinerman, "Connectivity threshold for random chordal graphs," to appear in *Graphs and Combin.*

[11] C. L. Monma, B. Reed and W. T. Trotter, "Threshold tolerance graphs," to appear in *J. Graph Theory.*

GRAPH TRANSFORMS:
A FORMALISM FOR MODELING CHEMICAL REACTION PATHWAYS

Mark Johnson

The Upjohn Company

Kalamazoo, Michigan

ABSTRACT

A goal of computer-assisted organic synthesis is the construction of a kit of reaction rules that enables the computer to suggest chemical paths of compounds leading to or from a compound of interest. Implicit in computer-aided organic synthesis the modeling of chemical reaction pathways. A metadigraphs is introduced as a graph-theoretic counterpart of a chemical reaction pathway and is defined to be a digraphs whose vertex set Γ *is a set of graphs. A graph transform* t *is defined such that for each ordered pair* (G, G') *of graphs one can determine if* t *can transform* G *into* G'*. A iransform kit is defined to be an ordered pair* $(T^i T^b)$ *of sets* T^i *of "inducing" transforms and* T^b *of "blocking" transforms. A set* Γ *of graphs and a kit* K *defines a unique metadigraph* $M(\Gamma, K)$*. The ordered pair* (Γ, K) *is referred to as a specification of the metadigraph* M *if* $M = M(\Gamma, K)$*. It is shown that every metadigraph* M *admits a specification of the form* $(V(M), K)$ *for some kit* K *where* $V(M)$ *denotes the vertex set of* M*.*

1. Introduction

Chemical graphs are labeled graphs in which letters assigned to the vertices denote the atoms of a molecule and letters assigned to the edges denote the bonds. They are widely used in chemistry as representations of molecular structures. A chemical reaction type can be viewed as a transform which, when applied to a compound, generates new compounds. For example, benzene undergoes hydroxylation to form hydroxybenzene. This idea has been made algorithmically explicit in a variety of ways with respect to chemical graphs in computer-assisted organic synthesis (1,10,11) and computer-assisted drug metabolism (6). The goal in these approaches is the construction of a kit of reaction rules that enables the computer to suggest chemical paths of compounds leading to or from a compound of interest. Recently, Koća et al. (10) have reexamined the approach by Ugi et al. (11) using graph-theoretic concepts. This paper develops a general graph-theoretic formalism analogous to the chemical concepts of reactions and kits of reaction rules, but freed from terminology and assumptions that limit the applications of this formalism to a particular scientific discipline. Such a mathematical formalism should eventually lead to improvements in computer-assisted organic synthesis.

Chemical pathways are diverse. They may consist of a single reaction or a complex biochemical pathway. Figure 1 presents part of the "shikimic acid" metabolic pathway (3). There we see cpd. 2 was hydroxylated to give cpd. 3. Similarly, cpd. 4 was hydroxylated to give cpd. 5 and cpd. 5 was further hydroxylated to give cpd. 6. In our analysis, "hydroxylation", which performs a particular structural change, will correspond to the graph-theoretic concept of a transform. In hydroxylation, a hydrogen atom is deleted and the atoms and bonds of an hydroxyl group are added. The graph elements (vertices or edges corresponding to the atoms and bonds) to be deleted by a transform will subsequently be indicated by assigning each such element $a-1$, and the graph elements to be added will be indicated by assigning each such element $a+1$. Note the asymmetry here. The reverse change, deleing an hydroxyl group and adding a hydrogen atom is called dehydroxylation. This asymmetry must be incorporated into our concept of a transform. In addition, we see that a transform acts on some compounds and not on others. In Figure 1, compounds 2,

4 and 5 were hydroxylated, but compounds 1, 3, and 6 were not. Those graph elements used to define the local environment where a structural change will take place, but which are neither to be added nor deleted, will subsequently be indicated by assigning each such element a 0. Finally, Figure 1 indicates that compound 2 is "deanimated" to give compound 5. The "deanimation" transform perform a different structural change than the "hydroxylation" transform. Thus, a solution to the problem of modeling chemical reaction pathways will require a specification of a broad class of possible transforms and a set of rules for deciding which transform is to operate where on each graph.

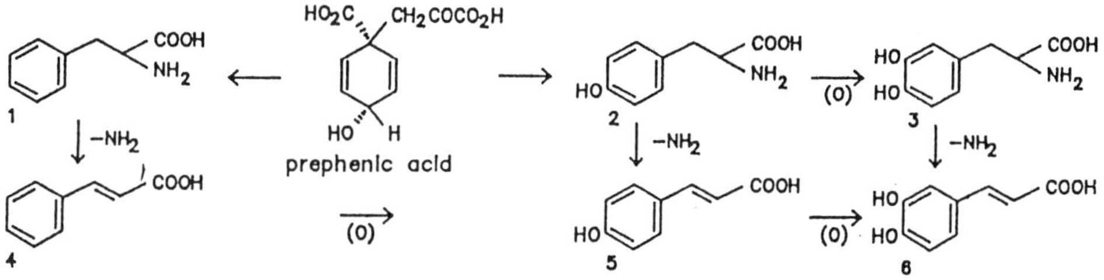

Figure 1. Part of the shikimic acid pathway

Figure 2 presents a graph-theoretic counterpart of a chemical reaction pathway. This figure should be viewed as involving two digraphs M_1 and M_2 whose vertices are themselves graphs. We shall refer to such digraphs as metadigraphs. The arcs of the metadigraphs are the graph-theoretic counterparts of chemical reactions. Those operations which convert one graph into another, the counterparts of "hydroxylation" and "deamination", will be called transforms. A transform kit K will consist of a set T^i of "inducing" transforms and set T^b of "blocking" transforms. Rules are given whereby an arc of a metadigraph can be considered to be an action of an "inducing" transform in T^i not blocked by a "blocking" transform in T^b. It will be seen that metadigraph M_1 in Figure 2 can be simply specified by an ordered pair $(\{P_1\}, K)$ where K consists of a single inducing transform and a single blocking transform. It will also be shown that any metadigraph M admits a specification of the form $(V(M), K)$ for some transform kit K where $V(M)$

denotes the set of graphs that comprise the vertex set of M.

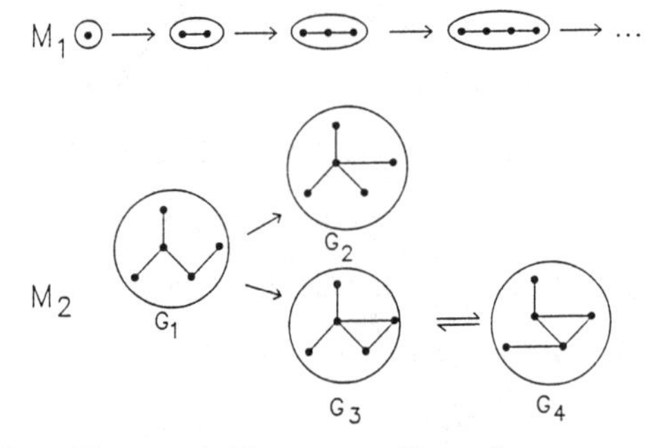

Figure 2. Two metadigraphs

2. Semitransforms and semiactions

The preceding notions will now be developed rigorously following the notation in Chartrand and Lesniak (4). A graph $C = (V, E)$ together with an assignment of the integers -1,0 and 1 to the elements (vertices and edges) of C, one integer per element, will be called a *semitransform* on C if zero edges (edges assigned 0's) are adjacent to only zero vertices, nonpositive (≤ 0) edges are adjacent only to nonpositive vertices, and nonnegative (≥ 0) edges are adjacent only to nonnegative vertices.

A semitransform will be illustrated by indicating negative, zero and positive edges with dashed, solid and dotted lines, and indicating negative, zero and positive vertices with x's, solid dots and circles, respectively. Figure 3 shows three semitransforms.

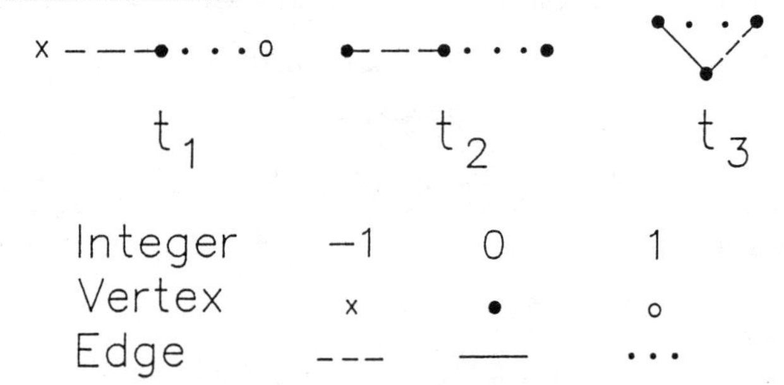

Figure 3. Some semitransforms

Let t be a semitransform on C. Four graphs will be commonly associated with t. Graph C is the *covering graph* of t. The nonpositive elements of t

define the *ingraph* I of t; the nonnegative elements define the *outgraph* O of t; and the zero elements define the *linking graph* L of t. The restrictions on the assignment function assure that I, O and L are well-defined. In Figure 3, the covering graph of t_3 is the cycle C_3, the ingraph and outgraph are both P_3 paths. The linking graph is the union of K_1 and P_2. The graphs C, I, O and L will be called the *defining* graphs of t. These letters, with appropriate superscripts and subscripts, will denote the defining graphs of a semitransform with corresponding superscripts and subscripts. Clearly, if any three of the defining graphs of a semitransform, when viewed as labeled subgraphs of C, are known or if the ingraph and outgraph are known, all of the defining graphs are completely determined. Define the *cardinality* $|G|$ of the graph $G = (V, E)$ by $|G| = |V| + |E|$. We obviously have

$$|I| + |O| = |C| + |L|, \tag{1}$$

$$p(I) + p(O) = p(C) + p(L), \tag{2}$$

and

$$q(I) + q(O) = q(C) + q(L), \tag{3}$$

where $p(G)$ and $q(G)$ denote the order and size of graph G.

There are many ways to form a semitransform having G for its ingraph and G' for its outgraph. To see this, let C be any common supergraph of G and G'. Let I and O denote labeled subgraphs of C that are isomorphic to G and G', respectively. Then the semitransform having I and O as the ingraph and outgraph is such a transform. Note that C will be the covering graph of that transform only if I and O cover C, i.e. every element of C is an element of either I or O. Let t be a semitransform on $C = (V, E)$ with assignment function $g : V \cup E \to \{-1, 0, 1\}$ and let t' be a semitransform on $C' = (V', E')$ with assignment function $h : V' \cup E' \to \{-1, 0, 1\}$. Then t and t' are isomorphic, and we write $t \simeq t'$, if there exists an isomorphism $f : V \to V'$ between C and C' such that $g(v) = h(f(v))$ for all $v \in V$ and $g(uw) = h(f(u)f(w))$ for all $uw \in E$.

Let t be a semitransform on C with assignment function g. Let C' be a labeled subgraph of C and let g' be the restriction of g to the elements of C'. The graph C' together with g' is a semitransform t' called a *subtransform* of t. If, in addition, t' has as many nonzero elements as t, then t' is a *reduction* of t. If t'' is any other semitransform isomorphic to t', then t'' is also called a subtransform (or reduction if such is the case) of t. The semitransform t_2 is a reduction of the semitransform t_3 in Figure 3. If t' is a reduction of t, then t is an *extension* of t'. Clearly, if t' is a reduction of t, then

$$|X| - |Y| = |X'| - |Y'|, \qquad (4)$$

$$p(X) - p(Y) = p(X') - p(Y') \qquad (5)$$

$$q(X) - q(Y) = q(X') - q(Y') \qquad (6)$$

where X and Y denote any pair of defining graphs of t and X' and Y' denote the corresponding pair of defining graphs of t'.

Let (G, G') be an ordered pair of graphs, and let t be a semitransform. We shall say (G, G') is a *semiaction* t if there exists an extension t' of t such that G and G' are isomorphic to the ingraph and outgraph of t'. The extension t' is called a *semioverlay* of (G, G') with respect to t. Clearly, every semitransform t is a semioverlay of (I, O) with respect to t itself where I and O denote the ingraph and outgraph of t. Let (G, G') be any semiaction of t having ingraph I and outgraph O. It follows directly from equations 5 and 6 that

$$p(G') = p(G) + p(O) - p(I), \qquad (7)$$

$$q(G') = q(G) + q(O) - q(I), \qquad (8)$$

$$|G'| = |G| + |O| - |I|. \qquad (9)$$

Example 1 Let (G, G') be any semiaction of t_1 in Figure 3. Let t be any overlay of (G, G') with respect to t_1 and let C, I, O, L denote the defining graphs of t. Then the linking graph L of t contains all the vertices and edges in I and O except for one edge, say uy, of I where $u \in V(L)$ and $y \notin V(L)$ and one edge of O adjacent to u, say uz, where $z \notin V(L)$. Define the function $f : V(I) \to V(O)$ by $f(v) = v$ if $v \neq y$ and $f(v) = z$ if $v = y$. Clearly, f defines an isomorphism between I and O. Thus, if (G, G') is a semiaction of t_1, then $G \simeq G'$. It might be pointed out that equations (7) and (8) imply that G and G' have identical orders and sizes since I and O have identical orders and sizes.

A graph G is transformable into G' by an *edge rotation* if G contains distinct vertices u, v, w such that $u\,v \in E(G)$, $uw \notin E(G)$ and $G' \simeq G - uv + uw$ (5). An edge rotation is an *edge shift*, if, in addition, $vw \in E(G)$ (9). It is a straightforward matter to show that G is transformable into G' by an edge rotation (edge shift) if and only if (G, G') is a semiaction of $t_2(t_3)$ in Figure 3.

3. Transforms and Actions

Let G and G' be two graphs. A *maximum common subgraph* of G and G' is a graph H of maximum cardinality which is isomorphic to a subgraph of both G and G'. A *minimum common supergraph* of G and G' is a graph H of minimum cardinality which is isomorphic to a supergraph of G and G'. (See Johnson (8) for other uses of maximum common subgraphs and minimum common supergraphs, referred to there as maximum intersections and minimum unions, for comparing graphs of arbitrary order and size.) A semioverlay t' of (G, G') with respect to a semitransform t is an *overlay* of (G, G') with respect to t if the linking graph of t' is a maximum common subgraph of G and G'. It follows directly from equation 1 that the linking graph of t' is a maximum common subgraph of G and G' if and only if the covering graph of t' is a minimum common supergraph of G and G'.

Figure 4 gives three overlays of graphs G_5 and G_6. Note that overlays o_1 and o_2 have isomorphic linking graphs but nonisomorphic covering graphs while overlays o_2 and o_3 have nonisomorhic linking graphs.

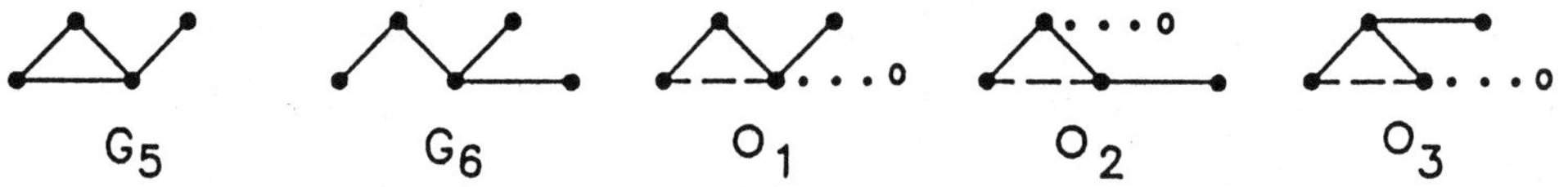

Various overlays of (G_5, G_6).

A semiaction (G, G') of a semitransform t is an *action* of t if t is a reduction of an overlay of (G, G'). A semitransform t is a *transform* if there exists an action of t. The following theorem establishes a consistency relationship between actions and overlays.

Theorem 1 A semioverlay of (G, G') with respect to t is an overlay with respect to t if and only if (G, G') is an action of t.

Proof By definition, it t' is an overlay of (G, G') with respect to t, then (G, G') is an action of t. To establish the only if relation, assume (G, G') is an action of t and t' is a semioverlay of (G, G') with respect to t By definition, there exists an overlay t'' of (G, G') with respect to t. Using equation 4 with $X = I$ and $Y = L$, we have

$$\begin{aligned} |G| &= |I'| \\ &= |L'| + |I'| - |L'| \\ &= |L'| + |I| - |L| \end{aligned} \tag{10}$$

and

$$\begin{aligned} |G| &= |I''| \\ &= |L''| + |I''| - |L''| \\ &= |L'| + |I| - |L|. \end{aligned} \tag{11}$$

Equating the right hand sides of equations (10) and (11), we have $|L'| = |L''|$. Since L' is a subgraph of both G and G' of cardinality $|L|$, L' must be a maximum common subgraph.

Not all semitransforms are transforms. In particular, the semitransform given by t_1 of Figure 3 is not. To see this, suppose (G, G') were an action of t_1. By definition, there exists an extension t of g_1 which is an overlay of $G, G')$. Let C, I, O, L be the defining graphs of t. Then $I \simeq G$ and $' \simeq O$. However, we showed in example 1 that $G \simeq G'$. Since t is an overlay, L is a maximum common subgraph of I and O, and consequently, $L = I = O$, i.e. t contains only zero elements. Since t_1 is a reduction of t, t_1 contains only zero elements, a contradiction. Thus t_1 cannot be a transform.

Let ST, SO, T and O denote the set of semitransforms, semioverlays, transforms and overlays. When defining semioverlays, it was shown that ST=SO. It has just been shown that T is a proper subset of ST. To see that O is a proper subset of T, note that the linking graph of t_2 in Figure 3 is the complement of K_3 while the maximum common subgraph of the ingraph and outgraph of t_2 is $P_2 \cup K_1$. Thus, t_2 is not an overlay. To show that t_2 and t_3 of Figure 3 are transforms, let $G \simeq K(1,3)$ and $G' \simeq P_4$. Let t be defined by t_4 of Figure 5. Let C, I, O and L be the defining graphs of t. Since $G \simeq I$ and $G' \simeq O, t$ is a semioverlay of (G, G'). Clearly, $L \simeq P_3 \cup K_1$ is a maximum common subgraph of G and G'. Thus, t is an overlay. Now we simply note that both t_2 and t_3 are reductions of t.

It should also be noted that not all semiactions of a transform are actions of that transform. To see this, let t' be defined by t_5 of Figure 5. Clearly, t' is a semioverlay of (P_4, P_4). Since t_2 and t_3 of Figure 3 are reductions of t', (P_4, P_4) is a semiaction of both t_2 and t_3. However, the linking graph of t' is $P_2 \cup P_2$, which is not a maximum common subgraph of P_4 and P_4. Thus, t' is not an overlay of t_3, and consequently not of t_2. It follows from Theorem 1 that (P_4, P_4) is not an action of either t_2 or t_3.

Figure 5. An overlay and a semioverlay with respect to t_3

Let $S(t)$ and $A(t)$ denote the sets of semiactions and actions of t. It would be nice to be able to prove that $A(t) = A(t')$ implies $t \simeq t'$, but I have been unable to do so. However, the proof of the following theorem is straight-forward.

Theorem 2 Let t be a transform and let t' be a reduction of t. Then t' is a transform. Moreover, $A(t) \subset A(t') \subset S(t')$.

3. Specifications of Metadigraphs

A *metadigraph* M is a, possibly infinite, digraph $D = (V, E)$ together with an injective function $g : V \to \Gamma_0$ that assigns a unique graph in Γ_0 to each vertex of D. Thus, a metadigraph can be considered to be a labeled digraph in which the set of labels is a set of unlabeled graphs. It follows that we can uniquely and unambiguously denote M by (Γ, E^g) where Γ is the image set $g(V)$ of V and $E^g = \{g(u)g(v)|uv \in E\}$, i.e. we can uniquely represent M by (Γ, E^g). Examples of two metadigraphs were given in Figure 2. Let M' be another metadigraph $D' = (V', E')$ with assignment function h, and let (Γ', E^h) denote its corresponding labeled graph representation. M is isomorphic to M' if there exists an isomorphism $f : V \to V'$ between D and D' such that $g(v) = h(f(v))$ for $v \in V$. Clearly, M is isomorphic to M' if and only if their corresponding labeled graph representations (Γ, E^g) and (Γ', E^h) are identical, i.e. $\Gamma = \Gamma'$ and $E^g = E^h$. Thus, we shall say M is a submetadigraph of M' if $\Gamma \subset \Gamma'$ and $E^g \subset E^h$.

An ordered pair (T^i, T^b) of, possibly infinite, sets of transforms is a *transform kit.* An action (G, G') of t is *blocked* by t' if (a) t' is an extension of t, (b) (G, G') is an action of t' and (c) $t' \in T^b$. Let $A(t|T^b)$ denote the actions of t that are not blocked by t' for any $t' \in T^b$. Clearly,

$$A(t|T^b) = A(t) \setminus \cup_t' A(t')$$

where the union is over those t' in T^b which are extensions of t. An ordered pair (G, G') of graphs is an *action* of $K = (T^i, T^b)$ is $(G, G') \in A(t|T^b)$

form some $t \in T^i$. A graph G' will be said to be *K-reachable* from G if either $G' \simeq G$ or there exists a sequence $G_1, \cdots, G_n$ such that $G \simeq G_1, G' \simeq G_n$ and (G_i, G_{i+1}) is an action of K for $i = 1, ..., n-1$. Let Γ be a set of graphs and let K be a transform kit. Define the metadigraph $M(\Gamma, K)$ as the metadigraph (Γ', E) where Γ' is the set of graphs which are K- reachable from a graph in Γ and where E is the subset of $\Gamma' \times \Gamma'$ whose members are actions of K. Given a metadigraphs M, we shall say (Γ, K) is a *specification* of M if $M(\Gamma, K) = M$.

Example 2 Let t_4 and t_5 be defined by Figure 6. Let $K = (\{t_4\}, \{t_5\})$ and let $M = (\Gamma, E)$ denote the metadigraph M_1 in Figure 2, i.e. $\Gamma = \{P_n | n = 1, ...\}$ and $E = \{(P_i, P_{i+1}) | i = 1, ...\}$. Then $(\{P_1\}, K)$ is a specification of M. To see this, we simply note that for every graph G and every vertex $v \in V(G), (G, G + u + uv)$ is an action of t_4 were $u \notin V(G)$. Similarly, $(G, G + u + uv)$ is an act of t_5 if and only if the deg $v > 1$. It follows that if (G, G') is an actio of K, then $G' \simeq G + u + uv$ where deg $v < 2$. Thus, if G is a path P_i, then $G' \simeq P_{i+1}$. Since every path is K-reachable from P_1, it follows that $(\{P_1\}, K)$ is a specification of M.

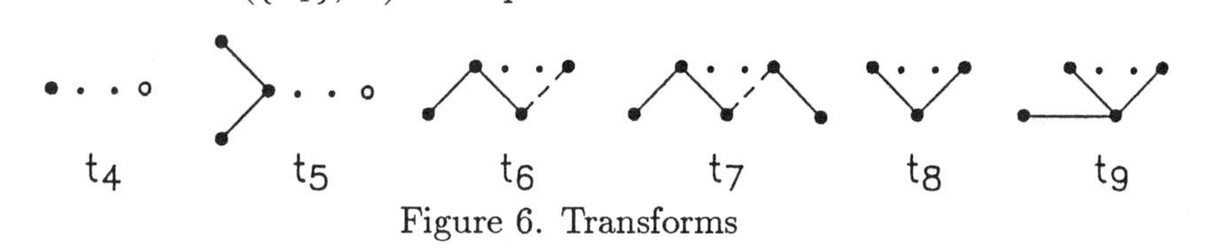

Figure 6. Transforms

Example 3 Let M be metadigraph M_2 in Figure 2, and let $t_i, i = 6, ..., 9$ be transforms defined by Figure 6. Define $T^i = \{t_6, t_8\}$ and $T^b = \{t_7, t_9\}$, and let K be the transform kit (T^i, T^b). Then $(\{G_1\}, K)$ is a specification of M. To see this, note that t_6 performs an edge shift of the form $G - uv + uw$ where v is adjacent in G to a vertex w with deg $w > 1$, while t_7 blocks those cases in which u is not a terminal vertex. Likewise, t_8 adds an edge between vertices u and v which are not adjacent, but which are both adjacent to another vertex w, while t_9 blocks those cases in which deg $w > 2$.

A theorem will now be proven which implies, as a corollary, that any metadigraph admits a specification for some K. The proof requires the following lemma.

Lemma 1 If t is a reduction of a semitransform t' for which $I \simeq I'$ and $O \simeq O'$, then $t \simeq t'$.

Proof To see that the reduction condition is essential, let t be defined by transform t_5 in Figure 5, and let t' be the transform for which $C' \simeq P_4 \simeq L'$. Since $L \simeq P_2 \cup P_2$, t and t' are not isomorphic. On the other hand, $I \simeq O \simeq P_4 \simeq I' \simeq O'$. Thus, the lemma fails without the reduction condition.

By definition, t is isomorphic to a reduction t'' of t' for which I'' and O'' are labeled subgraphs of I' and O'. Since $I'' \simeq I \simeq I'$ and $O'' \simeq O \simeq O$, it follows that $I'' = I'$ and $O'' = O'$. Thus, $t'' = t'$. Since $t \simeq t''$, transitivity implies $t \simeq t'$. Let $\mathcal{T}_0$ denote the set of all transforms. Theorem 3 and its corollary show how $\mathcal{T}_0$ can be used to prove that all metadigraphs admit a specification with respect to some transform kit.

Theorem 3 Let $K = (T^i, T^b)$ be the transform kit based on any partitioning $\{T^i, T^b\}$ of $\mathcal{T}_0$ such that every member of T^i is an overlay. Then (G, G') is an action of K if and only if there exists an overlay of (G, G') in T^i.

Proof Assume there exists an overlay t in T^i of (G, G'), i.e. $I \simeq G$ and $O \simeq G'$. If (G, G') is not an action of K, then the action (G, G') of t must be blocked by some $t' \in T^b$. This implies that I is a subgraph of I' and O is a subgraph of O'. By definition, (G, G') must be an action of t', i.e. t' is a reduction of an overlay of (G, G'). This implies that I' is a subgraph of G and that O' is a subgraph of G'. It follows that $I \simeq I'$ and $O \simeq O'$. By Lemma 1, $t \simeq t'$, a contradiction since T^i and T^b are disjoint sets.

On the other hand, assume (G, G') is an action of K. Then $(G, G') \in A(t|T^b)$ for some $t \in T^i$. Let t' be any extension of t such that (G, G') is an action of t'. If $t' \in T^b$, then $(G, G') \notin A(t|T^b)$. Thus, all extensions of t having (G,G') as an action must lie in T^i. But $(G,G') \in A(t|T^b)$ implies the existence of an overlay t' of (G,G') with respect to t. As an extension of t having (G,G') as an action, t' must lie in T^i, i.e. there exists an overlay of $(G,G') \in T^i$.

Corollary 1 For any metadigraph M, there exists a transform kit K such that $(V(M),K)$ is a specification of M.

Proof Write $M = (\Gamma, E)$. For every edge $GG' \in E$, form an overlay of (G,G') as noted in the section on semitransforms by letting the covering graph be a minimum common supergraph of G and G'. Let T^i denote the st of overlays so formed, and let $K = (T^i, \mathcal{T}_0 \backslash T^i)$. By Theorem 3, GG' is an action of K if and only if GG' is an edge of M. Since M and $M(\Gamma,K)$ have identical vertex set, $M = M(\Gamma,K)$.

As there generally exists many overlays of a pair (G,G') of graphs, there will generally be many kits K that satisfy the conditions of the preceding proof.

Acknowledgements

I would like to thank Gerald Maggiora for his encouragement, ideas an perspectives. I also want to thank the reviewers for their careful reading of the original manuscript and numerous helpful and detailed suggestions and clarifications. This work was supported by The Upjohn Company.

REFERENCES

[1] R. Barone and M. Chanon, "Computer-Aided Organic Synthesis (CAOS), In *Computer Aids to Chemistry*, Ed. by G. Vernin and M. Chanon, John Wiley and Sons, (1986) 19- 102.

[2] A. H. Beckett and D. A. Cowan, "Pitfalls in Drug-Metabolism Methodology," In *Drug Metabolism in Man*, ed. by J. W. Gorrod and A. H. Beckett, Taylor and Francis Ltd., London (1978) 237-257.

[3] J. D. Bu'lock, "The Biosynthesis of Natural Products: An Introduction to Secondary Metabolism," McGraw-Hill Pub. Co., New York (1965).

[4] G. Chartrand and L. Lesniak, "Graphs and Digraphs," Second Edition, Wadsworth and Brooks, Monterey, 1986.

[5] G. Chartrand, F. Saba and H.-B. Zou, "Edge Rotations and Distance between Graphs. *Casopis Pro Pest. Mat.* 110 (1985) 87-91.

[6] F. Darvas, "Predicting Metabolic Pathways by Logic Programming," *J. Mol. Graphics* 6 (1988) 80-86.

[7] S. Fujita, "Description of Organic Reactions Based on Imaginary Transition Structures. 1. Introduction of New concepts," *J. Chem. Inf. Comput. Sci.* 26 (1986) 205-212.

[8] M. Johnson, "Relating Metrics Lines and Variables Defined on Graphs to Problems in Medicinal Chemistry," In *Graph Theory and Its Applications to Algorithms and Computer Science*, ed. by Y. Alavi, G. Chartrand, L. Lesniak, D., R. Lick and C. E. Wall, John Wiley and Sons, Inc. (1985) 457-470.

[9] M. Johnson, "An Ordering of Some Metrics Defined on the Space of Graphs," *Czech. Math. J.* 37 (1987) 75-85.

[10] J. Koca, M. Kratochvíl, V. Kvasnička, L. Matyska and J. Pospíchal, "Synthon Model of Organic Chemistry an Synthesis Design," Springer-Verlag, Berlin, (1989) 207 p.

[11] I. Ugi, J. Bauer, J. Brandt, J. Friedrich, J. Gasteige, C. Jochum and W. Schubert, "New Applications of Computers in Chemistry," *Angew. Chem. Int. Ed. Engl.*, 18 (1979) 111-123.

[12] T. H. Varkony, D. H. Smith and C. Djerassi, "Computer-Assisted Structure Manipulation: Studies in the Biosynthesis of Natural Products," *Tetrahedron*, 34, (1978) 841-852.

A Note on Removable Pairs

H.A. Kierstead[1]
W.T. Trotter[2]

Department of Mathematics
Arizona State University

ABSTRACT

A long standing conjecture in the dimension theory for finite ordered sets asserts that every ordered set (of at least three points) contains a pair whose removal decreases the dimension at most one. Two stronger conjectures have been made:
(1) If (x, y) is a critical pair, then $dim(P) \leq 1 + dim(P - \{x, y\})$.
(2) For every $x \in P$, there exists $y \in P - \{x\}$ so that $dim(P) \leq 1 + dim(P - \{x, y\})$.

K. Reuter has disproved conjecture 1 by constructing a four–dimensional poset P containing a critical pair (x, y) so that $dim(P - \{x, y\}) = 2$. In this note, we construct for every $n \geq 5$ an n–dimensional poset P_n containing a critical pair (x, y) so that $dim(P_n - \{x, y\}) = n - 2$. Point y is a maximal point of P_n.

1. Preliminaries

Recall that the *dimension* of a finite ordered set P in the least positive integer t so that there exist t linear extensions $L_1, L_2, ..., L_t$ so that $P = L_1 \cap L_2 \cap ... \cap L_t$. An incomparable pair (x, y) is called a *critical* pair if any point less than x is less than y and any point greater than y is greater than x. The dimension of P is the least t for which there exist t linear extensions of P so that for every critical pair (x, y), there is at least one i for which $y < x$ in L_i. We refer the reader to the survey article [3] by D. Kelly and W.T. Trotter and the chapters [6], [7] by Trotter for additional background information on dimension theory.

1. Research supported in part by the Office of Naval Research.
2. Research supported in part by the National Science Foundation.

2. Removable Pairs

The following conjecture is one of the best known open problems in dimension theory and is a featured problem in **ORDER**. We believe the first reference to the conjecture is [1].

Conjecture 0 *If P is an ordered set having at least three points, then P contains a distinct pair (x, y) so that $dim(P) \leq 1 + dim(P - \{x, y\})$.*

A pair $x, y \in P$ for which $\dim(P) \leq 1 + \dim(P - \{x, y\})$ is called a *1–removable* pair, so that Conjecture 0 asserts that every poset contains a 1–removable pair.

The first reference to the following conjecture is apparently [5].

Conjecture 1 *Every critical pair is 1–removable.*

In [2], D. Kelly made the following conjecture which is also stronger than Conjecture 0.

Conjecture 2 *For every $x \in P$, there is a point $y \in P - \{x\}$ so that x, y is a 1-removable pair.*

K. Reuter [4] has disproved Conjecture 1 by constructing the ordered set shown in Figure 1. This ordered set P has dimension 4, (x, y) is a critical pair, and $\dim(P - \{x, y\}) = 2$. Note that y is a maximal point.

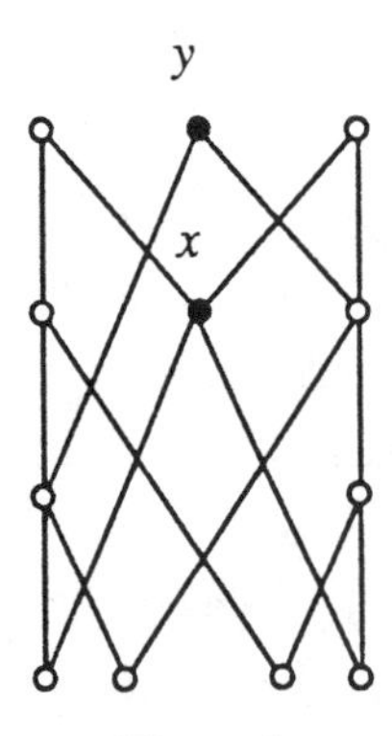

Figure 1

The purpose of this note is to show that Reuter's example is not an isolated phenomenon. To accomplish this, we will establish the following result.

Theorem *For every $n \geq 4$, there exists an n–dimensional ordered set P_n containing a critical pair (x, y) so that y is a maximal element in P_n, but (x, y) is not 1-removable, i.e., $dim(P - \{x, y\}) = n - 2$.*

Proof For $n = 4$, we have Reuter's example shown in Figure 1. For $n \geq 5$, the point set of P_n contains $4n - 4$ points labelled $\{a_i : 1 \leq i \leq n-2\} \cup \{b_i : 1 \leq i \leq n-2\} \cup \{c_i : 1 \leq i \leq n-2\} \cup \{d_i : 1 \leq i \leq n-2\} \cup \{x, y, z, w\}$. For all i, j with $1 \leq i, j \leq n-2$ and $i \neq j$, we have the cover relations $a_i <: b_j$ and $c_i <: d_j$. For each i with $1 \leq i \leq n-2$, we have $a_i <: y$, $c_i <: y$, $c_i < x$, $z <: b_i$, $w <: b_i$, $w < d_i$, $x <: b_i$, and $z < d_i$. We also have $w <: y$. We illustrate this definition with a diagram for P_n when $n = 5$.

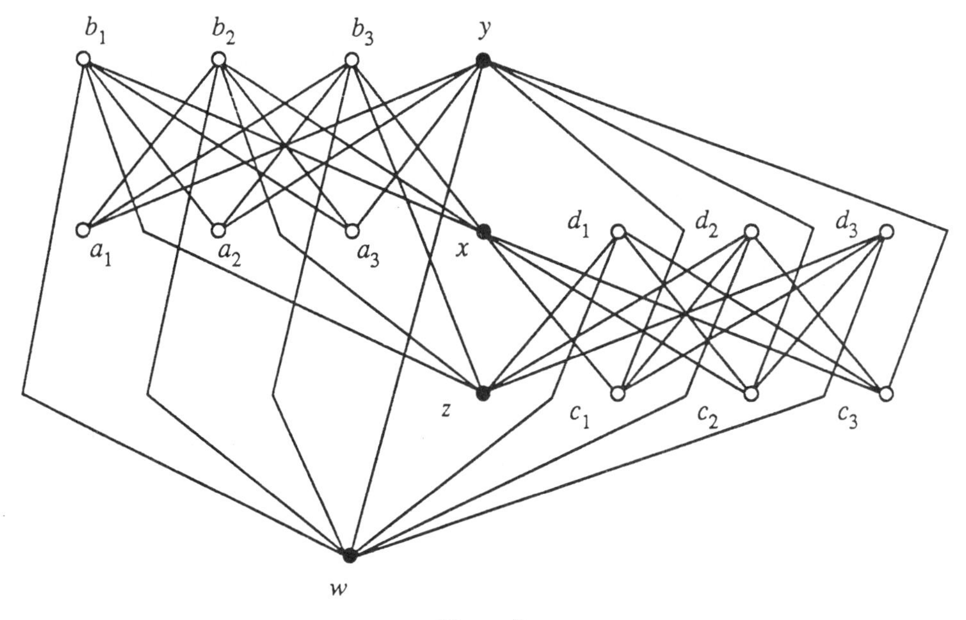

Figure 2

We first show that $\dim(P_n) \geq n$. To the contrary, suppose $\dim(P_n) \leq n - 1$, and let $L_1, L_2, ..., L_{n-1}$ be linear extensions whose intersection is P_n. Without loss of generality, we may assume that $b_i < a_i$ in L_i for $i = 1, 2, ..., n - 2$. Thus we must have $x > y$ in L_{n-1} and $z > y$ in L_{n-1}. However, this implies that for each $i = 1, 2, ..., n - 2$, there exists a unique $j_i \in \{1, 2, ..., n-2\}$ so that $c_i > d_i$ in L_{j_i}. Hence

$w > x$ in L_{n-1} also. But this implies that $w > x > y$ in L_{n-1} which is impossible since $w < y$ in P_n. The contradiction completes the proof that $\dim(P_n) \geq n$.

We observe that y is maximal element of P_n and that (x, y) is a critical pair. We now show that $\dim(P_n - \{x, y\}) \leq n - 2$. To accomplish this, consider the poset $Q_n = P_n - \{x, y\}$. In Q_n, we observe that z and w have duplicated holdings so that $\dim(Q_n - \{z\}) = \dim(Q_n)$. Let $Q'_n = Q_n - \{z\}$. We show that Q'_n has $n - 2$ linear extensions which intersect to give Q'_n. Let $A = \{a_1, a_2, ..., a_{n-2}\}$, $B = \{b_1, b_2, ..., b_{n-2}\}$, $C = \{c_1, c_2, ..., c_{n-2}\}$ and $D = \{d_1, d_2, ..., d_{n-2}\}$.

For $i = 1, 2$, L_i is any linear extension of Q'_n so that $A - \{a_i\} < C - \{c_i\} < w < d_i < c_i < b_i < a_i < B - \{b_i\} < D - \{d_i\}$. For $i = 3, 4, ..., n - 2$, L_i is any linear extension of Q'_n so that $w < C - \{c_i\} < d_i < c_i < D - \{d_i\} < A - \{a_i\} < b_i < B - \{b_i\}$. It is easy to see that any family constructed by these rules forms a realizer. With this observation, our proof is complete. ❑

We pause to note that the construction given in the preceding family for $\{P_n : n \geq 5\}$ does not work for $n = 4$. In this case, $\dim P_n = 4$, but $\dim(P_n - \{x, y\}) = 3$. Thus to handle the case $n = 4$, we need a special example, and Reuter's construction suffices.

3. Concluding Remarks

We view the results of this note as providing additional evidence as to the difficulty of Conjecture 0, but we are unable to decide whether our theorem argues for or against the conjecture. It is easy to see that the examples satisfy Conjecture 2, so at least this stronger form of the original conjecutre remains open.

REFERENCES

[1] K. Bogart and W.T. Trotter, Irreducible Posets with Arbitrarily Large Height Exist, *J. Comb. Theory (A)* **17** (1974), 337 – 344.

[2] D. Kelly, Removable Pairs in Dimension Theory, *Order 1* (1984) 217 – 218.

[3] D. Kelly and W.T. Trotter, Dimension Theory for Ordered Sets, *Proceedings of the Symposium on Ordered Sets*, I. Rival et. al., Reidel Publishing (1982), 171 – 212.

[4] K. Reuter, personal communication.

[5] W.T. Trotter, A Generalization of Hiraguchi's Inequality for Posets, *J. Comb. Theory (A)* **20** (1976), 114 – 123.

[6] W.T. Trotter, Graphs and Partially Ordered Sets, in *Selected Topics in Graph Theory II*, R. Wilson and L. Beineke, eds. Academic Press, (1983), 237 – 268.

[7] W.T. Trotter, Partially Ordered Sets, A chapter in *Handbook of Combinatorics*, R. Graham, M. Groetschel, L. Lovasz, eds., to appear.

On the Number of Maximal Independent Sets in Some (v,e)-Graphs

Felix Lazebnik
University of Delaware

ABSTRACT

Let $V(G)$ be the set of vertices of a simple undirected graph G and S be a subset of $V(G)$. S is an independent set *in G if no two vertices of S are joined by and edge of G. S is a* maximal independent set *(m.i.s.) in G if S is independent and is not a subset of any other independent set. Let $\mu(G)$ denote the number of m.i.s. of a graph G, and $\mu(v, e) = \max \{\mu(G) \mid G$ has v vertices and e edges$\}$. For $0 \leq e \leq v$, new bounds for $\mu(v,e)$ are found. For some subranges of the parameters, $\mu(v, e)$ is determined and extremal graphs are described. The results refine some known theorems from extremal graph theory and an upper bound for the running time of an algorithm of E. L. Lawler for determining the chromatic number of a graph.*

1. Introduction

The definitions in this paper are based on [Bo76]. All graphs we consider are undirected labelled graphs without loops and multiple edges. $V(G)$ and $E(G)$ denote sets of vertices and edges of G respectively. The number of elements of a finite set A is denoted by $|A|$. We write $v = v(G) = |V(G)|$ and $e = e(G) = |E(G)|$ and call G a (v, e)–graph. Let $\{x;y\}$ be an edge of G. Then by $G - \{x;y\}$ we mean the graph obtained from G by deleting $\{x;y\}$. By K_v, $K_v{}^c$, T_v and $K_{m,n}$ we denote correspondingly the complete graph on v vertices (any two vertices are joined by an edge), the completely disconnected graph on v vertices (no edges at all), a tree on v vertices, and the complete bipartite graph whose vertex classes contain m and n vertices. By $G + H$ we denote the disjoint union of graphs G and H. For a given $x \in V(G)$, by $N_G(x)$ we denote the set of all neighbors of x in G, i. e. the set of all $y \in$

1 This paper is based on a part of a Ph. D. Thesis written by the author under the supervision of Prof. H. S. Wilf at the University of Pennsylvania.

V(G) such that $\{x;y\} \in E(G)$. A set S, $S \subseteq V(G)$, is an *independent set* in G if no two vertices of S are joined by an edge of G. S is a *maximal independent set* (m.i.s.) in G if S is independent and is not a subset of any other independent set of G. Let $\mathcal{M}(G)$ denote the set of all m.i.s. of vertices in G, $\mu(G) = |\mathcal{M}(G)|$, $\mu(v) = \max\{\mu(G): |V(G)| = v\}$, $\mu(v, e) = \max\{\mu(G)$: G is a (v, e)–graph$\}$. Obviously, for any (v, e)–graph G, $\mu(G) \le \mu(v,e) \le \mu(v)$. A *clique* in G is a maximal complete subgraph of G. Let cl(G) denote the number of cliques of graph G. Let $cl(v) = \max\{cl(G): |V(G)| = v\}$, $cl(v,e) = \max\{cl(G)$: G is a (v,e)–graph$\}$. Let G^c denote the complement of graph G. It is easy to see that $\mu(G) = cl(G^c)$, $\mu(v) = c(v)$ and $\mu(v, e) = cl\,(v, v(v-1)/2 - e)$. The following problem was formulated by H. S. Wilf:

For the given pair of positive integers (v,e), find $\mu(v, e)$ or give a non–trivial upper bound of $\mu(v, e)$.

In this paper we present some partial results. Problems similar to this, but for different families of graphs, were considered by several authors. The value of cl(v) was determined by Miller and Muller [MiMu60] and independently by a different method by Moon and Mosser [MoMo65] in which they characterized the extremal graphs. The found that

$$\mu(v) = cl(v) = \begin{cases} 3^t & \text{if } v = 3t \ge 3; \\ 4 \cdot 3^{t-1} & \text{if } v = 3t+1 \ge 4; \\ 2 \cdot 3^t & \text{if } v = 3t+2 \ge 2 \end{cases} \tag{1.1}$$

and that the extremal graphs (for the number of cliques) are Turan graphs $T_t(v)$.

It turned out that the problem of finding cl(v) can be shown to be equivalent (Yao [Y76] attributes this result to D. E. Muller) to the problem of Katona on minimal separating systems:

Given the set $[n] = \{1, 2, \ldots, n\}$. Find the smallest number f(n) for which there exists a family of subsets of [n] $\{A_1, A_2, \ldots, A_{f(n)}\}$ with the following property: given any two elements $x, y \in [n]$ $(x \ne y)$, there exist k, ℓ such that $A_k \cap A_\ell = \emptyset$, and $x \in A_k$, $y \in A_\ell$.

Katona's problem was solved by Yao [Y76] and independently by a similar method by Cai [Ca83].

There were several papers in which the authors restricted their attention to a subset $\mathcal{F}$ of all the graphs on v vertices and determined either $cl(v, \mathcal{F}) = \max\{cl(G): G \in \mathcal{F}\}$ or $\mu(v, \mathcal{F}) = \max\{\mu(G): G \in \mathcal{F}\}$.

Hedman [He85,He] found the maximal number of cliques for the family $\mathcal{F}$ of all graphs on v vertices with the given clique number ω (the clique number ω of a graph G is the greatest number of vertices in a clique of G among all cliques of G).

Wilf [W86] found the larges number of m.i.s. vertices that any tree of v vertices can have. The same results were obtained by Cohen [Co84] and Sagan [Sa88] by different methods. Sagan's paper completely describes all extremal graphs. If we denoted this number by μ(v, Tree), then the result is

$$\mu(v, Tree) = \begin{cases} 2^{t-1}+1 & \text{if } v = 2t > 0; \\ 2^t & \text{if } v = 2t+1; \\ 1 & \text{if } v = 0 \end{cases} \tag{1.2}$$

Furedi [Fu87] gave a new proof of (1.1) and established an exact upper bound for μ(G) for a non–extremal graph G. In the same paper he found μ(v, Conn) = the maximum number of m.i.s. that a connected graph on v vertices can have (for v > 50) and described all extremal graphs. Independently, Griggs, Grinstead and Guichard [GGG88] determined μ(v, Conn) for all $v \geq 6$ and described all the extremal graphs. Their result is

$$\mu(v, Conn) = \begin{cases} 2\cdot 3^{t-1} + 2^{t-1} & \text{if } v = 3t > 6; \\ 3^t + 2^{t-1} & \text{if } v = 3t+1 > 6; \\ 4\cdot 3^{t-1} + 3\cdot 2^{t-2} & \text{if } v = 3t+2 > 6 \end{cases} \tag{1.3}$$

Harary and Lempel [HL74] studied the extremal graphs for the family of all graphs on v vertices with e edges. They developed some standard forms for such graphs and suggested a transformation which brings an extremal graph into this form. Similar results were obtained independently by the author. Unfortunately they have not helped much in finding μ(v, e).

Another motivation for the present work was an article by E. Lawler [La76] in which an algorithm for determining the chromatic number of a graph is discussed, and it is shown that its run time, in the worst case, is $0[ev(1+3^{1/3})^v]$ for graphs of e edges and v vertices. The appearance of $3^{1/3}$ derives from (1.1) because of the fact that a graph on v vertices has at most $3^{v/3}$ maximal independent sets. The fact that the graph has e edges is not used when the greatest number of maximal independent sets is estimated.

In Section 2 we give new bounds for μ(v, e) and determine μ(v,e) exactly for some ranges of v and e. The main results are in Theorems 2.1, 2.6 and Corollary 2.3.

2. Results

In this section we determine explicitly or find bounds for $\mu(v, e)$, for $0 \le e \le v$. We start with the following.

Theorem 2.1 *Let v, e be non-negative integers, $0 \le e \le v$, and $m(v, e) = 2^{v-e} \cdot 3^{(2e-v)/3}$. Then*

$$\mu(v, e) \le m(v, e) \tag{2.1}$$

The equality in (2.1) occurs if and only if $v/2 \le e \le v$ and $2e - v = 3t$ for some non-negative integer t. The only graph G, for which $\mu(G) = \mu(v, e) = m(v, e)$ is $G = (v - e)K_2 + tK_3$.

Proof We notice that for $e > v$, our upper bound $m(e, v)$ is worse than $\mu(v)$ given by (1.1). This explains the restriction $0 \le e \le v$. If $v = e = 0$, then $\mu(\varnothing) = m(0, 0) = 1$. If $v = 1$, $e = 0$, then $\mu(1, 0) = \mu(K_1) = 1 < m(1, 0) = 2/(3^{1/3})$. Let G be an extremal graph and $G_1, G_2, \ldots, G_n$ be connected components of G. Suppose G_i is a (v_i, e_i)–graph, $1 \le i \le n$. Then $\Sigma v_i = v$ and $\Sigma e_i = e$. Since $\mu(v, e) = \mu(G) = \Pi\mu(G_i)$ and $m(v, e) = \Pi\mu(v_i, e_i)$, Then in order to prove the theorem it is sufficient to show that for all i, $1 \le i \le n$, $\mu(G_i) \le m(v_i, e_i)$.

Lemma 2.2 *For any connected (v, e)–graph G, $\mu(G) \le m(v, e)$. $\mu(G) = m(v, e)$ if and only if $G = \varnothing$, $G = K_2$ or $G = K_3$.*

Proof It is enough to show that

$$\mu(v) \le m(v, e). \tag{2.3}$$

Then the first statement will be proved. Let $e = v + p$. Then

$$m(v, v+p) = 2^{-p} \cdot 3^{(v+2p)/3} = (9/8)^{p/3} 3^{v/3},$$

and (2.3) can be easily checked by using the table from Figure 1. Entries in the $\mu(v)$ column come from (1.1) and (1.2) (the only connected (v, e)–graphs with $e \le v - 1$ are trees, and $e = v - 1$.

e	v	μ (v)	m(v,e)
e = v + p, p≥0	v = 3t, t ≥1	3^t	$(9/8)^{p/3}3^t$
	v = 3t+1, t ≥1	$4 \cdot 3^{t-1}$	$3^{1/3}(9/8)^{p/3}3^t$
	v = 3t+2, t ≥0	$2 \cdot 3^t$	$3^{2/3}(9/8)^{p/3}3^t$
e = v−1	v = 2t, t ≥1	$2^{t-1}+1$	$(8/9)^{1/3}3^{2t/3}$
	v = 2t + 1, t ≥0	2^t	$(2/3^{1/3})3^{2t/3}$

Figure 1

Comparing the entries in the table on Figure 1, we conclude that equality occurs if and only if $v = 3t$, $p = 0$ or $v = 2$, $p = -1$. For $v = 3t$, $p = 0$, we have $e = 3t$, $\mu(3t) = 3^t$. As it follows from [MiMu60} and [MoMo65], the only extremal graph in this case is tK_3. This graph is connected for $t = 1$. Therefore $G = K_3$. For $v = 2$, $p = -1$, we have $e = 1$ and $G = K_2$. This proves the lemma. ❑

Thus (2.2) is true for each connected component of G and the bound (2.1) is proved. In order to get an equality in (2.1), each connected component of G has to be either K_3 or K_2. Suppose $G = tK_3 + sK_2$, for some non–negative integers t, s. then

$$3t + 2s = v \text{ and } 3t + s = e \tag{2.4}$$

The only solution of (2.4) is $s = v - e$, $t = (2e - v)/3$, and this concludes the proof of the theorem.

Corollary 2.3 *Let v, e be non–negative integers,* $v/2 \leq e \leq v$. *If* $2e - v = 3t + 1$, *then* $3^{-1/3}m(v, e) \leq \mu(v, e) < m(v, e)$. *If* $2e - v = 3t + 2$, *then* $\mu(v, e) = (1/2)(3^{1/3})m(v, e)$ *for* $v = 4, 6$; $(5/3^{5/3})m(v, e) \leq \mu(v, e) < m(v, e)$ *for* $v = 5, v \geq 7$.

Proof The upper bounds follow from Theorem 2.1. In the case $2e - v = 3t + 1$, the lower bound comes from the graph $tK_3 + (v - e - 1)K_2 + P_2$, where P_2 is a path with two edges. If $2e - v = 3t + 2$, then for $v = 4$, the only possible value of e is 4. In both cases $\mu(v, e) = (1/2)(3^{1/3})m(v, e)$. For $v = 5$ or ≥ 7, the lower bound comes from the graph $(t-1)K_3 + (v-e)K_2 + H$, where H is $(5, 5)$–graph shown on Figure 2. The lower bounds seem to be the best possible, but the author has been unable to prove it.

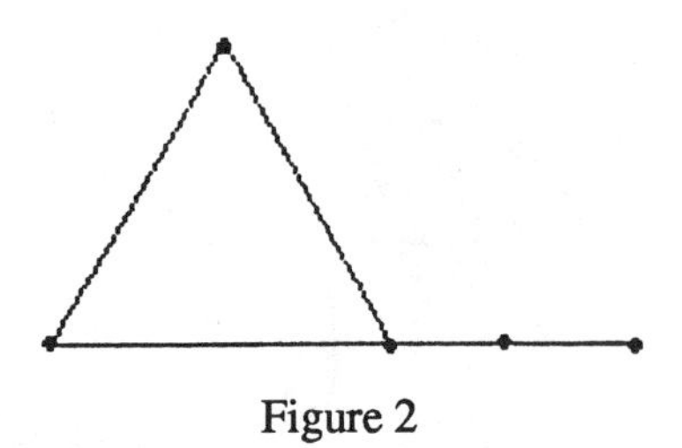

Figure 2

It turns out that for e, $0 \leq e \leq v/2$, the result of Theorem 2.1 can be substantially improved. The following lemma is the main step in this direction. It is also interesting on its own.

Lemma 2.4 *Let G be a graph and $\{x; y\}$ be an edge of G. Then*

$$\mu(G) \leq 2\mu(G - \{x; y\}). \tag{2.5}$$

The equality in (2.5) occurs if and only if $\{x; y\}$ is a connected component of G.

Proof The idea of the proof is to partition both $\mathcal{M}(G)$ and $\mathcal{M}(G-\{x;y\})$ into several classes and to compare numbers of elements in the corresponding classes. The description of the partitions is rather long, but the comparison will be easy. Figure 3 illustrates both stages of the proof. We divide $\mathcal{M}(G)$ into the following 7 classes some of which can be empty ($\dot{\cup}$ stands for the disjoint union of sets):

$\mathcal{M}_{x,1}(G) = \{M \in \mathcal{M}(G)\colon\ M = \{x\} \mathbin{\dot{\cup}} M',\ M' \neq \varnothing,\ M' \cap N_G(y) = \varnothing\}$;

$\mathcal{M}_{y,1}(G) = \{M \in \mathcal{M}(G)\colon\ M = \{x\} \mathbin{\dot{\cup}} M',\ M' \neq \varnothing,\ M' \cap N_G(x) = \varnothing\}$;

$\mathcal{M}_{x,2}(G) = \{M \in \mathcal{M}(G)\colon\ M = \{x\} \mathbin{\dot{\cup}} M',\ M' \cap N_G(y) \neq \varnothing\}$;

$\mathcal{M}_{y,2}(G) = \{M \in \mathcal{M}(G)\colon\ M = \{x\} \mathbin{\dot{\cup}} M',\ M' \cap N_G(x) \neq \varnothing\}$;

$\mathcal{M}_3(G) = \{M \in \mathcal{M}(G)\colon\ M \neq \varnothing,\ x \notin M,\ y \notin M\}$;

$$\mathcal{M}_x(G) = \begin{cases} \{x\}, \text{ if } \{x\} \in \mathcal{M}(G) \\ \varnothing, \text{ otherwise} \end{cases}$$

$$\mathcal{M}_y(G) = \begin{cases} \{y\},\ if\ \{y\} \in \mathcal{M}(G) \\ \varnothing,\ otherwise \end{cases}$$

It is easy to check that $\mathcal{M}(G)$ is the disjoint union of these classes. Similarily we divide $\mathcal{M}(G - \{x; y\})$ into the following 5 classes:

$\mathcal{M}_{x,y,1}(G - \{x; y\}) = \{M \in \mathcal{M}(G - \{x; y\}):\ M = \{x\} \dot\cup \{y\} \dot\cup M', M' \neq \varnothing\};$

$\mathcal{M}_{x,2}(G - \{x; y\}) = \{M \in \mathcal{M}(G - \{x; y\}):\ x \in M, M \cap N_{G-\{x;y\}}(y) \neq \varnothing\};$

$\mathcal{M}_{y,2}(G - \{x; y\}) = \{M \in \mathcal{M}(G - \{x; y\}):\ y \in M, M \cap N_{G-\{x;y\}}(x) \neq \varnothing\};$

$\mathcal{M}_3(G - \{x; y\}) = \{M \in \mathcal{M}(G - \{x; y\}):\ M \neq \varnothing, x \notin M, y \notin M\};$

$$\mathcal{M}_{x,y}(G - \{x; y\}) = \begin{cases} \{x, y\},\ if\ \{x, y\} \in \mathcal{M}(G - \{x; y\}) \\ \varnothing,\ otherwise \end{cases}$$

It is easy to check that $\mathcal{M}(G-\{x;y\})$ is the disjoint union of these classes.

The following bijections between some of these classes are obvious:

$f_{x,1}\ :\ \mathcal{M}_{x,1}(G) \to \mathcal{M}_{x,y,1}(G-\{x;y\}),\ f_{x,1}(\{x\} \dot\cup M') = \{x, y\} \dot\cup M';$

$f_{y,1}\ :\ \mathcal{M}_{y,1}(G) \to \mathcal{M}_{x,y,1}(G-\{x;y\}),\ f_{y,1}(\{y\} \dot\cup M') = \{x, y\} \dot\cup M';$

$f_{x,2}\ :\ \mathcal{M}_{x,2}(G) \to \mathcal{M}_{x,2}(G-\{x;y\}),\ f_{x,2}(M) = M;$

$f_{y,2}\ :\ \mathcal{M}_{y,2}(G) \to \mathcal{M}_{y,2}(G-\{x;y\}),\ f_{y,2}(M) = M;$

$f_3\ :\ \mathcal{M}_3(G) \to \mathcal{M}_3(G-\{x;y\}),\ f_3(M) = M;$

$f_x\ :\ \mathcal{M}_x(G) \to \mathcal{M}_{x,y}(G-\{x;y\}),\ f_x(\{x\}) = \{x, y\}$, if $\{x\}$ is m.i.s.;

$f_x(\{\varnothing\}) = \varnothing$, if $\{x\}$ is not m.i.s.;

$f_y\ :\ \mathcal{M}_y(G) \to \mathcal{M}_{x,y}(G-\{x;y\}),\ f_y(\{y\}) = \{x, y\}$, if $\{y\}$ is m.i.s.;

$f_y(\{\varnothing\}) = \varnothing$, if $\{y\}$ is not m.i.s.

(Notice that for each of these mappings the domain and the range are non–empty simultaneously.)

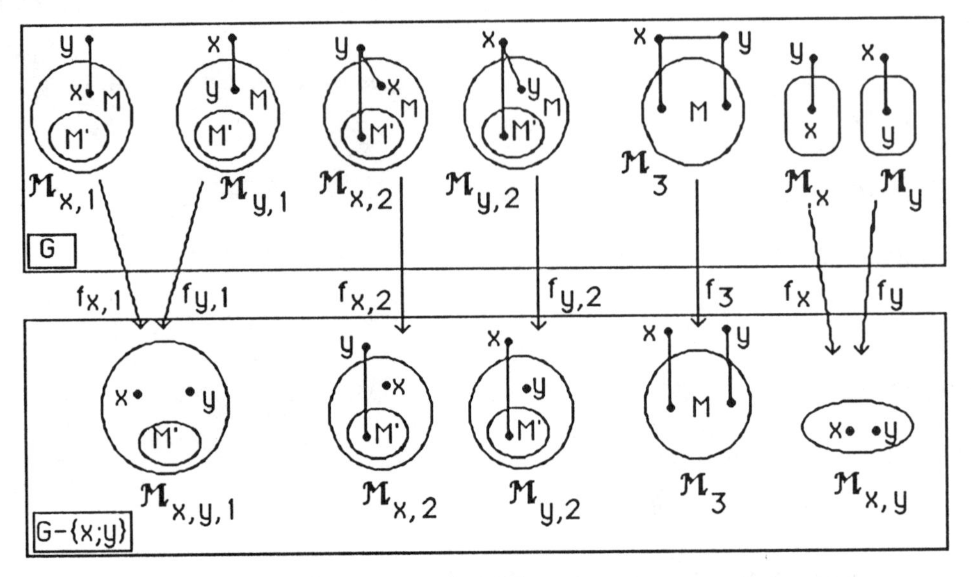

Figure 3

Finally we denote $|\mathcal{M}_{x,1}(G)| = |\mathcal{M}_{y,1}(G)| = |\mathcal{M}_{x,y,1}(G-\{x;y\})| = i_1$,

$$|\mathcal{M}_{x,2}(G)| = |\mathcal{M}_{x,2}(G-\{x;y\})| = i_{x,2},\ |\mathcal{M}_{y,2}(G)| = |\mathcal{M}_{y,2}(G-\{x;y\})| = i_{y,2},$$

$$|\mathcal{M}_3(G)| = |\mathcal{M}_3(G-\{x;y\})| = i_3,\ |\mathcal{M}_x(G)| = i_x\ (= 1 \text{ or } 0),$$

$$|\mathcal{M}_y(G)| = i_y\ (= 1 \text{ or } 0),\ |\mathcal{M}_{x,y}(G-\{x;y\})| = i_{x,y}\ (= 1 \text{ or } 0).$$

Then

$$2\mu(G-\{x;y\}) = 2(i_1 + i_{x,2} + i_{y,2} + i_3 + i_{x,y}),$$

$$\mu(G) = 2i_1 + i_{x,2} + i_{y,2} + i_3 + i_x + i_y.$$

Hence, $\mu(G) = 2\mu(G-\{x;y\}) - (i_{x,2} + i_{y,2} + i_3) - (2i_{x,y} - i_x - i_y)$. Obviously $i_{x,2} + i_{y,2} + i_3 \geq 0$. If at least one of i_x or $i_y = 1$, then $i_{x,y} = 1$, and $2i_{x,y} - i_x - i_y \geq 0$. If $i_x = i_y = 0$, then again $2i_{x,y} - i_x - i_y \geq 0$. Therefore we get

$$\mu(G) \leq 2\mu(G-\{x;y\}) \tag{2.6}$$

The equality sign in (2.6) occurs if and only if $i_{x,2} + i_{y,2} + i_3 = 0$ and $2i_{x,y} - i_x - i_y = 0$. The first of the equalities implies $i_{x,2} = i_{y,2} = i_3 = 0$. If $i_{x,2} = i_{y,2} = 0$, then vertices x and y have the same set of neighbors in G (each independent set of a graph is a subset of at least one m.i.s.). But $i_3 = 0$ implies that this set of neighbors is empty. Therefore the edge {x;y} is a connected component in G. If this is the only connected component of G, i. e. $G = K_2$, then $i_{x,y} = i_x = i_y = 1$ and $2i_{x,y} - i_x - i_y = 0$. If G has more than one connected component, then $i_{x,y} = i_x = i_y = 0$ and again $2i_{x,y} - i_x - i_y = 0$. The lemma is proved. ❑

Corollary 2.5 *If an extremal (v, e)–graph G has two isolated vertices, then it is a disjoint union of edges and isolated vertices.*

Proof Let G have a connected component H with 2 or more edges and two isolated vertices a and b. By deleting and edge {x;y} in H and joining vertices a and b we obtain a (v, e)–graph G'. Since $\mu(H + \{a\} + \{b\}) = \mu(H)$ and $\mu(\{a;b\} + (H - \{x;y\})) = 2\mu(H - \{x;y\})$, and by Lemma 2.4, $\mu(H) < 2\mu(H - \{x;y\})$, then $\mu(H + \{a\} + \{b\}) < \mu(\{a;b\} + (H - \{x;y\}))$. All other connected components (with the vertices in $V(G) - \{a, b\} - V(H)$) of G and G' are the same. Therefore $\mu(G) < \mu(G')$, which contradicts the extremality of G. ❑

The following theorem gives the exact value of $\mu(v,e)$ and describes the extremal graphs for $0 \leq e \leq v/2$.

Theorem 2.6 Let $0 \leq e \leq v/2$. Then $\mu(v,e) = 2^e$, and the only extremal graph is $eK_2 + (v - 2e)K_1$.

Proof The greatest number of vertices in a graph which are incident to e edges is 2e and this happens *only* if the graph is eK_2. Therefore if $v - 2e \geq 1$, then the statement of the theorem follows from Corollary 2.5. so we assume that $v = 2e$. Let G be extremal graph and $G_1, G_2, \ldots, G_n$ be connected components of G. Suppose $v(G_i) = v_i$ and $e(G_i) = e_i$, $1 \leq i \leq n$. Since for all i, $1 \leq e_i \leq v_i - 1$, then

$$\Sigma e_i = e \geq v - n \text{ and } n \geq v - e = 2e - e = e.$$

If G had two isolated vertices, then, due to Corollary 2.5, it would have at least 2e + 2 vertices and this is not the case.

Suppose G has no isolated vertices. Then each component must have at least one edge and $n \leq e$. So $n = e$, $G = eK_2$ and the theorem is proved.

The only case left is when G has only one isolated vertex. Then each of the remaining $n - 1$ components has at least one edge. If each of them has exactly one edge, then $G = eK_2 + K_1$ and $v(G) = 2e + 1$, but the latter is false. So there should be a component with at least two edges. It cannot have three edges, since in this case $e(G) \geq$

$3 + (n-2) = n + 1 > e$. Thus $G = P_2 + (e-2)K_2 + K_1$ (P_2 is a path with two edges), and $\mu(G) = 2 \cdot 2^{e-2} = 2^{e-1}$. But this is less than $2^e = \mu(eK_2)$ which contradicts the extremality of G. Therefore the theorem is proved.

Acknowledgement The author wishes to thank the referee for his numerous comments, suggestions and corrections which resulted in the improvement of the original version of this paper.

REFERENCES

[Bo78] B. Bollobas, Extremal Graph Theory. *Academic Press* (1978).

[Ca83] Cai Mao–cheng, Solution to Edmonds' and Katona's problems on families of separating subsets. *Discrete Math* **47** (1983) 13–21.

[Co84] D. I. A. Cohen, Counting stable sets in trees. *Seminaire Lotharingien de Combinatoire*, 10eme session, R. Konig ed, L'Institute de Recherche Mathematique, Avancee pub. Strasbourg, France (1984) d 48–52.

[Fu87] Z. Furedi, The number of maximal independent sets in connected graphs. *Journal of Graph Theory*, Vol. 11, No. 4 (1987) 463–470.

[GGG88] J. R. Griggs, C. M. Grinstead, D. R. Guichard, The number of maximal independent sets in a connected graph. *Discrete Math* **68** (1988) 211–220.

[HL74] F. Harary and A. Lempel, On clique–extremal (p, q)–graphs. *Networks* **4** (1974) 371–378.

[He] B. Hedman, Another extremal property of Turan graphs, to appear.

[He85] B. Hedman The maximal number of cliques in deuse graphs. *Discrete Math* **54** (1985) 161–166.

[La76] E. L. Lawler, A note on the complexity of the chromatic number problem. *Information Processing Letters*, v. 5, nom. 3 (1976) 66–67.

[MiMu60] R. E. Miller and D. E. Muller, A problem of maximum consistant set. *IBM Research Report RC–240*, J. T. Watson Res. Center, Yorktown Hights, N. Y. (1960).

[MoMo65] J. W. Moon and L. Moser, On clique graphs. *Israel J. Math* **3** (1965) 23–28.

[Sa88] B. Sagan, A note on independent sets in a tree, *SIAM J. Discrete Math.* **1** (1988) no. 1, 105–108.

[W86] H. S. Wilf, The number of maximal independent sets in a tree. *SIAM J. Alg. Disc. Math.* **7** (1986) 125–130.

[Y76] A. C. – C. Yao, On problem of Katona on minimal separating systems. *Discrete Math.* **15** (1976) 193–199.

On Critical and Cocritical Diameter Edge–invariant Graphs

Sin–Min Lee
San Jose State University

Ai–Yun Wang
Santa Clara University

ABSTRACT

The concepts of critical and cocritical diameter edge–invariant graphs are introduced. It is shown that every connected graph is an induced subgraph of a cocritical diameter edge–invariant graph.

1. Introduction

We can model a communication network by a graph $G = (V,E)$ consisting of a set V of vertices, or communication sites, and a set E of edges, or communication lines. The distance between two vertices x and y is defined as the length of the shortest (x,y)–path in G. The diameter $D(G)$ of a graph G is defined by $D(G) = \max\{d(x,y): x,y \in V(G)\}$. The diameter of a graph corresponds to the maximum number of links over which a message between two vertices of the network must travel. In networks where the time delay of signal degradation is approximately proportional to the length of the path being used, the diameter of the network is an important parameter which measures the efficiency and reliability of the network. ([1], [2], [3], [4])

Ore [14] considered graph G with $D(G) > D(G+e)$ for all e not in $E(G)$. Glivjak [5] in the sixties initiated the study of graphs G with property $D(G) < D(G \setminus e)$ for all e in $E(G)$. There is a vast literature devoted to the above two concepts. (See [6], [7], [8], [9], [18].)

Recently, the first author [12], [13] investigated the so–called diameter edge–invariant networks. A graph $G = (V,E)$ is said to be *diameter edge–invariant* if by deleting any edge e of G, its diameter does not change, i.e., $D(G) = D(G \setminus e)$ for all e in E. We will use an abbreviation "dei" for the diameter edge–invariant. The dei graph should be c–connected for $c \geq 2$, i.e., any two nonadjacent nodes of G are joined by c internally disjoint paths and cannot be separated by the removal of fewer than c vertices. However, c–connectedness is not sufficient for G to be dei; for example, $C_3 \times K_2$ is 3–connected but not diameter edge–invariant.

We shall denote the class of dei graphs by DEI. For any integer $k \geq 2$, we will denote the class of all dei graphs with diameter k by DEI(k).

A diameter edge–invariant graph $G = (V,E)$ is said to be *critical* (respectively *cocritical*), if the deletion of any vertex v in V results in a graph $G \setminus v$ which is not dei (respectively $G \setminus v$ is dei).

The potential importance of the cocritical graphs to the design of interconnection network is indicated by their high reliability.

Several methods of construction of dei graphs are proposed in [12]. It is the purpose of this paper to investigate the class of critical dei graphs and cocritical dei graphs that arise from such constructions.

Not every dei graph is critical dei or cocritical dei. The graph in Figure 1 (a) is in DEI(2). However, we see that $G \setminus A$ is in DEI(2) but $G \setminus B$ is not in DEI. Hence G is neither critical nor cocritical.

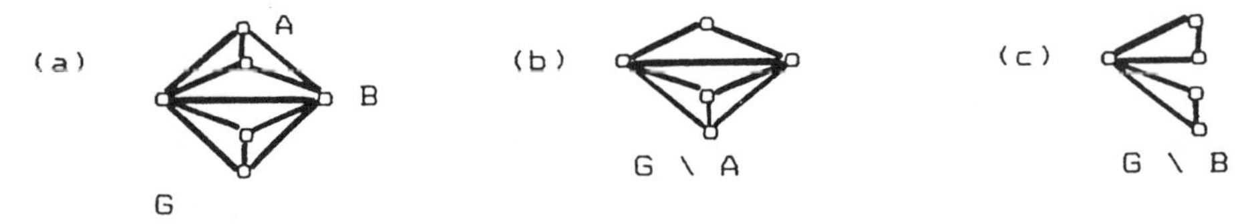

Figure 1

The second section of this paper is devoted to the construction of critical dei graphs by edge expansion of graphs. The critical dei graphs of diameter k with minimum number of nodes are completely described.

Section 4 presents the Sabidussi sum of graphs. This construction will lead to a large class of cocritical dei graphs. We show that every connected graph is an induced subgraph of a cocritical dei graph. The result not only indicates that there is an abundance of cocritical dei graphs but, as a consequence, also shows that it is impossible to have a characterization of cocritical dei graphs by forbidden subgraphs. For a fixed $k \geq 2$, we completely describe the structure of the cocritical dei graphs of diameter k with minimum number of nodes.

2. Critical dei graphs constructed by edge expansion of graphs.

Let *Gph* be the class of all undirected graphs. Denote by 2–*Gph* the class of all graphs of the form ⟨G; u,v⟩ where u, v in V(G), i.e. graphs with 2 terminals.

Given a directed graph G = (V,E) without loops and a mapping f: E→ 2–*Gph,* we construct a new undirected graph (G,f) which is called the *edge expansion* of G by f in [12] as follows:

if f((a,b)) = ⟨H; u,v⟩, then we replace the edge (a,b) in G by the graph H which identifies u with a and v with b.

An illustration of the construction of the edge expansion G from a mapping f is given in Example 1.

Example 1 Suppose G is given by Figure 2(a) and mapping f is given by Figure 2(b). Then (G,f) is shown in Figure 2(c).

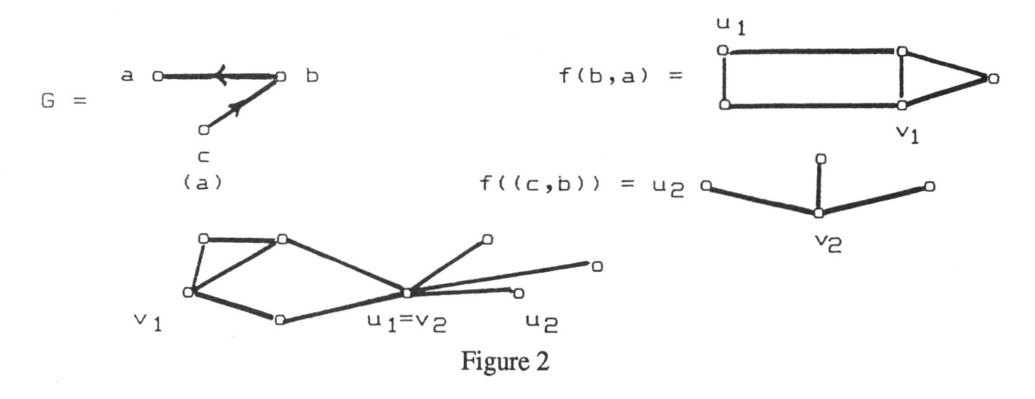

Figure 2

A graph G is a *minimum DEI(k)–graph* if it is a dei graph of diameter k with minimum number of nodes. It is clear that every minimum DEI(k)–graph is a critical dei graph.

The only minimum DEI(2)–graph is the "diamond" D of Klee and Quaife [10, 11], it is $P_3 + K_1$. There are two minimum DEI(3)–graphs J_1 and J_2 (Figure 3).

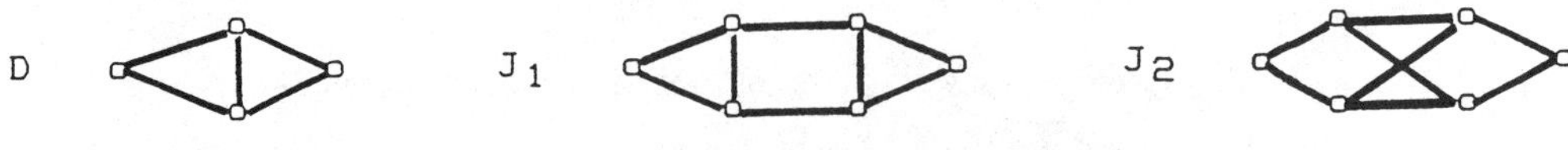

Minimum DEI(k)–graphs for $k = 2$ and 3.
Figure 3

The minimum DEI(k)–graphs are completely described by the first author in [12].

Theorem 1 *A graph G is a minimum DEI(k)–graph,*

(1) *for* $k = 2m$ *iff it is the edge expansion of* P_{m+1} *by f, where* $f(e_i) = \langle D; u,v\rangle$ *for* $i = 1, m$ *and* $f(e_i) \in \{ \langle D; u,v\rangle, \langle C_4; a,b\rangle \}$ *for* $1 < i < m$ *where*

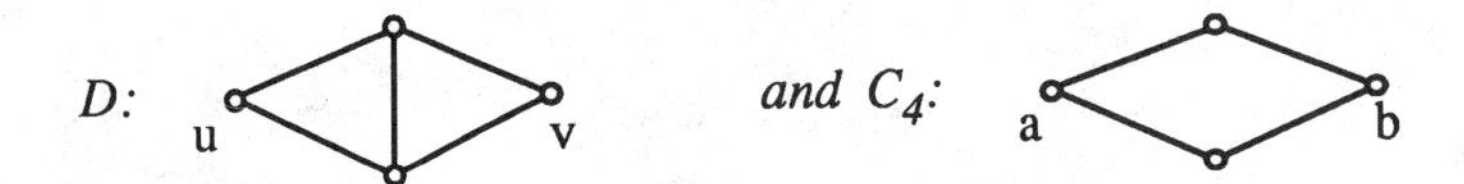

(2) *for* $k = 2m + 1$ *iff it is the edge expansion of* P_{m+1} *by f where*

(a) $m = 1$.

$f(e_1) = J_1$ *or* J_2

(b) $m > 1$, $P_{m+1} = (e_1, ..., e_m)$, *and for a fixed* k,

where $1 \leq k \leq m$,

$$f_k(e_i) = \begin{cases} J_1 \text{ or } J_2 & \text{for } i = k \\ D & \text{for } i = m \\ D \text{ or } C_4 & \text{otherwise.} \end{cases}$$

Example 2 The following configurations are minimum DEI(6) and DEI(7) graphs (Figure 4).

For $n = 3$ and $m \geq 3$, we denote by SF(n,m) the edge expansion of C_n by the complete graph K_m. The graph SF(n,m) is called the *generalized sunflower*. For $m = 3$, the graphs SF(n,3) for $n = 3, 4, 6$ are depicted in Figure 5. The graph SF (n,3) is called a sunflower graph.

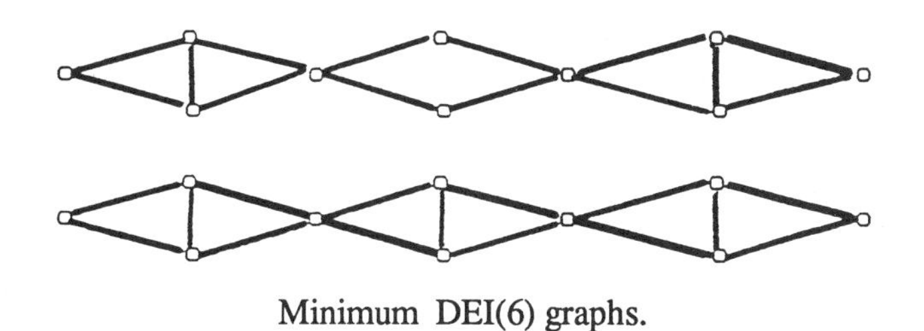

Minimum DEI(6) graphs.

(P_3, f_2)

(P_3, f_1)

Minimum DEI(7) graphs.

Figure 4

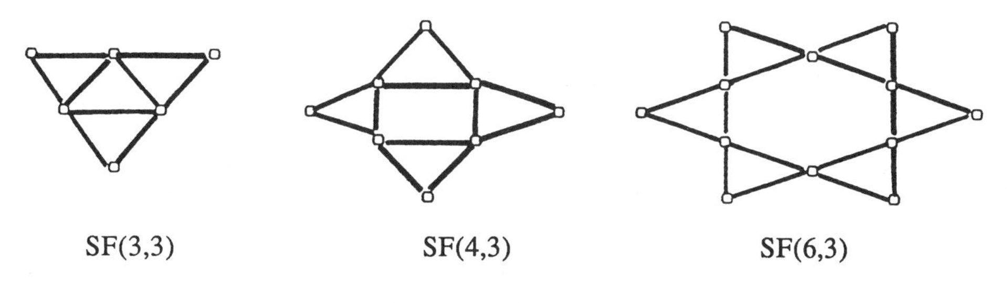

SF(3,3) SF(4,3) SF(6,3)

Sunflower Graphs.

Figure 5

Theorem 2 *The generalized sunflower* $SF(n,m)$ *is critical dei if and only if* $m = 3$, *and* n *is an even number greater than* 4.

Proof It is clear that the graph SF(n,m) is dei if and only if n is even.

If $m \geq 4$ and n is even, the graph SF(n,m) is not critical.

Assume $m = 3$ and $n = 4$, then the subgraph of SF(4,3) obtained by deleting an outer–vertex is diameter edge–invariant (Figure 6). Therefore, the graph SF(4,3) is not critical diameter edge invariant.

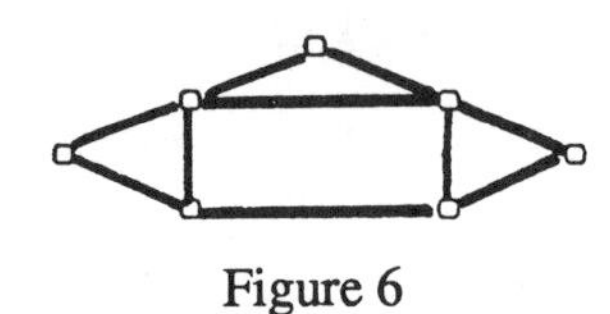

Figure 6

For $n = 2k \geq 6$, there are two types of vertex deleted subgraphs (Figure 7 for $k = 3$).

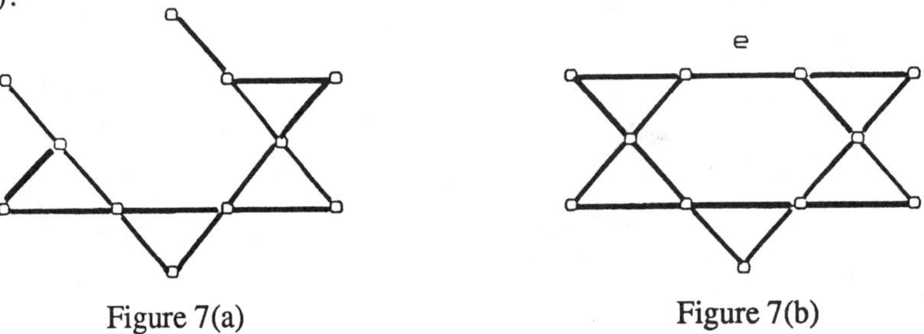

Figure 7(a) Figure 7(b)

We observe that Figure 7(a) is clearly not dei and the graph of Figure 7(b) has diameter $(n/2) + 1$. If we delete the edge e, then the resulting graph has diameter $n - 1$. Clearly $n - 1 > (n/2) + 1$ for $n = 2k \geq 6$. Therefore, SF(2k,3) is critical. ❑

Theorem 3 *If G is critical dei and f: E(G) → {Diamond}, then the edge expansion (G,f) is critical dei.*

Example 3 The graph which is constructed by the edge expansion of SF(6,3) by the diamond D is shown as follows (Figure 8):

Edge expansion of SF(6,3) by D.

Figure 8

3. Critical dei graphs constructed by Cartesian product.

The operation of forming Cartesian products affords the possibility of constructing many dei graphs [12], [13]. It may be worthwhile to note that not all such graphs are critical.

The Cartesian product G x H of two graphs is the graph with V(G) x V(H) and two vertices (g_1,h_1) and (g_2,h_2) are adjacent if $g_1 = g_2$ and $(h_1,h_2) \in E(H)$ or $(g_1,g_2) \in E(G)$ and $h_1 = h_2$.

The 2–dimensional grid is the Cartesian product of two paths (see Figure 9). It is clear that $P_2 \times P_n$ is not dei for any $n \geq 2$. But $P_m \times P_n$ is dei if $\min\{m,n\} \geq 3$.

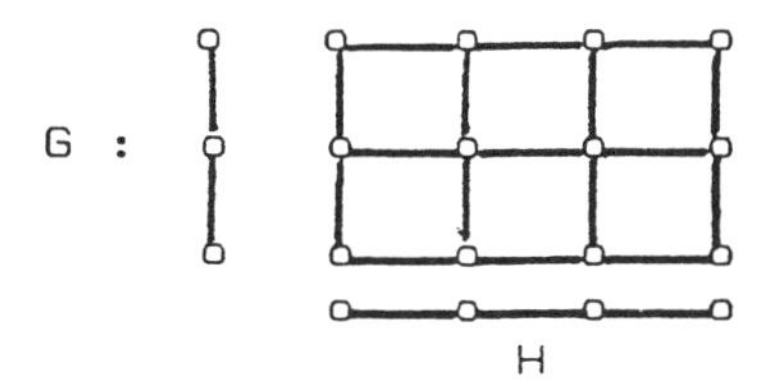

The products of two paths

Figure 9

The following theorem characterizes critical dei graphs which are 2–dimensional grids.

Theorem 4 *The 2–dimensional grid $P_n \times P_m$ is critical dei if and only if $(n,m) = (3,3)$ or $(3,4)$.*

Theorem 5 *The prism $C_n \times K_2$ is critical dei, if and only if $n > 3$.*

Since C_3 is a complete graph, one may conjecture that if G is a non–complete 2–connected graph, then $G \times K_2$ is critical diameter e–invariant.

However, the smallest counter–example is provided by the "diamond" D of Klee and Quaife. We see that $D \times K_2$ is dei. However, its two subgraphs which are depicted in Figure 10 are not both dei graphs.

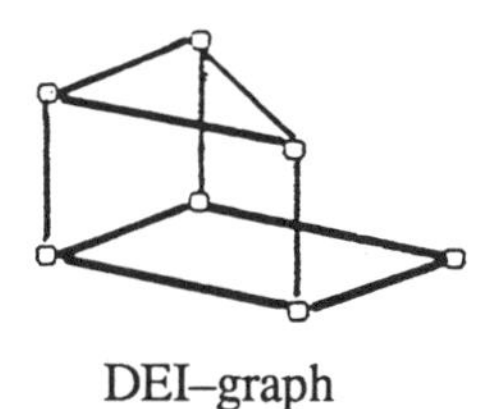

DEI–graph

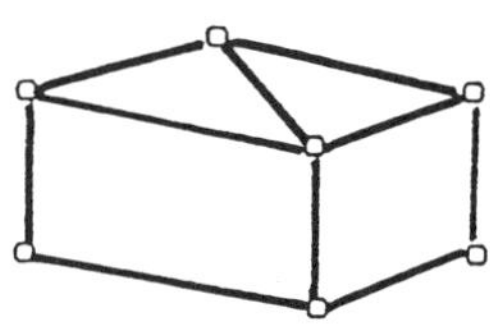

not DEI–graph

Figure 10

4. Cocritical dei graphs which are Sabidussi sum of graphs.

Let G be a connected graph and f: $V(G) \to Gph$ be a nontrivial mapping; we construct a graph $S^+(G,f)$ as follows:

$$V(S^+(G,f)) = \cup \{ f(v): v \in V(G) \}$$

$$E(S^+(G,f)) = \{ (x,y): x \in V(f(v)),\ y \in V(f(u)),\ (v,u) \in E(G) \}$$

The graph is connected and it is called the *Sabidussi sum* of { f(v): $v \in V(G)$ } in [12]. It is clear that $D(S^+(G,f)) = D(G)$, for $D(G) \neq 1$. For $D(G) = 1$, we have $D(S^+(G,f)) = 2$. This construction is a generalization of Sabidussi's composition of lexicographic product [15], [16], [17]. The Zykov sum G + H of two graphs can be considered as $S^+(K_2,f)$ where f(u) = G and f(v) = H. (Figure 11)

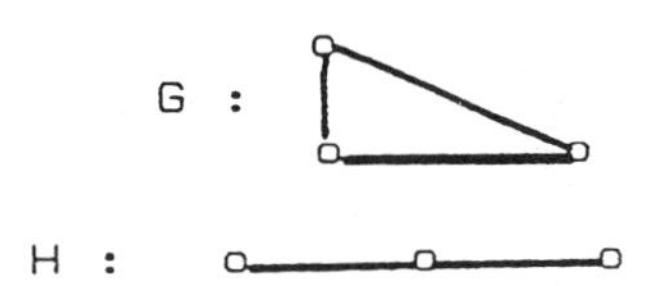

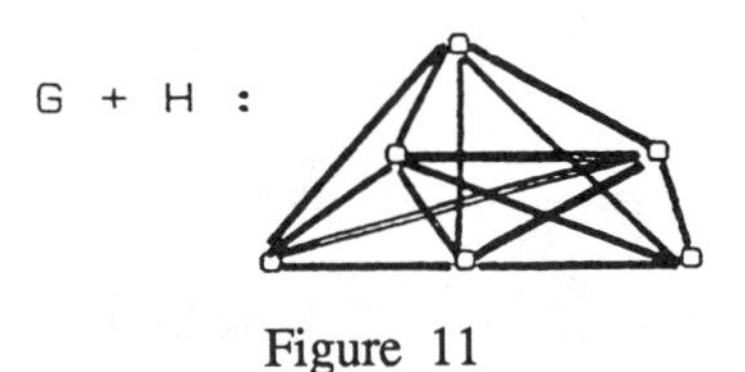

Figure 11

The lexicographic product of the graphs G_1 and G_2 is a graph denoted $G_1[G_2]$ whose vertex set is $V(G_1 \times G_2)$. For vertices $u = (u_1, v_2)$ and $v = (v_1, v_2)$ we have $uv \in E(G_1[G_2])$ if and only if either (a) $u_1 = v_1$ and $u_2v_2 \in E(G_2)$ or (b) $u_1v_1 \in E(G_1)$.

We see that $G_1[G_2] = S^+(G_1,f)$ where $f(v) = G_2$ for all v in $V(G_1)$.

Now we present the main innovation of this section.

Theorem 6 *For a 2–connected graph G, and f: $V(G) \to Gph$ with the property that f(v) is connected and $|V(f(v))| \geq 2$, the Sabidussi sum $S^+(G,f)$ is cocritical dei.*

We illustrate the above result by the following example:

Example 4 Let $C = C_4$ and f: $V(C_4) \to Gph$ be $f(i) = K_2$, then $S^+(G,f)$ is a graph which is cocritical dei. (Figure 12)

Cocritical dei graph with diameter 2.

Figure 12

Remark 1 The 2–connectedness in Theorem 6 is necessary. There exist connected graphs G such that f: $V(G) \to Gph$ with f(v) connected and $|V(f)| \geq 2$ but $S^+(G,f)$ is not cocritical dei. Let $G = P_3$, $f(v_1) = P_3$ and $f(v_2) = f(v_3) = P_2$ then the Sabidussi sum $S^+(G,f)$ (Figure 13) is not cocritical diameter e–invariant.

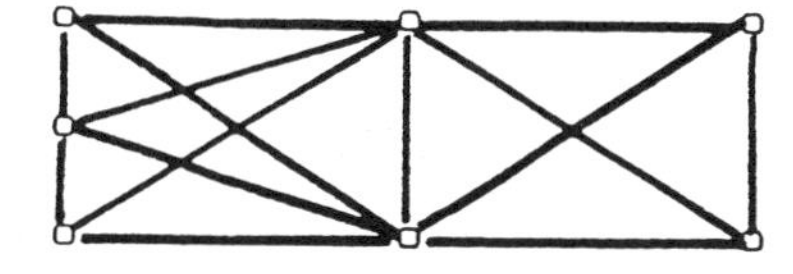

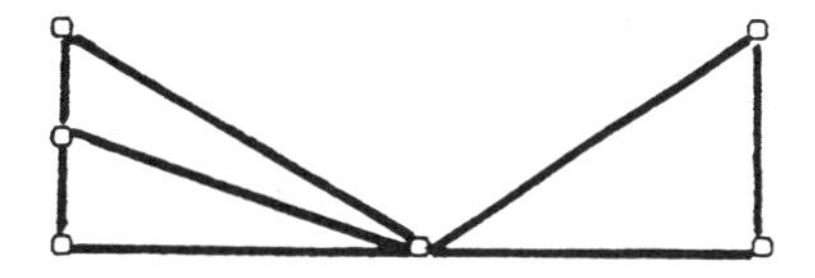

Figure 13

Remark 2 Let G be C_4 and $f(v_1) = f(v_2) = f(v_4) = K_2$ and $f(v_3) = K_1$, then $S^+(C_4,f)$ is not even cocritical (Figure 14). This example illustrates that the condition $|V(f(v))| \geq 2$ for all v is necessary.

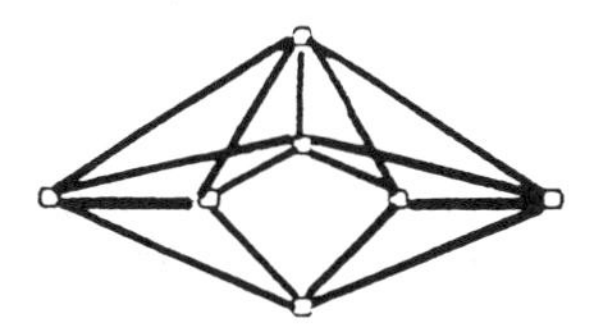

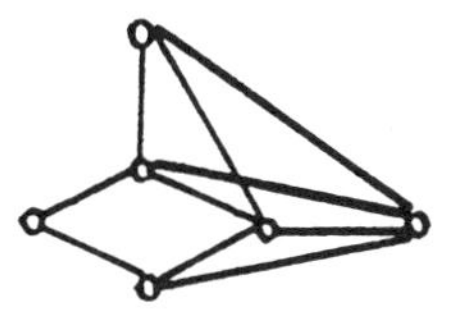

Figure 14

Corollary 7 *Every connected graph H is an induced subgraph of a cocritical dei graph.*

Proof If $|H| = 2$, the result is trivial true. Assuming $|H| \geq 3$, take any 2–connected graph G. Consider the lexicographic product G[H], then by Theorem 6 it is cocritical dei.❑

Our characterization of minimum cocritical DEI(k) graphs is

Theorem 8 *The minimum cocritical DEI graph with diameter k is given by $S^+(C_{2k},f)$ where $f(v) = P_2$ for all v in $V(C_{2k})$.*

Example 5 The following graphs (Figure 15) are minimum cocritical DEI graphs of diameter 2 and 3.

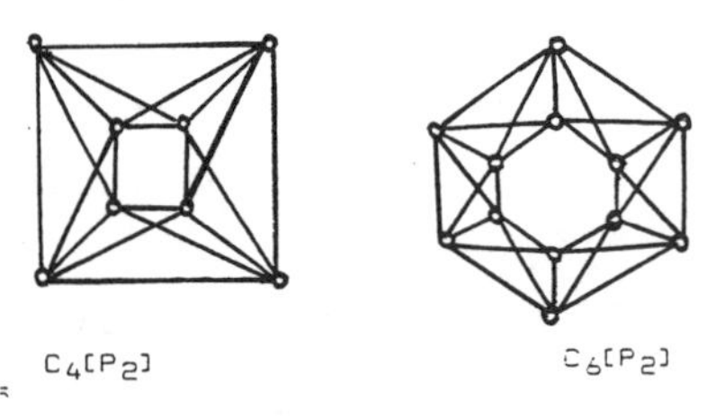

Figure 15

Acknowledgement We appreciate the referee for his helpful suggestions which improve the style of the paper.

REFERENCES

[1] J. C. Bermond and B. Bolobas, Diameters in graphs: a survey. *Congressus Numerantium* **32**, (1981), 3–27.

[2] J. C. Bermond, J. Bond, M. Paoli and C. Peyrat, Graphs and interconnection networks: Diameter and vulnerability, *London Math Society Lecture Notes Ser. 82*, Cambridge University Press, (1983), 1–30.

[3] J. C. Bermond, C. Delorme and G. Farhi, Large graphs with given degree and diameter III, *Annals of Discrete Mathematics* **13**, (1982), 23–32.

[4] F. R. K. Chung, Diameters of communication networks, in "Math. of information Processing", *Proc. Symposium in Applied Math.* **34**, ed. by Michael Anshel and William Gewirtz, American Math Society, (1986), 1–18.

[5] F. Glivjak, On certain classes of graphs of diameter two without superfluous edges, *Acta Fac. Rer. Nat. Univ. Comenian Math.* **21**, (1968), 39–48.

[6] F. Glivjak, On certain edge–critical graphs of a given diameter, *Mat. Casopis Sloven Akad Vied.* **25**, (1975), 249–263.

[7] F. Glivjak, On the impossibility to construct diametrically critical graphs by extensions, *Arch. Math. (Brno)* **11**, (1975), 131–137.

[8] D. Greenwill and P. Johnson, On subgraphs of critical graphs of diameter k, *Proc. 10th S–E Conf. Combinatorics, Graph Theory and Computing*, 465–467.

[9] J. Hartman and I Rubin, "On Diameter Stability of Graphs," *Theory and Applications of Graphs,* Eds. Y. Alavi and D. R. Lick, Springer Lecture Notes No. 642, (1987), 247–254 (1987).

[10] V. Klee and H Quaife, Minimum graphs of specified diameter, connectivity and valence 1. *Mathematics of Operations Research,* **1** (1976), 28– 31.

[11] V. Klee and H. Quaife, Classifications and enumeration of minimum (d,1,3)–graphs and minimum (d,2,3)–graphs, *J. Comb. Theory (B)* **23**, (1977), 83–93.

[12] Sin–Min Lee, Design of diameter e–invariant networks, *Congressus Numerantium* **65** (1988), 89–102.

[13] Sin–Min Lee and Rudy Tanoto, Three classes of diameter edge–invariant graphs. *Commentations Mathematicae Universatis Carolinae* **28**, (1987), 227–232.

[14] O. Ore, Diameters in graphs, *Journal of Comb. Theory* **5**, (1968), 75–81.

[15] G. Sabidussi, The composition of graphs, *Duke Math. Journal* **26**, (1959), 693–696.

[16] G. Sabidussi, The lexicographic product of graphs, *Duke Math. Journal* **28**, (1961), 573– 578.

[17] G. Sabidussi, Graph derivatives, *Math Zeitschrift* **76,** (1961), 385–401.

[18] A. A. Schone, H. L. Bodlaender and J. van Leeuwen, Diameter increase caused by edge deletion, *Journal of Graph Theory* **11** (1987), 409–427.

Facts and Quests on Degree Irregular Assignments

J. Lehel[1]

Computer and Automation Institute
Hungarian Academy of Sciences

ABSTRACT

We summarize the status of research on irregularity recently done by many people: G. Chartrand, R.J. Faudree, A. Gyárfás, K.S. Holbert, M.S. Jacobson, L. Kinch, J. Lehel, O.R. Oellermann, S. Ruiz, F. Saba, R.H. Schelp, H.C. Swart and T. Szônyi. The problem of irregular degree assignments is formulated in terms of set systems. The paper presents 27 conjectures and questions for particular structures together with the related facts and results.

Introduction

In [CJLORS], the following problem was proposed by Chartrand, Jacobson, Lehel, Oellermann, Ruiz and Saba. Assign positive integer weights to the edges of a simple connected graph of order at least 3 in such a way that the graph becomes irregular, i.e., the weight sums at each vertex are distinct. What is the minimum of the largest weight over all such irregular assignments?

The basic problem on irregular assignments is formulated here as a general extremal problem in terms of hypergraphs, i.e., set systems. Standard graph and hypergraph terminology will be assumed or may be found in [B].

A hypergraph H is a pair (V(H),E(H)), where V(H) is a finite non–empty set of elements called vertices, and E(H) is a finite collection of subsets of V(H) called (hyper)edges. A hypergraph is *vertex–distinguishable* if for each pair of vertices there is an edge of H which contains precisely one of them. We consider here simple hypergraphs which contain no multiple edges (that is the collection of E(H) is a set).

1 The work was done in part at the University of Louisville, Louisville, Kentucky, and was partially supported by the Hungarian Academy of Sciences under AKA grant 1–3–86–264.

Assign positive integer weights to the edges of a hypergraph H. This assignment is *irregular* if for all $x \in V(G)$ the weighted degree of x (the sum of the weights of the edges containing x) are distinct. The minimum of the largest weights assigned over all such irregular assignments of H is called the *irregularity strength* of H and is denoted by s(H). We say that s(H) is undefined (or infinite) if H is not vertex–distinguishable.

The basic problem of irregular assignments consists in determining or bounding the irregularity strength of finite set systems.

The material is arranged into sections according to the particular structure investigated or to the methods used as follows:

1. Irregularity Strength of Graphs
1.1 General bounds
1.2 Trees, cycles
1.3 Complete graphs

2. Regular and Dense Graphs
2.1 Regular graphs
2.2 Dense graphs
2.3 Graphs with s(G) = 2

3. Hypergraphs
3.1 Irregular hypergraphs
3.2 Irregularity strength
3.3 Complete hypergraphs

4. Irregular Embeddings
4.1 Consecutive degree sequences
4.2 F–degrees
4.3 Embeddings

5. Dual Strength
5.1 Graphs
5.2 Sum–distinct sequences
5.3 Projective planes

The text presents 22 propositions and 27 theorems (assembled from 15 manuscripts), furthermore, there are 27 problems which formulate the related conjectures and quests.

I am grateful to the many people for their explicit or indirect contribution to this paper. In particular I would like to thank Gary Chartrand, Paul Erdös, Ralph Faudree, András Gyárfás, Mike Jacobson, Lael Kinch, Dick Schelp and an anonymous referee.

Special thanks go to Dick, Mike, Ralph and Buck McMorris for their kind support making possible our collaboration at the University of Louisville and at Memphis State University in 1986 and 1987. Many problems presented here were proposed with Mike and Lael of Louisville while I spent three semesters at U of L.

1. Irregularity Strength of Graphs

The problem of studying $s(G)$, the irregularity strength of a graph G was proposed in [CJLORS]. Determining the strength proved to be rather hard, even for very simple graphs.

A graph in which each edge is assigned a positive integer weight will be called a network, and the largest weight s is the strength of the network (c.f. [CJLORS], [JL]). The same structure is Berge's s–graph which is a multigraph with edge multiplicity at most s (c.f. [B]).

An irregular network is one with distinct (weighted) degrees, and the irregularity strength $s(G)$ is the minimum strength of an irregular network with underlying graph G.

1.1. General Bounds Note that the graphs of order 2 (either a single edge or two isolated vertices) are the only minimal graphs which are not vertex–distinguishable. Obviously, $s(G)$ is finite if and only if G contains at most one isolated vertex and no connected component of order 2.

Any known lower bound on $s(G)$ of general use is based on the following observation.

Proposition 1 *([CJLORS]) If G contains n_i vertices of degree i, for some positive integer i, then $s(G) \geq (n_i + i - 1)/i$.*

A straightforward procedure which assigns non–unit weights to the edges of a spanning forest of the graph results in the following general upper bound.

Theorem 2 *(c.f. [CJLORS]) If G is a graph of order $n \geq 3$ with finite irregularity strength, then $s(G) \leq 2n - 3$.*

For connected graphs this bound can be lowered.

Theorem 3 *([JL]) If G is a connected graph of order $n \geq 4$ with finite irregularity strength then $s(G) \leq n-1$.*

Theorem 2 is sharp only for a 3–clique, while there is equality in Theorem 3 for every star.

Problem 4. Does there exist a real α, $0 < \alpha < 1$, such that $s(G) \leq \alpha n$ holds, if n is sufficiently large, for every (connected) graph G of order n containing no vertices of degree one?

Problem 5. Find the smallest real $\alpha = \alpha(d)$ such that $s(G) \leq \alpha n$ holds, if n is sufficiently large, for every (connected) graph G of order n containing no vertices of degree less than d.

One should note that the impact of connectedness on the irregularity strength of the graph is not at all clear.

1.2 Trees, cycles

Proposition 6 *(c.f. [CJLORS]) If T is a tree of order $n \geq 3$ then $s(T) \geq \lceil (n+2)/3 \rceil$, and the bound is sharp.*

The construction and the proof of Proposition 6 in [CJLORS] shows that the trees with $n' = \lceil (n+2)/3 \rceil$ vertices of degree 1 and n' vertices of degree 2 are expected to approach the lower bound. As far as we know no attempt of characterizing extremal trees (i.e. trees of order n with irregularity strength n') has been made.

Proposition 7 *([CJLORS]) If G is a unicyclic graph of order n (that is connected and has exactly one cycle) then $s(G) \geq n/3$, and the bound is tight.*

Proposition 8 *(c.f. [CJLORS]) For the path P_n with n vertices*

$$s(P_n) = \begin{cases} n/2 & \text{for } n \equiv 0 \pmod 4 \\ \lfloor n/2 \rfloor + 1 & \text{otherwise} \end{cases}.$$

One can verify that the irregularity strength of the complete binary tree is equal to (n + 1)/2, thus its strength is determined by the number of pendant vertices. This observation generates the following problems, the first is due to Chartrand (personal communication), the second one to Schelp ([CSS]).

Problem 9. Find the largest integer n(p) such that $s(T) = p$ holds for every tree T of order $n \leq n(p)$ with p vertices of degree 1 (e.g. $n(2) = 4$).

Problem 10. If T is a tree with no vertices of degree 2 then s(T) equals the number of vertices of degree 1.

Note that in the special case of a full binary tree (when it has no vertices of degree 2) Problem 10 is settled in [CSS].

Problem 11. Find the smallest positive real $\alpha = \alpha(D)$ such that $s(T) \leq \alpha$ holds, if n is sufficiently large, for every tree of order n with maximum degree $D \geq 2$.

Proposition 12 *([FJLS]) For the cycle C_n with n vertices*

$$s(C_n) = \begin{cases} \lceil n/2 \rceil & \text{for } n \equiv 1 \pmod 4 \\ \lceil n/2 \rceil + 1 & \text{otherwise} \end{cases}$$

Proposition 13 *([KLE]) If G is the disjoint union of cycles and paths of order at least 4 then $s(G) \leq n/2 + 10$, where n is the order of G.*

The exclusion of the P_3 components in the previous result is justified with the following.

Theorem 14 *([KLE]) Let G be the disjoint union of t paths of order 3. If G has n vertices, then*

$$\lceil (5n-1)/7 \rceil \leq s(G) \leq \lceil (5n-1)/7 \rceil + 1.$$

Kinch also conjectures that $s(G) = \lceil (5n-1)/7 \rceil$ holds for every n. If $G = tP_3$ as in Theorem 14 then $n = 3t$, $n_i = 2t$ and with $i = 1$ Proposition 1 gives the weaker lower bound $s(G) \geq 2t = 2n/3$. It is worth noting that this is the only example we know where Proposition 1 (and also its slight extension Proposition 44) gives no tight bound on the irregularity strength.

1.3. Complete Graphs Let K_n denote the complete graph (clique) of order n. In ([CJLORS]), $s(K_n) = s(K_{2k,2k}) = 3$ was proved for every n and for $k \geq 1$. These results are contained as special cases in the results mentioned in this section. Let $K_{p,q}$ denote the complete p x q bipartite graph. A complete multipartite graph is defined as the complement of some pairwise vertex disjoint cliques.

Theorem 15 *([FJLS]) If G is a complete k-partite graph with $k \geq 3$ having the same number of vertices in each vertex class, then $s(G) = 3$.*

In the case when G is a clique Theorem 15 yields the result in [CJLORS] that $s(K_n) = 3$ for every $n \geq 3$.

Theorem 16 *([FJLS])*

$$s(K_{p,q}) = \begin{cases} \lceil (q+p-1)/p \rceil & \text{for } q \geq 2p \\ 3 \text{ if either } p = q \text{ is even or } 1 < q/2 \leq p < q \end{cases}$$

The missing case when $p = q$ is odd was proposed by Chartrand et al. and settled by Gyárfás recently:

Theorem 17 *([GY])*

$$s(K_{2k+1,2k+1}) = 4.$$

2. Regular and Dense Graphs

The main conjectures are that every r–regular graph of order n has irregularity strength about n/r; and the irregularity strength of graphs of order n with minimum degree proportional to n does not depend on n. These questions are formulated in this section together with some related results.

2.1. Regular graphs Every r–regular graph G of order n satisfies $s(G) \geq (n + r - 1)/r$ by Proposition 1.

Proposition 18 *If G is regular, then $s(G) \geq 3$.*

The main problem in the background of all results pertaining to the irregularity strength of regular graphs is the following problem due to Jacobson.

Problem 19. Does there exist an absolute constant c such that $s(G) \leq n/r + c$ holds for every r–regular graph G of order n ?

One can find infinitely many regular graphs with irregularity strength indicated in the lower bound. The construction presented in [FJLS] is based on the vertex multiplication of a graph. Let $G^{(k)}$ denote the graph obtained from G by replacing each vertex of G by an independent set of k new vertices.

Theorem 20 *([FJLS]) Let G be an r–regular graph with n vertices obtained from the cycle C_t by vertex multiplication (that is $G = C_t^{(k)}$, $t \geq 3$, $n = tk$, $r = 2k$ and $k \geq 2$). Then $s(G) = \lceil n/r \rceil + 1$.*

Proposition 21 *([FJLS])* *Let G be a 2–regular graph of order n. Then* $\lfloor n/2 \rfloor \le s(G) \le \lfloor n/2 \rfloor + 2$. *Moreover, if G is the disjoint union of triangles, then*

$$s(G) = \begin{cases} \lceil n/2 \rceil + 1 & \text{for } n = (4k+1)3 \\ \lceil n/2 \rceil + 2 & \text{otherwise} \end{cases};$$

if G has no triangle components, then

$$s(G) = \begin{cases} \lceil n/2 \rceil & \text{for } n \equiv 1 \pmod 4 \\ \lceil n/2 \rceil + 1 & \text{otherwise} \end{cases}.$$

It has been observed by Jacobson that s(G) is about n/2 or less if G has a Hamiltonian cycle. This leads to the following.

Proposition 22 *([FJLS])* *If G is an r–regular graph of order n, with* $r \ge n/2$ *and* $n \ge 3$, *then* $s(G) \le \lceil n/2 \rceil + 1$.

A result in [J] yields a similar theorem: *If G is a 2–connected r–regular graph of order n, with* $r \ge n/3$, *then* $s(G) \le \lceil n/2 \rceil + 1$.

The disjoint union of t cliques of order p is denoted by tK_p. The irregularity strength of these structures was determined (up to an additive constant) with Kinch.

Theorem 23 *([FJKL])* *If* $p \ge 3$, $r = p - 1$ *and* $n = tp$, *then* $G = tK_p$ *is an r–regular graph of order n having irregularity strength* $s(G) \le \lceil (n+1)/r \rceil + 2$ *for every t.*

Theorem 24 *([FL])* *If G is a regular graph of order n having finite irregularity strength, then* $s(G) < n/2 + 9$.

Problem 25. Does there exist a constant c such that $s(G) \le n/3 + c$ holds for every 3–regular graph of order n ?

2.2. Dense graphs A graph is considered here dense if the degree of each vertex is proportional to the order of the graph. The main problem concerning dense graphs is the following:

Problem 26. Let α be a real with $0 < \alpha < 1$. Does there exist a constant $c = c(\alpha)$ such that $s(G) \le c$ for every graph G with a minimum degree $\delta(G) \ge (1 - \alpha)n$?

The upper bound in the next theorem improves on the general bound $s(G) \le 2n - 3$ of Theorem 2, however, it still depends on n.

Theorem 27 *([FJKL]) Let $0 < \alpha < 1$ and let G be a $(1-\alpha)n$–regular graph of order n. Then $s(G) \leq (1/(\lfloor 1/\alpha \rfloor - 1))\, n + 5$ holds for sufficiently large n.*

More is known about the irregularity strength of graphs of order n having minimum degree $n - t$, with t constant.

Theorem 28 *([FJKL]) If t is a fixed positive integer and n is sufficiently large, then $s(G) \leq 3$ holds for each graph of order n with minimum degree $n - t$.*

In some sense this theorem gives a sharp result on the minimum degree of a graph of irregularity strength 3 or less, since for fixed $t \geq 0$ the strength of infinitely many graphs of minimum degree $n - t$ is equal to 3 (e.g. regular graphs according to Proposition 18).

Problem 29. Find the largest integer $t = t(n)$ such that $s(G) \leq 3$ holds, if n is sufficiently large, for every graph G of order n with minimum degree $n - t$.

Theorem 30 *([FJKL]) Let G be a graph of order n and minimum degree $\delta(G) = n - t$. If $1 \leq t \leq \sqrt{n/18}$, then $s(G) \leq 3$.*

If $t = t(n) > 2n/3$, it follows that $s(G) > 4$ for every $(n-t)$–regular graph G of order n. By lowering $2n/3$ (or by pushing up the bound $\sqrt{n/18}$ in Theorem 30) one could try to find the threshold between the minimum degree of graphs of irregularity strength 3 and 4.

Problem 31. Find the smallest integer $t = t(n)$ such that there exist infinitely many graphs of order n with minimum degree $n - t$ and having irregularity strength at least 4.

Problem 32. Find the smallest constant c such that $s(G) \leq c$ holds for every graph G of order n having minimum degree $n-3$.

2.3. Graphs with $s(G) = 2$ Some extremal problems pertaining to the number of edges in a graph of irregularity strength 2 were thoroughly investigated in [FGYS]. Note that in most of the cases of sharp results the non–constructive part of the proof is based on a supply–demand type counting argument developed by Gyárfás (c.f. [GY],[GYA]). Let n and m denote, respectively, the number of vertices and edges of a graph G.

Theorem 33 *([FGYS]) If G has irregularity strength 2, then* $m \geq \lceil (n^2 - 1)/8 \rceil$, *moreover,* $m \geq (n^2 - 1)/8 + 1$ *of* $n \equiv 3 \pmod 4$*. Furthermore, there exist graphs for which equality holds.*

Theorem 34 *([FGYS]) If G has irregularity strength 2, then*

$$m \leq \binom{n}{2} - (n-1)/4.$$

A related result is the following:

Theorem 35 *([FGYS]) Let* $N \subset K_n$ *be subgraph of at most* $n/2$ *vertices and at least* $(n-1)/4$ *edges. Then* $s(K_n - N) = 2$.

The results above have the following corollary: $\lceil (n-1)/4 \rceil$ *is the minimum number of edges which need to be removed from* K_n *in order to decrease its irregularity strength from 3 to 2.*

The question of determining the irregularity strength of $K_n - F_k$, where F_k is a matching of k edges, was proposed in [FJKL].

Proposition 36 *([FJKL]) For every* $k \geq 0$ *and* $n \geq 3$, $s(K_n - F_k) \leq 3$.

Theorem 37 *([GYA])* $s(K_n - F_k) = 2$ *if and only if* $n = 4t \pm 1$ *or* $4t$, *and* $k = t$.

A graph whose vertex set is the union of a clique and an independent (or stable) set is called a split graph. Then $G = K_n - K_p$ is a complete split graph $(n \geq p)$. It is not hard to verify that $s(K_n - K_p) = 3$ for $n \geq 2p^2 - 2p + 2$.

Proposition 38 *(c.f. [FGYS])* $s(K_n - K_p) = 2$ *if and only if* $2p - 1 \leq n \leq 2p^2 - 2p + 1$.

3. Hypergraphs

The starting point is that there do exist non–trivial irregular hypergraphs (H is irregular if the degrees of each vertex are distinct or equivalently, if $s(H) = 1$). The rank of a hypergraph is the maximum cardinality of its hyperedges. A hypergraph is called r–uniform if each edge contains r vertices.

3.1. Irregular hypergraphs

Proposition 39 *([GYJKLS]) If* $r \geq 3$ *and* $n \geq r + 3$, *then there exist irregular r–uniform hypergraphs of order n.*

Faudree, Jacobson and Schelp proposed the question of the irregularity behavior of a typical (random) hypergraph.

Theorem 40 *([GYJKLS]) Almost all r–uniform hypergraphs are irregular for $r \geq 6$.*

For the remaining cases Erdös and Gyárfás ask (personal communication):

Problem 41. Is it true that almost all 3– and 4–uniform hypergraphs have 2 vertices with the same degree?

Problem 42. Does it have positive probability that a 5–uniform hypergraph is irregular?

An $\langle r \times m \rangle$–hypergraph is an r–partite, r–uniform hypergraph with m vertices in each vertex class.

Theorem 43 *([GYJKLS]) For either fixed $m \geq 2$ or fixed $r \geq 6$ almost all $\langle r \times m \rangle$–hypergraphs are irregular.*

3.2. Irregularity strength The lower bound given in [JL] (and essentially in Proposition 1) generalizes to hypergraphs.

Proposition 44 *([GYJKLS]) If H has n_i vertices of degree i, then*

$$s(G) \geq \max \left\{ \left(\left(\sum_{k=i}^{j} n_k \right) + i - 1 \right) / j \ : \ i \leq j \right\} .$$

Theorem 45 *([GYJKLS]) If H is a vertex distinguishable hypergraph of rank r, then $s(H) \leq r|V(H)| - r + 1$.*

Proposition 46 *([GYJKLS]) The dual of a complete graph having n edges is a 2–regular hypergraph of order n with irregularity strength $2n - o(n)$.*

Most likely, there is no absolute constant c with $s(H) \leq c|V(H)|$. However, examples with high irregularity strength are unknown.

Problem 47. Find hypergraphs of order n with irregularity strength at least 2n.

3.3 Complete Hypergraphs Denote by $K_n^{(r)}$ the complete r–uniform hypergraph of order n and by $K\langle r \times m \rangle$ the complete r–partite, r–uniform hypergraph with m vertices in each vertex class.

For the most part the hypergraphs in each of the two mentioned classes have irregularity strength 2. Note that this is equivalent to showing the existence of an irregular hypergraph of the given type.

Proposition 48 *([GYJKLS])* *For* $r \geq 3$,

$$s(K_n^{(r)}) = \begin{cases} n & \text{if } n = r+1 \\ 3 & \text{if } n = r+2 \\ 2 & \text{if } n \geq r+3 \end{cases} .$$

Theorem 49 *([GYJKLS])* *For* $r \geq 3$ *and* $m \geq 2$,

$$s(K\langle r \times m\rangle) = \begin{cases} 3 & \text{if } r+m \leq 6 \\ 2 & \text{otherwise} \end{cases} .$$

4. Irregular Embeddings

This section contains problems and results pertaining to the existence of irregular networks under various restrictions or with additional constraints.

4.1. Consecutive degree sequences In [FJKL], irregular complete networks were studied with degree sequence consisting of consecutive or almost consecutive integers. (A network on a complete graph is called a complete network.)

Theorem 50 *([FJKL])* *If* $D = (d, d-1, \ldots, d-n+1)$ *is a sequence of* n *consecutive integers with even sum, then there exists a complete network of strength* $\lceil d/(n-1) \rceil$ *with degree sequence* D *for each* $n \geq 5$ *and* $d > 2(n-1)$.

Note that $\lceil d/(n-1) \rceil$ is obviously the smallest possible strength a network can have with the degree sequence D.

Let $a \geq 0$, $a+1 \leq b \leq a+2$ and $b+n-2 \leq c \leq b+n-1$. Then $D = (a, b, b+1, \ldots, b+n-3, c)$ with even sum is called a quasi–consecutive sequence. If D is the degree sequence of some irregular network G of order n, then obviously $s(G) \geq \lceil c/(n-1) \rceil$

Theorem 51 *([FJKL])* *If* D *is a quasi–consecutive sequence of at least 5 integers, then there exists a network with degree sequence* D *and having irregularity strength*

$$s = \begin{cases} \lceil c/(n-1) \rceil & \text{if } c \neq n-1 \\ 2 & \text{if } c = n-1 \end{cases} .$$

4.2. F–degrees Chartrand, Holbert, Oellermann and Swart in [CHOS] introduced the generalized degree of a graph as follows. Let F and G be graphs; the F-degree of a vertex x in G is the number of subgraphs of G isomorphic to F that contain x. A graph is called *F–irregular* (F–regular) if the F–degrees of the vertices of G are all distinct (equal).

Theorem 52 *([CHOS]) There exist $K_{1,n}$–irregular and K_{n+1}–irregular graphs for every $n \geq 2$.*

Erdös, Székely and Trotter (c.f. [CHOS]) have the remark that there are infinitely many K_3–irregular graphs.

Problem 53. Do there exist infinitely many $K_{1,n}$–irregular and K_n–irregular graphs?

Chartrand et al. in [CHOS] have the following:

Problem 54. Does there exist an F–irregular graph for every connected graph F of order at least 3?

4.3. Embeddings On the analogy of the embedding of a simple graph into a regular one, investigated by König ([K]), Chartrand has proposed the problem of the irregular embedding of networks (personal communication, c.f. [CJLORS]).

Proposition 55 *([CJLORS]) If G is a network of strength at least 2, then there exists an irregular network H containing G as an induced subnetwork and such that s(H) = s(G).*

It was also mentioned in [CJLORS] that if G is bipartite, then H might be bipartite as well. This result is an instance of the following general problem:

Problem 56 [CJLORS]. Given a network G of strength at least 2, and possessing a specified graphical property P, does there exist an irregular network H having property P such that H contains G as an induced subnetwork and s(H) = s(G) ?

This question has an affirmative answer in the case when P is the property of being k–chromatic.

Theorem 57 *([JLE]) If G is a k–chromatic network of strength at least 2, then there exists an irregular k–chromatic network H with s(H) = s(G) such that H contains G as an induced subnetwork.*

The following general problem on the F–irregular embedding of graphs contains the question on the existence of infinitely many F–irregular graphs (Problems 53 and 54).

Problem 58. Given the graphs F and G (F is connected and has at least 3 vertices). Does there exist an F–irregular graph H which contains G as an induced subgraph?

Note that Problem 58 has a positive answer when $F = P_3$, a path on 3 vertices (Jacobson, personal communication).

If $F = K_r$ then the F–irregular embedding of a graph with maximum clique r can be related in a natural way to the irregular embedding of a conform r–uniform hypergraph. Here we formulate a slightly weaker problem (for hypergraph terminology, see [B]).

Problem 59. Let H be an r–uniform hypergraph. Does there exist an irregular r–uniform hypergraph containing H as a subhypergraph?

5. Dual Strength

If H* stands for the dual of the hypergraph H, and H is simple then obviously H* is vertex–distinguishable. So it is quite natural to introduce s*(H), the dual notion of the irregularity strength; it is the minimum of the largest label used in any *irregular labelling of the vertices* with (not necessarily distinct) positive integers such that for each $e \in E(H)$ the sum of the labels of the vertices of e are distinct. Then $s^*(H) = s(H^*)$, and it is called the dual (irregularity) strength of H.

5.1. Graphs The study of the dual strength has been initiated in [GYJKLS]. The basic problem pertaining to the dual strength of graphs is: how large s*(G) can be compared to the number of edges. The following conjecture is due to Gyárfás (personal communication).

Problem 60. Does there exist a constant c such that $s^*(G) \leq cm$ for every graph G with m edges?

Note that the question above is open even for bipartite graphs. The case of trees, however, is an easy corollary of the following observation due to Gyárfás (personal communication).

Proposition 61 *If G is a graph of order n such that every induced subgraph contains a vertex of degree at most k, then* $s^*(G) \leq kn + O(1)$.

For a complete graph the following holds true.

Proposition 62 *([GYJKLS]) If G is an n–clique, then*

$$s^*(G)/\binom{n}{2} \to 2 \quad \text{as } m \text{ approaches infinity.}$$

For the dual strength of the complete bipartite graph Gyárfás and Schelp [GYS] have the following conjecture settled in the case of $p = 3$.

Problem 63. Let $p \leq q$ and $G = K_{p,q}$. Is it true that

$$s^*(G) = \begin{cases} p(q+1)/2 & \text{for } p \text{ even} \\ q(p+1)/2 & \text{for } q \text{ even, } p \text{ odd ?} \\ q(p+1)/2 + (p-1)/2 & \text{for } p \text{ and } q \text{ odd} \end{cases}$$

5.2. Sum–distinct sequences The notation of a B_2–sequence was introduced in additive number theory by Erdös and Turán ([ET]) to denote a sequence $a_1 < a_2 < \ldots < a_m$ of positive integers with distinct pairwise sums. They ask the following question.

Problem 64 ([ET]). Is there a constant c such that $a_m \geq m^2 + cm$?

Partial results to this well studied extremal problem on B_2–sequences have some interesting corollaries for the dual strength of complete graphs (see Propositions 46 and 62). Note that since cliques contain no loops, the notion of a B_2–sequence is to be modified by considering only sums of distinct elements.

Positive integers $a_1 < \ldots < a_m$ form a sum–distinct sequence if all the 2^m subsums are distinct. An example is the sequence $a_i = 2^{i-1}$, $i = 1, \ldots, m$. A problem due to Erdös and Moser, still open since 1955, is the following. (For answering any of these problems Erdös offers a good price.)

Problem 65 ([E]). Prove or disprove the existence of an absolute constant c such that $a_m > c2^m$ holds for every sum–distinct sequence.

This question is clearly related to the problem of estimating the dual strength of a totally complete hypergraph of order m.

The concept of a sum–distinct sequence is extended to more than one sequence and their generation by 'greedy' procedures in [KL].

Two sequences $A = \{a_1, a_2, ...\}$ and $B = \{b_1, b_2, ...\}$ are said to be compatible sum–distinct sequences if both A and B are sum–distinct, and the subsums of A differ from that of B.

Denote F(n) the nth Fibonacci number, that is $F(0) = F(1) = 1$ and if $n \geq 2$, then $F(n) = F(n-1) + F(n-2)$.

Proposition 66 *([KL]) Create two compatible sum–distinct sequences by the first–fit procedure as follows. Start with $a_1 = 1$ and $b_1 = 2$; if elements a_i and b_i are defined for $i = 1, 2, ..., k$, then choose the smallest positive integer x as a_{k+1}, then the smallest positive integer y as b_{k+1} such that $\{a_i : 1 \leq i \leq k\} \cup \{x\}$ remains sum–distinct and compatible with $\{b_i : 1 \leq i \leq k\}$, and $\{b_i : 1 \leq i \leq k\} \cup \{y\}$ remains sum–distinct and compatible with $\{a_i : 1 \leq i \leq k+1\}$. Then $a_i = F(2i-1)$ and $b_i = F(2i)$ for $i = 1, 2, ...$.*

We say that t sum–distinct sequences are compatible if they are pairwise compatible ($t > 1$). No formula is known so far for the compatible sequences created by the first–fit procedure for $t \geq 3$. Note that in [KL] it is shown that by introducing a minor change in the procedure the sequence of generalized Fibonacci numbers are obtained.

For $n = 2$ one can verify easily the following:

Proposition 67 *The first–fit procedure creates t compatible sum–distinct pairs with maximum element $\lfloor 5t/2 \rfloor$.*

Let s(t,n) denote the minimum of the largest element of t compatible sum–distinct sequences.

Theorem 68 *([KLE]) $s(t,n) \geq (15t-1)/7$ and the bound is sharp (up to an additive constant) for every t.*

As in the case of $t = 1$, one can propose the following general extremal problem.

Problem 69. Determine the order of magnitude of s(t,n).

Problem 70. Is there a constant $c = c(n)$ such that $s(t,n) \leq ct$ for every t?

5.3. Projective planes A projective plane of order q is considered as a $(q + 1)$–uniform, regular hypergraph defined by the $q2 + q + 1$ lines as edges. For the terminology pertaining to infinite geometries see [H]. Jacobson asked (personal communication) whether the finite projective plane has large irregularity strength compared to the number of points. It is more convenient to formulate the results and the problems in terms of the dual strength.

Theorem 71 *([LSZ]) If P is a projective plane of order q and $q \geq 3$, then $s^*(P) \leq q^2$.*

Note that the lower bound from Proposition 44 is linear in q. For special q there are sharper results.

Proposition 72 *([LSZ]) If P is the Fano plane $(q = 2)$, then $s^*(P) = 5$. If P is the projective plane of order 3, then $6 \leq s^*(P) \leq 8$.*

Proposition 73 *([LSZ]) If P is a Desarguesian projective plane of square order q, then $s^*(P) \leq q\sqrt{q} + 0(q)$.*

The following result improves on the lower bound $s^*(P) \geq q + 2$ resulting by proposition 44, if q is sufficiently large.

Theorem 74 *([LSZ]) If P is a projective plane of order q, then $s^*(P) \geq 7q/4$ for sufficiently large q.*

Problem 75. Disprove that $s^*(P) \leq 0(q)$ for every projective plane P of order q.

There are further structures of geometric nature for which the problem of determining the dual strength seems to have some interest, for instance:

Problem 76. Determine the dual strength of uniform interval hypergraphs.

REFERENCES

[B] C. Berge, Graphs and hypergraphs (North Holland, Amsterdam) (1973).

[CEO] G. Chartrand, P. Erdös, O. R. Oellermann, How to define an irregular graph, *College Math. J.* **19** (1988) 36-42.

[CHOS] G. Chartrand, K. S. Holbert, O. R. Oellermann, H. C. Swart, F–degrees in graphs, *Ars Combinatoria* **24** (1978) 133-148.

[CJLORS] G. Chartrand, M. S. Jacobson, J. Lehel, O. R. Oellermann, S. Ruiz, F. Saba, Irregular networks, to appear in the Proceedings of the 250th Anniversary Conference on Graph Theory, Fort Wayne, IN, 1986.

[CL] G. Chartrand, L. Lesniak, Graphs and Digraphs, 2nd Edition, Wadsworth and Brooks/Cole, (1986).

[CSS] L. A. Cammack, R. H. Schelp, G. C. Schray, Irregularity strength of full d–ary trees, In preparation.

[E] P. Erdös, Problems and results in additive number theory, in "Colloque sur la Théorie des Nombres, Bruxelles 1955", Liége et Paris, (1956), 127-137.

[ET] P. Erdös, P. Turán, On a problem of Sidon in additive number theory and some related problems, *J. London Math Soc.* **16** (1941) 212-215.

[FJLS] R. J. Faudree, M. S. Jacobson, J. Lehel, R. H. Schelp, Irregular networks, regular graphs and integer matrices with distinct row and column sums, to appear in *Discrete Math.*

[FGYS] R. J. Faudree, A. Gyárfás, R. H. Schelp, On graphs of irregularity strength 2, to appear in the Proceedings of the Hungarian Conference in Eger 1987.

[FL] R. J. Faudree, J. Lehel, Bound on the irregularity strength of regular graphs. to appear in the Proceedings of the Hungarian Conference in Eger 1987.

[FJKL] R. J. Faudree, M. S. Jacobson, L. Kinch, J. Lehel, Irregularity of dense graphs. Submitted.

[GY] A. Gyárfás, The irregularity strength of $K_{m,m}$ is 4 for odd m, *Discrete Math.* **71** (1988) 273-274.

[GYA] A. Gyárfás, The irregularity strength of $K_n - mK_2$, to appear in *Utilitas Mathematica.*

[GYJKLS] A. Gyárfás, M. S. Jacobson, L. Kinch, J. Lehel, R. H. Schelp, Irregularity strength of uniform hypergraphs. Submitted.

[GYS] A. Gyárfás, R. H. Schelp, A matrix labelling problem. Submitted.

[H] J. W. P. Hirschfeld, Projective Geometries over Finite Fields, Clarendon Press, Oxford, (1979).

[J] B. Jackson, Hamiltonian cycles in regular 2–connected graphs, *Journal of Combinatorial Theory,* **B 29** (1980) 27-46.

[JL] M. S. Jacobson, J. Lehel, Upper bound on the irregularity strength of a simple graph. Submitted.

[JLE] M. S. Jacobson, J. Lehel, Irregular embeddings. In preparation.

[KL] L. Kinch, J. Lehel, Sum–distinct sequences and Fibonacci numbers. Submitted.

[KLE] L. Kinch, J. Lehel, The irregularity strength of tP_3. Submitted.

[K] D. König, Theorie der endlichen und unendlichen Graphen, Leipzig, (1936) (Reprinted Chelsea, New York, 1950)

[LSZ] J. Lehel, T. Szônyi, Irregular labeling of the finite projective planes. Unpublished.

Neighborhood Unions and Graphical Properties

Linda M. Lesniak
Drew University

ABSTRACT

If S is a set of vertices of a graph G, then the neighborhood $N(S)$ of S is defined to be the set of all vertices of G which are adjacent to at least one vertex of S. A survey is presented of recent results of the following form:

Given a graph G, if $|N(S)|$ is at least k for every subset S of $V(G)$ of a specified type, then G has property P.

We look, for example, at a variety of hamiltonian properties P, where the sets S are all pairs of nonadjacent vertices and also where the sets S are ***all*** *pairs of vertices.*

1. Introduction

Graph theory literature is rich with sufficient conditions of the following type:

If $\deg v \geq k$ for every vertex v of a graph G, then G has property P.

Perhaps one of the best-known is a result of Dirac [8] giving a sufficient condition for a graph to be hamiltonian.

Theorem 1 *If G is a graph of order $p \geq 3$ such that $\deg v \geq p/2$ for every vertex v, then G is hamiltonian.*

In many instances, minimum degree conditions have been improved by results involving degree sums of nonadjacent vertices. (See [6] for an excellent survey.) Ore [25], for example, extended Dirac's result in this fashion.

Theorem 2 *If G is a graph of order $p \geq 3$ such that $\deg u + \deg v \geq p$ for each pair u, v of nonadjacent vertices, then G is hamiltonian.*

Theorem 1 was yet generalized further by Bondy [4].

Theorem 3 *If G is a 2–connected graph of order p such that $\deg u + \deg v + \deg w \geq 3p/2$ for each triple u, v, w of mutually nonadjacent vertices, then G is hamiltonian.*

A condition of the form " $\deg v \geq k$ for every vertex v of G" could alternatively be described as "$|N(S)| \geq k$ for every subset S of $V(G)$ with $|S| = 1$", where $N(S)$ denotes the *neighborhood of the set* S and is defined to be the set of all vertices of G which are adjacent to at least one vertex of S. This rather more complicated way of describing a minimum degree condition suggests obvious variations obtained by changing the sets S under consideration. Such results, unlike Theorems 2 and 3, would reflect overlapping neighborhoods of the vertices in S.

Although this idea of a neighborhood condition is not new, only a few such results appeared in the literature until recently. Hall [21] showed that if G is a bipartite graph with partite sets V_1 and V_2, then V_1 can be matched to a subset of V_2 if and only if $|N(S)| \geq |S|$ for every nonempty subset S of V_1. Anderson [2] showed that if G is a graph of even order p satisfying

$$|N(S)| \geq (p + 2|S| + 1)/4$$

for every nonempty subset S of $V(G)$, then G has a 1–factor. Woodall [28] established that if G is a 2–connected graph of order p satisfying

$$|N(S)| \geq (p + |S| - 1)/3$$

for every nonempty subset S of $V(G)$ and

$$\deg v \geq (p + 2)/3$$

for every vertex v, then G is hamiltonian. Posa [26] showed that for a fixed integer k, if G is a graph such that $|N(S) - S| \geq 2|S| - 1$ for each set S of at most k vertices, then G contains a path of length $3k - 2$.

In this paper we present a survey of the many recent neighborhood union results. In all cases, these results are distinct from known results based on degree conditions. For hamiltonian properties we will compare neighborhood union and degree sum results. For the sake of brevity, we leave other such comparisons as an exercise for the reader.

The following notations will be useful in presenting examples. If G and H are two graphs with disjoint vertex sets, then $G \cup H$ denotes the *union* of G and H and is the graph with vertex set $V(G) \cup V(H)$ and edge set $E(G) \cup E(H)$. The union of n copies of G is denoted by nG. The *join* of G and H, denoted $G + H$, is the graph with vertex set $V(G) \cup V(H)$ and edge set

$$E(G) \cup E(H) \cup \{ uv \mid u \in V(G) \text{ and } v \in V(H)\}.$$

For all other terms and notation not specifically defined here, see [7].

2. All Pairs of Nonadjacent Vertices

In this section we examine the relationship between the cardinality of the union of the neighborhoods of an arbitrary pair of nonadjacent vertices of a graph and various hamiltonian properties.

A graph G is *hamiltonian* (*traceable*) if it has a cycle (path) containing all of the vertices of G. Such a cycle (path) is called a *hamiltonian cycle* (*hamiltonian path*). A graph G is *hamiltonian–connected* if each pair of vertices of G are the endvertices of a hamiltonian path in G.

Ore's original paper on a degree sum condition for a graph to be hamiltonian contained a similar condition (as a corollary) for a graph to be traceable.

Theorem 4 *If G is a graph of order p such that* $\deg u + \deg v \geq p - 1$ *for each pair u, v of nonadjacent vertices, then G is traceable.*

In a later paper [24] he established a similar result for hamiltonian–connected graphs.

Theorem 5 *If G is a graph of order p such that* $\deg u + \deg v \geq p + 1$ *for each pair u, v of nonadjacent vertices, then G is hamiltonian–connected.*

If G is a (noncomplete) graph with $p \geq 2$ vertices such that $|N(u) \cup N(v)| \geq k$ for each pair u, v of nonadjacent vertices, then necessarily $0 \leq k \leq p - 2$. Since the graph $K_1 \cup K_{p-1}$ is disconnected with $k = p - 2$, it is clear that such a neighborhood condition will not force any connectivity conditions. However, if G is traceable, hamiltonian, or hamiltonian–connected, then G is necessarily connected, 2–connected, or 3–connected, respectively. Thus, in the case of hamiltonian properties, a minimum connectivity condition as well as a neighborhood condition must be assumed.

In [14], the following "hamiltonian type" neighborhood conditions were established.

Theorem 6 *If G is a 2–connected graph of order p such that* $|N(u) \cup N(v)| \geq (2p - 1)/3$ *for each pair u, v of nonadjacent vertices, then G is hamiltonian.*

Corollary *If G is a connected graph of order p such that* $|N(u) \cup N(v)| \geq (2p - 2)/3$ *for each pair u, v of nonadjacent vertices, then G is traceable.*

Theorem 7 *If G is a 3–connected graph of order p such that* $|N(u) \cup N(v)| > 2p/3$ *for each pair u, v of nonadjacent vertices, then G is hamiltonian–connected.*

If G is the graph $3K_n + K_2$, then G is 2–connected of order $p = 3n + 2$. Furthermore, $|N(u) \cup N(v)| = 2n = (2p - 1)/3 - 1$ for each pair u, v of nonadjacent vertices. However, G is not hamiltonian. Thus, for graphs of order $p = 3n + 2$, Theorem 6 is best possible.

If H is the graph obtained by taking three copies of K_n $(n \geq 3)$, and joining corresponding vertices in each copy by an edge, then H is a 2–connected graph of order $p = 3n$ which satisfies $|N(u) \cup N(v)| = 2n = 2p/3$ for each pair u, v of nonadjacent vertices. Thus Theorem 6 states that H is hamiltonian. However, each vertex of H has degree $n + 1 = p/3 + 1$, and so neither of Theorems 2 nor 3 applies.

Similar constructions exist for hamiltonian–connected graphs. In particular, if G is the graph $3K_n + K_3$, then G is 3–connected of order $p = 3n + 3$ and G satisfies the condition that $|N(u) \cup N(v)| = 2n + 1 = 2p/3 - 1$ for each pair u, v of nonadjacent vertices. However, G is not hamiltonian–connected. On the other hand let H be the graph obtained from three disjoint copies of K_n, $n \geq 10$, where the vertices of each copy are ordered 1, 2, ..., n. Add edges between the copies as follows. The first vertex of each copy is adjacent to the n–th, first, and second vertices of the other two copies. The second vertex of each copy is adjacent to the first, second, and third vertices of each copy. In general, vertex i of each copy is adjacent to vertices $i - 1$, i, and $i + 1$ of the other two copies. Then H is a 3–connected graph of order $p = 3n$ in which each vertex has degree $n + 5 = p/3 + 5$. Since $|N(u) \cup N(v)| \geq 2n + 3 = 2p/3 + 3$ for each pair u, v of nonadjacent vertices, Theorem 7 may be applied to conclude that G is hamiltonian–connected. However, Theorem 5 gives no information since $2p/3 + 10 < p + 1$.

The neighborhood condition for a graph to be traceable can be decreased substantially if we assume that G is 2-connected [14].

Theorem 8 *If G is a 2–connected graph of order p such that* $|N(u) \cup N(v)| \geq (p - 1)/2$ *for each pair u, v of nonadjacent vertices, then G is traceable.*

If G is the graph $3K_n + K_1$, $n \geq 2$, then G is connected, has order $p = 3n + 1$ and satisfies $|N(u) \cup N(v)| = 2n - 1 \geq (p - 1)/2$ for each pair u, v of nonadjacent vertices. Since G is not traceable we see that the connectivity condition of Theorem 8 cannot be dropped. Secondly, if we consider the complete bipartite graph $K_{n-2,n}$, $n \geq 4$, we have a 2–connected graph of order $p = 2n - 2$ satisfying $|N(u) \cup N(v)| \geq n - 2 = (p - 2)/2$ for each pair u, v of nonadjacent vertices, and this graph is not traceable. Thus the neighborhood condition of Theorem 8 is sharp. Finally, if H is the graph $3K_n + K_2$, $n \geq 2$, then H is 2–connected of order $p = 3n + 2$ and $|N(u) \cup N(v)| = 2n \geq (p - 1)/2$ for each pair u, v of nonadjacent vertices. Thus Theorem 8 states that H is traceable. Since, however, the degree sum of any pair of nonadjacent vertices is $2n + 2 < p - 1$, Theorem 4 cannot be applied.

Theorem 8 suggests that neighborhood conditions may be relaxed somewhat by imposing other conditions (higher connectivity, for example) on the graphs under consideration. Results of this type were obtained in [11], where a minimum degree condition was added.

Theorem 9 *For a fixed integer* $t \geq 2$, *let* G *be a 2–connected graph of order* p *with* $\delta(G) \geq t$. *If* $|N(u) \cup N(v)| \geq p - t$ *for each pair* u, v *of nonadjacent vertices, then* G *is hamiltonian.*

Corollary *For a fixed integer* $t \geq 2$, *let* G *be a connected graph of order* p *with* $\delta(G) \geq t - 1$. *If* $|N(u) \cup N(v)| \geq p - t$ *for each pair* u, v *of nonadjacent vertices, then* G *is traceable.*

If G is the 2–connected graph $3K_{t-2} + K_2$, $t \geq 3$, of order $p = 3t - 4$, then G is 2–connected and satisfies the neighborhood condition of Theorem 9. However, $\delta(G) = t - 1$ and G is not hamiltonian.

Ore's degree sum condition guarantees the existence of *one* hamiltonian cycle in a graph. In [22] Ore's result was generalized to a condition giving multiple edge–disjoint hamiltonian cycles.

Theorem 10 *Let* G *be a graph of order* p *and minimum degree* δ *such that* $p \geq 2\delta^2$. *If* $\deg u + \deg v \geq p$ *for each pair* u, v *of nonadjacent vertices, then* G *contains* $\lfloor (\delta - 1)/2 \rfloor$ *edge–disjoint hamiltonian cycles.*

A neighborhood condition for multiple hamiltonian cycles was established in [19]. The *edge–connectivity* $\kappa_1(G)$ of a nontrivial graph G is the minimum number of edges whose removal from G results in a disconnected graph.

Theorem 11 *Let* k *be a fixed positive integer. Then there is a constant* $c = c(k)$ *such that if* G *is a graph of sufficiently large order* p *satisfying*

(1) $|N(u) \cup N(v)| \geq (2p + c) / 3$ *for each pair* u, v *of nonadjacent vertices,*

(2) $\delta(G) \geq 4k + 1$,

(3) $\kappa_1(G) \geq 2k$, *and*

(4) $\kappa_1(G - v) \geq k$ *for every vertex* v,

then G *contains* k *edge–disjoint hamiltonian cycles.*

Any theorem that gives a sufficient condition for a graph G to have k edge–disjoint hamiltonian cycles and is based on a neighborhood condition must have the types of restrictions listed in Theorem 11, as we shall see. However, only restrictions (3) and (4) are necessarily sharp.

(1) If G_1 is the graph $3K_n + K_2$, then G_1 is a graph of order $p = 3n + 2$ satisfying (2), (3), and (4) for p sufficiently large. Furthermore, $|N(u) \cup N(v)| = 2(p - 2) / 3$ for each pair u, v of nonadjacent vertices. However, G_1 is not hamiltonian.

(2) If G_2 is the graph $(K_{2k-1} \cup K_{p-2k-1}) + K_2$, then G_2 is a graph of order p satisfying (1), (3), and (4), and $\delta(G) = 2k$ for p sufficiently large. However, G_2 does not contain k edge–disjoint hamiltonian cycles; otherwise, each such cycle would necessarily contain a hamiltonian path in the subgraph K_{2k-1}, implying that K_{2k-1} has at least $k(2k - 2)$ edges.

(3) If G_3 is the graph obtained from two copies of the complete graph K_n by adding $2k - 1$ mutually nonadjacent edges between the copies, then G_3 is a graph of order $p = 2n$ satisfying (1), (2), and (4) and with $\kappa_1(G) = 2k - 1$ for p sufficiently large. However, it is clear that G_3 contains at most $k - 1$ edge–disjoint hamiltonian cycles.

(4) Let G_4 be the graph obtained from two copies of the complete graph K_n by adding edges from a fixed vertex v in the first copy of K_n to all of the vertices of the second copy of K_n, and edges from a second vertex in the first copy of K_n to $k - 1$ vertices in the other copy of K_n. Then G_4 is a graph of order $p = 2n$ satisfying (1), (2), and (3) for p sufficiently large, and $\kappa_1(G - v) = k - 1$. However, G contains at most $k - 1$ edge–disjoint hamiltonian cycles.

A graph G of order $p \geq 3$ is *pancyclic* if contains a cycle of length r for $r = 3, 4, \ldots, p$. If each vertex of G lies on a cycle of every possible length, then G is *vertex–pancyclic.*

In [5] and [3] it was shown that the degree sum conditions in Theorems 2 and 3 are, in fact, "nearly sufficient" conditions for a graph to be pancyclic.

Theorem 12 *If G is a graph of order $p \geq 3$ such that $\deg u + \deg v \geq p$ for each pair u, v of nonadjacent vertices, then either G is pancyclic or else p is even and G is the graph $K_{p/2\,,\,p/2}$.*

Corollary *If G is a graph of order $p \geq 3$ such that $\deg u + \deg v \geq p + 1$ for each pair u, v of nonadjacent vertices, then G is pancyclic.*

Theorem 13 *If G is a 2–connected graph of order $p \geq 6$ such that $\deg u + \deg v + \deg w \geq 3p/2$ for each triple u, v, w of mutually nonadjacent vertices, then either G is pancyclic or else p is even and G is the graph $K_{p/2,\,p/2}$.*

Corollary *If G is a 2–connected graph of order $p \geq 6$ such that $\deg u + \deg v + \deg w > 3p/2$ for each triple u, v, w of mutually nonadjacent vertices, then G is pancyclic.*

Just as a slight increase in Ore's degree sum condition gives a sufficient condition for a graph to be pancyclic, so too with the neighborhood condition of Theorem 6 [11].

Theorem 14 *If G is a 2–connected graph of order p with $|N(u) \cup N(v)| \geq (2p + 5)/3$ for each pair u, v of nonadjacent vertices, then G is pancyclic.*

We have seen that, in the case of the properties of being hamiltonian and traceable, an imposed minimum degree condition allows the neighborhood condition to be relaxed. It is conjectured [11] that this is true for the property of being vertex pancyclic.

Conjecture *For a fixed integer $t \geq 2$, let G be a 2–connected graph of order p with $\delta(G) \geq t + 1$. If $|N(u) \cup N(v)| \geq p - t$ for each pair u, v of nonadjacent vertices of G, then G is vertex pancyclic.*

The conjecture has been verified for the case $t = 2$ [11].

Theorem 15 *If G is a 2–connected graph of order p with $\delta(G) \geq 3$ and such that $|N(u) \cup N(v)| \geq p - 2$ for each pair u, v of nonadjacent vertices, then G is vertex pancyclic.*

We note that if G is the 2–connected graph obtained from $K_{p-t-1} + \overline{K}_t$, $t \leq p/2$, by adding a new vertex which is adjacent to each vertex of the $\overline{K}_t$, then $\delta(G) = t$ and $|N(u) \cup N(v)| \geq p - t$ for each pair u, v of nonadjacent vertices. But G is not vertex pancyclic.

We next turn our attention to neighborhood conditions guaranteeing cycles and paths of at least a specified order [13].

Theorem 16 *Let G be a graph of order p satisfying $|N(u) \cup N(v)| \geq s$ for each pair u, v of nonadjacent vertices.*

(1) If G is 2–connected, then G contains a path of order at least $3s/2 + 2$ or (if $p < 3s/2 + 2$) G is traceable.

(2) If G is 2–connected, then G contains a cycle of order at least $s + 2$ or (if $p < s + 2$) G is complete. If, in addition, s is odd and $p > s + 2$, then G contains a cycle of order at least $s + 3$.

(3) *If G is connected, then G contains a path of order at least* $s + 2$ *or (if* $p < s + 2$*) G is complete. If, in addition, s is even and* $p > s + 2$*, then G contains a path of order at least* $s + 3$.

(4) *If G is connected and* $s \geq 3$*, then G contains a cycle of order at least* $(s + 2)/2$ *or (if* $p < s + 2$*) G is complete.*

(5) *If G is not connected, then at most one component of G has fewer than* $(s + 2)/2$ *vertices and if there is one, then it is complete. Every other component has at least* $(s + 2)/2$ *vertices, and is either complete or contains a path of order at least* $s + 2$ *and (if* $s \geq 3$*) a cycle of order at least* $(s + 2)/2$..

The results of Theorem 16 are sharp, as the following examples indicate. In each case, the graph described has $|N(u) \cup N(v)| \geq s$ for each pair u, v of nonadjacent vertices.

(1) and (2): For s even, let G_1 be the graph $mK_{s/2} + K_2$, where $m \geq 4$. Then G_1 is 2–connected. Clearly, a longest path on G_1 has order $3s/2 + 2$ and a longest cycle has order $s + 2$.

(3) For s even, let G_2 be the graph $mK_{(s+2)/2} + K_1$, where $m \geq 3$. Then G_2 is connected and a longest path in G_2 has order $s + 3$.

(4) For even $s \geq 4$, let G_3 be the graph obtained from $m \geq 1$ copies $H_1, H_2, \ldots, H_m$ of $K_{(s+2)/2}$ by adding an edge joining a vertex of H_i and a vertex of H_{i+1} for $i = 1, 2, \ldots, m - 1$. Then G is connected, and a longest cycle in G_3 has order $(s + 2)/2$.

It is unknown if the results of Theorem 16 can be improved by imposing higher connectivity restrictions. For example, if G is a m–connected graph of sufficiently large order for which $|N(u) \cup N(v)| \geq s$, does G necessarily contain a cycle of order 2s ? The complete bipartite graph $K_{s, p-s}$ shows this would be best possible for $m \leq s \leq p/2$.

A *matching* in a graph G is a set of independent edges, i.e., a set of edges such that no two are adjacent. The *edge–independence number* $\beta_1(G)$ of G is the maximum number of edges in a matching in G. The next theorem [13] answers the questions: If $|N(u) \cup N(v)| \geq s$ for each pair u, v of nonadjacent vertices of a graph G, then how large is $\beta_1(G)$? Does the answer change if a connectivity condition is imposed?

Theorem 17 *Let G be a graph of order p such that* $|N(u) \cup N(v)| \geq s$ *for each pair u, v of nonadjacent vertices.*

(1) *If* $s \leq (p-2)/2$ *then* $\beta_1(G) \geq s$.

(2) *If* $s \geq (p-1)/2$ *and p is odd, then* $\beta_1(G) \geq (p-3)/2$.

(3) *If* $s > 2\lfloor p/3 \rfloor - 2$ *and* p *is odd, then* $\beta_1(G) = (p-1)/2$, *unless* $p = 5$ *and* $s = 1$.

(4) *If* $s \geq (p-1)/2$ *and* p *is odd and* G *is connected, then* $\beta_1(G) = (p-1)/2$.

(5) *If* $s \geq p/2$ *and* p *is even, then* $\beta_1(G) \geq (p-2)/2$.

(6) *If* $s > 2(p-1)/3 - 1$ *and* p *is even and* G *is connected, then* $\beta_1(G) = p/2$.

(7) *If* $s \geq p/2$ *and* p *is even and* G *is 2–connected, then* $\beta_1(G) = p/2$.

Certainly, $\beta_1(G) \leq \lfloor p/2 \rfloor$ for every graph G of order p. The following examples demonstrate the sharpness of the remaining parts of Theorem 17. Each graph satisfies the condition $|N(u) \cup N(v)| \geq s$ for all pairs u, v of nonadjacent vertices.

(1) The graph $K_{s,p-s}$, $s \leq p/2$, has edge–independence number s and indicates that the conclusion in (1) is best possible even for graphs with high connectivity.

(2) The graph $3K_n$ of order $p = 3n$ has edge–independence number $(p-3)/2$ for any odd integer n, and shows that (2) is best possible for values of s as large as $2\lfloor p/3 \rfloor - 2$.

(3) The graph $K_{1,4}$ is the excluded case in (3).

(5) The graph $K_1 \cup K_n$ (n odd) of order $p = n + 1$ has edge–independence number $(p-2)/2$, and shows that (5) is best possible for values of s as large as $p - 2$.

(6) The graph $3K_{n-1} + K_1$ (n even) shows that (6) is *not* true for $s = 2(p-1)/3 - 1$.

If G is a graph of even order p and M is a matching in G with $p/2$ edges, then M is called a *perfect matching* of G. We see from the previous theorem that if a connected graph G of even order p satisfies $|N(u) \cup N(v)| > 2(p-1)/3 - 1$, then G contains a perfect matching. We can, if fact, say more [19].

Theorem 18 *Let* m *be fixed positive integer. Then there is a constant* $c = c(m)$ *such that if* G *is a graph of sufficiently large even order* p *satisfying*

(1) $|N(u) \cup N(v)| \geq (2p + c)/3$ *for each pair* u, v *of nonadjacent vertices,*

(2) $\delta(G) \geq 2m$, *and*

(3) $\kappa_1(G) \geq m$,

then G *contains* m *edge–disjoint perfect matchings.*

Condition (3) is certainly sharp since any graph consisting of two odd order disjoint subgraphs with $m - 1$ edges between them contains at most $m - 1$ perfect matchings. Although condition (1) may not be sharp, it is of the correct order of magnitude as indicated by Theorem 17 (6). Similarly, if G has m perfect matchings, then $\delta(G) \geq m$ and so a condition like (2) is necessary.

The final result of this section involves the relationship between the neighborhood unions of pairs of nonadjacent vertices and "Menger path systems". For positive integers m and d, let $P_{d,m}$ denote the property that between each pair x, y of vertices of a graph G there are m vertex disjoint (except for x and y) paths, each of length at most d. For integers $d \geq 3$ and $m \geq 1$, let G_1 be the graph obtained from a path of order $d + 2$ by first replacing each of the first two and last two vertices with a copy of $K_{3t/2}$, t even, then replacing each of the remaining vertices with a copy of K_t, and finally adding all possible edges between "consecutive" complete graphs. If G is the graph $G_1 + K_{m-1}$, then G is an m–connected graph of order $p = (d + 4)t + m - 1$. Furthermore, $|N(u) \cup N(v)| \geq 5(p - m + 1)/(d + 4) + m - 3$ for each pair u, v of nonadjacent verices. However, if x is a vertex from the "first" copy of $K_{3t/2}$ and y is a vertex from the "last" copy of $K_{3t/2}$, then every $x - y$ path of length at most d contains a vertex of K_{m-1}. Consequently, G does not have property $P_{d,m}$. This example indicates that the neighborhood condition in the next theorem [18] is of correct order of magnitude.

Theorem 19 *Given integers $d \geq 2$ and $m \geq 1$, if G is an m-connected graph of sufficiently large order p with $|N(u) \cup N(v)| > 5p / (d + 2) + 2m$ for each pair u, v of nonadjacent vertices, then G satisfies $P_{d,m}$.*

3. All Sets of t Independent Vertices, $t \geq 2$

In Section 2 we examined sufficient conditions of the form $|N(u) \cup N(v)| \geq k$ for all sets S of two nonadjacent vertices, i.e., for all sets S of two independent vertices. An obvious question to ask is whether such conditions exist for sets S of t independent vertices, $t > 2$. It is not surprising that the number of known results here is small. In three instances of graphical properties, however, complete solutions have been obtained. The first of these generalizes Theorem 6 [20].

Theorem 20 *If G is a t–connected graph of order $p \geq 3$ and each set S of $t \geq 1$ independent vertices satisfies*

$$|N(S)| > t(p - 1)/(t + 1)$$

then G is hamiltonian.

This result is "nearly best possible" as is indicated by the graph G defined to be $(t + 1)K_q + K_t$. Then G is a t-connected graph of order $p = t + (t + 1)q$.

Furthermore, for each set S of t independent vertices, $|N(S)| = t + t(q - 1) = t(p-t)/(t+1)$. However, G is not hamiltonian. We note that the case $t = 1$ in Theorem 20 is Dirac's theorem.

If $|N(S)|$ is "large enough" for sets S of t independent vertices of a graph G, then intuitively G contains a large number of well–distributed edges. Consequently, it is reasonable to expect the existence of a large complete subgraph in G. Such a result was established in [1].

Theorem 21 *Let t and m be fixed integers, $t \geq 1$ and $m \geq 3$. If G is a graph of sufficiently large order p such that each set S of t independent vertices satisfies*

$$|N(s)| > (m-2)p/(m-1)$$

then K_m is a subgraph of G.

If G is the complete $(m-1)$–partite graph with $k \geq t$ vertices in each partite set (i.e., G is the Turan graph), then G has order $p = (m-1)k$ and every set S of t independent vertices satisfies

$$|N(S)| = (m-2)k = (m-2)p/(m-1).$$

However, K_m is not a subgraph of G. We note that the case $t = 1$ in Theorem 21 follows from the classic extremal result of Turan [27].

Theorem 21 was extended in [12], where a neighborhood union condition for independent t–sets was given that insures multiple copies of a complete graph of a fixed order.

Theorem 22 *Let t, m, n be fixed integers, $t \geq 1$, $m \geq 3$, $n \geq 1$. If G is a graph of sufficiently large order p such that each set S of t independent vertices satisfies*

$$|N(S)| \geq \frac{(m-2)p+n}{m-1}$$

then nK_m is a subgraph of G.

For fixed integers $t \geq 1$, $m \geq 3$, $n \geq 1$, let G be the graph $H + K_{n-1}$, where H is the complete $(m-1)$–partite graph with $k \geq t$ vertices in each partite set. Then G has order $p = k(m-1) + n - 1$. Furthermore, every set S of t independent vertices satisfies

$$|N(S)| = (m-2)k + n - 1 = \frac{(m-2)p + n - 1}{m-1}.$$

But G does not contain nK_m as a subgraph. Thus the result in Theorem 22 is sharp.

In Section 2 we looked at neighborhood conditions based on all pairs of independent vertices and then, in this section, considered a generalization to all sets of t independent vertices. Very recently, a different direction from "all pairs of nonadjacent vertices" has been investigated by Lindquester [23]. For vertices u and v of a graph

G, let d(u, v) denote the length of a shortest u – v path in G. Then u and v are nonadjacent if and only if $d(u, v) \geq 2$.

Theorem 23 *If G is a 2–connected graph of order p such that* $|N(u) \cup N(v)| \geq (2p-1)/3$ *for each pair u, v of vertices with* $d(u, v) = 2$*, then G is hamiltonian.*

Corollary *If G is a 2–connected graph of order p such that* $|N(u) \cup N(v)| > (2p-4)/3$ *for each pair u, v of vertices with* $d(u, v) = 2$*, then G is traceable.*

Certainly Theorem 6 and its corollary follow from these two results. Moreover, Theorem 23 is an improvement of Theorem 6. For example, let I_r denote a graph with 2r vertices and r independent edges. Consider the graph G which consists of $I_r \cup K_{4r}$ plus a matching between the vertices of I_r and 2r of the vertices of K_{4r}. Then G has order $p = 6r$ and for vertices u, v with $d(u, v) = 2$, we have $|N(u) \cup N(v)| = 2p/3 > (2p-1)/3$. Thus G is hamiltonian by Theorem 23. However, G does not satisfy the neighborhood condition of Theorem 6.

4. All Pairs of Vertices

The results of this section parallel, in some sense, those of Section 2. Here, however, we are considering neighborhood conditions of the type

$|N(u) \cup N(v)| \geq k$ for *all* pairs u, v of vertices.

The first results involve hamiltonian properties [9].

Theorem 24 *If G is a 2–connected graph of order p such that* $|N(u) \cup N(v)| \geq p/2$ *for each pair of u, v of vertices, then G is hamiltonian.*

Corollary *If G is a connected graph of order p such that* $|N(u) \cup N(v)| \geq (p-1)/2$ *for each pair u, v of vertices, then G is traceable.*

Certainly p/2 cannot be lowered in Theorem 24 since, in fact, Dirac's minimum degree condition $\delta(G) \geq p/2$ is sharp. The graph $2K_{(p-1)/2} + K_1$ (p odd) shows that the connectivity condition cannot be relaxed.

In light of Theorem 24 one might hope that

$$|N(u) \cup N(v)| \geq (p+1)/2 \text{ for each pair } u, v \in V(G) \rightarrow G \text{ is hamiltonian–connected,}$$

since we know that

$$\delta(G) \geq (p+1)/2 \rightarrow G \text{ is hamiltonian–connected}$$

and this minimum degree condition is best possible. At the present, the best result known is the following [15].

Theorem 25 *There is a constant* c *such that if* G *is a 3–connected graph of order* p *with* $|N(u) \cup N(v)| \geq p/2 + c$ *for each pair* u, v *of vertices, then* G *is hamiltonian–connected.*

Similarly, $\delta(G) \geq (p + 1) / 2$ implies that G is pancyclic, and we have a corresponding neighborhood result [15].

Theorem 26 *There is a constant* d *such that if* G *is a 2–connected graph of order* p *with* $|N(u) \cup N(v)| \geq p/2 + d$ *for each pair* u, v *of vertices, then* G *is pancyclic.*

For shorter paths and cycles, the next two theorems [9] summarize the known results.

Theorem 27 *Let* G *be a connected graph of order* p *such that* $|N(u) \cup N(v)| \geq s$ *for each pair* u, v *of vertices and for some* $3 \leq s < p/2$. *Then* G *contains a path of order at least* $2s - 1$ *and a cycle of order at least* s. *Furthermore,* $\beta_1(G) \geq s - 1$.

Theorem 28 *Let* G *be a 2–connected graph of order* p *such that* $|N(u) \cup N(v)| \geq s$ *for each pair* u, v *of vertices and for some* $3 \leq s \leq p/2$. *Then* G *contains a path of order at least* $2s$ *and a cycle of order at least* $2s - 2$. *Furthermore,* $\beta_1(G) \geq s$.

For $s, t \geq 3$, let G_1 be the graph $K_2 + tK_{s-2}$. Then G_1 is 2–connected, has order $p = t(s - 2) + 2$, satisfies $|N(u) \cup N(v)| \geq s$ for each pair u, v of vertices, and contains a cycle of order $2s - 2$ but no longer cycle. Thus the cycle result of Theorem 28 cannot be improved for $3 \leq s \leq (p + 4)/3$. For $3 \leq s \leq p/2$, let G_2 be the graph $K_s + \overline{K}_{p-s}$. Then G_2 is a 2–connected graph of order p and satisfies $|N(u) \cup N(v)| \geq s$ for each pair u, v of vertices. However, G_2 has no path of order $2s + 2$ and $\beta_1(G_2) = s$. Thus, only an improvement of 1 on the order of the path is possible in Theorem 28.

Theorem 28 gives a sufficient condition for a graph to contain a perfect matching. This was extended in [16].

Theorem 29 *For a fixed positive integer* k, *let* G *be a graph of even order* p *such that*

(1) $\delta(G) \geq k + 1$,

(2) $\kappa_1(G) \geq k$, *and*

(3) $|N(u) \cup N(v)| \geq p/2$ *for each pair* u, v *of vertices.*

If p *is sufficiently large, then* G *contains* k *edge–disjoint perfect matchings.*

Certainly condition (2) is necessary since any graph consisting of two odd order disjoint subgraphs with $k-1$ edges between them contains at most $k-1$ perfect matchings. For even p sufficiently large, the complete bipartite graph $K_{p/2-1,p/2+1}$ satisfies conditions (1) and (2), but not (3), and contains no perfect matchings. Finally, let G be any graph obtained by indentifying one vertex of a copy of $K_{p/2}$ with one vertex of another copy of $K_{p/2}$ and then adding a vertex x of degree k so that in the resulting graph, x is adjacent to the only vertex of degree $p-1$. Then for $p \equiv 0 (\text{mod } 4)$ and p sufficiently large, G satisfies (2) and (3) but not (1), and the maximum number of edge–disjoint perfect matchings in G is $k-1$.

The final all pairs neighborhood condition in this section is a sufficient condition for a graph to have property $P_{d,m}$ [17], i.e., for a graph to have the property that between each pair x, y of vertices there are m vertex disjoint (except for x and y) paths, each of length at most d.

Theorem 30 *Given integers $d \geq 2$ and $m \geq 1$ there is a constant $c = c(d, m)$ such that if G is an m-connected graph of order p with $|N(u) \cup N(v)| \geq 4p/(d+2) + c$ for each pair u, v of vertices, then G satisfies $P_{d,m}$.*

Given integers $d \geq 2$ and $m \geq 2$ ($d \equiv 2 (\text{mod } 4)$), let H be the graph obtained from a cycle of order $d+2$ by first replacing the first, second, fifth, sixth, ninth, tenth, etc. vertices with disjoint copies of K_n for some $n \geq 2$ and then adding all possible edges between "consecutive" complete graphs (including K_1) on the cycle. Define G to be the graph $H + K_{m-2}$. Then G is an m–connected graph of order

$$p = (n+1)(d+2)/2 + m - 2.$$

For all vertices u, v of G, $|N(u) \cup N(v)| \geq m - 2 + 2n + 1 =$

$$4p/(d+2) + (md - 2m + 2 - 3d)/(d+2).$$

However, G does not have property $P_{d,m}$. Thus, the order of magnitude of the previous theorem is correct.

There are relatively few known results of the type

$$|N(S)| \geq k \text{ for every set } S \text{ of } t \text{ vertices of } G \rightarrow$$
$$G \text{ has property } P$$

for $t > 2$. In the case of hamiltonian graphs we have the following extension of Theorem 24 [9].

Theorem 31 *Given a positive integer t there is a constant $c = c(t)$ such that if G is a 2–connected graph of order p with $\delta(G) \geq t$ and $|N(S)| \geq p/2 + c$ for each set S of t vertices, then G is hamiltonian.*

Some preliminary work has been done with neighborhood conditions based on all t–sets of vertices and the property $P_{d,m}$. Given integers m, t, and d with $2 \leq t \leq d-1$, there are constants c_1 and c_2 depending only on m and d, and examples of m–connected graphs G of order p which satisfy

$$(1) \qquad |N(S)| \geq \max \begin{cases} 4p/(d + 4 - t) + c_1 \\ 5p(t - 1)/(d + 2)(t + 1) + c_2 \end{cases}$$

for each set S of t vertices. These graphs "just miss" satisfying property $P_{d,m}$. However, it is believed that (1) is of the correct order of magnitude for the desired sufficient condition for $P_{d,m}$.

5. Neighborhood Closures

In [6], a property P defined for all graphs of order p was called *k–degree stable* if, for all graphs G of order p, whenever G + uv has property P and deg u + deg v ≥ k, then G has property P. It was shown, for example, that the property of being hamiltonian is p–degree stable. The *k–degree closure* $D_k(G)$ of a graph G of order p is the graph obtained from G by recursively joining pairs of nonadjacent vertices whose degree sum is at least k (in the resulting graph at each stage) until no such pair remains. Since for $p \geq 3$ the complete graph of order p is hamiltonian, we have the following sufficient condition for a graph to be hamiltonian.

Theorem 32 *If G is a graph of order $p \geq 3$ such that $D_p(G)$ is complete, then G is hamiltonian.*

We note that Theorems 1 and 2 are immediate corollaries of this stronger degree result. However, the graph H described after the statement of Theorem 7 indicates that the neighborhood condition of Theorem 6 is distinct from the result in Theorem 32.

Analogously, let P be a property dfefined for all graphs G in a class Γ and let k be an integer. Then P is *k-neighborhood stable* in Γ if whenever G + uv has property P, where $G \in \Gamma$ and $|N(u) \cup N(v)| \geq k$, then G has property P [10]. The *k–neighborhood closure* $N_k(G)$ of a graph G is the defined to be the graph obtained from G by recursively joining pairs of nonadjacent vertices u, v for which the cardinality of the neighborhoods of u and v is at least k (in the resulting graph at each stage) until no such pair remains.

Theorem 33 *Let p, s, δ be integers satisfying $p \geq 2s \geq 2$. If Γ denotes the class of graphs of order p with $\delta(G) \geq \delta$ (connected graphs if $p = 2s$), then the property of containing sK_2 is $(2s - 1 - \delta)$–neighborhood stable in Γ.*

Corollary 1 *Let p, s, and δ be integers satisfying $p \geq 2s \geq 2$. If G is a graph of order p with $\delta(G) \geq \delta$ and $N_{2s-1-\delta}(G)$ is complete, then G contains sK_2 provided* $n \geq 2s+1$ *or G is connected.*

Corollary 2 *Let p and s be positive integers satisfying $p \geq 2s$. If G is a graph of order p without isolated vertices for which $N_{2s-2}(G)$ is complete, then G contains sK_2 provided $p \geq 2s+1$ or G is connected.*

The corresponding result for degree closure states that if G is a graph of order $p \geq 2s \geq 2$ and $D_{2s-1}(G)$ is complete, then G contains sK_2 (see [6]). If G is the graph obtained from two copies A and B of $K_s - e$, $s \geq 3$, where two vertices of degree $s-2$, one from A and one from B, are joined by an edge and the remaining two vertices of degree $s-2$ are joined by a path of length 2, $N_{2s-2}(G)$ is complete and consequently Theorem 33 gives that G contains sK_2. However, since $\Delta(G) = s-1$, the $(2s-1)$–degree closure $D_{2s-1}(G)$ is not complete and so the degree closure yields no information. By adding vertices to G, maintaining a maximum degree of at most $s-1$, we obtain graphs G' of all orders $p \geq 2s+1$ for which $N_{2s-2}(G')$ is complete but $D_{2s-1}(G')$ is not complete and is, in fact, just G'.

Theorem 34 *Let Γ denote the class of 2–connected graphs of order $p \geq 6$. Then the property of containing the cycle C_k, $6 \leq k \leq p$, is $(p-2)$–neighborhood stable in Γ.*

Corollary *Let G be a 2–connected graph of order $p \geq 6$. If $N_{p-2}(G)$ is complete, then G contains the cycle C_k for $k = 6, 7, \ldots, p$.*

We note that the two previous results do not hold for values of $k < 6$.

The corresponding result for degree closure states that for integers p and k satisfying $p \geq s \geq 5$, if $D_{2p-s}(G)$ is complete, then G contains C_k (see [6]). For even $p \geq 8$, let G be the graph obtained from two copies A and B of $K_{p/2}$ by adding $p/2$ independent edges from the vertices of A to the vertices of B. Then $N_{p-2}(G)$ is complete and so, by Theorem 34, G contains cycles of lengths $6, 7, \ldots, p$. However, since G is $p/2$–regular, $D_{2p-s}(G)$ is complete only if $s = p$.

REFERENCES

[1] N. Alon, R.F. Faudree, Z. Furedi, A Turan–like condition and cliques in graphs. *Ann. NY Acas. Sci*, to appear.

[2] I. Anderson, Sufficient conditions for matchings. *Proc. Edin. Math. Soc.*, 18 (1972) 129 – 136.

[3] D. Bauer and E. Schmeichel, Pancyclic graphs and a degree condition of Bondy. Preprint.

[4] J.A. Bondy, Longest paths and cycles in graphs of high degree. *Research report CORR 80 – 16*, University of Waterloo, Waterloo, Ontario.

[5] J.A. Bondy, Pancyclic graphs I. *J. Combinatorial Theory* 11B (1971) 80 – 84.

[6] J.A. Bondy and V. Chvatal, A method in graph theorey. *Discrete . Math.* 15 (1976) 111 – 135.

[7] G. Chartrand and L. Lesniak, *Graphs & Digraphs*, Prindle, Weber, and Schmidt, Boston, 1986.

[8] G.A. Dirac, Some theorems on abstract graphs, *Proc. London Math. Soc.* 2 (1952) 69 – 81.

[9] R.J. Faudree, R.J. Gould, M.S. Jacobson, L.M. Lesniak, Neighborhood unions and a generalization of Dirac's theorem. Preprint.

[10] R.J. Faudree, R.J. Gould, M.S. Jacobson, L.M. Lesniak, Neighborhood closures for graphs. *Colloquia Mathematica Societatis Janos Bolyai* 52 (1987) 227 – 237.

[11] R.J. Faudree, R.J. Gould, M.S. Jacobson, L.M. Lesinak, Neighborhood unions and highly hamiltonian graphs. Preprint.

[12] R.J. Faudree, R.J. Gould, M.S. Jacobson, L.M. Lesniak, On a neighborhood condition implying the existence of disjoint complete graphs, *European J. Combinatorics,* to appear.

[13] R.J. Faudree, R.J. Gould, M.S. Jacobson, R.H. Schelp, Extremal problems involving neighborhood unions. *J. Graph Theory* 11 (1987) 555 – 564.

[14] R.J. Faudree, R.J. Gould, M.S. Jacobson, R.H. Schelp,Neighborhood unions and hamiltonian properties in graphs, *J. Combinatorial Theory B*, to appear.

[15] R.J. Faudree, R.J. Gould, L.M. Lesniak, Generalized degree conditions and hamiltonian properties in graphs. Preprint.

[16] R.J. Faudree, R.J. Gould, L.M. Lesniak, Neighborhood conditions and edge–disjoint perfect matchings. Preprint.

[17] R.J. Faudree, R.J. Gould, L.M. Lesniak, Neighborhood conditions and Menger path systems. Preprint.

[18] R.J. Faudree, R.J. Gould, R.H. Schelp, Menger path systems. Preprint.

[19] R.J. Faudree, R.J. Gould, R.H. Schelp, Neighborhood conditions and edge–disjoint hamiltonian cycles, *Congressus Numerantium* 59 (1987) 55 – 68.

[20] P. Fraisse, A new sufficient condition for hamiltonian graphs. *J Graph Theory* 10 (1986) 405 – 410.

[21] P. Hall, On representation of subsets. *J. London Math. Soc.* 10 (1935) 26 – 30.

[22] H. Li, Edge disjoint cycles in graphs. *J. Graph Theory,* to appear.

[23] T.E. Lindquester, The effects of distance and neighborhood union conditions on hamiltonian properties in graphs, *J. Graph Theory,* to appear.

[24] O. Ore, Hamiltonian connected graphs. *J. Math. Pures. Appl.* 42 (1963) 21 – 27.

[25] O. Ore, Note on hamilton circuits. *Amer. Math. Monthly* 67 (1960) 55.

[26] Posa, Hamiltonian circuits in random graphs. *Discrete Math.* 14 (1976) 359 – 364.

[27] P. Turan, Eine Extremalaufgabe aus der Graphtheorie. *Mat. Fiz. Lapok* 48 (1941) 436 – 452.

[28] D.R. Woodall, A sufficient condition for hamiltonian circuits. *J. Combinatorial Theory* B25 (1978) 184 – 186.

An Algorithm to Decide if a Cayley Diagram is Planar

H. Levinson
Rutgers University

ABSTRACT

Given a presentation of a group on generators and defining relators, its associated Cayley diagram may or may not be planar. An algorithm is given which produces the set of all possible rotation sequences for embeddings, in the form of a tree. Only the defining relators are used, and a form of the word problem must be assumed solvable. The Cayley diagram is planar if and only if some "consistent" subset of these rotation sequences exists. Although Maschke in 1896 gave all finite Cayley diagrams, these results apply to any finite presentation with solvable word problems (most of which are infinite groups).

1. Introduction

The construction of the Cayley diagram of a presentation of a group is equivalent to solving the word problem for the presentation. Because of this, we must assume an effective solution of the word problem for a presentation in order to decide whether its Cayley diagram is planar. In particular, we shall confine out attention to those (finite) presentations for which the defining relators, considered as cyclic words, contain no proper segment equal to the group identity. We shall assume this condition throughout this article and not restate it again explicitly each time it is needed.

If we are confronted with a presentation P in which some defining relator R contains two or more segments $S_1,...,S_k$ which are themselves the identity, we may replace R by $S_1,...,S_k$ and obtain another presentation P' satisfying our condition and whose Cayley diagram is identical to that of P. For example, $P = \langle a,b;\ a^3 = a^4ba^{-1}b^{-1}a^3 = 1\rangle$ has $aba^{-1}b^{-1}$ and a^3 both 1. $Q = \langle a,b;\ a^3 = a^3 = aba^{-1}b^{-1} = 1\rangle$ has the same Cayley diagram as P.

2. Fundamental regions and effective construction.

We define the fundamental region F(P, g) at a point g of a Cayley diagram for $P = \langle x_1,...,x_n;\ R_1 = ... = R_k = 1\rangle$ as follows. Construct a colored directed cycle corresponding to each distinct cyclic permutation of the defining relators $R_1,...,R_k$. Each such cycle has a distinguished vertex from which the succession of edges, in some direction, is the sequence of generators, starting from the first, in the cyclic permutation of the relator it represents. Identify all these distinguished vertices together as one point g. If any two edges having the same direction and color are incident at the same point, coalesce them. When there are no more edges to coalesce, the remaining graph is our fundamental region. It is, in fact, a point g of the Cayley diagram C(P) of P with an incident cycle starting at g for each cyclic permutation of each defining relator of P. That F(P, g) is well defined results from the uniqueness of C(P) which, in turn, results from the group given by P being well defined.

The classic construction of C(P) starts with a tree T corresponding to the free group on $x_1,...,x_n$, and continues by identifying all points of T in equivalence classes of words which are "consequences" of the defining relators, $R_1,...,R_k$. The problem with this method is to discern which words are "consequences". A consequence is a juxtaposition of (a finite number of) conjugates of the defining relators and their inverses. After free cancellation (removing all adjacent pairs $x \cdot x^{-1}$), the remaining word may not be easily recognized. Moreover, there may be some very short words which only result from very large aggregates of conjugates of defining relators. "The word problem is not effectively decidable" means that there are presentations in which it is impossible to obtain an upper bound on the length of a string of conjugates of defining relators which collapses to a given word, should such a string exist.

We shall construct C(P) differently. Call a point p *saturated* if there is an entrant and a salient edge at p for each "color" $x_1,...,x_n$. In F(P, g), g is saturated. If a point p in F(P, g) is not saturated identify the g of a copy of F(P, g) with p, and coalesce adjacent edges having the same color and direction, as in the construction of F(P, g). We call this process *saturating p with F(P, g)*. Starting with F(P, g) and saturating each unsaturated point, including all newly introduced points, with F(P, g) results in C(P). We refer to this as the F(P, g) construction of C(P). Although this task is, in general, infinite, the condition we prescribe in the introduction gives us an advantage.

Theorem 1 *Each (finite) diagram obtained during the F(P, g) construction of C(P), in which all points are saturated, is a subgraph of C(P).*

Proof Suppose there are two distinct points p and q of the saturated part of a diagram resulting during the F(P, g) construction of C(P) which will be identified in C(P). Then there is some finite sequence of saturations of points, starting with p, which results in p being identified with q, via edge identifications. However the identification of two saturated points will result in edge identifications leading to the collapse of some part of some F(P, g). This means that some cycle of F(P, g) has a proper segment equal to the group identity, which contradicts our introductory assumption on P.❑

Thus F(P, g) saturation, coupled with our assumption on the defining relators of P, gives us an effective method for constructing any finite portion of C(P). In addition, the identifications of edges during the process affects algebraic cancellation in a geometric manner.

3. Embedding fundamental regions.

We shall show that is planar embeddings of F(P, g) exist and satisfy certain compatibility conditions, then any subgraph of C(P) is planar. This has already been exploited in the case where each generator appears exactly twice among all the symbols of all the defining relators of P. In this very special case, if C(P) is planar, it is point–symmetric planar or weakly point–symmetric planar and an algorithm to decide planarity has been given in [2]. This result made crucial use of the Edmonds–Heffter embedding technique [1].

The algorithm given in [2] has now been generalized using an idea due to A. Barnasconi of the Universidad Técnica Federico Santo Maria, Valparaiso, Chile. He suggested that the different occurrences of the generators in the defining relators be subscripted to distinguish them from each other. Figure 1 gives the algorithm which results from exploiting the adjacencies of generators at g prescribed by their adjacencies in the defining relators of P. If F(P, g) has a planar embedding, the Edmonds–Heffter embedding technique will produce it. Since the algorithm yields *all* embedding schemas (rotation systems), any particular embedding will appear. The algorithm produces a forest Φ of rooted trees of occurrences of generators within the defining relators. If there is some path, or some "feasible" collection of paths, in Φ which contains an instance of each occurrence of each generator in each defining relator of P, then F(P, g) may have a planar embedding. (The meaning of "feasible" will become clear later.) In such a planar embedding, should it be realizable, each relator cycle does *not* necessarily bound its own disk.

The sequence of generators is not necessarily the succession of edges appearing at a given vertex. Rather, a list of *successive inner corners of relator cycles* is what is produced by the algorithm. These corners may be determined as follows from a sequence, or collection of sequences, given by the forest. Number the entries of a complete list of the defining relators and their distinct cyclic permutations. Next to each subscripted generator with exponent +1 in the forest, write the number of (the cyclic permutation of) the

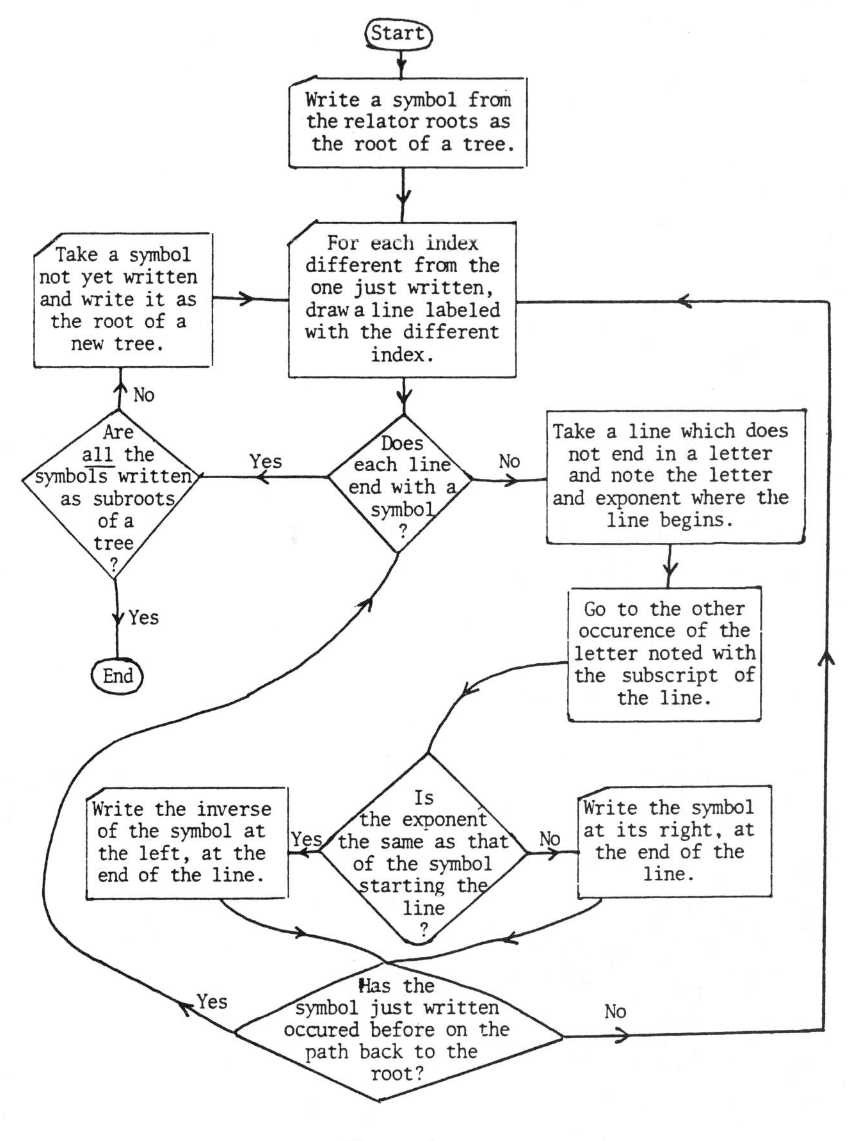

Figure 1

Take the set of all roots of the defining relators and subscript the distinct occurrences of the generators in them. (i.e. ab is a root of $(ab)^{13}$.) A symbol will be considered distinct from its inverse.

defining relator in the numbered list, which starts with that numbered generator. Next to each generator with exponent −1, write the number of the relator which ends with that numbered generator. This induced list of relator numbers gives the relator corners as they appear around g in an embedding of F(P, g). Some may be superimposed on others.

$$P = \langle a, b, c;\ abc = (ab)^3 = (ac)^3 = 1\rangle$$

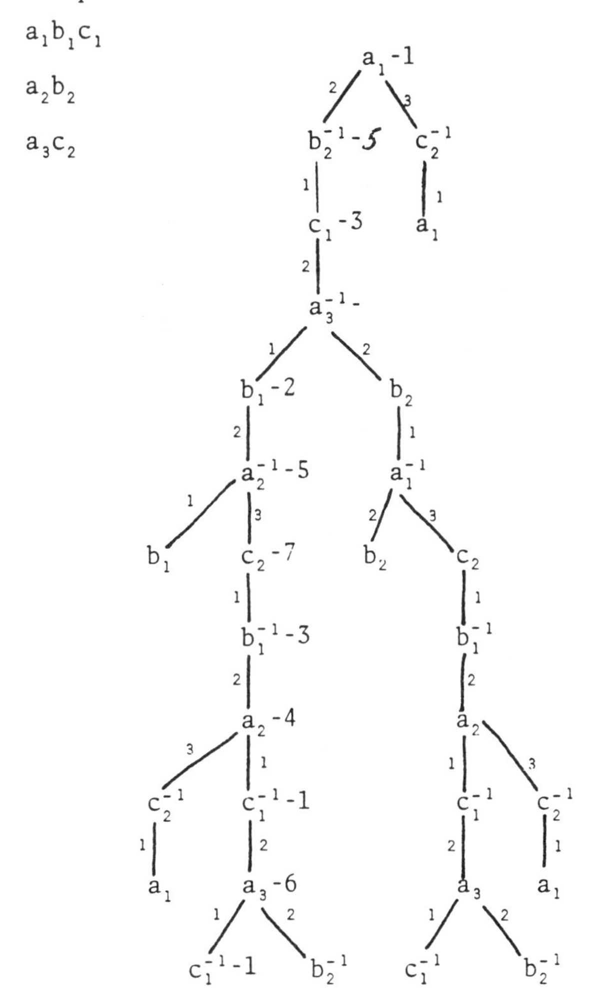

Numbered cyclic permutations:

1 - abc
2 - bca
3 - cab
4 - $(ab)^3$
5 - $(ba)^3$
6 - $(ac)^3$
7 - $(ca)^3$

Figure 2

Figure 2 shows an example of a presentation for which the algorithm gives a single tree. The subscripted generators along a particular path have been numbered according to the directions given in the previous paragraph. Since numbers for all (7) distinct cyclic permutations of each defining relator appear on this list, there may be a planar embedding of F(P, g) which realizes this sequence of generators. A planar embedding of F(P, g) is given in Figure 3. Here the "sequence" involves jumps. From a_1, jump clockwise to b_2^{-1} to realize 4–$(ab)^3$. From b_2^{-1}, go counterclockwise to c_1 to realize 3–cab, etc. The list of relator cycles is also shown separately in Figure 3 since they are hard to discern individually within F(P, g).

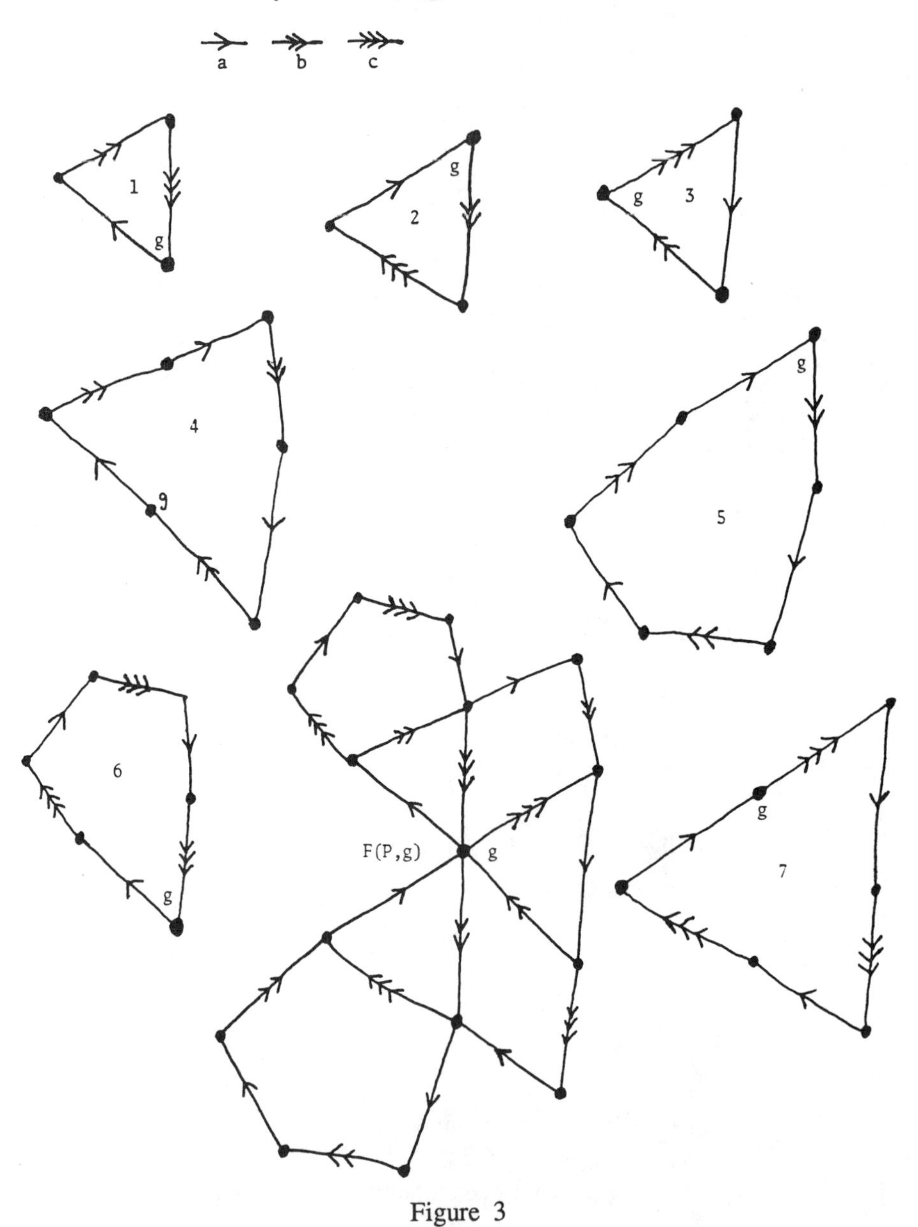

Figure 3

To produce a planar embedding, should one exist, construct cycles corresponding to the numbered defining relators, each with a distinguished vertex at the start of the leading generator of the relator it represents. Embed the first relator of the relator list induced by the path, or collection of paths, chosen in the forest. Next, superimpose the leading generator in the second relator from the list on the appropriate generator emanating from g in what has already been embedded. Unless there are adjacent edges of the same color and direction to be merged, there is now a choice of embedding the second relator clockwise or counterclockwise in various disks. Embed it whichever way does not give improper crossings with edges already embedded. (If several ways are possible, make separate embedding diagrams for all, each of which is to be continued, if possible.) Embed successive relators from the induced relator list similarly to the way in which the second was embedded. If it is impossible to continue at any step without introducing improper crossings, then that particular line of embedding, reflecting previous choices of cycle embeddings, is unfeasible. Should there be no way of continuing this embedding process to a complete embedding of $F(P, g)$, then the path, or collection of paths, chosen from the forest does not result in a planar embedding, but some higher genus embedding. (Remember that if there is *some* planar embedding of $F(P, g)$, the forest will yield it.)

If there is a collection of paths from the forest in which all the distinct cyclic permutations of the relators occur, it may be the case that the induced relator lists from these paths may be realized by a planar embedding. If so, we shall call the system of paths *feasible*. If the paths are relator–wise disjoint, the corresponding regions of $F(P, g)$ they represent will be distinct blocks, and g will be a cut point. (all points in $C(P)$ will be cut points.) In this case, any planar block may be embedded in any disk of any other, provided only that g be on the boundary of that disk (in order to identify the distinguished vertices of all the blocks.) Figure 4 shows an example of a presentation which yields a forest of paths. A collection of paths is indicated which gives the embedding of $F(P, g)$ shown. (The presentation in Figure 4 is of the free group on 3 generators, *not* freely presented.)

It may seem that this algorithm is itself so complicated that one might just as well try all possible sequences of edges about a point by "brute force". However, reflection will show that $\frac{(2n-1)!}{2}$ sequences of edges must be examined for a presentation on n generators. In most cases, finding feasible sequences from the forest produced by the algorithm reduces the search for planar embeddings to a much smaller order of magnitude. Although the computational complexity of the algorithm presented is exponential, it is better than sheer brute force. Furthermore, it is reasonable to conjecture that the problem itself is only solvable in exponential time.

$$P = \langle a, b, c, d, e, f:\ adf^{-1} = bed^{-1} = cfe^{-1} = 1 \rangle$$

$$adf^{-1} \cdot fe^{-1}c = ade^{-1}c$$

$$ad \cdot d^{-1}be \cdot e^{-1}c = abc$$

Since abc is a consequence of the defining relators of P, adding it as a "new" relator won't change the Cayley diagram and each generator will appear at least twice. The subscripted roots are:

$$a_1d_1f_1^{-1} \qquad b_1e_1d_2^{-1} \qquad c_1f_2e_2^{-1} \qquad a_2b_2c_2$$

The forest is:

a_1	d_1	f_1^{-1}	b_1^{-1}	a_1^{-1}	a_2	b_2^{-1}	d_1^{-1}
2	2	2	2	2	1	1	2
c_2^{-1}	b_1	e_2^{-1}	c_2	b_2	f_1	e_1	e_1^{-1}
1	2	1	1	1	2	2	2
f_2	a_2^{-1}	d_2^{-1}	e_2	d_2	c_1^{-1}	c_1	f_2^{-1}
1	1	1	1	1	2	2	1
a_1	d_1	f_1^{-1}	b_1^{-1}	a_1^{-1}	a_2	b_2^{-1}	d_1^{-1}

The paths above the horizontal bracket form a compatible system yielding the planar embedding of F(P, g) given below.

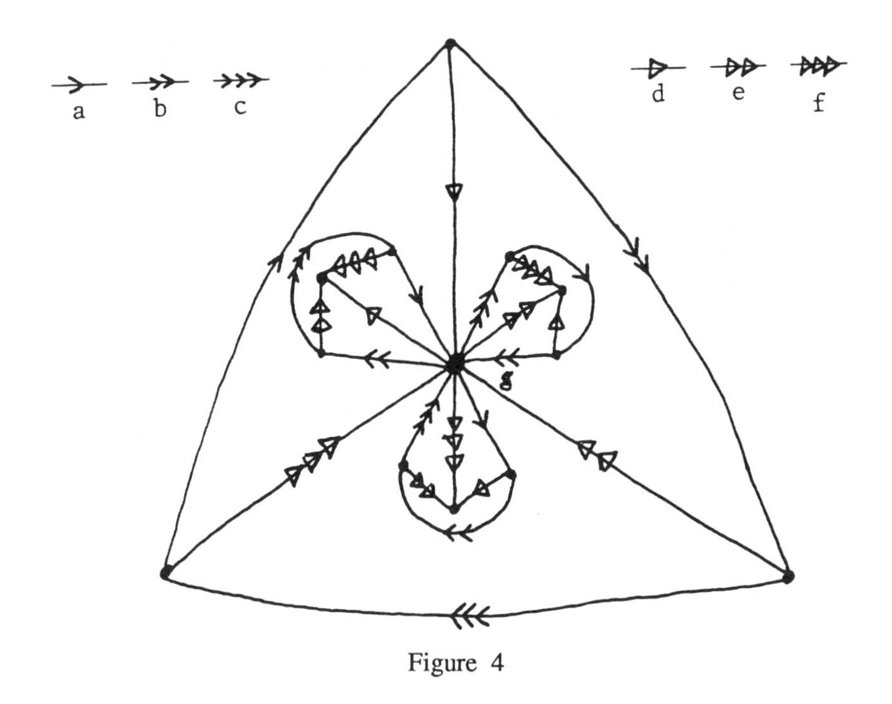

Figure 4

3. Compatible systems of embeddings.

We will now proceed from (local) planarity of F(P, g) to (global) planarity of C(P). We must devise a (finite) test to ensure that the planarity of an embedding of F(P, g) may be extended at each successive point saturation by F(P, g) to planar embeddings of the resulting regions of C(P).

Let us denote the sequence of successive edges about g, in a planar embedding of F(P, g) by s. We will not stipulate clockwise or counterclockwise, and we consider s as a cyclic word in the generators and inverses of generators of P, obviating the representation of cyclic permutations of s. We call the points in an embedding of F(P, g) whose sequences of edges are subsequences of s, *points of type s*. If there are points in our embedding of F(P, g) which are not of type s, we must find other new planar embeddings of F(P, g) such that their sequences are subsequences of the new sequences about g of these embeddings. Further, each "new" embedding must be inspected to discover if all points have sequences of edges about them which are of types already present. If not, we continue adding types (different embeddings of F(P, g)), if possible, until sequences exist about g to type all points present. Should there be no system of embeddings of F(P, g) for which each point in the embeddings is of one of the types of the embeddings, C(P) is not planar, although F(P, g) may be planar.

Should there exist a system of embedding for which each point is of the type of one of the embeddings, we must check each one to see if each point may be saturated using an embedding of F(P, g) of its type such that no types of embeddings conflict with each other on their intersections. If such is the case, we call the system of embeddings of F(P, g) *compatible.*

Theorem 2 *If a presentation P has a compatible system of planar embeddings of its fundamental region F(P, g), then its Cayley diagram may be embedded in the plane.*

Proof We simply note that at each stage of the F(P, g) construction of C(P), each embedding of F(P, g) used to saturate an unsaturated point p, fits into the disks about p without improper crossings, as guaranteed by our compatibility assumption.❑

In the two examples shown, each point in the embedding of F(P, g) is of the same type as the sequence about g. It is easy to check that each respective embedding forms a self–compatible system, and in this case the Cayley diagrams for the presentations have point–symmetric embeddings.

REFERENCES

[1] J. Edmonds, A combinatorial representation for polyhedral surfaces, *Notices Amer. Math. Soc.* 7 (1960) 646.

[2] H. Levinson, Planarity of Cayley diagrams: weak point–symmetry, *Proc. 10th S-E Conf. Combinatorics, Graph Theory & Computing* (1979) 679–697.

CRITICALLY N-CONNECTED DIGRAPHS

W. Mader
Universität Hannover
Institut für Mathematik
Federal Republic of Germany

ABSTRACT

A graph or digraph G is critically n-connected, if it is n-connected, but for every vertex x of G, $G - x$ is not n-connected. In 1972, Chartrand, Kaugars, and Lick proved in [1] that every critically n-connected, finite (undirected) graph has a vertex of degree at most $\frac{3n}{2} - 1$. Generalizing this result it was shown in [5] that every critically n-connected, noncomplete, finite graph has a fragment of order at most $\frac{n}{2}$, where a fragment of a graph G of connectivity number n is an induced subgraph F of G of order less than $|G| - n$ such that F has exactly n neighbours in G. In this paper we will study the corresponding problems for directed graphs. We shall survey the results on critically n-connected digraphs and shall point out some open questions. We shall much simplify some proofs for results in [13] and shall prove some new results. Furthermore, we give counterexamples to a conjecture in [14].

1. Introduction

First we need some definitions and notation. A digraph has neither loops nor multiple edges of the same direction. An edge (x, y) of a digraph D is called *symmetric*, if also the edge (y, x) from y to x belongs to D. A digraph without symmetric edges is called *antisymmetric*. For a digraph D, $\underline{D}$

denotes the multigraph arising from D by disregarding the orientation of the edges. The digraph $\overleftrightarrow{G}$ arises from the graph G by replacing every edge of G with a pair of oppositely directed edges; in particular, $\overleftrightarrow{K}_n$ denotes the complete digraph on n vertices. For a digraph, the term "connected" always means "strongly connected", and we use the terms "path", "circuit", and "component" always in the directed sense. A (simple) path P from vertex x_o to vertex x_n is written $x_o, x_1, ..., x_n$ and is called an x_o, x_n -path; for $0 \leq i \leq j \leq n, P[x_i, x_j]$ denotes the subpath of P from x_i to x_j. For $A, B \subseteq V(D)$, an *A,B - path* P is an a,b-path for certain (not necessarily distinct) $a \in A$ and $b \in B$ such that $V(P-a) \cap A = \emptyset$ and $V(P-b) \cap B = \emptyset$ holds. An x,y-path P and an x',y'-path P' are *openly disjoint*, if $P \neq P'$ holds and $z \in P \cap P'$ implies $z = x = x'$ or $z = y = y'$. A (forward) *chord* of a path $x_0, x_1, ..., x_n$ in a digraph D is an edge $(x_i.x_j)$ of D such that $j \geq i+2$ holds; if there is none, the path is called *chordless*. For $X \subseteq V(D)$, the subdigraph $D(X)$ induced by X is $D-(V(D)-X)$ and for a subdigraph $H \subseteq D$, set $D(H) := D(V(H))$. An *induced subdigraph* of D is a subdigraph of the form $D(X)$. A subdigraph H of a digraph D is called a *subcomponent of* D, if there is a component C of D so that $H = D(X)$ holds for an X maximal in $(\{Y \subsetneqq V(C) : Y \neq 0$ and $D(Y)$ connected$\}, \subseteq)$. Obviously, every vertex of a connected, finite digraph D of order at least 2 is contained in a subcomponent of D. For a subdigraph $F \subseteq D$ or $F \subseteq V(D)$, we define $N^+(F;D) := \{y \in V(D) : y \notin F$ and there is an $x \in F$ such that $(x,y) \in E(D)$ holds$\}$, and $N^-(F;D)$ is defined dually. For $x \in V(D)$, set $N^\epsilon(x;D) := N^\epsilon(\{x\});D)$ for $\epsilon = +,-$ and $N(x;D) := N^+(x;D) \cup N^-(x;D)$, and $d^\epsilon(x;D) := |N^\epsilon(x;D)|$ denotes for $\epsilon = +$ the *outdegree* and for $\epsilon = -$ the *indegree of* x in D. For a digraph D, $D = \emptyset$ means $V(D) = \emptyset$ and $x \in D$ means $x \in V(D)$. For graphs or digraphs, $G_1 \cong G_2$ means that G_1 and G_2 are isomorphic. $\overline{K}_n$ denotes the graph of order n without edges and let L_m be an undirected circuit of length $m \geq 3$. For disjoint graphs G and H, the sum $G+H$ arises from

$G \cup H$ by adding all edges between G and H. If G and H are not disjoint, let $G + H$ be any graph isomorphic to $G' + H'$, where G' and H' are disjoint copies of G and H, respectively. Instead of $G_1 + G_2 + ... + G_k$ with $G_1 \cong G_2 \cong ... \cong G_k$, we write kG_1.

A digraph D is *n-connected* for a non-negative integer n, if it has more than n vertices, and for every $V' \subseteq V(D)$ with $|V'| < n, D - V'$ is connected. By Menger's Theorem, D is n-connected, iff for every pair of vertices x, y in D, there are n openly disjoint x,y-paths. The *connectivity number* $\kappa(D)$ of a digraph D is the largest n for which D is n-connected. (For the infinite digraphs we shall consider, such a maximal n will always exist.) In particular, $\kappa(\overleftrightarrow{K}_{n+1}) = n$ holds. A digraph D is *k-critically n-connected* for integers $0 \leq k \leq n$ (or *(n,k)-critical*), if $\kappa(D - V') = n - |V'|$ holds for every $V' \subseteq V(D)$ with $|V'| \leq k$. In particular, for an (n,k)-critical digraph D, we have $\kappa(D) = n$. A digraph D is called *k-critical*, if it is $(\kappa(D), k)$-critical, it is called *critically n-connected*, if it is $(n,1)$-critical, and it is called *critically connected*, if it is $(1,1)$-critical. A *fragment of sign* $\epsilon \in \{+,-\}$ of a digraph D of (finite) connectivity number n is an induced subdigraph $F \neq \emptyset$ of D such that $|N^{\epsilon}(F;D)| = n$ and $V(F) \cup N^{\epsilon}(F;D) \subsetneq V(D)$ holds. A *positive fragment* is one of sign $+$ and a *negative fragment* is one of sign -, and a *fragment* is a positive or a negative fragment. If F is a positive fragment of D, then $\overline{F} := D - V(F) \cup N^{+}(F;D))$ is a negative fragment of D. Of course, every non-complete digraph of finite connectivity number has at least one positive fragment and at least one negative fragment.

Let $\mathbb{N}$ and $\mathbb{N}_n$ denote the set of positive integers and $\{m \in \mathbb{N} : m \leq n\}$, respectively. $\mathbb{Z}_n$ denotes the integers modulo n. For a set M and $n \in \mathbb{N}$, define $(\binom{M}{n}) := \{' \subseteq M : |M'| = n\}$. Let us consider now some examples for critically n-connected digraphs.

2. Examples

(1) Let $m \geq n$ be positive integers and let $K_{m,m}$ be the complete bipartite graph with bipartition A, B into the independent vertex sets A

and B. Choose an n-factor F of $K_{m,m}$. The digraph $\vec{F}$ may arise from F by orienting all edges of F from A to B. Then the digraph $D := (A \cup B, E(\vec{F}) \cup (B \times A))$ is critically n-connected. Since $K_{m,m}$ is m-connected, we see that *for every* $n \geq 1$, *there are critically n-connected digraphs D so that the connectivity number of* $\underline{D}$ *is arbitrarily large.* In particular, there is no value $f(n)$ such that for all critically n-connected, finite digraphs D, $\min_{x \in V(D)} (d^+(x;D) + d^-(x;D)) \leq f(n)$ holds.

(2) Let $n \geq 1$ and $m \geq 4$ be integers and choose $k \in \{3, ..., m-1\}$. Consider disjoint copies $D_1, ..., D_m$ of $\overleftrightarrow{K}_n$. The digraph D' arises from $\bigcup_{i \in \mathbb{Z}_m} D_i$ by adding all edges $\bigcup_{i \in \mathbb{Z}_m} V(D_i) \times V(D_{i+1})$. Then D' is critically n- connected. Every D_i is a positive and a negative fragment of D' is critically n- connected. Every D_i is a positive and a negative fragment of D' and every $x \in D$ has $d^+(x;D') = d^-(x;D') = 2n-1$. If we add the further edges

$$\bigcup_{\substack{i \in \mathbb{Z}_m \\ i \neq 1,k}} V(D_{i+1}) \times V(D_i) \text{ to } D',$$

getting D in this way, D remains critically n-connected. Now, D has exactly two positive fragments of order at most n (namely D_2 and D_{k+1}) and exactly two negative fragments of order at most n (namely D_1 and D_k), and these fragments have order n. In general, therefore, *in a critically n-connected, finite digraph we cannot find a vertex of outdegree or indegree less than* $2n-1$ *nor a fragment of order less than* n.

(3) Let $m \geq n \geq 2$ be integers. Consider a complete digraph $\overleftrightarrow{K}_m$ and subsets $S_1, ..., S_k \in \binom{V(\overleftrightarrow{K}_m)}{n}$ such that $V(\overleftrightarrow{K}_m) = \bigcup_{i=1}^{k} S_i$ holds. Let $C_1, ..., C_k$ be disjoint copies of $\overleftrightarrow{K}_n$ with $C_i \cap \overleftrightarrow{K}_m = \emptyset$ for $i \in \mathbb{N}_k$ and let f_i be a bijection of S_i to $V(C_i)$ for $i \in \mathbb{N}_k$. The digraph D may arise from $\overleftrightarrow{K}_m \cup \bigcup_{i \in \mathbb{N}_k} C_i$ by adding all edges $\bigcup_{i \in \mathbb{N}_k} \{(x, f_i(x)) : x \in S_i\}$

and $\bigcup_{i \in \mathbb{N}_k} V(C_i) \times V(\overleftrightarrow{K}_m)$. The digraph D is critically n-connected, but $d^+(x; D) \geq m$ for all $x \in D$. Hence, *for every* $n \geq 2$, *there are critically n-connected, finite digraphs* D *with arbitrarily large* $\min_{x \in D} d^+(x; D)$.

It is not an imperfection of our construction that we had to assume $n \geq 2$ in example 3, because it was shown in [13] that every critically connected, finite digraph has two vertices of outdegree 1. The proof given in [13] has become too complicated, because it took its rise as a by-product of the proof of the general case (see [14]). Se we shall give a simple proof here.

Theorem 1 Let C be a subcomponent of a critically connected digraph D of order at least 3. Then $m := |D - V(C)$ is finite and at least 2 and $D - V(C)$ has a chordless, hamiltonian path $P : x_1, x_2, ..., x_m$ such that $N^+(C; D) = \{x_1\}$ and $N^-(C; D) = \{x_m\}$ holds.

Proof Since D is critically connected, but $\kappa(C) \geq 1$ or $|C| = 1$ holds, we get $m \geq 2$. Of course, $N^\epsilon := N^\epsilon(C; D) \neq \emptyset$ holds for $\epsilon \in \{+, -\}$ and there is an N^+, N^- -path $P : x_1, \cdots, x_k$ in $D - V(C)$. Since $D(C \cup P)$ is connected and C is a subcomponent of D, P must be a hamiltonian path of $D - V(C)$, hence $m = k \in \mathbb{N}$. Since this is true for every N^+, N^- -path in $D - V(C)$, we see that $N^+ = \{x_1\}$ and $N^- = \{x_m\}$ holds and that P has no chord.

Remark Since, by Theorem 1, a subcomponent of a critically connected digraph D of order at least 3 is a positive (and a negative) connected fragment of D, Theorem 1 implies immediately Theorem 1 of [13].

Corollary 1 Every critically connected, finite digraph D of order at least 4 has disjoint, chordless paths P_1 and P_2, say, from x_i^+ to x_i^- for $i = 1, 2$ such that $x_i^+ \neq x_i^-$ and $N^\epsilon(D - V(P_i); D) = \{x_i^\epsilon\}$ holds for $i = 1, 2$ and $\epsilon = +, -$.

Proof We may assume that D is not a circuit. Then there is a subcomponent C of D with $|C| \geq 2$. By Theorem 1, there is an x_2^+, x_1^- -path P_1 with

$V(P_1) = V(D) - V(C)$ which has the properties described in Corollary 1. If there is a subcomponent of D containing P_1, then Theorem 1 provides an x_2^+, x_2^- -path P_2 disjoint from P_1 with the desired properties. So we may assume that every path P from $N^+(x_1^-; D) \cap C$ to $N^-(x_1^+; D) \cap C$ contains $V(C)$. There is such a path P_2 in C, say, an x_2^+, x_2^- -path. Then as above, P_2 has no chord and $N^\epsilon(P_1; D) = \{x_2^\epsilon\}$ holds for $\epsilon = +, -$, since $N^\epsilon(C; D) = \{x_1^\epsilon\}$ is true for $\epsilon = -, +$. Since $|C| \geq 2$, we have $x_2^+ \neq x_2^-$, and Corollary 1 is proved.

Since $d^+(x_i^+; D) = d^-(x_i^-; D) = 1$ holds for the endvertices of the x_i^+, x_i^- -path in Corollary 1, we get immediately as a slight improvement of Corollary 2 in [13].

Corollary 2 Every critically connected, finite digraph D of order at least 4 has 4 distinct vertices x_1, x_2, y_1, y_2 such that $d^+(x_i; D) = d^-(y_i; D) = 1$ holds for $i = 1, 2$.

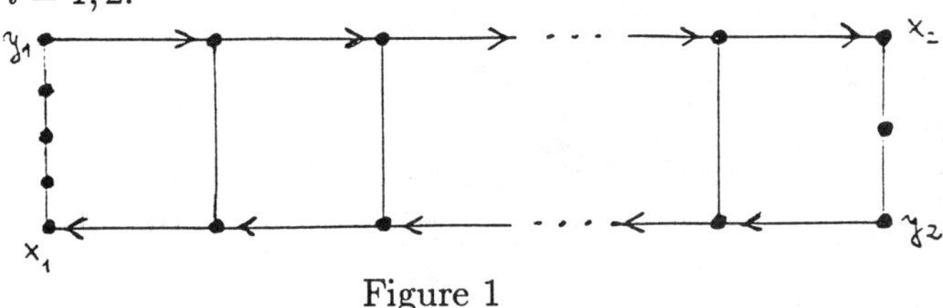

Figure 1

The second construction in example 2 for $n = 1$ shows that *for every integer* $m \geq 4$, *there are critically connected digraphs of order* m *containing exactly two vertices of outdegree* 1 *and exactly two vertices of indegree 1.* For $m = 3$ this is not true, because the circuit of length 3 is the only critically connected digraph of order 3.) Figure 1 displays a further critically connected digraph D with exactly two vertices x_1, x_2 of outdegree 1 and exactly two vertices y_1, y_2 of indegree 1 such that the vertex set $\{x_1, x_2, y_1 y_2\}$ is independent in D. (In figure 1, a pair of symmetric edges is drawn as an undirected edge.) It is not so by chance that D contains symmetric edges. For it is shown in [13] and easily derived from Corollary 1 that *every antisymmetric, critically connected, finite digraph* D *with exactly two vertices of outdegree 1 must contain disjoint edges* (x_1, y_1) *and* (x_2, y_2) *such that* $d^+(x_i; D) = d^-(y_i; D) = 1$ *holds for* $i = 1, 2$. It is even possible to characterize all antisymmetric, critically connected, finite digraphs with exactly two vertices of outdegree 1. They are built up from almost 20 blocks in an analogous way as shown in Figure 2.

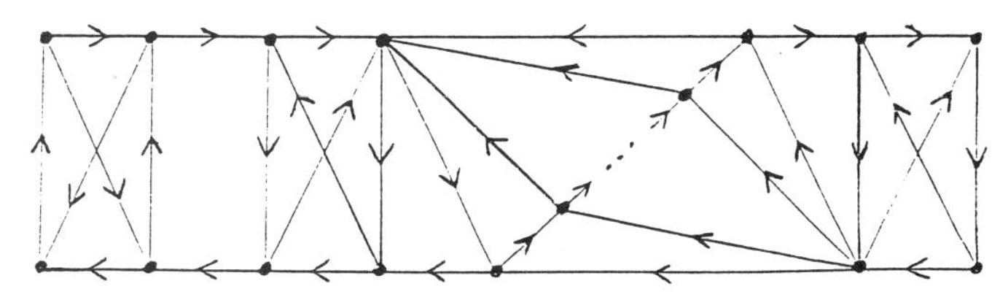

Figure 2

Let us turn now to critically n-connected digraphs for $n \geq 2$. Let us first have a glance at the undirected case. Chartrand et al. showed in [1] that every critically n-connected, non-complete, finite graph has a vertex of degree at most $\frac{3n}{2} - 1$. It was later proven in [5] that such a graph has a fragment F of order at most $\frac{n}{2}$. Of course, every vertex of such a fragment has degree at most $\frac{3n}{2} - 1$. As example 2 shows, in general, we cannot find a fragment of order less than n in a critically n-connected digraph. Example 3 shows that the minimal order of a positive fragment in a critically n-connected, finite graph is not bounded above by a function of n, because every vertex x of a positive fragment F of D satisfies $d^+(x; D) \leq |F| - 1 + \kappa(D)$. But there is a "small" positive or negative fragment in a critically n-connected, non- complete, finite digraph.

Theorem 2 [14]. Every critically n-connected, non-complete, finite digraph has a fragment of order at most n.

This had been conjectured by Y. O. Hamidoune in [2] and was proved for digraphs with vertex transitive automorphism group (which digraphs, of course, are 1-critical) in [4].

One of the main steps in the proof of Theorem 2 is the following lemma. For this we need further definitions. A set $\mathcal{F}$ of positive fragments of a digraph is called *non-crossing*, if for all $F, F' \in \mathcal{F}, F \cap F' \neq \emptyset$ implies $F \subseteq F'$ or $F' \subseteq F$. If F is a positive fragment and G is any fragment of D, then $F \leq G$ means $F \subseteq G$, if G is positive, and $V(F) \subseteq V(G) \cup N^-(G; D)$, if G is negative.

Lemma 1 [14]. Let G be a fragment of a finite digraph D of connectivity number N and let $\mathcal{F}$ be a non-empty, non-crossing set of positive fragments F' of D with $F' \leq G$. Define $\mathcal{F}_o := \{F \in \mathcal{F} : F \text{ minimal in } (\mathcal{F}, \subseteq)\}$ and assume

$$V(G) \cap \bigcup_{F \in \mathcal{F}_o} V(F) \subseteq \bigcup_{F \in \mathcal{F}} N^+(F; D).$$

Then $|\overline{G}| < n$ holds or there is an $F \in \mathcal{F}$ with $|F| \leq n$.

It is still more intricate to get such a system $\mathcal{F}$ we need for the application of lemma 1. Theorem 2 immediately implies

Corollary 3 [14. Every critically n-connected, finite digraph D contains a vertex x with $d^+(x; D) < 2n$ or $d^-(x; D) < 2n$.

This was also conjectured in [2]. Example 2 shows that this upper bound $2n - 1$ for $\min_{x \in D} \{d^+(x; D), d^-(x; D)\}$ is best possible, and example 3 shows that for $n \geq 2$, there is not always a vertex x in a critically n-connected, finite digraph D such that $d^+(x; D) < 2n$ holds.

For antisymmetric digraphs we get a better upper bound for the minimum degree, because an antisymmetric digraph F contains a vertex x with $d^+(x; F) \leq \frac{1}{|F|}\binom{|F|}{2}$ and a vertex y with $d^-(y; F) \leq \frac{|F|-1}{2}$. Hence Theorem 2 implies

Corollary 4 [14]. Every antisymmetric, critically n-connected, finite digraph D contains a vertex x with $d^+(x; D) \leq \lfloor \frac{3n-1}{2} \rfloor$ or $d^-(x; D) \leq \lfloor \frac{3n-1}{2}$.

The following example shows that this bound is also best possible.

Example (4). For $n \in \mathbb{N}$ and $m \geq 3$, define $H_n := (\mathbb{Z}'_n\{(x, x+i) : x \in \mathbb{Z}_n$ and $i = 1, ..., \lfloor \frac{n-1}{2} \rfloor\})$ and consider m disjoint copies $D_1, ..., D_m$ of H_n. The digraph $D := (\bigcup_{i \in \mathbb{Z}_m} V(D_i), \bigcup_{i \in \mathbb{Z}_m} E(D_i) \cup \bigcup_{i \in \mathbb{Z}_m} V(D_i) \times V(D_{i+1}))$ is critically n-connected and antisymmetric, and for every $x \in D, d^+(x;D) = d^-(x;D) = \lfloor \frac{3n-1}{2} \rfloor$ holds.

A digraph D is *n-edge-connected*, if $D - E'$ is connected for every $E' \subseteq E(D)$ such that $|E'| < n$, and it is called *minimally n-edge-connected*, if it is n-edge-connected, but for every $e \in E(D), D - e$ is not n-edge-connected, finite digraph D contains a vertex x such that $d^+(x;D) = d^-(x;D) = n$ holds. This result is now easily derived from Theorem 2 by the following construction.

Let D be a minimally n-edge-connected, finite digraph. We assign to every $x \in D$ a complete digraph $K(x)$ of order $d^+(x;D) + d^-(x;D)$ such that $K(x) \cap K(y) = \emptyset$ for $x \neq y$. We replace now every $x \in D$ with $K(x)$ and every $(x,y) \in E(D)$ with an edge from $K(x)$ to $K(y)$ in such a way that all edges corresponding to the edges of D are disjoint. The resulting digraph is critically n-connected, and an application of Corollary 3 easily provides

Corollary 5 [7, 14]. Every minimally n-edge-connected, finite digraph D has at least two vertices x such that $d^+(x;D) = d^-(x;D) = n$ holds.

It was conjectured in [11] that Corollary 5 is true also for minimally n-(vertex-) connected digraphs. This conjecture seems not derivable from Theorem 2 and remains open. - As in the undirected case, Theorem 2 and Corollaries 1 to 5 have no counterparts for infinite digraphs.

It is probable that in a critically n-connected, finite digraph, not only one outdegree or indegree is less than $2n$. But let us first consider the undirected case. It was proved in [3] that every critically n- connected, finite graph $(n \geq 2)$ has at least two vertices of degree at most $\frac{3n}{2} - 1$ and that for $n \geq 3$, there are critically n-connected, finite graphs of arbitrarily large order having exactly two such vertices. Generalizing this result, it was proved in [10] that every critically n-connected, non-complete, finite digraph has two disjoint fragments of order at most $\frac{n}{2}$.

Let us now turn to the directed case. The examples given in [14] suggested the conjecture that every critically n-connected, finite digraph D has at least four distinct pairs $(x, \epsilon) \in V(D) \times \{+, -\}$ with $d^{\epsilon}(x; D) < 2n$. This is true for $n = 1$, as stated in Corollary 2, but for $n \geq 2$, this conjecture is wrong, as the following examples show.

Examples (5) For integers $n \geq 2$ and $m \geq 3$, let $G_o, G_1, ..., G_m$ be disjoint graphs such that $G_o = (\{a\}, \emptyset)$ and $G_i \cong K_n$ for $i \in \mathbb{N}_m$ hold. Define $G := \bigcup_{i=1}^{m}(G_{i-1} + G_i)$ and choose any $b \in G_m$. The digraph D may arise from $\overleftrightarrow{G}$ by deleting the edges $\{(x, b) : x \in G_{m-1}\}$ and adding the edges $\{(a, x) : x \in \bigcup_{i=2}^{m} G_i\}$. The digraph D is n-connected. Take a copy $\overline{D}$ of D which is disjoint from D. The vertex $\overline{x} \in \overline{D}$ and the graph $\overline{G}_i$ may correspond to $x \in D$ and G_i, respectively. The digraph D_1 may arise from $D \cup \overline{D}$ by adding all edges $V(G_m) \times V(\overline{G}_m)$ and $V(\overline{G}_m) \times V(G_m)$. For $n = 2$ and $m = 3$, this construction is illustrated in Figure 3. *The digraph* D_1 *is critically n-connected. It has no vertex of outdegree less than* $2n$ *and exactly two vertices of indegree less than* $2n$, namely the vertices a and $\overline{a}$.

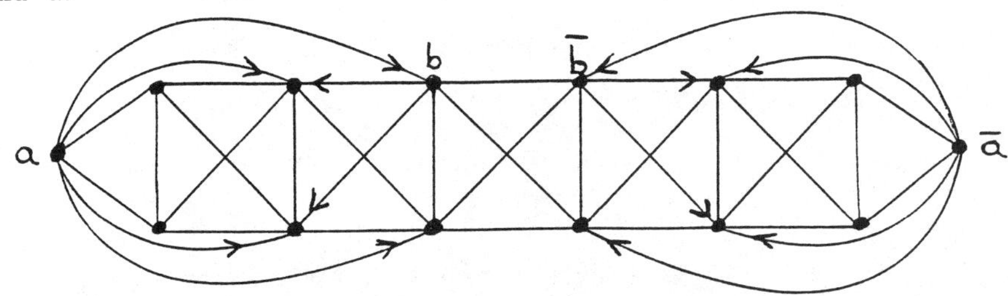

Figure 3

If we take in the above construction a disjoint copy of the dual D' of D instead of $\overline{D}$, we get a critically n-connected digraph D_2 with exactly two outdegrees or indegrees less than $2n$, one outdegree n and one indegree n.

(6) For integers $n \geq 2$ and $m \geq 2$, let $D_o, D_1, ..., D_m$ be disjoint digraphs such that $D_o = (\{a\}, \emptyset), D_1 \cong D_m \cong \overleftrightarrow{K}_{3n-1}$, and $D_i \cong \overleftrightarrow{K}_{4n-2}$ for $i = 2, .., m-1$ hold. For $i \in \mathbb{N}_m$, let A_i, A'_i, B_i, B'_i be a partition of $V(D_i)$ such that for $i \in \mathbb{N}_m$, $|A_i| = |A'_i| = n$, for $i \in \mathbb{N}_{m-1}, |B_i| = n - 1$, and $B_m = \emptyset$ hold. (Hence, $B'_i = \emptyset$ and for $i = 2, ..., m, |B'_i| = n - 1$ holds.)

Define $A_o = A'_o = \{a\}$. The digraph D may arise from $\bigcup_{i=0}^{m} D_i$ by adding the edges $\bigcup_{i=1}^{m-1} B_i \times B'_{i+1}$ and $\bigcup_{i \in \mathbb{Z}_{m+1}} A_{i+1} \times A'_i$. For $n = 2$ and $m = 3$, this digraph D is displayed in Figure 4, where the edges of the complete digraphs D_1, D_2, D_3 are omitted. *The digraph D is critically n-connected, but every vertex of D has outdegree and indegree at least $2n$ with the only exception of the vertex a having $d^+(a; D) = d^-(a; D) = n$.*

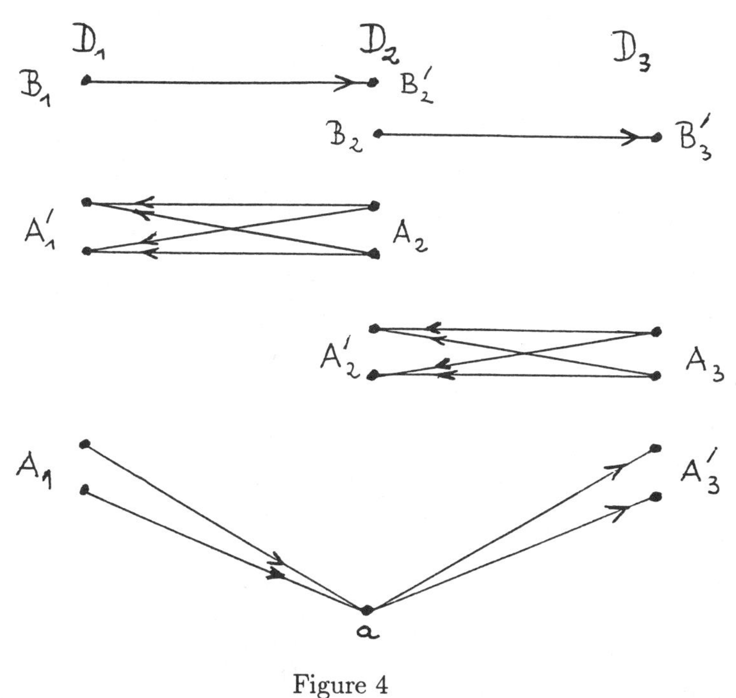

Figure 4

This example shows that in a critically n-connected, finite digraph D, there are not always two vertices x such that $\min\{d^+(x; D), d^-(x; D)\} < 2n$ holds. But the following conjecture should be true.

Conjecture 1 For every critically n-connected, finite digraph D, there are at least two pairs $(x, \epsilon) \in V(D) \times \{+, -\}$, such that $d^\epsilon(x; D) < 2n$ holds.

Example 6 also shows that in a critically n-connected, non-complete, finite digraph, one cannot always find two disjoint fragments of order at most n. The following conjecture, obviously, is stronger than conjecture 1 and, by example

5, in a sense, best possible.

Conjecture 2 If a critically n-connected, non-complete, finite digraph does not contain a negative fragment of order at most n, then it has two disjoint positive fragments of order at most n.

It was shown in [15] that there is no (n,n)-critical, finite graph G but K_{n+1}. This follows by considering a spanning tree T of G and deleting $\{x_1, ..., x_n\}$ from G, where x_i is an endvertex of $T - \{x_1, ..., x_{i-1}\}$. It was proved in [8] that every 2-critical graph is finite, but in [12], for every $n \in \mathbb{N}$, an example of an infinite, (n,n)-critical digraph was constructed. (In example 7 we shall give further constructions for infinite, (n,n)-critical digraphs.) Therefore, the question arises, if there are finite, (n,n)-critical digraphs for $n \geq 2$ besides $\overleftrightarrow{K}_{n+1}$. Applying Corollary 2, this question was answered in the negative in [13].

Theorem 3 [13] For $n \geq 2$, there is no finite, (n,n)-critical digraph besides $\overleftrightarrow{K}_{n+1}$.

The real reason for this fact had not become quite clear in the proof given in [13]. So we shall give here a proof more in the spirit of the proof in the undirected case, which makes this result intuitively clear. We need the following

Lemma 2 Let z be a vertex of a 2-connected digraph D and suppose there are disjoint subdigraphs H_1 and H_2 of $D - z$ such that for $i = 1, 2$, the digraph H_i contains vertices $x_i^+ \neq x_i^-$ with the following property: $N^\epsilon((D - z) - V(H_i); D - z) = \{x_i^\epsilon\}$ and $z \in N^\epsilon(x_i^\epsilon; D)$ for $\epsilon = +, -$.

Then $D - \{x_1^+, x_2^-\}$ is connected.

Proof Consider any $x \in D - \{x_1^+, x_2^-, z\}$. We will show that there is an x, z-path in $D' := D - \{x_1^+, x_2^-\}$. Since D is 2-connected, there are openly disjoint x, z-paths P_1 and P_2 in D. We may assume that $x_1^+ \in P_1$ and $x_2^- \in P_2$ holds. If $x \notin H_2$, then $x_2^- \in P_2$ implies $x_2^+ \in P_2$, since $N^+(D - z - V(H_2); D - z) = \{x_2^+\}$. But then we get an x, z-path $P_2[x, x_2^+] \cup (x_2^+, z)$ in D', since $x_2^+ \neq x_2^-$ and $z \in N^+(x_2^+; D)$ holds. So we may assume $x \in H_2$. Then $x_1^+ \in P_1$ implies $x_2^- \in P_{\underline{1}}$, since $x_1^+ \notin H_2$ and $N^-(D - z - N(H_2); D - z) = \{x_2^-\}$ holds. But $x_2 \in P_1$ contradicts $V(P_1) \cap V(P_2) = \{x, z\}$. In an analogous way, we see that there is a z, x-path in D'. Hence, D' is connected.

Proof of Theorem 3 It is enough to prove Theorem 3 for $n = 2$, since an (n,n)-critical digraph provides a (2,3)-critical one by delcting any n - 2 vertices. Let us suppose there is a (2,2)-critical, finite digraph D with $|D| \geq 4$. Since $D - x$ is critically connected for every $x \in D$ and since the only critically connected digraph of order 3 is a circuit of length 3, we must have $|D| \geq 5$. Choose $z \in D$. By Corollary 1, $D - z$ has two disjoint x_i^+, x_i^--paths P_i of length at least 1 such that $N^\epsilon(D - z - V(P_i); D - z) = \{x_i^\epsilon\}$ and $d^\epsilon(x_i^\epsilon; D - z) = 1$ holds for $\epsilon = +, -$ and $i = 1, 2$. Since D is 2-connected, $d_i^\epsilon; D - z) = 1$ holds for $\epsilon = +, -$ and $i = 1, 2$. Since D is 2-connected, $d^\epsilon(x_i^\epsilon; D - z) = 1$ implies $z \in N^\epsilon(x_i^\epsilon; D)$ for $\epsilon = +, -$ and $i = 1, 2$. So we can apply lemma 2 and get a contradiction.

Instead of deleting vertices, we may delete edges or vertices and edges. The case of deleting at least two edges is trivial: There are no graphs or digraphs besides L_m such that deleting any two edges, the connectivity number decreases by 2. (Consider any edge $e = [x, y]$ or $e = (x, y)$ of G and a system $P_1, ..., P_{\kappa(G)}$ of openly disjoint x, y-paths in G. Then for every $e' \in E(G) - \bigcup_{i=1}^{\kappa(G)} E(P_i), \kappa(G - \{e, e'\}) \geq \kappa(G) - 1$ holds). So only the case of mixed deletion remains. A graph or digraph G is called *vertex-edge-critical* or *v-e-critical*, if $\kappa(G) \geq 2$ and for all $v \in V(G)$ and $e \in E(G - v), \kappa(G - \{v, e\}) = \kappa(G) - 2$ holds. It is easy to see (in the same way as just before) that there are no infinite v-e-critical graphs or digraphs. The finite, v-e-critical graphs have been characterized in [6]: *The v-e-critical graphs are exactly the graphs* $G_{m,k,1} := k\overline{K}_m + 1L_{m+2}$, *where* $m \geq 1, k, 1$ *are non-negative integers such that* $\kappa(G_{m,k,1})^{\geq 2}$ *holds.* (The proof given in [6] was not so easy, but using Theorem 4 in [9], we get a straightforward proof.)

Of course, the digraphs $\overleftrightarrow{G}_{m,k,1}$ are v-e-critical. So the question arises, if there are still further ones. In an analogous way as in (a_2) on page 277 in [6], one can see that every vertex of a v-e-critical digraph of connectivity number n has outdegree and indegree equal to n. Using results of [11], one can show that every v-e-critical digraph must have a symmetric edge, but I do not know, if every edge must be symmetric. But this is the case for $n = 2$, as seen in the following way. Let D be a v-e- critical digraph of connectivity number 2. For $x \in D, D - x$ is minimally connected and, therefore, has two vertices of indegree and outdegree 1, by Corollary 5. Hence, the 4 edges incident to x are symmetric and $D = \overleftrightarrow{L}_{|D|}$ holds.

For every cardinal $\aleph_\alpha$ and every $n \in \mathbb{N}$, an (n,n)-critical digraph of order $\aleph_\alpha$ was constructed in [12]. There is not much hope to able to describe the structure of all (n,n)-critical digraphs for $n \geq 2$ precisely, but we can say something about the degrees in such digraphs. The examples of (n,n)-critical digraphs of order $\aleph_\alpha$ given in [12] have in every vertex outdegree equal to $\aleph_\alpha$ and in some vertices indegree equal to $\aleph_\alpha$. The next result shows that for $n \geq 3$, it was not so by chance that every vertex had outdegree equal to $\aleph_\alpha$.

Proposition 1 Let D be an (n,n)-critical digraph of order $\aleph_\alpha$ for an integer $n \geq 2$. Then for every $x \in V(D)$, $d^+(x; D) + d^-(x; D = \aleph_\alpha$ holds. For $n \geq 3$, every vertex of D has outdegree $\aleph_\alpha$ or every vertex of D has indegree $\aleph_\alpha$.

Proof For the first and for the second assertion, it is enough to consider the case $n = 2$ and $n = 3$, respectively. Choose $n \in \{2, 3\}$ and let D be an n-connected digraph of order $\aleph_\alpha$ which contains vertices x,y such that $^-(x; D) < \aleph_\alpha$ and $d^+(y; D) < \aleph_\alpha$ holds, where x=y for n = 2 and $x \neq y$ for n = 3 is assumed. For every pair of vertices $a \in N^-(x; D)$ and $b \in N^+(y; D)$, there is an a,b-path $P_{a,b}$ in $D - \{x, y\}$, since $D - \{x, y\}$ is connected. Since $V' := \bigcup_{-} V(P_{a,b})$ has less than $\aleph_\alpha$ elements,

$$a \in N^-(x; D)$$

$$b \in N^+(y; D)$$

there is a $z \in V(D)-(V' \cup \{x,y\})$. If $D' := D-\{x,y,z\}$ were not connected, there were a partition $A \neq \emptyset, B \neq \emptyset$ of $V(D')$ such that $N^+(A; D') = \emptyset$. Then $N^+(A; D) = \{x, y, x\} = N^-(B; D)$ holds, since D is n-connected. But this implies $N^-(x; D) \cap A \neq \emptyset$, say $a \in N^-(x; D \cap A$, and $N^+(y; D) \cap B \neq \emptyset$, say $b \in N^+(y; D) \cap B$. But $P_{a,b} \subseteq D'$ holds, contradicting $N^+(A; D') = \emptyset$. Hence, D' is connected and D is not (n,n)-critical. From this, Proposition 1 follows.

Obviously, Proposition 1 is not true for $n = 1$. The examples given in [12] show that not at all every vertex of an infinite, (n,n)-critical digraph must have infinite out-degree and infinite indegree. *For every* $n \geq 2$, *we construct now even an example of an infinite, (n,n)-critical digraph where every vertex has indegree n.* (Of course, such a digraph cannot exist for $n = 1$.) Such a digraph must be countable and have infinite outdegree in every vertex by Proposition 1. Subsequently, we shall give an example which shows that there is no integer $f(n)$ such that $\min_{z \in D}\{d^+(z; D), d^-(z; D)\} \leq f(n)$ holds for every countable, (n,n)-critical digraph D.

Example 7 Let n be an integer at least 2 and set $C_o := \{z \in \mathbb{Z} : z \equiv 0 (\text{mod } n+1)\}$. First we consider the digraph $D_o := (\mathbb{Z}, \{(z, z+i) : z \in \mathbb{Z}$ and $i \in \mathbb{N}_{n-1}\} \cup \{(z, z+n+1) : z \in \mathbb{Z} - C_o\})$. The digraph D_o has exactly one hamiltonian path and for every $T \subseteq \mathbb{Z}$ with $|T| \leq n-2, D_o - T$ is still hamiltonian. If $D_o - T$ is not hamiltonian for a $T \in (\mathbb{Z}_{n-1})$, then T is an interval of $\mathbb{Z}$, say, $T = \{z_o + i : i \in \mathbb{N}_{n-1}\}$, and $D_o - (T \cup \{z_o\})$ or $D_o - (T \cup \{z_o + n\})$ is hamiltonian. Let $C_1, C_2, \ldots$ be a partition of C_o such that for every $i \in \mathbb{N}, C_i$ is not bounded below, and let $z_o := \mathbb{Z}, z_1, z_2, \ldots$ be disjoint countable sets. For $k \in \mathbb{N}$, let f_k be a bijection of $((\bigcup_{i<k} z_i)_n)$ onto z_k. For $i \in \mathbb{N}$, we define now inductively digraphs

$$D_i := (V(D_{i-1}) \cup z_i, E(D_{i-1}) \cup \bigcup_{M \in \binom{V(D_{i-1})}{n}} \{(x, f_i(M) : x \in M\}).$$

In $\tilde{D} := \bigcup_{i \in \mathbb{N}} D_i$, the vertices of C_o have indegree $n-1$, but all other vertices have indegree n. For every $x \in C_o$, we define exactly one further incoming edge in such a way that the arising digraph D becomes n-connected. For $x \in V' := \bigcup_{i \in \mathbb{N}} z_i$, define $B(x) := \{z \in \mathbb{Z} :$ there is a z,x-path in $\tilde{D} - (\mathbb{Z}^{i \in \mathbb{N}}\{z\})\}$. For all $x \in V', B(x)$ is finite and $|B(x)| \geq n$ holds. Choose $i \in \mathbb{N}$. We define now a bijection $h_i : Z_i \to C_i$. Write Z_i in a sequence $z_1, z_2, \ldots$. Since $B(z_1)$ is finite and C_i is not bounded below, $\{x \in C_i : x < \min B(z_1)\}$ is not empty and has a maximum which we denote by $h_i(x_1)$. Assume $h_i(z_i), \ldots, h_i(z_m)$ are defined. Then define $h_i(z_{m+1}) := \max\{x \in C_i - \{h_i(z_1), \ldots, h_i(z_m)\} : x < \min B(z_{m+1})\}$. Of course, $h_i : Z_i \to C_i$ is injective. But h_i is also subjective, because for every $x \in C_i$, min $B(f_i(\{x+1, \ldots, x+n\}) = x+1$ holds. Define $E_i := \{(z, h_i(z)) : z \in Z_i\}$ and $D := \tilde{D} \bigcup_{i \in \mathbb{N}} E_i$. Then $d^-(z; D) = n$ for every $z \in D$, and D is an antisymmetric digraph.

We shall prove now that D is (n,n)-critical. Since for every $T \in \binom{V(D)}{n}$, $D - T$ has a vertex of indegree 0, it is enough to show that D is n-connected. Choose $T \in \binom{V(D)}{n-1}$. We will show that $D' := D - T$ is connected. First we prove

(α) All vertices of $D_o - T$ belong to the same component of D'.

Proof of (α) Consider $x' < x$ in $D_o - T$. Since $N^+(x; D) \cap z_1$ is infinite and for all $y \in N^+(x; D) \cap Z_1, h_1(y) < x$ holds, there is a path (of length 2) in D' from x to a vertex $x'' < x'$ in $D_o - T$. This implies (α), if $D_o - T$ is hamiltonian. Suppose that $D_o - T$ is not hamiltonian. Then $T \subseteq \mathbb{Z}$ and T is an interval, say $T = \{z_o + i : i \in \mathbb{N}_{n-1}\}$, and $D_o - (T \cup \{z_o\})$ or $D - (T \cup \{z_o + n\})$ has a hamiltonian path P. Since P belongs again to the same component of D', it suffices to show that there is a $z_{o'}(z_o+n)$-path in D'. If z_o is not on P, then $(z_o+n) \in P$ and there is a path from z_o to an $x < z_o$ in D', as we have seen above. But $x \in P$ holds and, therefore, a $z_{o'}(z_o+n)$ - path exists in D'. Assume z_o+n is not on P. Then z_o is on P. If $(z_o-1, z_o+n) \in E(D_o)$ holds, then $D_o-(T\cup\{z_o\})$ is hamiltonian and we return to the case just considered. Hence we may assume $(z_o-1, z_o+n) \notin E(D_o)$, i.e. $z_o+n \in C_o$. So there is an $i \in \mathbb{N}$ with $z_o+n \in C_i$ and a $z \in Z_i$ with $h_i(z) = z_o+n$. Then $z_o+n < \min B(z)$ and so $B(z) \subseteq V(P)$ holds. Since $(z, z_o+n) \in E(D')$, we get a $z_{o'}(z_o+n)$-path in D' via any $y \in B(z)$.

Since for every $x \in V'$, there are infinitely many disjoint edges from $N^+(x;D)$ to $\mathbb{Z}$, and n disjoint $\mathbb{Z}, N^-(x;D)$-paths in D, (α) immediately implies that D' is connected. We will show now that by addition of further edges we get an (n,n)-critical digraph from D which has also infinite indegree in every vertex. Consider any $i \in \mathbb{N}$ and any $x \in Z_i$. Defining $F(x) := \{f_{i+1}(M) : x \in M \in \binom{N^-(x;D)\cup\{x\}}{n}\}$, we can form $E(x) := \{(f_j(F(x)), x) : j \geq i+2\}$, since $|F(x)| = n$ holds. The digraph $\overline{D} := (V(D), E(D) \cup \bigcup_{x\in V'} E(x) \cup \bigcup_{x\in\mathbb{Z}} \{(y,x) : y \in V' - N^+(x;D)\})$ is antisymmetric and has infinite indegree in every vertex. We will show that it is still (n,n)-critical. Choose any $T \in \binom{V(D)}{n}$. We define vertex sets $T_o := T \subseteq T_1 \subseteq T_2 \subseteq \cdots$ inductively by $T_i := T_{i-1} \cup \{x \in V' : N^-(x;D) \subseteq T_{i-1}\}$ for $i \in \mathbb{N}$. Then $\overline{T} := \bigcup_{i\in\mathbb{N}} T_i$ is an infinite, proper subset of $V(D)$. Suppose there is an edge $(y,x) \in E(\overline{D})$ such that $y \in V(D) - \overline{T}$ and $x \in \overline{T} - T$ holds. Then $x \in V'$, say $x \in Z_i$ for an $i \in \mathbb{N}$. Since $N^-(x;D) \subseteq \overline{T}$ for every $x \in \overline{T} - T$, we see $(y,x) \in E(\overline{D} - E(D)$, hence $(y,x) \in E(x)$. Hence, there is a $j \geq i+2$ such that $f_jF(x)) = y$ holds. But $x \in \overline{T} - T$ implies $N^-(x;D) \subseteq \overline{T}$, therefore, also $f_{i+1}(M) \in \overline{T}$ for all $M \in \binom{N^-(x;D)\cup\{x\}}{n}$. So $F(x) \subseteq \overline{T}$ and hence $y = f_j(F(x)) \in \overline{T}$ follows. This contradiction shows $N^-(\overline{T} - T; \overline{D} - T) = \emptyset$. So $\overline{D}$ is (n,n)-critical.

Using this last idea, *for every cardinal* $\aleph_\alpha$ *and every* $n \in \mathbb{N}$, *it is possible to construct an (n,n)-critical digraph of order* $\aleph_\alpha$ *where every vertex has outdegree and indegree equal to* $\aleph_\alpha$. For this, one has to take for D a digraph built as is [12], but with the following modification: In every step, split z_i into $\aleph_\alpha$ many disjoint subsets $z_{i\beta}$ of cardinality $\aleph_\alpha$ and for every β, use a bijection $f_{i\beta}$ of $\left(\bigcup_{j \lesssim i} z_j\right)$ onto $z_{i\beta}$.

It remains open, if for any cardinal $\aleph_{\alpha'}$ there are (2,2)-critical digraphs of order $\aleph_\alpha$ which have vertices of outdegree less than $\aleph_\alpha$ and vertices of indegree less than $\aleph_\alpha$.

REFERENCES

[1] G. Chartrand, A. Kaugars, and D. R. Lick, "Critically n-connected graphs," *Proc. Amer. Math. Soc.* 32 (1972), 63-68.

[2] Y. O. Hamidoune (Mohamedou Yahya Ould Elmoctar), Contribution à l'étude de la connectivité d'un graphe, Thèse d'Etat, Paris 1980.

[3] Y. O. Hamidoune, "On critically h-connected simple graphs," *Discrete Math.* 32 (1980), 257-262.

[4] Y. O. Hamidoune, "Quelques problèmes de connexité dans les graphes orientés," *J. Combinatorial Theory Ser.* B 30 (1981), 1-10.

[5] W. Mader, "Eine Eigenschaft der Atome endlicher Graphen," Arch. Math. 22 (1971), 333- 336.

[6] W. Mader, "1-Faktoren von Graphen," Math. Ann. 201 (1973), 269-282.

[7] W. Mader, "Ecken vom Innen-und Aubengrad n in minimal n-fach kantenzusammenhängenden Digraphen," Arch. Math. 25 (1974), 107-112.

[8] W. Mader, "Endlichkeitssätze für k-kritische Graphen," Math. Ann. 229 (1977), 143-153.

[9] W. Mader, "Zur Struktur minimal n-fach zusammenhängender Graphen," Abh. Math. Sem. Univ. Hamburg 49 (1979), 49-69.

[10] W. Mader, "Disjunkte Fragmente in kritisch n-fach zusammenhängenden Graphen," *Europ. J. Combinatorics* 6 (1985), 353-359.

[11] W. Mader, "Minimal n-fach zusammenhängende Digraphen," *J. Combinatorial Theory Ser.* B 38 (1985), 102-117.

[12] W. Mader, "On infinite, n-connected graphs," to appear in the Proceedings of the Workshop on "Cycles and rays" in Montreal 1987.

[13] W. Mader, "On critically connected digraphs," to appear in the Journal of Graph Theory.

[14] W. Mader, "Ecken von Kleinem Grad in kritisch n-fach zusammenhängenden Digraphen," submitted to the Journal of Combinatorial Theory Series B.

[15] St. Maurer and P. Slater, "On k-critical, n-connected graphs," *Discrete Math.* 20 (1977), 255-262.

Walks in Covering Spaces

Bennet Manvel* and Richard Osborne
Colorado State University

ABSTRACT

Many topological problems on covering spaces can be restated as problems involving homomorphic images of graphs or directed graphs. In this paper we consider homomorphisms between directed graphs which are k–regular. The principal question is: which homomorphisms between such digraphs allow the lifting of all walks? We cannot answer that question, but we have found some surprising conditions on lifting homomorphisms. In the last section we briefly discuss some related questions, including: which digraphs have the same number of walks of every length?

1. Introduction

We begin with basic definitions. For this paper, a *digraph* may have loops and multiple arcs. A *homomorphism* between digraphs is any onto map sending vertices to vertices and arcs to arcs, preserving directions. (We do not require that the homomorphism be a local isomorphism, as is sometimes done when considering coverings of graphs, e.g. [2].) In the case of multiple arcs in the image, it is necessary to specify the image arc, rather that just the vertex map. We will be working with homomorphisms which are onto. If there is a homomorphism from D* to D then D* is said to *cover* D. A *walk* in a digraph is an alternating sequence of vertices and arcs $v_0, x_1, v_1, ..., x_n, v_n$ in which each arc x_i is directed from v_{i-1} to v_i. A digraph is *k–regular* if there are exactly k arcs into each vertex and k arcs out of each vertex. We call a digraph *regular* if it is k–regular for some k. The following definition is central to our discussion.

Definition Given a homomorphism h from D* to D, and a walk W then W *lifts* if there is some walk W* in D* such that h(W*) = W. We say that W* is a *lifting* of W.

Because homomorphisms allow identification of vertices, some walks in the digraph D, covered by D*, may not lift to walks in D*. If every walk in D lifts to a walk in D*, we say that the homomorphism h has the *lifting property*. (A topologist would say it has the path lifting property, but a topologist's path is a graph theorist's walk. We will not use the term path in this paper, to avoid confusion.)

When does a homomorphism h between digraphs D* and D have the lifting property? Why would we like to know? We cannot answer the first question, although the rest of this paper gives some rather surprising partial results. It is easier to answer the second question, by citing applications to topology and computer science.

Covering spaces and liftings are central to topology. It turns out that in the case of undirected walks, lifting occurs only when you have a local isomorphism, i.e. when the homomorphism is what is called a covering projection. The case of directed walks is more interesting, and has applications in the study of 3-manifolds. For more details, see [5].

In a more applied direction, finite state machines are just digraphs with certain functions (transition and output) attached. They are intended to model computers, while the functions model programs. Given two finite state machines D* and D, D* can be programmed to emulate anything that D can do if and only if there is a homomorphism h: $D^* \rightarrow D$ with the lifting property. That is so because D* must be able to perform any sequence of operations which D can perform, in the same order.

So which digraph homomorphisms have the lifting property? The question for arbitrary triples (D*, D, h) is interesting, but seems extremely difficult. We focus our attention here on the case where both D* and D are regular. Two special classes of digraphs are particularly interesting.

The *deBruijn graph* $dB_{m,n}$ has m^n vertices, labelled with the m–ary n–tuples, and the vertex $e_1e_2...e_n$ has an arc going to the vertex $f_1f_2...f_n$ if and only if $e_i = f_{i-1}$ for $i = 2$ to n. Thus $dB_{m,n}$ is m–regular, for each n.

The *Kautz graph* $Kz_{m,n}$ is defined in a similar way, except that its vertices are labelled with m–ary n-tuples in which consecutive terms must be different. (We use Kz for Kautz in deference to the customary use of K for complete). Since there are $m(m-1)^{n-1}$ such n–tuples, the binary Kautz graphs have only two vertices, and are not interesting. Ternary Kautz graphs are 2–regular, and will turn up in our examples. In general, the m–ary Kautz graphs are (m–1)–regular. Of course the Kautz graph $Kz_{m,n}$ is a subgraph of the deBruijn graph $dB_{m,n}$.

In Figure 1 we offer some examples of digraphs and homomorphisms. In each case the digraph D has its vertices labelled, and the labelling of the digraph D* indicates the intended homomorphism h. In Figures 1a and 1b, D* is the Kautz graph $Kz_{3,3}$ and D is the deBruijn graph $dB_{2,2}$. In Figures 1c and 1d, D* is the deBruijn graph $dB_{2,3}$

and D is the deBruijn graph $dB_{2,1}$. The homomorphism of Figure 1d does not have the lifting property. For example, the walk 01001 does not lift. All of the other homomorphisms do have the lifting property, for reasons we will now explain.

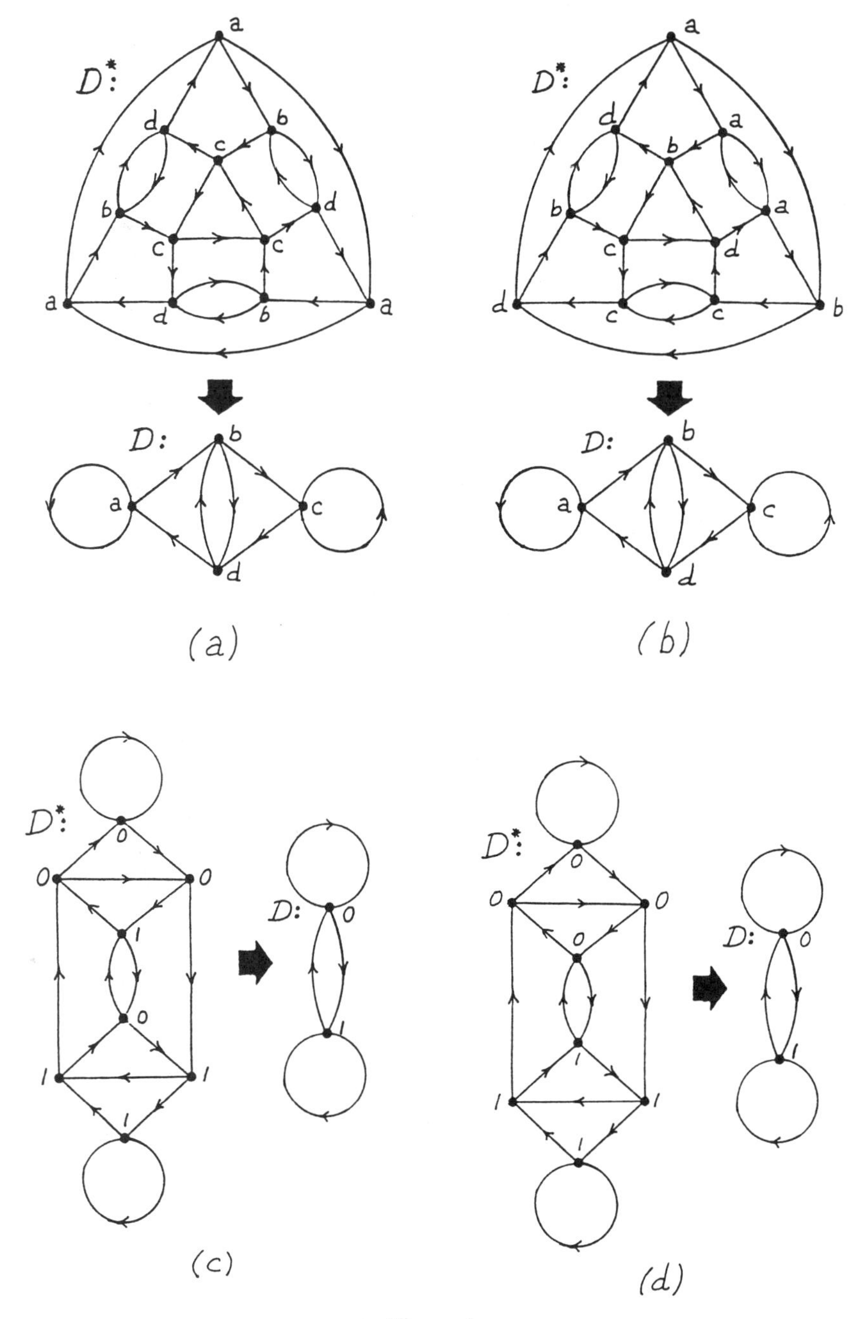

Figure 1

The *out–neighborhood* of a vertex v in a digraph is the set of vertices adjacent from v, and the *in–neighborhood* is the set of vertices adjacent to v. We say that a homomorphism h from D* to D is *out–regular* if the out–neighborhood of each vertex v of D* is mapped onto (not just into) the out–neighborhood of h(v), and similarly for *in–regular*. If h is both in– and out–regular then it is *regular*.

Lemma 1 *If h: D* → D is either out–regular or in–regular, then h has the lifting property.*

Proof Say h is out–regular, and we seek a lifting of the walk $v_0, x_1, v_1, ..., x_n, v_n$. Since h is onto, there is a vertex u_0 on D* such that $h(u_0) = v_0$. By out–regularity, some vertex u_1 adjacent from u_0 is mapped to v_1, and so on. Thus the walk lifts. If h is in–regular, we use the same argument starting with v_n instead of v_0. ❑

Since the homomorphism of Figure 1a is regular and that of Figure 1b is out–regular, they have the lifting property, by the Lemma. The homomorphism of Figure 1c is neither in– nor out–regular, so we need another argument to show lifting in that case. The following observation is obvious.

Lemma 2 *If homomorphisms h*: D** → D* and h: D* → D have the lifting property, then so does hh*: D** → D.*

In view of this lemma, all we need to do to prove that the homomorphism h of Figure 1c has the lifting property is exhibit two homomorphisms with lifting whose composition is h. That is done in Figure 2, where the first homomorphism, indicated with the labels a, b, c, d, is in–regular and the second, indicated with the lables 0 and 1, is out–regular.

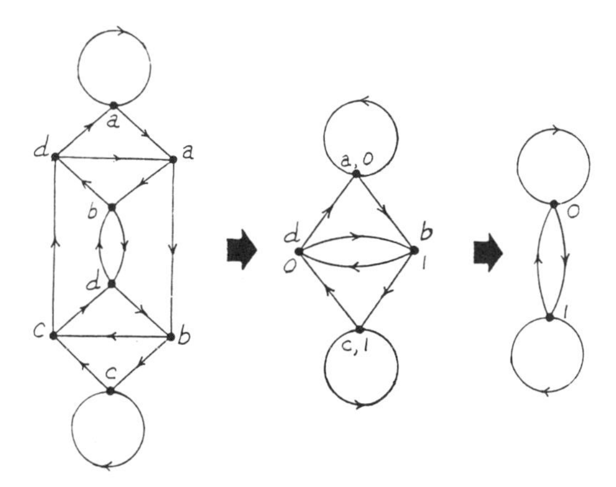

Figure 2

Since preparing this paper, we have been able to prove the existence of a function f(n) such that h has the lifting property if every walk of length at most f(n) lifts, where n the size of D*. We are still working on reducing the size of f(n), and will publish those results in a later paper. Here we present, in the next section, some observations about liftings, particularly in regular digraphs. In the final section we offer some ideas for further exploration.

2. Conditions for Lifting

For the remainder of this paper we will assume that the digraphs we are dealing with are k–regular (with k fixed within each homomorphism) and connected. (Because of the regularity, any sort of connectivity implies strong connectivity.) Several of our results depend on the following obvious lemma.

Lemma 3 *If W is any walk in a strongly connected digraph, almost all long walks contain W as a subwalk.*

Proof Say W is of length n and the diameter of the digraph is d. Given any vertex v of D, there is a walk starting at v of length at most d + n, containing W. Thus the number of walks of length d + n beginnning at v and not containing W is less than the total number of (d + n)–walks from v. Denote these numbers by N_v and M_v, respectively, and let R denote the maximum of the ratios N_v/M_v among all vertices v. Note that R < 1. To build a long walk, first build a (d + n)–walk. The number of ways to do that without including a copy of the walk W is at most R times the total number of ways to build a walk. Such a (d + n)–walk ends at some vertex. To continue and make a walk of length 2(d + n), go to an adjacent vertex and build another (d + n)–

walk. No matter what that adjacent vertex was, the number of ways to continue without including a copy of W is at most R times the total number of ways to continue without restriction. By proceeding in this way we see that the number of walks of length $k(d + n)$ including no copy of W is at most R^k times the total number of walks of that length. Since $R < 1$, almost every long walk contains a copy of the given walk W. ❑

For any homomorphism $h: D^* \to D$ there must be some walk W_m in D which has the minimum number of liftings, m. The homomorphism has the lifting property if and only if $m > 0$.

Lemma 4 *If h is a homomorphism from D* to D, D and D* are k regular, and W_m has the minimum number of liftings, m, then every walk containing W_m has exactly m liftings.*

Proof Clearly no walk has fewer than m liftings. Suppose, on the other hand, that some walk containing W_m has more than m liftings. Say that W is such a walk which is minimal, so that removing the vertex at either end of W results either in a walk with m liftings, or a walk not containing W_m. Remove an endvertex u which results in a walk W^- which has only m liftings, and contains W_m. Then the endvertex u is one of k vertices which could have been chosen to extend W^-. One of those k extensions, W, has more than m liftings, and the total number of liftings of those extensions is km. Thus, by the pigeonhole principle, some extension has fewer than m liftings, which contradicts the definition of m. ❑

Definition If two different walks W^* and W^{**} in D^* have the same beginning and ending vertices and both map to walk W under a homomorphism h, then (W^*,W^{**}) form a *split* in D^*.

Theorem 1 *A homomorphism between connected k–regular digraphs has the lifting property if and only if it has no splits.*

Proof Say that W does not lift. Then nothing containing W can lift. By Lemma 3, almost every long walk contains W, so almost no long walks lift. In D^*, there are k^r walks of length r, each a lifting of some walk in D. Hence, for r sufficiently large, if $k > 1$, then some walk in D must have more than n^2 liftings in D^*, where D^* has order n. But if D^* has more than n^2 liftings of any walk, two must have common starting and ending vertices, so there is a split. That leaves the case $k = 1$. But if $k = 1$, every homomorphism lifts and there can be no splits.

Suppose, on the other hand, that h has the lifting property, and suppose there is a split (W^*, W^{**}). Since D^* is connected, there is a closed walk W including a walk

W_m with the minimum number of liftings and also including the homomorphic image of W^*. Placing t copies of W end–to–end we obtain a walk W^t which includes the homomorphic image of the split t times, and includes W_m. Thus on the one hand W^t has at least 2^t liftings and on the other hand, by Lemma 4, it has only m liftings. Since t is arbitrary and m is fixed, this is a contradiction. ❑

Although this theorem gives a necessary and sufficient condition for lifting, it is not clear how useful that condition is, since there is no obvious way to check for the existence of a split. The next lemma and theorem may be more useful.

Lemma 5 *Under a homomorphism between connected k–regular digraphs, if the number of liftings of a walk is bounded, then every walk has the same number of liftings.*

Proof The number of liftings is always an integer, so say that M is the least upper bound on the number of liftings, and W_M has exactly M liftings. Then every walk containing W_M has at most M liftings. By k–regularity, that implies in turn (as in the proof of Lemma 4) that every walk containing W_M has exactly M liftings. But D is strongly connected, so some walk contains both W_M and W_m, a walk with the minimum number of liftings. Hence, by Lemma 4, M = m, and every walk has the same number of liftings. ❑

Theorem 2 *A homomorphism between connected k–regular digraphs has the lifting property if and only if every walk has the same number of liftings.*

Proof Certainly constant lifting implies lifting. On the other hand, Lemma 5 implies that every walk has the same number of liftings unless the number of liftings of some walk is arbitrarily large. But if D* has n vertices and some walk has more than n^2 liftings, there must be a split. By Theorem 1, that implies we do not have lifting. ❑

In view of Lemmas 3 and 4, Theorem 2 is not very surprising. On the other hand, it has two immediate corollaries, obtained by applying it to 0–walks and 1–walks, which are not intuitively obvious.

Corollary 2.1 *A homomorphism between connected k–regular digraphs D* and D can have the lifting property only if the number of vertices of D divides the number of vertices of D*.*

If h is homomorphism between D* and D, with lifting, we call an arc e of D* *busy* if h: $(D^* - e) \rightarrow D$ does not have lifting.

Corollary 2.2 *In any homomorphism between connected k–regular graphs with lifting, every arc is busy.*

This corollary brings out just how strongly global, rather than local, the property of lifting is in k–regular digraphs.

3. Related Problems

Work on liftings has suggested several directions for further investigation, which are interesting in their own right.

First, consider a question which arises from applying Theorem 2 to homomorphisms from 2–regular digraphs to the deBruijn graph $dB_{2,1}$. Since $dB_{2,1}$ is complete, with loops, any labelling of the vertices of a digraph D^* with two labels, say 0 and 1, defines a homomorphism onto $dB_{2,1}$. If the homomorphism has lifting, then by Theorem 2 it has constant lifting. In particular, all walks which are strings of 0's or strings of 1's must lift exactly the same number of times. Thus the subgraph of D^* induced by the vertices in D^* labelled 0, the *0–subgraph*, is a digraph with exactly k vertices, k arcs, k directed 2–walks, and so on, for some k. It is somewhat surprising that such a digraph can exist, but a glance at the 0–subgraph in Figure 1c gives a non–trivial example. Trivial examples include directed cycles and functional (or antifunctional) digraphs.

The general question of which digraphs have exactly k walks of length n, for every n, is somewhat complicated. The results will appear in full in [4], but we summarize them here. The first observation to make is that a digraph with constant walk numbers must have at least one directed cycle, but no two directed cycles are joined by a directed path. Any digraph with such a cycle structure will have walk numbers which are eventually periodic. (Note the similarity of this result to theorems on iterated line digraphs in Beineke [1], and Hemminger [3].) By reducing unilateral components off of the cycle to directed trees, numbers can be generated which determine the number of walks of every length. The conditions for eventually constant walk numbers are relatively neat, and are derived using eigenvalues of circulant matrices. Making the initial numbers constant as well is somewhat messier.

A second area for further study is to examine certain natural ways to define homomorphisms, particularly for the deBruijn and Kautz graphs, and see what can be said about the lifting property in each case.

As was pointed out before, any map onto a complete digraph with loops is a homomorphism -- every arc automatically maps to an arc. If we view the homomorphism as a labelling of a k–regular digraph D*, the conditions for such a labelling to have lifting include: all vertex labels must appear the same number of times; the subgraphs induced by the vertices with a single label must have constant walk numbers; the bipartite subgraph with arcs joining vertices of two different labels must have constant walk numbers. Another image graph which gives interesting homomorphisms is the complete graph without loops. In that case each homomorphism corresponds to a labelling of D* which is a proper vertex coloring. The same necessary conditions for lifting apply. A third interesting image digraph has a single vertex, and two directed loops. In this case a homomorphism is just a labelling of the arcs with two labels. In fact this digraph can be considered as the smallest deBruijn graph, and is of interest in studying factorings of homomorphisms, which we will discuss in a moment.

The connection between digraphs and their line digraphs gives other interesting homomorphisms. If D* = L(D), then the natural corresponcdence between vertices of D* and arcs of D gives two natural homomorphisms. The *forward* homomorphism maps v, in D*, to the head of the arc in D corresponding to v. The *backward* homomorphism maps v to the tail of the same arc. The deBruijn graphs and Kautz graphs are examples of sequences of iterated line digraphs, and so give examples of such line digraph homomorphisms. The homomorphisms in Figure 2, for example, are respectively backward and forward.

The labellings of the vertices of the deBruijn and Kautz graphs with n–tuples present many more opportunities for interesting homomorphisms. Each vertex is labelled $x_1x_2...x_n$, with x_i from Z_m. Every function from $Z_m{}^n$ to Z_m defines a homomorphism from $dB_{m,n}$ to $dB_{m,1}$. Conversely, by using sums of products such as $(x_1)(x_2-1)(x_3-1)$ (which gives the value 1 at the vertex 100 and is 0 elsewhere), a function can be defined for any homomorphism. The question of lifting then becomes a question about the functions. For example, it is not hard to see which such functions will give in–regular or out–regular homomorphisms. Unfortunately, this formulation does not yield any obvious new theorems, although it does generate some nice examples. It is interesting to note that the natural homomorphism which maps each vertex in $dB_{2,n}$ with coordinates $x_1x_2...x_n$ to the vertex in $dB_{2,(n-1)}$ with coordinates $(x_1+x_2)(x_2+x_3)\ldots(x_{n-1}+x_n)$ is neither the foward nor the backward homomorphism obtained from the line digraph map.

Another approach which may offer answers to some questions about homomorphisms and liftings is factoring. It was already observed in Lemma 2 that the composition of maps which lift is a map which lifts. Given a homomorphism h from D* to D, it may be possible to find another digraph E so that there is a homomorphism from D* to E and another from E to D, whose composition is h. If there is no such homomorphism, h might be considered prime. Of course a similar definition might be made, with lifting demanded of each homomorphism. Then it is really only necessary to study prime homomorphisms with lifting, rather than all homomorphisms.

Clearly a great deal is still to be learned about homomorphisms and lifting of walks in digraphs.

REFERENCES

[1] Beineke, L.W., On derived graphs and digraphs, in *Beitrage zur Graphentheorie* (ed. H Sachs et al), Teubner–Verlag, Leipzig (1968) 17 – 23.

[2] Fellows, Mike, On mutual and common covers and the complexity of covering, this Proceedings.

[3] Hemminger, R.L., Digraphs with periodic line digraphs, *Studia Sci. Math. Hungar.* 9 (1974) 27 – 31.

[4] Iyer, H., B. Manvel, and R. Osborne, Digraphs with constant walk numbers, in preparation.

[5] Osborne, Richard, Lifting directed paths, in preparation.

Chordal and Interval Multigraphs

Terry A. McKee*
Wright State University

ABSTRACT

By allowing multiple edges to represent multiple intersection, much of the work done with intersection graphs can be upgraded to intersection multigraphs. In particular, starting from subtrees of trees and subpaths of paths leads to notions of chordal and interval multigraphs. We present efficient characterizations and applications of these concepts.

1. Introduction

Chordal (and interval) graphs may be characterized as intersection graphs of families of subtrees of a tree (respectively, subpaths of a path). We propose their intersection multigraph analogs, where two subtrees or subpaths having k vertices in common correspond to two vertices being joined by k parallel edges in the intersection multigraph. Because such intersection multigraphs carry more information about the original tree or path representations, they can be more appropriate for applications. In [5] we use our characterizations of these multigraphs to determine which partially ordered sets have upper bound multigraphs (i.e., intersection multigraphs of the sets of upper bounds of elements) which are chordal or interval.

Intersection graphs are simply the underlying graphs of the analogous intersection multigraphs. But the characterization problem for such intersection multigraphs is quite different from (yet, as we shall see, closely related to) that for the corresponding intersection graphs. For instance, the complete tripartite graph $K_{1,1,2}$ (i.e., a rectangle with one diagonal) is a subpaths–of–a–path intersection graph, but not multigraph. A second parallel diagonal would have to be added before four subpaths could intersect in

* The author thanks the Western Michigan University Department of Mathematics & Statistics for its exuberant hospitality during his sabbatical stay there. The research was partially supported by Office of Naval Research contract N00014–88–K–0163.

precisely that fashion. (Yet adding parallel edges can sometimes destroy being such an intersection multigraph).

For any multigraph M with vertex set V(M), let $M\downarrow$ be the underlying (simple) graph. For each pair $u, v \in V(M)$, let $\mu(u,v)$ be the number of parallel edges joining u with v; if u is adjacent to v, then $\mu(u,v)$ is called the *multiplicity* of the multiple edge. A *clique* of M is a simple subgraph of M whose vertex set induces an inclusion–maximal complete subgraph of $M\downarrow$ (i.e., a clique of $M\downarrow$). (Hence a clique of a multigraph contains exactly one edge from each bundle of parallel edges joining its vertices.) For $u, v \in V(M)$, let $\kappa(u,v)$ be the number of cliques of $M\downarrow$ which contain $\{u, v\}$. A *simplicial vertex* of M is a vertex in a unique clique of $M\downarrow$.

A multigraph M is said to satisfy the $\mu \geq \kappa$ *condition* if, for every pair $u, v \in V(M)$, $\mu(u,v) \geq \kappa(u,v)$. Thus M satisfies the $\mu \geq \kappa$ condition if and only if all the cliques of $M\downarrow$ can be simultaneously identified with edge–disjoint cliques of M. Any edges left over, i.e., $\mu(u,v) - \kappa(u,v)$ parallel edges joining each pair $u, v \in V(M)$, are called *residual edges* of M. The submultigraph of M consisting of residual edges is called the *residual submultigraph*, or *residuum*, of M, denoted rM. (We use r^2M for r(rM), etc.) The residuum plays a key role in relating the theories of intersection graphs and multigraphs.

2. Chordal Multigraphs

We define a *subtrees–of–a–tree* representation of a graph G to be a tree T whose vertices are subsets of V(G) such that:

(a) For each $v \in V(G)$, $T_v = \{A \in V(T) : v \in A\}$ induces a subtree of T; and

(b) For each $u, v \in V(G)$, u and v are adjacent in G if and only if $T_u \cap T_v$ is nonempty.

A subtrees–of–a–tree representation of a multigraph M is defined similarly, except replacing (b) with

(b') For each $u, v \in V(M)$, $\mu(u,v) = |T_u \cap T_v|$.

Notice that, by (b) or (b'), each vertex of T induces a complete subgraph of M or $M\downarrow$.

Lemma *In any subtrees–of–a–tree representation of a multigraph M, each vertex set of a clique of M must equal one or more vertices of the tree.*

Proof We first show that whenever vertices v_1, v_2, v_3, form a triangle in $M\downarrow$, then some vertex of the representing tree T must contain them all. By (b), each two v_i, v_j are in a common vertex A_{ij} in T. If each such A_{ij} failed to contain the third vertex of the triple, then, by (a), there would be a unique vertex of G at the intersection of the three paths pairwise connecting A_{12}, A_{13}, and A_{23}, and this vertex would contain all three v_i's. Similarly, the vertex set of every complete subgraph of $M\downarrow$ must be contained in a common vertex of T. ❑

A *chord* of a circuit in a graph is an edge which joins vertices in, but not consecutive along the circuit. *Chordal graphs* are characterized by every circuit of length at least four having at least one *chord*. (See Chapter 4 of [2] for an extensive survey of chordal graphs, called there "triangulated graphs" and sometimes elsewhere "rigid circuit graphs".) Chordal graphs were characterized (independently) by Walter, Gavril, and Buneman [2, Theorem 4.8] as those graphs having subtrees–of–a–tree representations. This motivates our defining a multigraph to be a *chordal multigraph* if it has a subtrees–of–a–tree representation.

Theorem 1 *A multigraph M is a chordal multigraph if and only if:*
(1) M satisfies the $\mu \geq \kappa$ condition; and
(2) $M\downarrow$ is a chordal graph.

Proof Suppose first M has a subtrees–of–a–trees representation T. Since T is also a subtrees–of–a–tree representation of $M\downarrow$, condition (2) follows immediately. If vertices u and v of M are in κ common cliques of $M\downarrow$, then by the Lemma these are in at least κ common vertices of T, so $|T_u \cap T_v| \geq \kappa$, and so condition (1) follows from (b').

Conversely, suppose conditions (1) and (2) hold. Suppose M+ is obtained from M as follows: For each residual edge of M, add a new simplicial vertex (simply) adjacent to precisely the two endvertices of that edge. Repeat if necessary until M+ satisfies $\mu(u,v) = \kappa(u,v)$ for all $u, v \in V(M+)$. Let $G = M+\downarrow$. It is easy to check that (2) implies that G is also chordal.

Claim If T is the subtrees–of–a–tree representation for the graph G produced in Golumbic's proof of Theorem 4.8 of [2], then T is also a subtrees–of–a–tree representation for the multigraph M+. (Then removing the elements of $V(M+) - V(M)$ out of the vertices of T will leave the desired subtrees–of–a–tree representation for M itself.)

We show the Claim inductively, with the case when G is complete as basis. (T then has V(G) as its unique vertex.) For the inductive step, suppose G is not complete and that the Claim is true for all smaller graphs. Because G is chordal, there is [2, Lemma 4.2] a simplicial vertex a. (Our notation is chosen to be consistent with proof of Theorem 4.8 of [2].) Put A = N[a], the closed neighborhood of a. Thus A is a clique of G, and so must be a vertex of T by our Lemma. Letting N(u) be the open neighborhood of vertex u, put $U = \{u \in A : N(u) \subset A\}$ and $Y = A - U$. Put $G' = G - U$ and let B be any clique of G' which contains Y. The cases $B = Y$ and $B \neq Y$ are handled separately.

Suppose $B = Y$ and put $M' = M+ - U$. Note that M' satisfies (1) and (2) and that $G' = M'\downarrow$. Since U is nonempty, G' is smaller than G, and so by the inductive hypothesis, the subtrees–of–a–tree representation T' for G' produced as in Theorem 4.8 of [2] is also a subtrees–of–a–tree representation for M'. Define T to be the same as T' except with vertex B replaced by A, and verify that T is a subtrees–of–a–tree representation for M by comparing the multiplicities with the number of common vertices of T for each of the six possible types of vertex pairs from the partition U, Y, $V(M+) - A$ of $V(M+)$.

Now suppose $B \neq Y$ and put $M' = M+ - U$ except also remove one edge from between every two members of Y. Proceed as in the $B = Y$ case, except define T to be the same as T' except for adding the one new vertex A and one new edge between A and B. ❑

For comparison with Theorem 2 below, note that we have essentially proved an equivalent version of Theorem 1 with condition (2) replaced by:

(2') For M+ defined from M by adding a new simplicial vertex for each edge of rM (repeating if necessary until $\mu(u,v) = \kappa(u,v)$ for all u, v), $M+\downarrow$ is a chordal graph.

Corollary *If M is a chordal multigraph, then every circiut of length at least four must have at least two (possibly parallel) chords.*

Proof One chord exists by condition (2) of Theorem 1. The existence of the second follows by induction on the length of the circuit. The basis (length four) case follows from condition (1). For lengths greater than four, use a smaller circuit subtended by the first chord. ❑

3. Interval Multigraphs

We can define a *subpaths–of–a–path* representation of a graph or a multigraph exactly as we did a subtrees–of–a–tree representation, except that the tree T will be a path P such that each $P_v = \{A \in V(P) : v \in A\}$ is a subpath of P, with adjacency corresponding to the intersection of the P_v's in the expected manner.

Interval graphs are defined as those graphs having a subpaths–of–a–path representation and have been characterized in many ways; see Section 1.3 and Chapter 8 of [2] or Section 3.4 of [6]. The earliest characterization was that of Lekkerkerker and

Boland [2, Theorem 8.4], as being both chordal and having no 'asteroidal triples'. This latter statement can be viewed as a sort of generalized chordality: Given any closed walk passing through three distinguished "anchor" vertices, there must exist a chord joining one of the anchors to some vertex of the walk lying between (or at one of) the other two anchors.

We define a multigraph to be an *interval multigraph* if and only if it has a subpaths–of–a–path representation. It should be noted that, unlike chordal multigraphs, the family of interval multigraphs is not closed under duplicating edges; i.e., increasing an already positive multiplicity in an interval multigraph may produce a non interval multigraph.

Theorem 2 *A multigraph M is an interval multigraph if and only if:*

(1) M, rM, r^2M, ... each satisfies the $\mu \geq \kappa$ condition; and

(2) For M+ defined from M by adding a new simplicial vertex for each clique of rM, r^2M, ... ((simply) adjacent to precisely the vertices of that clique), M+$\downarrow$ is an interval graph.

Proof Suppose first M has a subpaths–of–a–path representation P. Since this is also a subtrees–of–a–tree representation, M is a chordal multigraph and so, by Theorem 1, satisfies $\mu \geq \kappa$. By the Lemma, each clique of M corresponds to a vertex of P. Removing these vertices from P by contracting edges of P leaves a subtrees–of–a–tree representation for rM. So rM is a chordal multigraph and so, by Theorem 1, satisfies $\mu \geq \kappa$. Thus and similarly, condition (1) holds. Finally, adding the vertices of V(M+) − V(M) into the appropriate vertices of P produces a subpaths–of–a–path representation for M+$\downarrow$, thereby proving (2).

Conversely, suppose conditions (1) and (2) hold. By (2), M+$\downarrow$ has a subpaths–of–a–path representation P, where the vertices of P will be precisely the vertex sets of the cliques of M+$\downarrow$. By (1) and the construction of M+, each pair $u, v \in V(M)$ will occur in exactly $\mu(u,v)$ of the vertices of P. Hence, removing the elements of V(M+) − V(M) out of the vertices of P will leave a subpaths–of–a path representation for M itself. ❑

4. Algorithms, Applications, and Alterations

Known algorithms and results concerning chordal graphs (namely, Corollary 4.6 and Theorem 4.17 of [2]) can be combined with Theorem 1 to recognize chordal multigraphs efficiently. Theorem 1 also allows Theorem 2 to be equivalently stated with condition (1) replaced with:

(1') Each of M, rM, r^2M, ... is a chordal multigraph.

This allows the recognition procedure for chordal multigraphs to be combined with known algorithms and results for interval graphs (namely, Theorem 8.3 of [2]) to recognize interval multigraphs efficiently.

Corollary *Chordal [and interval] multigraphs can be recognized in time, proportional to the order [times the maximum multiplicity occurring].*

Chordal and interval graphs serve as models in a wide variety of applications; Section 8.4 of [2] and Section 3.4 of [6] survey many of these. (When interval graphs are defined in terms of intervals of the real line, subpaths–of–a–path represent[illegible] can be obtained by taking left endpoints as vertices, or by using representatives of equivalence classes of points as vertices of P, where two points are equivalent whenever they are in precisely the same intervals.)

One of the leading applications in [1] was the construction of the "genealogy" of medieval manuscripts by tracing errors (or other distinctive features) which were made in handcopying and might be repeated (or possibly corrected) in later recopying. (Reconstructing evolutionary trees is a second example in [1], using the same ideas; both arose from real–life concerns.) The manuscripts can be viewed as the vertices of a rooted tree, with the errors corresponding to certain subtrees of the tree. Thus the intersection (multi)graph of errors must be a chordal (multi)graph.

As an oversimplified example, suppose there are four errors designated A, B, C, D and five manuscripts having the following combinations of errors: A; A & B; A & B & D; B & C; and C & D. The intersection graph of errors is the complete tripartite graph $K_{1,1,2}$ and is chordal. But the intersection multigraph ($K_{1,1,2}$ plus a residual edge between A and B) is not a chordal multigraph: By the Corollary to Theorem 1, a second chord is needed for the circuit A, B, C, D; the missing edge predicts a missing manuscript. (In fact, our Lemma implies there must be a manuscript with errors B, C and D.) The point is, if all one has is the intersection graph, all looks well; but by having the intersection multigraph, an inconsistency (or incompleteness) is seen.

Suppose we were to start with an actual chordal multigraph M which did represent errors in manuscripts. Suppose also that each manuscript has at least two errors (or other distinctive features). Each manuscript would then correspond to a clique of $M+\downarrow$ (using the notation of the proof of Theorem 1). Forming the "clique (multi)graph" (i.e., the intersection (multi)graph of the family of cliques) of $M+\downarrow$ would produce the intersection (multi)graph of the original family of manuscripts, each now viewed as a set of errors. (This "manuscript (multi)graph" would not have been similarly producible if we had started with a chordal graph rather than multigraph.) This then imparts structure to the manuscripts–as–sets–of–errors (multi)graph, as well as to the errors–as–sets–of–manuscripts (multi)graph.

A similar example concerns the detection of a linear structure underlying developmental psychology; see the bottom of page 120 in [6] for discussion and primary references. Children would correspond to vertices of what was hypothesized as a subpaths–of–a–path representation, with certain traits or characteristics as subpaths. If the hypothesis were correct, the intersection (multi)graph of traits would be an interval (multi)graph. Using the multigraph model, the appropriate multigraph criterion of Theorem 2 gives a more stringent test of the hypothesis. Missing edges now correspond to unrepresented types of children (i.e., to undiscovered combinations of traits).

There are also obvious alterations to our results made by defining "subthings–of–a–thing representations" and seeking corresponding characterization theorems having clauses involving (1) the $\mu \geq \kappa$ condition for, and the (2) underlying graph of a multigraph and its residuum. For instance, the "things" could easily be caterpillars or stars as well as paths or trees, or they could be mixed, as in subpaths–of–a–tree representations. But the problem would still remain of going from knowledge of such representations to suitable conditions for the intersection multigraphic characterizations. (The similar problem for intersection graphic characterizations is discussed in [3]; a somewhat unsatisfactory approach for intersection multigraphic characterizations is mentioned at the end of [4].)

As long as the "things" in a subthings–of–a–thing representation are varieties of trees, we can simultaneously characterize *edge intersection multigraphs*, where the subgraphs are considered as sets of edges instead of sets of vertices. Namely, the edge intersection multigraph is obtained from the (vertex) intersection multigraph by subtracting from it the edges of the (underlying) intersection graph. (This corresponds to each tree having exactly one more vertex than edge.) Thus of the four alternatives determined by the vertex/edge and simple/multi choices, the (vertex) intersection multigraph directly determines each of the others, while the standard (vertex) intersection (simple) graph determines none of the others.

In closing, we mention that our Lemma corresponds to saying that any family of subtrees–of–a–tree satisfies the *Helly Property*: Whenever all the subtrees of any subfamily have pairwise nonempty intersections, there is a vertex in common to all the subtrees of the subfamily; see Proposition 4.7 of [2]. If we define a *Helly multigraph* to be the intersection multigraph of a Helly family, then the arguments we made above can be rephrased as follows.

Theorem 3 *A multigraph M is a Helly multigraph if and only if M satisfies the $\mu \geq \kappa$ condition.*

As a corollary, every graph underlies a Helly multigraph, and so every graph would be a "Helly graph". As a corollary of Theorems 1, 2 and 3, we can sharpen the role of the $K_{1,1,2}$ example used at the beginning of this paper. A chordal (or interval or Helly) graph G is a chordal (or respectively, interval or Helly) multigraph if and only if each edge of G is in a unique clique, and so if and only if G contains no induced subgraph isomorphic to $K_{1,1,2}$.

REFERENCES

[1] P. Buneman, A characterization of rigid circuit graphs, *Discrete Math.* **9** (1974) 205 – 212.

[2] M.D. Golumbic, *Algorithmic Graph Theory and Perfect Graphs*, Academic Press (New York) 1980.

[3] T.A. McKee, Intersection properties of graphs, to appear.

[4] T.A. McKee, Foundations of intersection graph theory, to appear.

[5] T.A. McKee, Upper bound multigraphs for posets, to appear.

[6] F.R. Roberts, *Discrete Mathematical Models with Applications to Social, Biological and Environmental Problems*, Prentice–Hall (Englewood Cliffs NJ) 1976.

The Expected Number of Symmetries in Locally–Restricted Trees I

Kathleen A. McKeon
Connecticut College

ABSTRACT

Exact formulas are derived for the number of symmetries in several types of unlabeled trees with vertices of restricted degree. The trees are d–trees whose vertices have degree at most d and (1,d)–trees whose vertices have degree 1 or d. These results together with similar results for the number of such trees provide formulas for the expected number of symmetries in these trees.
These trees give rise to significant examples in polymer chemistry. For example, (1,4)–trees represent the alkanes and 4–trees represent the carbon skeletons of alkanes. The expected number of symmetries in such trees is important in the study of collections of molecular species formed during some chemical reaction process.

1. Introduction

The enumeration of trees is an important problem in graph theory with a distinguished history as well as applications to theoretical chemistry. The first major work in this area was performed by Cayley who determined exact formulas for the number of labeled trees [C89], the number of rooted trees [C57] and the number of free trees [C75, C81]. These results were extended and an asymptotic analysis of the numbers was provided by Pólya [P37] and Otter [O48].

Cayley's work [C75] was motivated by the problem of enumerating isomers of alkanes, compounds of carbon and hydrogen atoms which have valencies of 4 and 1 respectively. The alkanes have the general formula C_kH_{2k+2} and can be represented by (1,4)–trees. They are the best documented family of chemical compounds and provide a model for much of chemical theory [GoK73]. Generalizing (1,4)–trees, we have (1,d)–trees, which give rise to other meaningful examples in polymer chemistry. There is a correspondence between (1,d)–trees and d–trees that also has chemical significance.

While 4–trees correspond to the carbon skeletons of alkanes [GoK73], d–trees in general correspond to skeleton polymers, i.e., polymer molecules that have been stripped of their reactive end groups [GoT76].

The problem addressed in this paper, the enumeration of symmetries in (1,d)–trees and d–trees for $d = 3, 4$, is also motivated by chemistry. In the study of collections of molecular species, it is almost always the average of some property over an appropriate class of trees that is required. In computing such an average, it is necessary to assign weights to the various trees in the class so as to reflect the (not usually equal) proportions in which they are formed by the chemical process involved. The proper assignment of weights to the trees often involves the orders of their automorphism groups [GoL75]. Consequently, chemists are interested in the orders of the automorphism groups of large trees of various species such as (1,d)–trees and d–trees.

The tool used to do the counting is a two–variable generating function, an approach that seems to have originated in the work of Etherington [Et38]. For a given class C of trees, let $t(x,y)$ be the generating function in two variables x and y such that the coefficient of $y^m x^n$ is the number of trees T in C of order n in which m is the logarithm base 2 of the order of the automorphism group of T. In $t(x,2)$, the coefficient of x^n is the sum of the orders of the automorphism groups of all such trees.

The technique used to do the counting was developed by Pólya [P37], perfected by Otter [O48] and described as a twenty step algorithm for counting various types of trees by Harary, Robinson and Schwenk [HRS75]. The generating functions $t(x,y)$ and $t(x,2)$ satisfy functional equations from which recurrence relations for their coefficients are determined.

In this paper, the technique is illustrated and results given for (1,3)–trees. Exact formulas are determined for the number of symmetries in both planted and free unlabeled (1,3)–trees. These results together with similar results for the number of (1,3)–trees provide formulas for the expected number of symmetries in these trees. A study of the asymptotic behavior of the number of symmetries in such trees will appear in Part II of this series. A brief sketch of the general approach is provided in the short research announcement [KMPR].

2. Generating Functions

While equations are given for (1,3)–trees only, the method has been used to enumerate symmetries in four types of trees: d–trees and (1,d)–trees for $d = 3, 4$ [Mc87] and may be applied for higher values of d.

We begin by defining a generating function that counts symmetries in planted trees of the specified type. In general, the results for planted trees provide a means for obtaining the results for free trees. However, removing the root of a planted (1,3)–tree leaves a binary tree that has the same automorphism group as the original planted (1,3)–tree. Thus, when counting symmetries in planted (1,3)–trees we are also counting symmetries in binary trees.

For the planted trees of each type, a two–variable logarithmic generating function is defined as follows:

$$T(x,y) = \sum_{n=1}^{\infty} \sum_{m} T_{m,n} y^m x^n \tag{2.1}$$

For d–trees, $T_{m,n}$ is the number of planted trees T of order $n+1$ in which $m = \log_2|\Gamma(T)|$, where $\Gamma(T)$ denotes the automorphism group of T. Every (1,d)–tree, planted or free, has 2 modulo (d–1) vertices. This is taken into account in the definition of $T(x,y)$ for (1,d)–trees. In this example, since every planted (1,3)–tree has an even number of vertices, $T_{m,n}$ is defined to be the number of planted (1,3)–trees on $2n$ vertices (or binary trees on $2n-1$ vertices) with 2^m symmetries.

The values which m may assume in the sum (2.1) depend on both d and the type of tree. Since an automorphism of a rooted tree must leave the root fixed, the order of the automorphism group of a planted (1,3)–tree is of the form 2^m where m is an integer ranging from 0 to $n-1$.

Note that when $y = 1$ is substituted in (2.1), $T(x,1)$ counts the number of planted trees of the specified type. Substituting $y = 2$ in (2.1) results in a one–variable generating function which counts symmetries in planted trees of the specified type. Let S_n be the total number of symmetries in all planted (1,3)–trees on $2n$ vertices.

$$S_n = \sum_{m} T_{m,n} 2^n \tag{2.2}$$

and

$$T(x,2) = \sum_{n=1}^{\infty} S_n x^n \tag{2.3}$$

Similarly, $t(x,y)$ can be defined for free trees. However, we actually only work with $t(x,2)$. Thus, we define

$$t(x,2) = \sum_{n=1}^{\infty} s_n x^n \tag{2.4}$$

which counts symmetries in free, i.e., unrooted trees. For (1,3)–trees, s_n is the total number of symmetries in all free (1,3)–trees on $2n$ vertices.

3. Functional Relations

To obtain the exact formulas for the number of symmetries in these trees, functional relations satisfied by $T(x,y)$, $T(x,2)$ and $t(x,2)$ are now derived.

First observe that rooted and planted trees of a specified type can be formed from planted trees of that type. A rooted tree in which the root has degree k is formed by taking a collection of k planted trees and identifying their roots to form the root of the new tree. Adding a new vertex adjacent to the root of this rooted tree results in a planted tree in which the degree of the vertex adjacent to the root is $k+1$. Based on this observation, relations expressing $T(x,y)$ in terms of $T(x,y)$, $T(x^2,y^2)$, and $T(x^3,y^3)$ are derived.

Theorem 3.1 *The generating functions $T(x,y)$ and $T(x,2)$ which count symmetries in planted (1,3)–trees satisfy*

$$T(x,y) = x + \frac{1}{2}T(x,y)^2 + (y - \frac{1}{2})\, T(x^2,y^2) \tag{3.1}$$

and

$$T(x,2) = x + \frac{1}{2}T(x,2)^2 + \frac{3}{2}T(x^2,4) \tag{3.2}$$

Proof The vertex adjacent to the root of a planted (1,3)–tree has degree 1 or 3. The term x counts the symmetries in a planted K_2, the only planted (1,3)–tree in which the vertex adjacent to the root has degree 1.

To count symmetries in those trees in which the vertex adjacent to the root has degree 3, two cases must be considered. Suppose T is the planted (1,3)–tree formed from the planted (1,3)–trees T_1 and T_2 in the manner described above. If $T_1 \neq T_2$, then we have $|\Gamma(T)| = |\Gamma(T_1)|\,|\Gamma(T_2)|$. Then $\frac{1}{2}(T(x,y)^2 - T(x^2,y^2))$ counts symmetries in this case. If $T_1 = T_2$, then we have $|\Gamma(T)| = 2|\Gamma(T_1)|^2$ since the two branches T_1 and T_2 can be permuted. This case is handled by $yT(x^2,y^2)$ with the factor of y accounting for the additional factor of 2 in the group order. Now (3.2) is obtained by substituting $y = 2$ in (3.1). ❑

Using the following lemma which relates the order of the automorphism group of a free tree to the orders of the automorphism groups of the vertex and edge–rooted versions of the tree, $t(x,2)$ is expressed in terms of $T(x,2)$, $T(x^2,4)$ and $T(x^3,8)$.

Lemma 3.2 *For any tree T,*

$$|\Gamma(T)| = \sum_{T_1} |\Gamma(T_1)| - \sum_{T_2} |\Gamma(T_2)| + |\Gamma(T_3)| \tag{3.3}$$

where the first sum is taken over all different vertex–rooted versions T_1 of T and the second sum is taken over all different edge–rooted versions T_2 of T. If T has a symmetry edge, an edge whose vertices are interchanged by some automorphism of T, then $T_3 = T$. If T does not have a symmetry edge, then T_3 is the empty graph and $|\Gamma(T_3)| = 0$.

Proof This lemma is a variation of a lemma due to Otter [HP73]. As in the proof of Otter's lemma, the vertex and edge–rooted versions of T can be paired such that the paired vertex and edge–rooted versions of T have the same automorphism group. Recall that an automorphism of a graph must leave the root fixed while an automorphism of an edge–rooted graph must leave the vertices of the root edge fixed. For each vertex v that is not in the center of T, match the version of T that is rooted at v with the edge–rooted version that is rooted at the first edge on the path from v to the center of T.

If T has a symmetry edge, the center of T consists of two vertices, say u and v. Since the edge uv is a symmetry edge of T, rooting T at u is equivalent to rooting T at v. Hence if the version of T that is rooted at the vertex v is paired with the version of T that is rooted at the edge uv, then the difference of the two sums in (3.3) is 0 and $T_3 = T$. Thus, (3.3) holds in this case.

If T does not have a symmetry edge two cases must be considered. If the center of T consists of two vertices u and v, match the version of T that is rooted at v with the version of T that is rooted at the edge uv. In this case and the case that the center of T consists of just one vertex u, there is one vertex–rooted version of T that cannot be paired with an edge–rooted version. This is the tree that results from rooting T at the vertex u which is in the center of T. Since T does not have a symmetry edge, the vertices in the center of T are all fixed points of the automorphisms of the unrooted tree T. Hence this extra vertex–rooted version of T has the same automorphism group as T and (3.3) holds in this case also. ❑

This lemma can be extended to a statement about the generating functions that count symmetries by multiplying (3.3) by x^n and summing over all trees of the appropriate order. Summing the result over all $n \geq 1$ gives $t(x,2)$ on the left side. The first sum on the right side gives the series that counts symmetries in rooted trees and the second sum gives the series that counts symmetries in edge–rooted trees while $|\Gamma(T_3)|x^n$ sums to the series that counts symmetries in trees with a symmetry edge. For (1,3)–trees we have the following functional relation.

Theorem 3.3 *The generating function* $t(x,2)$ *for symmetries in free* $(1,3)$*–trees is given by*

$$t(x,2) = \frac{1}{2x}T(x,2)^2 - \frac{1}{3x}T(x,2)^3 + \frac{3}{2x}T(x^2,4) + \frac{13}{3x}T(x^3,8).$$

Proof First we determine an expression for the series that counts symmetries in rooted (1,3)–trees. As previously described, this expression can be found by using planted (1,3)–trees to build rooted (1,3)–trees. The series for rooted (1,3)–trees is equal to

$$T(x,2) + \frac{3!}{x}T(x^3,8) + [\frac{2}{x}T(x^2,4)\,T(x,2) - T(x^3,8)]$$

$$+ [\frac{1}{3!\,x}(T(x,2)^3 - 3\,T(x^2,4)\,T(x,2) + 2\,T(x^3,8))]$$

Symmetries in rooted (1,3)–trees in which the root has degree 1, i.e., in planted (1,3)–trees are counted by $T(x,2)$. To count symmetries in rooted (1,3)–trees in which the root has degree 3, three cases must be considered. Suppose T is the rooted tree formed from the planted (1,3)–trees T_1, T_2 and T_3. The second term of (3.5) counts symmetries in the case that all three trees are the same. The case that exactly two of the three trees are the same and the case that all three are different are handled by the first and second bracketed terms of (3.5) respectively.

A tree rooted at an edge can be formed by identifying the edges incident to the roots of two planted trees. That edge is the root edge of the edge–rooted tree. When the two trees which are combined are the same, that edge is a symmetry edge. Thus,

$$\frac{1}{2x}(T(x,2)^2 + T(x^2,4))$$

counts symmetries in edge–rooted (1,3)–trees and

$$\frac{2}{x}T(x^2,4)$$

counts symmetries in (1,3)–trees that have a symmetry edge. Combining (3.5), (3.6) and (3.7) as in Lemma 3.2 and using the functional relation (3.2) to simplify gives equation (3.4). ❑

4. Recurrence Relations

From the functional equation (3.1) recurrence relations for $T_{m,n}$, the coefficient of $y^m x^n$ in $T(x,y)$ can now be determined. Note that throughout this section the subscripts on the variables are always non–negative integers. Otherwise one can assume the value of the variable is zero.

The only planted (1,3)–tree with the identity group as its automorphism group occurs when $n = 1$. Otherwise, there is at least one pair of end vertices that can be permuted. Thus, $T_{0,1} = 1$ and $T_{0,n} = 0$ if $n \geq 2$. For $n \geq 2$ and $1 \leq m \leq n-1$, $T_{m,n}$, the number of planted (1,3)–trees on $2n$ vertices (binary trees on $2n-1$ vertices) with 2^m symmetries is as follows:

$$T_{m,n} = T_{(m-1)/2,\, n/2} - \frac{1}{2}T_{m/2,\, n/2} + \begin{cases} 0, \text{ if } m = n-1 \\ \frac{1}{2}\sum_{k=1}^{n-1} \sum_{i} T_{i,k}T_{m-i,n-k}, \text{ if } m \neq n-1 \end{cases}$$

Recall that S_n is the coefficient of x^n in $T(x,2)$ and let B_n and C_n be the coefficients of x^{2n} and x^{3n} in $T(x^2,4)$ and $T(x^3,8)$ respectively. Then as a consequence of equation (3.4), s_n, the coefficient of x^n in $t(x,2)$, can be expressed in terms of S_n, B_n and C_n. For $n \geq 2$,

$$s_n = \frac{1}{2}\sum_{k=1}^{n} S_k S_{n-k+1} - \frac{1}{3}\sum_{i=1}^{n-1}\sum_{j=1}^{n-i} S_i S_j S_{n-i-j+1} + \frac{3}{2}B_{(n+1)/2} + \frac{13}{3}C_{(n+1)/3}$$

Let T_n be the number of planted (1,3)–trees on $2n$ vertices (binary trees on $2n-1$ vertices) and let t_n be the number of free (1,3)–trees on $2n$ vertices. Then equations for T_n and t_n [BaKP81] can be combined with (4.1) and (4.2) respectively to give formulas for the expected number of symmetries in these trees. That is, the expected number of symmetries in a planted (1,3)–tree on $2n$ vertices (binary tree on $2n-1$ vertices) is S_n/T_n and the expected number of symmetries in a free (1,3)–tree on $2n$ vertices is s_n/t_n.

5. Numerical Results

Values of $T_{m,n}$, S_n and s_n were computed using the CDC Cyber 750 in the Computer Laboratory at Michigan State University. The computation of these numbers was limited by the available accuracy and storage restrictions. Another limiting factor was the time required to compute the values using the recurrence relations. In the case of (1,3)–trees, the Fortran programs used to compute S_n for $n \leq 50$ took 52 seconds while an additional 300 seconds were required to compute S_{68}. Table 1 contains values of $T_{m,n}$ for $n = 6$ to 13. Table 2 contains values of S_n and s_n for $n = 6$ to 25.

Table 1. Coefficients of T(x,y) for Planted (1,3)–trees

n	m	$T_{m,n}$
6	1	1
	2	2
	3	2
	4	1
7	1	1
	2	4
	3	3
	4	3
8	1	1
	2	6
	3	7
	4	7
	5	1
	6	0
	7	1
9	1	1
	2	9
	3	14
	4	14
	5	6
	6	1
	7	1
10	1	1
	2	12
	3	28
	4	28
	5	21
	6	4
	7	3
	8	1
11	1	1
	2	16
	3	50
	4	58
	5	54
	6	17
	7	8
	8	3
12	1	1
	2	20
	3	85
	4	119
	5	126
	6	61
	7	27
	8	9
	9	2
	10	1
13	1	1
	2	25
	3	135
	4	239
	5	273
	6	187
	7	80
	8	32
	9	8
	10	3

Table 2. Coefficients of T(x,2) and t(x,2) for (1,3)–trees

n	S_n	s_n
6	42	24
7	90	168
8	354	240
9	758	608
10	2290	920
11	6002	5680
12	18410	6104
13	51310	18416
14	154106	43008
15	449322	148152
16	1384962	325608
17	4089174	980840
18	12475362	2421096
19	37746786	7336488
20	116037642	19769312
21	355367310	58192608
22	1097869386	164776248
23	3393063162	502085760
24	10546081122	1427051544
25	32810171382	4261678656

Values of T_n and t_n and the ratios S_n/T_n and s_n/t_n were computed using the Macintosh II microcomputer. Table 3 contains values of S_n/T_n, the expected number of symmetries in a planted (1,3)–tree on 2n vertices (binary tree on $2n-1$ vertices), and s_n/t_n, the expected number of symmetries in a free (1,3)–tree on 2n vertices, for $n = 6$ to 25.

Table 3. Expected Number of Symmetries in Planted and Free (1,3)–trees

n	Planted	Free
6	7.000	12.000
7	8.182	42.000
8	15.391	40.000
9	16.478	55.273
10	23.367	51.111
11	28.995	153.514
12	40.820	92.485
13	52.197	136.415
14	70.723	162.294
15	92.644	268.391
16	127.002	287.640
17	166.017	406.988
18	222.731	474.911
19	295.100	665.743
20	395.295	829.041
21	525.569	1114.097
22	702.244	1435.383
23	935.721	1973.833
24	1250.072	2522.478
25	1667.160	3367.375

REFERENCES

[BaKP81] C. K. Bailey, J. W. Kennedy and E. M. Palmer, Points by degree and orbit size in chemical trees I, *Theory and Applications of Graphs* (G. Chartrand et al, eds.) Wiley, New York (1981) 27–43.

[BaKP83] C. K. Bailey, J. W. Kennedy and E. M. Palmer, Points by degree and orbit size in chemical trees II, *Discrete Appl. Math.*, **5** (1983) 157–164.

[C57] A. Cayley, On the theory of the analytical forms called trees, *Philos. Mag.* (4) **13** (1857) 172–176 = *Math. Papers*, Vol. 3, 242–246.

[C75] A. Cayley, On the analytical forms called trees, with application to the theory of chemical combinations, *Rep. Brit. Assoc. Advance. Sci.* **45** (1875) 257–305 = *Math. Papers,* Vol. 9, 427–460.

[C81] A. Cayley, On the analytical forms called trees, *Amer. J. Math.* **4** (1881) 266–268 = *Math. Papers*, Vol. 11, 365–367.

[C89] A. Cayley, A theorem on trees, *Quart. J. Pure Appl. Math.* **23** (1889) 376–378 = *Math. Papers*, Vol. 13, 26–28.

[Et38] I. M. H. Etherington, On non–associative combinations, *Proc. Roy. Soc. Edinburgh* **59** (1938/39) 153–162.

[GoK73] M. Gordon and J. W. Kennedy, The graph–like state of matter, Part 2, LGCI schemes for the thermodynamics of alkanes and the theory of inductive inference, *J. Chem. Soc.*, Farady II **69** (1973) 484–504.

[GoL75] M. Gordon and C. G. Leonis, Combinatorial shortcuts to statistical weights and enumeration of chemical isomers, *Proc. 5th British Combinatorial Conf.* (1975) 231–238.

[GoT76] M. Gordon and W. B. Temple, The graph–like state of matter and polymer science, Chapter 10 in *Chemical Applications of Graph Theory* (A. T. Balaban, ed.), Academic, London (1976).

[HP73] F. Harary and E. M. Palmer, *Graphical Enumeration*, Academic, New York (1973).

[HRS75] F. Harary, R. W. Robinson and A. J. Schwenk, Twenty–step algorithm for determining the asymptotic number of trees of various species, *J. Austral. Math. Soc. Ser. A* **20** (1975) 483–503.

[KMPR] J. W. Kennedy, K. A. McKeon, E. M. Palmer and R. W. Robinson, Asymptotic number of symmetries in locally–restricted trees, to appear.

[Mc87] K. A. McKeon, Enumeration of symmetries in locally–restricted trees, Dissertation, Michigan State University, (1987).

[O48] R. Otter, The number of trees, *Ann. of Math.* **49** (1948) 583–599.

[P37] G. Pólya, Kombinatorische Anzahlbestimmungen für Gruppen, Graphen und chemische Verbindungen, *Acta Math.*, **68** (1937) 145–254.

MULTIDIMENSIONAL BANDWIDTH IN RANDOM GRAPHS

Zevi Miller

Miami University

ABSTRACT

Let $[n]^k$ be the graph of the k-dimensional grid of side n. That is, $[n]^k$ has as its vertices the lattice points in k-dimensional space with coordinates between 0 and n, and as its edges those pairs of lattice points differing by exactly 1 in one coordinate. Let G be a random graphs with fixed edge probability and having $N = |[n]^k| = (n+1)^k$ vertices. Given a one to one map of vertices $f : G \to [n]^k$, let $[f] = max[dist_{[n]^k}(f(x), f(y)) : xy \epsilon E(G)]$ and $B(G) = min_f|f|$. Under the assumption $\Omega(k) = n \leq e^{e^{o(k)}}$, we show that $p[diam([n]^k) - C_1 \ log(log \ N) \leq B(G) \leq diam(diam([n]^k) - C_2 \frac{\log(\log \ N)}{\log \ k}] \to 1$ as $k \to \infty$, where C_1 and C_2 are constants.

1. Introduction

Let G and H be graphs with $|G| = |H|$, and $f : V(G) \to V(H)$ a one to one map of the vertex set of G to that of H. Let $|f| = \max[\text{dist}_H(f(x), f(y)) : xy\epsilon E(G))]$ (where dist_H denotes distance in H), and let $B(G, H) = \min_f |f|$.

The investigation of $B(G, H)$ for certain classes of "host graphs" H, apart from its intrinsic interest as a graph theory problem is motivated by issues in VLSI and parallel computation. when H is a path the parameter $B(G, H)$, denoted $b(G)$, is known as the *bandwidth* of G.

The problem of determining $b(G)$ is NP-complete, even when G is a tree of maximum degree three $[GGJK]$. This has prompted research into the probabilistic analysis of $b(G)$. The probable behavior of $b(G)$ for random graphs with fixed edge probability p was investigated in [KMc], where it was shown that for almost all graph on n points we have $b(G) = n - (2 + 2^{\frac{1}{2}} + 0(1))\frac{\log(n)}{\log(\frac{1}{1-p})}$. A similar result is proved in [EHW]. In [Tu] the success of some well known heuristics for $b(G)$ is explained from a probabilistic point of view.

We now consider the case when H is a grid of arbitrary dimension. As notation, let $[n]^k$ denote the graph with vertex set $V = \{(x_1, x_2, ..., x_k) : 0 \leq x_i \leq n,\ x_1$ integers] and edges consisting of those unordered pairs $[x, y]$ from V for which $\sum_{i=1}^{d} |x_i - y_i| = 1$. Thus $V([n]^k)$ may be identified with the lattice points in k-dimensional space having coordinates between 0 and n, where two points are joined by an edge when they are "adjacent" as lattice points. Note that $[n]^k$ has $(n+1)^k$ points and diameter kn. The NP-completeness of finding $B(G, [n]^k)$ when $k = 2$ was shown independently in [Mil], [BCo], and [BSu], and the proofs extend readily to arbitrary dimension. Bounds for $B(G, [n]^k)$ may be derived from [Ro] and [RoSn], and other work by these and other authors.

Given certain upper and lower bounds for n in terms of k, our purpose in this paper is to give reasonably tight upper and lower bounds for $B(G, [n]^k)$ which are valid with probability approaching 1 as $k \to \infty$.

2. Grid Bandwidth in Random Graphs

We take as our probability model the set $G(N, p)$ of all labelled graphs on $N = (n+1)^k$ points, with labels from $\{1, 2, ..., N\}$, having edge probability p where p is fixed. We set $q = 1 - p$, and we denote an element of $G(N, p)$ by G. Under this model the probability that two points i and j are joined by an edge in G is p, and the two events $i_1 j_1 \epsilon E(G)$ and $i_2 j_2 \epsilon E(G)$ are independent for distinct pairs of points $i_1 j_1$ and $i_2 j_2$. For applications of this model to many different problems in graphs theory the reader is referred to [Bo] and [Pa].

In this section we will study the probable behavior of $B(G, H)$, where $G \in G(N, p)$ and $H = [n]^k$. As H will always have this meaning, we abbreviate

$B(G, H)$ by $B(G)$ from now own.

We use the standard notation $O(s(n)), o(s(n))$, and $\Omega(s(n))$ to denote any function $g(n)$ for which the absolute ration $|\frac{g(n)}{s(n)}|$ is bounded above by a constant, approaches 0, and is bounded below by a constant respectively as n approaches ∞.

We will need the following facts for later use. The first gives standard bounds on binomial coefficients, and the second follows readily from Taylor's theorem.

Lemma 1

a) $(\frac{n}{k}) \leq \binom{n}{k} \leq (\frac{ne}{k})^k$ for any $k \leq n$.

b) Suppose $r(k)$ is a function satisfying $r(k) = o(k)$. Then $(1 + \frac{r(k)}{k})^k \sim \exp(O(r(k)))$ as $k \to \infty$.

We now introduce some notation. A point of $[n]^k$ will be called an *extreme point* if each of its coordinates is 0 or n. Note that there are 2^k such points in $[n]^k$. For each extreme point v let $\underline{D_i(v)}$ be the set of points in $[n]^k$ at distance i from v and let $\underline{Nb(v,r)} = \bigcup_{i=0}^{r} D_i(v)$. We call the sets $D_i(v)$ the *diagonals* at v and $Nb(v,r)$ the *r-neighborhood* about v, and note that by symmetry $|Nb(v,r)| = |Nb(w,r)|$ for any two extreme points v and w. For $i \leq n$ we can identify $D_i(v)$ with the set of nonnegative integer sequences $(x_1, x_2, ..., x_k)$ such that $\sum_{j=i}^{k} x_j = i$, so $|D_i(v)| = \binom{i+k-1}{k-1}$ for $i \leq n$. As $|D_i(v)|$ is independent of v, we denote it by $\underline{D_i}$. Also for $r \leq n$ we then have $|Nb(v,r)| = \sum_{i=o}^{r} \binom{i+k-1}{k-1} = \binom{r+k}{k}$.

Let $r \geq 0$ be and consider a sequence of sets $(S_0, S_1, ..., S_r)$ where S_i is a subset of $G \epsilon G(n,p)$ of size D_i, and $S_i \cap S_j = \phi$ for $i \neq j$. The set $\bigcup_{i=0}^{r} S_i$ will be called an *r-corner* of G (or just corner when r is understood). Thus an r-corner in G has the same size as an r-neighborhood about an extreme point in $[n]^k$, and we will refer to the sets S_i as the *diagonals* of the corner. Now let $v(1), v(2), ..., v(2^k)$ be a fixed ordering of the extreme points of $[n]^k$. A sequence $(S^{(1)}, S^{(2)}, ..., S^{(2^k)})$ of r-corners of G will then be called an *r-system* if the set systems $\bigcup_{j=0}^{2^k} S_j$ and $\bigcup_{j=0}^{2^k} Nb(v(j), r)$ (with edge sets $[S^{(j)} : 1 \leq j \leq 2^k]$ and $[Nb(v(j), r) : 1 \leq j \leq 2^k]$ respectively) are isomorphic by an isomorphism under which the induced map on edges sends $S^{(j)}$ to $Nb(v(j), r)$ for each j. As motivation for these definitions note that if $f : G \to [n]^k$ is an embedding, then the inverse image of the collection $\{Nb(v, r) : v$ extreme $\}$ must be an r-system while the inverse image of each $Nb(v, r)$ must be an r-corner of that r-system.

We will assume now that the ordering $v(i)$ of the extreme points is such that for any odd i the pair $\{v(i), v(i+1)\}$ are antipodal (i.e. at distance kn). Let S and T be two r-corners of G (with diagonal sets $\{S_i\}$ and $\{T_i\}$ respectively) such that each point $v \epsilon S_i$ is nonadjacent in G to all points in $\bigcup_{j=0}^{r-1} T_j$. (It follows symmetrically that any point $v \epsilon T_i$ is nonadjacent in G to all points in $\bigcup_{j=0}^{r-i} S_j$). We then call S and T *opposite corners*. An r-system will then be termed a *good r-system* if each pair $\{S^{(i)}, S^{(i+1)}\}$ with i odd are opposite corners. If S is a good r-system of G, then the required set of nonadjacencies among the 2^{k-1} opposite corner pairs of S will be referred to as the *nonadjacencies implied by* S.

The significance of a good r-system is seen in the following lemma.

Lemma 2 $B(G) \leq$ diam$([n]^k) - r - 1 \leftrightarrow G$ has a good r-system.

Proof Observe that points $s, t \epsilon [n]^k$ satisfy dist$(s, t) \geq kn - r$ iff there are antipodal extreme points $v(i)$ and $v(i+1)$ such that (without loss of generality) $s \epsilon D_j(v)$ and $t \epsilon \bigcup_{t=0}^{r-j} D_t(v(i+1))$ for some $j \leq r$.

Now let $f : G \to [n]^k$ be an embedding. It follows that $|f| \leq kn - r - i$ if and only if for each extreme point $v(i) \epsilon [n]^k$ (say with i odd) the two sets $f^{-1}(Nb(v(i), r))$ and $f^{-1}(Nb(v(i+1), r))$ are opposite corners in G. Thus $|f| \leq kn - r - 1$ if and only if the sequence $(f^{-1}(Nb(v(1), r)),$

$f^{-1}(Nb(v(2),r)),\ldots,f^{-1}(Nb(v(2^k),r)))$ is a good r-system in G.

We may now proceed to our theorem.

Theorem Let $N=(n+1)^k$. Assume $\Omega(k)=n\le e^{e^{o(k)}}$. Then there are constants C_1 and C_2 such that for p fixed and $G\epsilon G(N,p)$ we have $P[\mathrm{diam}([n]^k)-C_1\log(\log N)\le B(G)\le \mathrm{diam}([n]^k)-C_2\frac{\log(\log N)}{\log k}]\to$ as $k\to\infty$.

Toward proving this theorem we begin with a lemma.

Lemma 3 The number of nonadjacencies implies by an r-system for $r\le n$ is $2^{k-1}\binom{2k+r}{r}$.

Proof Consider an opposite corner pair S and T (with diagonal sets $\{S_i\}$ and $\{T_i\}$ respectively) belonging to an r-system. For a given $i\le r$ we have $|\bigcup_{m=0}^{r-1}T_m|=\sum_{m=0}^{r-i}\binom{m+k-1}{k-1}=\binom{r-i+k}{k}$. Thus the number of nonadjacencies between S_i and $\bigcup_{m=0}^{r-i}T_m$ is $\binom{i+k-1}{k-1}\binom{r-i+k}{k}$. Hence the number of implied nonadjacencies among S and T is

$$\sum_{i=0}^{r}\binom{i+k-1}{k-i}\binom{r-i+k}{k}=\sum_{i=0}^{r}\binom{i+k-1}{i}\binom{r-i+k}{r-i}=\binom{2k+r}{r}.$$

Since there are 2^{k-1} pairs of opposite corners in an r-system, the lemma follows.

As further notation, for $G\epsilon G(n,p)$ let $N_r(G)$ be the number of r-systems in G. Also if $n=rs$ then let $\binom{n}{rX_s}$ denote the multinomial coefficient $\binom{n}{r,r,\ldots r}$ where the r appears s times in the bottom.

Proof of Theorem First we show that $P[B(G)>\mathrm{diam}([n]^k)-$ C $\log(\log N)]\to 1$ as $k\to\infty$.

First observe that with $r=$ C $\log(\log N)$ and k sufficiently large we have

$$E(N_r)=\left(2^k\binom{r^N+k}{k}\right)\left(\frac{2^k\binom{r+k}{k}}{\binom{r+k}{k}\times 2^k}\right)\left(\frac{\binom{r+k}{k}}{\binom{k-1}{k-1}\binom{1+k-1}{k-1}\ldots\binom{r+k-1}{k-1}}\right)^{2^k}(2^k)!q^{\binom{r+2k}{r}2^{k-1}}$$

To see this, note that since $r = \log(k) + \text{loglog}(n) = o(k)$ and $n = \Omega(k)$ it follows that for k sufficiently large the r-corners making up an r-system are pairwise disjoint. Hence the number of points in an r-system for large k is $2^k\binom{r+k}{k}$. Thus the first factor on the right is the number of ways of choosing the ground set of $2^k\binom{r+k}{k}$ vertices for our r-system. The second factor is the number of ways to partition these vertices into a collection of 2^k corners. The third factor is the number of ways to partition each of the corners into the diagonals $S_0, S_1, S_2, ..., S_r$, these partitions being independent. The fourth factor is the number of different orderings of the 2^K corners. Thus the product of the first four factors is the number of r-systems. The exponent of q is the number of disallowed edges in a good r-system by lemma 3, so the fifth factor is the probability that an r-system is a good r-system. Thus the full product is the expectation of N_r.

We now estimate $E(N_r)$. First we will bound the multinomial coefficients by powers of binomial coefficients. The one with $2^k\binom{r+k}{k}$ as the top is clearly at most $\binom{2^k\binom{r+k}{k}}{\binom{r+k}{k}}^{2^k}$. The one with $\binom{r+k}{k}$ as the top may be written as the product $\prod_{j=0}^{r}\binom{\binom{j+k}{k}}{\binom{j+k-1}{k-1}}$ using the fact that for any integer $j \geq 0$ we have $\sum_{i=0}^{j}\binom{i+k-1}{k-1} = \binom{j+k}{k}$. Now some analysis of binomial coefficients shows that the biggest among the $r+1$ factors in this product is $\binom{\binom{r+k}{k}}{\binom{r+k-1}{k-1}}$. Hence the multinomial coefficient is bounded above by $\binom{\binom{r+k}{k}}{\binom{r+k-1}{k-1}}^{r+1}$.

Now substituting for the multinomial coefficients by the above bounds, and using lemma 1 to upper bound all binomial coefficients, one can show that if $r >$ C log(logN) for a suitable constant C, then $E(N_r) \to 0$ as $k \to \infty$. We omit details of this analysis here, as they will appear later in the full paper. Now using lemma 2 and Markov's inequality we have $P[B(G) \leq \text{diam}([n]^k) - r - 1] = P[N_r \geq 1] \leq E(N_r) \to 0$ as $k \to \infty$ for $r >$ C log(logN). Hence $P[B(G) > \text{diam}([n]^k) -$ C log(logN)$] \to 1$ as $k \to \infty$, proving half of our theorem.

To prove that $P[B(G) \leq \text{diam}([n]^k) - C_2 \frac{\log(\log N)}{\log k}] \to 1$ we will apply the second moment method.

Let R be the set of r-systems. For each $A \epsilon R$ let

$$Z_A(G) = \begin{cases} 1 & \text{if } A \text{ is a good r-system} \\ 0 & \text{otherwise} \end{cases}$$

Note that $N_r(G) = \sum_{A \epsilon R} Z_A(G)$.

We begin by estimating the variance of N_r. Let F be a fixed member of R. Then

$$\begin{aligned} \text{Var}(N_r) &= E(N_r^2) - E(N_r)^2 \\ &= \textstyle\sum_{A \epsilon R} \sum_{B \epsilon R} (E(Z_A Z_B) - E(Z_A)E(Z_B)) \\ &= |R| \textstyle\sum_{A \epsilon R} (E(Z_F Z_A) - E(Z_F)E(Z_A)). \end{aligned}$$

To estimate $E(Z_F Z_A)$, note first that since Z_F and Z_A are 0-1 valued random variables we have $E(Z_F Z_A) = P(Z_F Z_A = 1) = P(Z_A = 1 | Z_F = 1).P(Z_F = 1)$.

By lemma 2 we have $P(Z_F = 1) = q^{\binom{2k+r}{r} 2^{k-1}}$. Now given that F is a good r-system the probability that A is a good r-system is just the probability that A has the required nonadjacencies not already implied by F. This probability is $q^{\binom{2k+r}{r} 2^{k-1} - c(A)}$, where $c(A)$ is the number of nonadjacencies implied by jointly by A and F. Thus $E(Z_F Z_A) = q^{\binom{2k+r}{r} 2^k - c(A)}$.

We can now develop the above expression a bit further. Let $h = 2^k \binom{r+k}{k}$, and let v_t be the set of elements of R having t points in common with F. Also let $c(t)$ be the maximum number of nonadjacencies implied jointly by F and any member of v_t. Then continuing from above we have

$$\begin{aligned}\mathrm{Var}(N_r) &= |R|q^{\binom{2k+r}{r}2^k}\textstyle\sum_{A\epsilon R}(q^{-c(A)}-1)\\ &\le |R|q^{\binom{2k+r}{r}2^k}\sum_{t=2}^{h}|\nu_t|q^{-c(t)}\\ &= |R|^2q^{\binom{2k+r}{r}2^k}\sum_{t=2}^{h}\frac{|\nu_t}{|R|}q^{-c(t)}\\ &= |E|(N_r)^2\sum_{t=2}^{h}\frac{|\nu_t}{|R|}q^{-c(t)}\end{aligned}$$

We therefore have

$$\frac{\mathrm{Var}(N_r)}{E(N_r)^2}\le\sum_{t=2}^{h}\frac{|\nu_t|}{|R|}q^{-c(t)} \qquad (A1)$$

It remains to estimate the right side of the last expression. First note that we may write $\nu_t = N_1N_2$, where N_1 = number of ways to choose h points from $\{1,2,...,N\}$ of which t are in the ground set of F, and N_2 = number of ways to arrange any ground set counted in N_1 into a member of $|R|$.

We have $N_1 = \binom{h}{t}\binom{N-h}{h-t}$, while $N_2 = \frac{|R|}{\binom{N}{h}}$. Therefore from (A1) we get

$$\frac{\mathrm{Var}(N_r)}{E(N_r)^2}\le\sum_{t=2}^{h}\frac{\binom{h}{t}\binom{N-h}{h-t}}{\binom{N}{h}}q^{-c(t)} \qquad (A2)$$

We recognize the coefficients of the $q^{-c(t)}$ in this sum as the terms of the hypergeometric distribution. In the course of approximating the hypergeometric by the Poisson distribution one can derive the following inequality [Bo pp. 7-8].

$$\frac{\binom{h}{t}\binom{N-h}{h-t}}{\binom{N}{h}}\le\frac{(\frac{h^2}{N})^t\mathrm{ext}(-\frac{h^2}{N})}{t!}(1-\frac{t}{N})^{t-h}\mathrm{ext}(\frac{ht}{N})$$

(This is obtained by first approximating the hypergeometric by the binomial distribution b(t; h,p) with success probability $p=\frac{h}{N}$, and then the binomial by the Poisson distribution with mean $\lambda = ph = \frac{h^2}{N}$).

Next we need an estimate for $c(t)$. Consider then a set $Q\epsilon\nu_t$ at which the maximum defined by $c(t)$ is realized. Thus $c(t)$ is the number of nonadjacent pairs xy, x, y ϵ Q $\cap$ F, implied jointly by Q and F. Now $c(t) = \frac{1}{2}\sum_{x \epsilon Q\cap F} {}^{d(x)}$, where $d(x)$ is the number of nonadjacencies implied jointly by Q and F in which x is a member. Since $Q\epsilon R$ we have $d(x) \leq \binom{r+k}{k}$, with equality occurring precisely when there is a corner pair S, T in Q for which $x = S_0$ and the points counted in $d(x)$ comprise all of T. Thus $c(t) \leq \frac{1}{2}t\binom{r+k}{k}$.

Now using Chebyshev's inequality and (A2) we therefore have $P(B > \mathrm{diam}([n]^k) - r - 1) = P(N_r = 0)$

$$\leq \frac{\mathrm{Var}(N_r)}{E(N_r)^2}$$

$$\leq \exp(-\frac{h^2}{N})\sum_{t=2}^{h}\frac{(\frac{\;}{h^2}N)^t}{t!}(1-\frac{t}{N})^{t-h}\exp\frac{(}{ht}N)q^{-t\binom{r+k}{k}}$$

$$\leq \exp(-\frac{h^2}{N})(1-\frac{h}{N})^{-h}[\exp\{\frac{h^2}{N}\exp\frac{(}{h}N)q^{-\binom{r+k}{k}}\} - 1 - \frac{h^2}{N}\exp\frac{h}{N}q^{-\binom{r+k}{k}}] \quad (A3)$$

Now using lemma 1 and the assumed upper bound on n one can show that if $r \leq C\frac{\mathrm{loglog}(N)}{\log(k)}$ for an appropriate constant C, then the right side of (A3) approaches 0 as $k \to \infty$. Again the details will appear later in the full paper. It follows that almost every graph satisfies $B(G) \leq \mathrm{diam}([n]^k) - C\frac{\mathrm{loglog}(N)}{\log(k)}$ and the theorem is proved.

Remark J. Spencer [Sp] points out the following. From random graph theory we know that almost every graph on N points has essentially $\frac{N}{\log N}$ disjoint independent sets of size logN. We can use these sets as r-corners in constructing a good r-system. The value of r is determined by the requirement logN $\sim \binom{r+k}{k} \sim r^k$ for fixed k as $n \to \infty$ (where $N = (n+1)^k$). Thus for fixed k we have $B(G) \leq \mathrm{diam}([n]^k)-$ (logN)$^{1/k}$ for almost all graphs.

Now for $n \geq e^{e^{0(k^{1+\epsilon})}}$ this upper bound on $B(G)$ is an improvement on the one presented here. In fact it is less than the lower bound we have, which shows that the hypothesis $n \leq e^{e^{o(k)}}$ in our theorem cannot be relaxed very much. But under this hypothesis the upper bound of our theorem in the range $n \geq e^{k^{1+\epsilon}}$ is an improvement on the one above.

REFERENCES

[1] [BSu] P. Bertolazzi, I. H. Sudborough, "The grid embedding problem is NP-complete even for edge length 2", *Technical Report* (1983), EE/CS Dept., Northwestern University.

[2] [BCo] S. Bhatt, S. Cosmadakis, "The complexity of minimizing wire lengths in VLSI layouts", manuscript (1983).

[3] [Bo] B. Bollobas, "Random Graphs", *Academic Press* (1985).

[4] [EHW] P. Erdös, P. Hell, P. Winkler, "Bandwidth versus bandsize", manuscript (1985).

[5] [GGJK] M. Garey, R. Graham, D. Johnson, D. Knuth, "Complexity results for bandwidth minimization", SIAM J. of Applied Mathematics 34 (1978) 477-495.

[6] [KMc] Y. Kuang, C. McDiarmid, "On the bandwidth of a random graph", manuscript (1985).

[7] [Mil] Z. Miller, J. Orlin, "NP-completeness for minimizing maximum edge length in grid embeddings", *J. of Algorithms* 6, (1985) 10-16.

[8] [Pa] E. M. Palmer, *Graphical Evolution: An Introduction to the Theory of Random Graphs*, John Wiley & Sons (1985).

[9] [Ro] A. Rosenberg, "On embedding graphs in grids", *IBM Research Report RC* 7559 (#2668) (1970).

[10] [RoSn] A. Rosenberg, L. Snyder, "Bounds on the costs of data encodings", *Mth. Systems Theory* 12 (1978) 9-39.

[11] [Sp] J. Spencer, private communication.

[12] [Tu] J. Turner, "Probabilistic analysis of bandwidth minimization algorithms", *Proc. 15th Annual ACM Symposium on Theory of Computing* (1983) 467-476.

THE LAPLACIAN SPECTRUM OF GRAPHS

Bojan Mohar

University of Ljubljana

Jadranska 19, 61111 Ljubljana, Yugoslavia

ABSTRACT

The paper is essentially a survey of known results about the spectrum of the Laplacian matrix of graphs with special emphasis on the second smallest Laplacian eigenvalue λ_2 *and its relation to numerous graph invariants, including connectivity, expanding properties, isoperimetric number, maximum cut, independence number, genus, diameter, mean distance, and bandwidth-type parameters of a graph. Some new results and generalizations are added.*

1. Introduction

The Laplacian matrix of a graph and its eigenvalues can be used in several areas of mathematical research and have a physical interpretation in various physical and chemical theories. The related matrix - the adjacency matrix of a graph and its eigenvalues were much more investigated in the past than the Laplacian matrix. The reader is referred to the monographs [CDS, CDGT]. However, in the author's opinion the Laplacian spectrum is much more natural and more important than the adjacency matrix spectrum. It is the aim of this survey paper to explain where this belief comes from.

The work supported in part by the Research Council of Slovenia, Yugoslavia. Part of the work was done while the author was a Fulbright Scholar at the Ohio State University, Columbus, Ohio.

We shall use the standard terminology of graph theory, as it is introduced in most textbooks on the theory of graphs (e.g., [Wi]). Our graphs are unoriented, but they may have loops and multiple edges. We also allow *weighted graphs* which are viewed as a graph which has for each pair u, v of vertices, assigned a certain *weight* $a_{u,v}$. The weights are usually real numbers and they must satisfy the following conditions:

(i) $a_{u,v} = a_{vu}, v, u \in V(G)$, and

(ii) $a_{vu} \neq 0$, if and only if v and u are adjacent in G.

Usually the additional condition on the non-negativity of weights is assumed:

(iii) $a_{uv} \geq 0, v, u \in V(G)$.

It will be clear from the context or otherwise explicitly specified if a graph is weighted. Unweighted graphs can be viewed as a special case of weighted graphs, by specifying, for each $u, v \in V(G)$, the weight a_{uv} be equal to the number of edges between u and v. The matrix $A = A(G) = [a_{uv}]_{u,v \in V(G)}$, is called the *adjacency matrix* of the graph G. We shall use the same name for the matrix of weights if the graph is weighted.

Let $d(v)$ denote the degree of $v \in V(G), d(v) = \sum_u a_{uv}$, and let $D = D(G)$ be the diagonal matrix indexed by $V(G)$ and with $d_{vv} = d(v)$. The matrix $Q = Q(G) = D(G) - A(G)$ is called the *Laplacian matrix* of G. It should be noted at once that loops have no influence on $Q(G)$. The matrix $Q(G)$ is sometimes called the *Kirchhoff matrix* of G due to its role in the well-known Matrix-Tree theorem (cf. §4)which is usually attributed to Kirchhoff. Another name, the *matrix of admittance*, comes from the theory of electrical networks (admittance = conductivity). It should be mentioned here that the rows and columns of graph matrices are indexed by the vertices of the graph, their order being unimportant. The matrix $Q(G)$ acts naturally on the vector space $\ell^2(V(G))$. For any vector $x \in \ell^2(V(G))$ we denote its coordinates by $x_v, v \in V(G)$.

Throughout the paper we shall denote by $\mu(G, x)$ the characteristic polynomial of $Q(G)$. Its roots will be called the *Laplacian eigenvalues* (or sometimes just *eigenvalues*) of G. They will be denoted by $\lambda_1 \le \lambda_2 \le \cdots \le \lambda_n (n = |V(G)|)$, always enumerated in increasing order and repeated according to their multiplicity. We shall use the notation $\lambda_k(G)$ to denote the k-th smallest eigenvalue of the graph G (counting multiplicities). The letter n will always stand for the order of G, so $\lambda_n(G)$ will be the maximal eigenvalue of $Q(G)$.

Let G be a given graph. Orient its edges arbitrarily, i.e. for each $e \in E(G)$ choose one of its ends as the *initial* vertex, and name the other end the *terminal* vertex. The *oriented incidence matrix* of G with respect to the given orientation is the $|V| \times |E|$ matrix $C = [c_{ve}]$ with entries

$$c_{ve} = \begin{cases} +1, & \text{if } v \text{ is the terminal vertex of } e, \\ -1, & \text{if } v \text{ is the initial vertex of } e, \\ 0, & \text{if } v \text{ and } e \text{ are not incident.} \end{cases}$$

It is well known that

$$Q(G) = CC^t \tag{1.1}$$

independent of the orientation given to the edges of G (cf., e.g., [Bi]). It should be noted that (1.1) immediately implies the formula (2.1) since the inner product $(Q(G)x, x)$ is equal to $(CC^t x, x) = (C^t x, C^t x)$.

The Laplace differential operator Δ is one of the basic differential operators in mathematical physics. One looks for non-trivial solutions of $\Delta\phi = \lambda\phi$ on a certain region Ω. By discretizing the Laplace equation one gets the Laplacian matrix Q of the discretisized space (usually a graph). We mention that, by this correspondence, the oriented incidence matrix C, as defined above, corresponds to the gradient operator, and so (1.1) has a clear physical interpretation.

In Section 2 we review the basic spectral properties of $Q(G)$. The next section presents the results on the spectra of graphs obtained by means of some operations on graphs, including the disjoint union, Cartesian product and the join of graphs, deleting or inserting an edge, the complement, the line graph, etc. Section 4 is devoted to the renowned application of $Q(G)$, the Matrix-Tree-Theorem, which expresses the number of spanning trees of a graph in terms of its non-zero eigenvalues.

There are many problems in physics and chemistry where the Laplacian matrices of graphs and their spectra play the central role. Some of the applications are mentioned in Section 5. It is worth noting that the physical background served as the idea of a well known algorithm of W. T. Tutte [T] for testing planarity and constructing "nice" planar drawings of 3-connected planar graphs.

The second smallest Laplacian eigenvalue λ_2 plays a special role. Recently its applications to several difficult problems in graph theory were discovered (e.g., the expanding properties of graphs, the isoperimetric number, and the maximum cut problem). Section 6 presents these applications, including the relation of λ_2 to the diameter and the mean distance of a graph. In addition, a relation of λ_2 to the independence number, genus, and bandwidth-type invariants is presented. The structure of the eigenvectors corresponding to λ_2 is discussed in the next section. The last section covers a few other results on $Q(G)$ and its applications.

There are some new results in this paper. Many of them are more or less trivial and have probably been known to researchers in the field, although not published before. The results surveyed in the paper are biased by the viewpoint of the author. We apologize to all who feel that their work is missing in the references, or has not been emphasized sufficiently in the text.

2. Basic Properties

The following properties were established by several authors [K3, V, AnM] for the case of unweighted graphs. The proofs carry over to the weighted case if all the weights are non-negative.

Theorem 2.1 Let G be a (weighted) graph with all weights non-negative.

Then:

(a) $Q(G)$ has only real eigenvalues,

(b) $Q(G)$ is positive semidefinite,

(c) its smallest eigenvalue if $\lambda_1 = 0$ and a corresponding eigenvector is $(1, 1, ..., 1)^t$. The multiplicity of 0 as an eigenvalue of $Q(G)$ is equal to the number of components of G.

We mention that the positive semidefiniteness of $Q(G)$ follows from the next useful expression for the inner product $(Q(G)x, x)$ which holds also in the weighted case:

$$(Q(G)x, x) = \sum_{vu \in E} a_{vu}(x_v - x_u)^2 \tag{2.1}$$

Let $\lambda_1 \leq \lambda_2 \leq \cdots \leq \lambda_n$ be the eigenvalues of $Q(G)$ in increasing order and repeated according to their multiplicity. So, $\lambda_1 = 0$, and $\lambda_2 > 0$ if and only if G is connected. The following bounds for the eigenvalues are known.

Theorem 2.2 Let G be a graph of order n. Then:

(a) [F1] $\lambda_2 \leq \frac{n}{n-1} \min\{d(v); v \in V(G)\}$.

(b) [AnM] $\lambda_n \leq \max\{d(u) + d(v); uv \in E(G)\}$. If G is connected then the equality holds if and only if G is bipartite semiregular.

(c) [K3] If G is a simple graph then $\lambda_n \leq n$ with equality if and only if the complement of G is not connected.

(d) $\sum_{i=1}^{n} \lambda_i = 2|E(G)| = \sum_{v} d(v)$.

(e) [F1] $\lambda_n \geq \frac{n}{n-1} \max\{d(v); v \in V(G)\}$.

(f) [MM, p. 168] $\lambda_n \geq \max\{\sqrt{(d(v) - d(u))^2 + 4a_{uv}^2}; v, u \in V(G), v \neq u\}$.

Let G be a (weighted) graph and $V_1 \cup V_2 \cup ... \cup V_k$ a partition of its vertex set. This partition is said to be *equitable* if for each $i, j = 1, 2, ..., k$ there is a number d_{ij} such that for each $v \in V_i$ there are exactly d_{ij} edges between v and vertices in $V_j(\sum_{u \in v_j} a_{vu} = d_{ij}, v \in V_i)$. The name equitable partition was introduced by Schwenk. There are several other terms used for the same thing (e.g., divisor [CDS], coloration degree refinement, etc.).

Theorem 2.3 Let $V_1 \cup V_2 \cup ... \cup V_k$ be an equitable partition of G with parameters $d_{ij}(i, m = 1, 2, ..., k)$, and let $B - [b_{ij}]_{i,j=1,...,k}$ be the matrix defined by

$$b_{ij} \begin{cases} -d_{ij}, & \text{if } i \neq j \\ (\sum_{s=1}^{k} d_{is}) - d_{ii}, & \text{if } i = j. \end{cases}$$

If λ is an eigenvalue of B then λ is also an eigenvalue of $Q(G)$.

Proof Let $Bx = \lambda x, x = (x_1, ..., x_k)^t$. Let $y = (y_v)_{v \in V(G)}$ be defined by: if $x \in V_i$ then let $y_v = x_i$. Now it is not too difficult to verify that $Q(G)y = \lambda y$. Let $v \in V_i$ be any vertex of G. Then

$$(Qy)_v = d(v)y_v - \sum_u a_{vu}y_u = (\sum_{j=1}^{k} d_{ij})x_i - \sum_{j=1}^{k} d_{ij}x_j =$$
$$= (Bx)_i = \lambda x_i = \lambda y_v.$$

Note We may view $D = [d_{ij}]$ as the matrix of a weighted *directed* graph with k vertices. Then B is just its Laplacian matrix.

Let us mention briefly that equitable partitions of vertices arise in many important situations. For example, if $p : \tilde{G} \rightarrow G$ is a graph covering projection (in the sense of topology) then the fibres $p^{-1}(v), v \in V(G)$, form an equitable partition of $\tilde{G}$. The corresponding matrix B is just $Q(G)$, and this shows that the Laplacian spectrum of $\tilde{G}$ contains the spectrum of G. Many examples of equitable partitions of a graph G are obtained by taking, as the classes of a partition, the orbits of some group of automorphisms of the graph G.

3. Operations on Graphs and the Resulting Spectra

Many published works relate the Laplacian eigenvalues of graphs with the eigenvalues of graphs obtained by means of some operations on the graphs we start with. The first result is obvious.

Theorem 3.1 Let G be the disjoint union of graphs $G_1, G_2, ..., G_k$. Then

$$\mu(G, x) \prod_{i=1}^{k} \mu(G_i, x).$$

Let G be a (weighted) graph and let $G' = G + 3$ be the graph obtained from G by inserting a new edge e into G (possibly increasing the multiplicity of an existing edge). Then $Q(G')$ and $Q(G)$ differ by a positive semidefinite matrix of rank 1. It follows by the Courant-Weyl inequalities (see, e.g., [CDS, Theorem 2.1]) that the following is true.

Theorem 3.2 The eigenvalues of G and $G' = G + e$ interlace:

$$0 = \lambda_1(G) = \lambda_1(G') \leq \lambda_2(G) \leq \lambda_2(G') \leq \lambda_3(G) \leq \ldots \leq \lambda_n(G) \leq \lambda_n(G').$$

We notice that $\sum_{i=1}^{n}(\lambda_i(G') - \lambda_i(G)) = 2$ by Theorem 2.2(d), so that at least one inequality $\lambda_i(G) \leq \lambda_i(G')$ must be strict.

By inserting more than one edge we may loose interlacing of the eigenvalues. Nevertheless, there is an important result on λ_2.

Theorem 3.3 Let $G = G_1 \bigoplus G_2$ be a factorization of a graph G. Then
(a) [F1] $\lambda_2(G) \geq \lambda_2(G_1) + \lambda_2(G_2)$.
(b) $\max\{\lambda_n(G_1), \lambda_n(G_2)\} \leq \lambda_n(G) \leq \lambda_n(G_1) + \lambda_n(G_2)$.

Note In Theorem 3.3, graphs may be weighted with non-negative weights. In that case the factorization means that the uv-weight in G is the sum of the uv-weights in G_1 and G_2.

Proof (b) $G = G_1 \bigoplus G_2$ means that $Q(G) = Q(G_1) + Q(G_2)$. Then

$$\begin{aligned}\lambda_n(G) &= \max_{||x||=1}(Q(G)x, x) = \max_{||x||=1}[(Q(G_1)x, x) + (Q(G_2)x, x)] \leq \\ &\leq \max_{||x||=1}(Q(G_1)x, x) + \max_{||x||=1}(Q(G_2)x, x) = \lambda_n(G_1) + \lambda_n(G_2).\end{aligned}$$

The other inequality follows from a similar consideration.

Corollary 3.4 [F1] If G_1 is a spanning subgraph of G_2 then $\lambda_2(G_1) \leq \lambda_2(G_2)$.

Fiedler [F1] derived also a result about the Cartesian products of graphs.

Theorem 3.5 [F1] The Laplacian eigenvalues of the Cartesian product $G_1 \times G_2$ of graphs G_1 and G_2 are equal to all the possible sums of eigenvalues of the two factors:

$$\lambda_i(G_1) + \lambda_j(G_2),\ i = 1, ..., |V(G_1)|, j = 1, ..., |V(G_2)|.$$

By applying Theorem 3.5 we can easily determine the spectrum of "lattice" graphs. The $m \times n$ lattice graph is just the Cartesian product of paths, $P_m \times P_n$. The spectrum of P_k is [AnM]

$$\ell_i^{(k)} = 4\sin^2(\frac{\pi i}{2k}),\quad i = 0, 1, ..., k-1$$

so $P_m \times P_n$ has eigenvalues

$$\lambda_{i,j} = \ell_i^{(m)} + \ell_j^{(n)} = 4\sin^2(\frac{\pi i}{2m}) + 4\sin^2(\frac{\pi i}{2n}).$$

The next two results were first observed by Kel'mans [K1,K2].

Theorem 3.6 [K1,K2] If $\overline{G}$ denotes the complement of the graph G then

$$\mu(\overline{G}, x) = (-1)^{n-1} \frac{x}{n-x} \mu(G, n-x)$$

and so the eigenvalues of $\overline{G}$ are $\lambda_1(\overline{G}) = 0$, and

$$\lambda_{i+1}(\overline{G}) = n - \lambda_{n-i+1}(G),\quad i = 1, 2, ..., n-1.$$

Note Theorem 3.6 has a generalization to the weighted case, see [MP], if we define the weights of $\overline{G}$ to be $a'_{uv} = 1 - a_{uv} (u \neq v)$.

Corollary 3.7 [K1,K2] Let $G_1 * G_2$ denote the join of G_1 and G_2, i.e. the graph obtained from the disjoint union of G_1 and G_2 by adding all possible edges $uv, u \in V(G_1), v \in V(G_2)$. Then

$$\mu(G_1 * G_2, x) = \frac{x(x - n_1 - n_2)}{(x - n_1)(x - n_2)} \mu(G_1, x - n_2)\mu(G_2, x - n_1).$$

where n_1 and n_2 are orders of G_1 and G_2, respectively.

Let G be a simple unweighted graph. The *line graph* $L(G)$ of G is the graph whose vertices correspond to the edges of G with two vertices of $L(G)$ being adjacent if and only if the corresponding edges in G have a vertex in common. The *subdivision graph* $S(G)$ of G is obtained from G by inserting, into each edge of G, a new vertex of degree 2. The *total graph* $T(G)$ of G has its vertex set equal to the union of vertices and edges of G, and two of them being adjacent if and only if they are incident or adjacent in G.

Theorem 3.8 [K3] Let G be a d-regular simple graph with m edges and n vertices. Then

(a) $\mu(L(G), x) = (x - 2d)^{m-n}\mu(G, x)$

(b) $\mu(S(G), x) = (-1)^m(2 - x)^{m-n}\mu(G, x(d + 2 - x))$

(c) $\mu(T(G), x) = (-1)^m(d + 1 - x)^n(2d + 2 - x)^{m-n}\mu(G, \frac{x(d+2-x)}{d+1-x})$.

The part (a) of Theorem 3.8 was also obtained by Vahovskii [V]. Theorem 3.8(a) can be proved also for bipartite semiregular graphs. Recall that a graph G is *r,s)-semiregular* if it is bipartite with a bipartition $V = U \cup W$ such that all vertices in U have degree r and all vertices in W have degree s.

Theorem 3.9 If G is a simple (r,s)-semiregular graph then

$$\mu(L(G), x) = (-1)^n(x - (r + s))^{m-n}\mu(G, r + s - x).$$

Proof Orient the edges of G in the direction from U to W ($U \cup W$ is a semiregular bipartition) and let C be the oriented incidence matrix of G with respect to this orientation. Then

$$CC^t = Q(G) \text{ and } C^tC = 2I + A(L(G)).$$

The line graph of an (r,s)-semiregular graph is $(r+s-2)$-regular, hence $Q(L(G)) = (r+s-2)I - A(L(G))$. It is well-known that the matrices CC^t and C^tC have the same eigenvalues with the exception of the possible eigenvalue 0. It follows that $\mu(G,x)$ and the characteristic polynomial of $(r+s)I - Q(L(G))$ have the same non-zero roots (including their multiplicities). The proof is finished by observing that the difference between the dimensions of $Q(L(G))$ and $Q(G)$ is $m-n$ and the fact that the leading coefficient of the characteristic polynomial is equal to 1.

Note a) If G is (r,s)-semiregular then $\lambda_n(G) = r+s$ and this eigenvalue corresponds, by the formula of Theorem 3.9, to the eigenvalue 0 of $\mu(L(G))$.

b) Let $\varphi(.,.)$ denote the characteristic polynomial of the adjacency matrix of the graph. It is clear from the proof of Theorem 3.9 that $\varphi(L(G),x),x) = (x+2)^{m-n}\mu(G,x+2)$ for any bipartite graph G.

Subdivision graphs, with many vertices subdividing each edge of the original graph, and their spectra are particularly important in the study of thermodynamic properties of crystalline solids (cf. §5). This practical problem led B. E. Eichinger and J. E. Martin [EM] to devise an algorithm for computing the Laplacian eigenvalues of a subdivided graph by applying numerical linear algebraic methods only to the matrix of the unsubdivided graph.

4. The Matrix-Tree-Theorem The most renowned application of the Laplacian matrix of a graph is in the well-known Matrix-Tree- Theorem. This result is usually attributed to Kirchhoff [Ki].

Theorem 4.1 **(Matrix-Tree-Theorem).** Let u, v be vertices of a graph G, and let $Q_{(uv)}$ be the matrix obtained from $Q(G)$ by deleting the row u and the column v. The absolute value of the determinant of $Q_{(uv)}$ is equal to the number of spanning trees $\kappa(G)$ of the graph G.

Corollary 4.2 The number $\kappa(G)$ of spanning trees of the graph G of order n is equal to

$$\frac{1}{n}(-1)^{n-1}\mu'(G,0) = \frac{1}{n}\lambda_2(G)\lambda_3(G)\cdots\lambda_n(G).$$

A generalization of the Matrix-Tree Theorem was obtained by Kel'mans [K3] who gave a combinatorial interpretation to all the coefficients of $\mu(G,x)$ in terms of the numbers of the certain subforests of the graph. This result has been obtained even in greater generality (for weighted graphs) by fiedler and Sedlácek [FS].

Theorem 4.3 [FS,K3] If $\mu(G,x) = x^n + c_1x^{n-1} + \cdots + c_{n-1}x$ then

$$c_i = (-1)^i \sum_{\substack{s \subset V \\ |S|=n-i}} \kappa(G_S)$$

where $\kappa(H)$ is the number of spanning trees of H, and G_S is obtained from G by identifying all points of S to a single point.

In [K4] graphs are compared by their polynomial $\mu(G,x)$ with application to the number of spanning trees. Kel'mans and Chelnokov [KC] consider the problem of determining the graphs with extreme number of spanning trees (minimal, or maximal) in the family of graphs with a given number of vertices and edges. They make use of Corollary 4.2 Constantine [Co] further generalizes the results of [KC].

5. Physical and Chemical Applications

The Laplace differential operator Δ is one of the basic differential operators in mathematical physics. There are two boundary problems connected with this operator. In each of them one has to look for non-trivial solutions of $\Delta\phi = \lambda\phi$ on Ω. If we add the boundary condition $\phi_{|\partial\Omega} = 0$ we get the *Dirichlet problem*. The same equation with the *Neumann condition* at the boundary describes the vibration of a membrane which does not have its boundary fixed. The same two problems have been studied on Riemannian manifolds. It also makes sense to consider Riemannian manifolds without boundary, in which case there is no distinction between both problems (see, e.g.

[Cha]). The discretization of these problems gives rise to the Laplacian matrix of a graph (possibly infinite) and the eigenvalue problem for this matrix. The first problem we mention is the vibration of a membrane. It is described by the Laplace equation

$$\Delta z = -\lambda z, \ z = 0 \ \text{ on } \ \Gamma \tag{5.1}$$

where Γ is a simple closed curve in the z-plane. Discrete analogue of Δ is the Laplacian matrix of a graph which discretizes the region where the equation (5.1) is studied, cf. [CDS, p. 257].

Fisher [Fi] discusses a discrete model of a vibrating membrane where interaction occurs only between neighbouring atoms (vertices of a graph). The discretization of the vibration of a membrane in this model leads to the Laplacian matrix of the graph with its eigenvalues corresponding to the characteristic frequencies of the membrane. (It seems that the author of [Fi] realizes, because of the regularity of his "lattice" graphs, the connection with adjacency matrix eigenvalues, and addresses the general problem to $A(G)$.) The Laplacian problem on graphs in this interpretation determines the so-called *combinatorial drum* [CDS, p. 256]. It means vibrating of a drum membrane without boundary.

Viewing a graph as a system of vertices joined by elastic springs representing its edges, and observing a kinematic system which vibrates in the xy-plane tends to its equilibrium (stationary) state, led W. T. Tutte [T] to a very interesting algorithm for convex straight-line embeddings of 3-connected planar graphs in the plane. First, one has to select a non-separating induced cycle C of the given 3-connected graph G. Let $(x_1, y_1), (x_2, y_2), \cdots, (x_k, y_k)$ be the vertices of a convex k-gon in the plane where k is the length of C. These points will be the coordinates of vertices of C in the constructed embedding. There are unique solutions x, y of

$$Q(G)x = x^o \quad \text{and} \quad Q(G)y = y^o \tag{5.2}$$

where the components of $x = (x_v)_{\in V}$ corresponding to vertices on C are equal to their coordinates x_i, and the other components of x are unknowns. The vector x^o on the right side of (5.2) has all coordinates corresponding to vertices not on C equal to 0 and the others unknown. Similar holds for y and y^p. It turns out [T] that the solution x, y of (5.2) determines the coordinates of a planar convex, straight-line embedding of G if and only if G is planar.

The Laplacian matrix appears also in the theory of electrical currents and flows - the incidence matrices C and $Q = CC^t$ can be found in the famous Kirchhoff laws. As a reference we give the classical Kirchhoff's paper [Ki].

C. Maas showed in [Ma] that the Laplacian eigenvalues of the underlying graph determine the kinematic behaviour of a liquid flowing through a system of communicating pipes. It turns out that the second smallest eigenvalue λ_2 (see also §6) determines the basic behaviour of the flow (e.g., whether the flow is of periodic, or aperiodic type). Let G be a graph representing a system of beads as vertices and edges representing the mutual interactions between these beads. Then the potential, or the kinematic energy of such a system is a quadratic form which can be expressed (cf. (2.1)) by the use of the Laplacian matrix of G. Many related physical quantities have the same relation to $Q(G)$. Eichinger, *et al.* [E1, E2, E3, E4, EM] for example showed that the eigenvalues of the Laplacian matrix of a molecular graph determine the distribution function of the so-called radius of gyration of the molecule, and that the non-zero eigenvalues and their eigenvectors can be used efficiently to compute the scattering functions for Gaussian molecules. See [GS] for some additional references. It is worth mentioning that the asymptotic behaviour of the distribution function of the radius of gyration of a molecule depends mostly upon the magnitude and multiplicity of λ_2 [E1].

6. λ_2 - The Algebraic Connectivity of Graphs

The second smallest Laplacian eigenvalue λ_2 of graphs is probably the most important information contained in the spectrum of a graph. This eigenvalue is related to several important graph invariants, and it has been extensively investigated. Most of the results are consequences of the well- known Courant-Fischer principle which states that

$$\lambda_2(G) = \min_{\substack{x \perp 1 \\ x \neq 0}} \frac{(Q(G)x, x)}{x, x} \tag{6.1}$$

where $0 = (0, 0, ..., 0)^t$, and $1 = (1, 1, ..., 1)^t$ is an eigenvector of $\lambda_1 = 0$. Fiedler [F2] obtained another expression for λ_2.

6.1 Proposition [F2] Let G be a weighted graph with non-negative weights a_{uv}. Then

$$\lambda_2(G) = 2n \min_{x \in \Phi} \frac{\sum\limits_{uv \in E(G)} a_{uv}(x_u - x_v)^2}{\sum\limits_{u \in V} \sum\limits_{u \in V} (x_u - x_v)^2} \tag{6.2}$$

where Φ is the set of all non-constant vectors $x \in \ell^2(V)$.

It can be shown easily, using the fact that $\lambda_n(G) = |V(G)| - \lambda_2(\overline{G})$, that a similar formula holds for the maximal eigenvalue of a graph:

$$\lambda_n(G) = 2n \max_{x \in \Phi} \frac{\sum\limits_{uv \in E(G)} a_{uv}(x_u - x_v)^2}{\sum\limits_{u \in V} \sum\limits_{u \in V} (x_u - x_v)^2} \tag{6.2'}$$

Fiedler [F1, F3] calls the number $\lambda_2(G)$ the *algebraic connectivity* of the graph G. This is influenced by its relation to the classical connectivity parameters of the graph - the *vertex connectivity* $\nu(G)$ and the *edge connectivity* $\eta(G)$.

Theorem 6.2 [F1] Let G be a graph of order n and with maximal valency $\Delta(G)$, and denote by $\omega = \frac{\pi}{n}$. Then

(a) $\lambda_2(G) \leq \nu(G) \leq \eta(G)$,

(b) $\lambda_2(G) \geq 2\eta(G)(1 - \cos\omega)$, and

(c) $\lambda_2(G) \geq 2(\cos\omega - \cos 2\omega)\eta(G) - 2\cos\omega(1 - \cos\omega)\Delta(G)$.

It was discovered recently that graphs with large λ_2 (with respect to the maximal degree) have some properties which make them to be very useful objects in several applications. It is important that λ_2 imposes reasonably good bounds on several properties of graphs which are, for an explicit graph, very hard to compute. We shall mention three such applications. It should be noted that, in all of them, the graph invariants, on which λ_2 imposes non-trivial bounds, can be viewed as measures of connectivity.

Concentrators and expanders are graphs with certain high connectivity properties. They are used in the construction of switching networks that exhibit high connectivity, in the recent parallel sorting of Ajtai, Komlós, and Szermerédi [AKS], in the construction of linear-sized tolerant networks which arise in the study of fault tolerant linear arrays [AC], for the construction of the so-called superconcentrators which are extensively used in the theoretical computer science (e.g., the study of lower bounds in the algorithmic complexity (cf. [Va]), in the establishment of time space tradeoffs for computing various functions [Ab, JJ, To], the construction of graphs that are hard to pebble [LT, Pi, PTC], the construction of low complexity error-correcting codes, etc.), etc. Tanner [Ta] was probably the first who realized that the concentration and expanding properties of a graph can be analyzed by its (adjacency) eigenvalues. He observed that a small ratio of the subdominant adjacency eigenvalue to the dominant eigenvalue implies good expansion properties. Alon [A2] and Alon and Milman [AM1, AM2] followed Tanner's approach, but later they realized [A1, AM3, AGM] that the Laplacian spectrum of a graph (in particular the second smallest eigenvalue) appears more naturally in the study of expanding properties of graphs. [Ro] is an overview article about superconcentrators, and it includes also an exposition of some of the eigenvalue methods which we are trying to summarize here. In [AM3] the authors present several inequalities of the isoperimetric nature relating λ_2 and several other quantities in graphs. These results have analytic analogues [GM] in the theory of Riemannian manifolds where the role of λ_2 is played by the smallest positive eigenvalue of the

Laplacian differential operator on the Riemannian manifold (cf. also [Cha]).

The basic lemma of [AM3] is the following equality. Let A and B be subsets of $V(G)$ at distance ρ (this is the minimal distance between a vertex in A and a vertex in B), and let F be the set of edges which do not have both ends in A and do not have both ends in B. Then

$$|F| \geq \rho^2 \lambda_2(G) \frac{|A||B|}{|A| + |B|} \tag{6.3}$$

In particular, when $B = V \backslash A$, then $F = \delta A = \delta B$ (the *coboundary* of A, or of B) is the set of edges with one end in A and the other end outside A. In this case $\rho = 1$ and (6.3) implies

$$|\delta A| \geq \lambda_2(G) \frac{|A|(n - |A|)}{n} \tag{6.4}$$

A refinement of (6.3) is derived in [M3]. If A and B are subsets of $V(G)$ at distance $\rho > 1$ then

$$(\rho - 1)^2 < \frac{\lambda_n(G)}{4\lambda_2(G)} \frac{(n - |A| - |B|)(|A| + |B|)}{|A||B|} \tag{6.5}$$

This paper [AGM] considers the expansion properties of graphs and their applications. The main eigenvalue based lemma of [AGM] gives a lower bound on the number of neighbours of a set $X \subseteq V$. If $N(X)$ is the set of those neighbours of vertices of X which do not lie in X, then

$$|N(X)|^2 - 2(n - 2|X| - \alpha n)|N(X)| - 4|X|(n - |X|) \geq 0 \tag{6.6}$$

where $\alpha = \frac{1}{2}(1 + \frac{\Delta}{\lambda_2})$, and Δ is the maximum valency in G.

In [A1] expanders and graphs with large λ_2 are related. Expanders can be constructed from graphs which are *c-magnifiers* ($c \in \mathbb{R}^+$). These are graphs which are highly connected according to the following property. For every set X of vertices of G with $|X| \leq \frac{n}{2}$ the neighbourhood $N(X)$ of X contains at least $c|X|$ vertices. In [A1] it is shown that a graph G is $\frac{2\lambda_2}{\Delta + 2\lambda_2}$-magnifier and, conversely, if G is a c-magnifier then $\lambda_2(G) \geq \frac{c^2}{4+2c^2}$. The first result is based on (6.4), while the second one is a discrete version of the Cheeger inequality [Che] from the theory of Riemannian manifolds.

A strong improvement over the Alon's discrete version of the Cheeger inequality was obtained by the author [M2] in connection with another problem. The *isoperimetric number* $i(G)$ of a graph G is equal to

$$i(G) = \min\{\frac{\delta X}{|X|}; X \subset V, 0 < |X| \leq \frac{|V|}{2}\}.$$

This graph invariant is very hard to compute, and even obtaining any lower bounds on $i(G)$ seems to be a difficult problem. It is shown in [M2] that

$$i(G) \geq \frac{\lambda_2(G)}{2} \tag{6.7}$$

and, moreover, a strong discrete version of the Cheeger inequality holds [M2]:

$$i(G) \leq \sqrt{\lambda_2(2\Delta - \lambda_2} \tag{6.8}$$

where Δ is, as usual, the maximal degree in G. The reader is also referred to [M1]. We mention that these results are very important since they yield efficient checking procedures of several graph properties. For example, if $\lambda_2(G) = 2$ then we know by (6.7) that $i(G) \geq 1$. If, moreover, we find a cut X with $|\delta X| = |X|$ then we can conclude that $i(G) = 1$.

Besides the expansion properties and the isoperimetric numbers of graphs, an eigenvalue based inequality can be used of the max-cut problem [MP] (also the weighted case) which is known to be NP-hard. It is shown in [MP] that the number of edges MC(G) in a maximal cut in a graph G is bounded above by

$$MC(G) \leq \frac{n\lambda_n(G)}{4}. \tag{6.9}$$

Notice that $\lambda_n(G)$ is related to the second smallest eigenvalue of the complement of G (cf. Theorem 3.6).

The second eigenvalue is also related to some other graph invariants. One of the most interesting connections is its relation to the diameter and the mean distance of graphs. There is a lower bound

$$\operatorname{diam}(G) \geq \frac{4}{n\lambda_2(G)} \tag{6.10}$$

This bound was obtained by Brendan McKay [McK] but its proof appeared for the first time in [M3].

To get an upper bound one may use the inequality (6.5) which gives rise to an eigenvalue-based upper bound on the diameter of a graph [M3]:

$$\operatorname{diam}(G) \leq 2\lceil \sqrt{\frac{\lambda_n(G)}{\lambda_2(G)}} \sqrt{\frac{a^2-1}{4a}} + 1 \rceil \lceil \log_a \frac{n}{2} \rceil \tag{6.11}$$

where γ is any real number which is > 1. For any particular choice of n, λ_n, and λ_2 one can find the value of γ which imposes the lowest upper bound on the diameter of the graph. See [M3] for details. A good general choice is $a = 7$.

In [M3] another upper bound on the diameter of a graph is obtained

$$\operatorname{diam}(G) \leq 2\lceil \frac{\Delta + \lambda_2(G)}{4\lambda_2(G)} \ln(n-1) \rceil. \tag{6.12}$$

This improves a bound of Alon and Milman [AM3]. It should be noted that Chung, Faber and Manteuffel [CFM] found another bound:

$$\operatorname{diam}(G) \leq \lceil \frac{\log(n-1)}{\log((\lambda_n + \lambda_2)/(\lambda_n - \lambda_2))} \ln(n-1) \rceil.$$

In [M3], some bounds on the mean distance $\overline{\rho}(G)$ are derived. Recall that the mean distance is equal to the average of all distances between distinct vertices of the graph. A lower bound is

$$(n-1)\overline{\rho}(G) \geq \frac{2}{\lambda_2(G)} + \frac{n-2}{2} \tag{6.13}$$

and an upper bound, similar to (6.12), is

$$\overline{\rho}(G) \leq \frac{n}{n-1} \lceil \frac{\Delta + \lambda_2(G)}{4\lambda_2(G)} \ln(n-1) \rceil. \tag{6.14}$$

There is also an upper bound on $\overline{\rho}(G)$ related to the inequality (6.11). Cf. [M3].

Some inequalities relating graph invariants to the spectrum of the adjacency matrix of a graph can as well be formulated in terms of the Laplacian spectrum - usually obtaining even stronger results this way. As an example we extend a Hoffman-Lovász' bound [CDS, Lo] on the *independence number* $\alpha(G)$ of a graph. They proved that a d-regular graph G has $\alpha(G) \leq n(1 - d/\lambda_n)$.

Let G be a graph of order n with vertices of degrees $d_1 \leq d_2 \leq \cdots \leq d_n$. Set

$$e_r = \frac{1}{r}(d_1 + d_2 + \cdots + d_r), \quad 1 \leq r \leq n.$$

Assume we have an independent set R of vertices of size r. Define a vector $x \in \ell^2(V(G))$ by setting

$$x_v = \begin{cases} 0, & v \in R \\ 1, & v \notin R \end{cases}$$

By (6.2'),

$$\lambda_n(G) \sum_{u \in V} \sum_{v \in V} (x_u - x_v)^2 \geq 2n \sum_{uv \in E} (x_u - x_v)^2$$

which reduces to $\lambda_n r(n-r) \geq n|\delta R| \geq nre_r$. It follows that

$$r \leq \frac{n(\lambda_n - e_r)}{\lambda_n} \tag{6.15}$$

Theorem 6.3 If r_0 is the smallest number r for which (6.15) fails then

$$\alpha(G) \leq r_0 - 1. \tag{6.16}$$

It is interesting that large graphs of bounded genus and with bounded maximal degree have small λ_2.

Theorem 6.4 Let G be a graph of order n, with maximal vertex degree Δ and genus g. In $n > 18(g+2)^2$ then

$$\lambda_2(G) \leq \frac{6(g+2)\Delta}{\sqrt{n/2} - 3(g+2)}.$$

Proof Boshier [Bo] proved that under the hypothesis of the theorem

$$i(G) \le \frac{3(g+2)\Delta}{\sqrt{\frac{n}{2}} - 3(g+2)}$$

where $i(G)$ is the isoperimetric number of G. The inequality (6.7) now completes the proof.

It would be interesting to have a "direct" proof of Theorem 6.4. We believe that such a proof would yield even better inequality,, between λ_2 and the genus of a graph, than outlined above.

C. Maas [Ma] studied extensively how can $\lambda_2(G)$ change if we delete or insert an edge into G. He derived several upper and lower bounds on this change, some of them being quite non- obvious. In [Ma] there is also a result which was obtained independently by R. Merris [Me2]. See also [F4].

Theorem 6.5 For a tree T, $\lambda_2(T) \le 1$, with equality if and only if T is a star.

Three interesting notions are introduced in [Ma], depending on λ_2 and the corresponding eigenspace: *permeability* of a graph, the *well-connectedness* of pairs of vertices, and a measure to relate how *good position* have particular vertices. It is shown on some examples that the notions introduced behave in correspondence with our intuitive notion of permeability, well connected pairs of vertices, and a good (strategic, central) position in a network. Merris [Me2] also found that a kind of a central position in a tree can be defined by using λ_2 and the corresponding eigenvectors.

Let us finish this section with a few words about infinite graphs. Let G be a locally finite countable graph with bounded vertex degrees. Its Laplacian matrix $Q(G)$ gives rise to a self-adjoint linear operator on the Hilbert space $\ell^2(V)$. Its spectrum $\sigma(G)$ is called the Laplacian spectrum of G. It is worth mentioning that the role of λ_2 is replaced by $\lambda = \inf \sigma(G)$. It might happen that $\lambda = 0$ for a connected graph G. Indeed, this is true for any graph of polynomial growth. More details can be found in [BMS, M1, MW], especially about the relation between λ, the growth, and the isoperimetric number of G.

7. Characteristic Valuations

An eigenvector of $\lambda_2(G)$ is called a *characteristic valuation* of G. The characteristic valuations, especially their sign structure, have been studied by Fiedler [F2], and Merris and Grone [Me2, GM1, GM2]. We shall collect here only the most interesting results of these papers.

Theorem 7.1 [F2, F4] Let G be a connected weighted graph with non-negative weights, and $y = (y_v)_{v \in V(G)}$ a characteristic valuation of G. For $r \geq 0$ let

$$S(r) = \{v \in V(G) | y_v \geq -r\}.$$

Then the subgraph induced on $S(r)$ is connected.

A similar result holds for $r \leq 0$ and $S'(r) = \{v | y_v \leq -r\}$. It is an interesting corollary to Theorem 7.1 that if $c \geq 0$ is such a constant that $y_v \neq c$ for all vertices v of G, and $S_1 = \{v | y_v > c\}, S_2 = V(G) \backslash S_1$, then the subgraphs of G induced on S_1 and S_2 are both connected.

Theorem 7.2 [F2] Let G be a connected graph, $y = (y_v)_{v \in V(G)}$ a characteristic valuation of G, and v is a cut-vertex of G. Denote by $G_1, G_2, ..., G_r$ the components of $G - v$. Then:

(a) If $y_v > 0$ then exactly one G_i contains a vertex u with $y_u < 0$. All vertices in other components G_j have positive y-value.

(b) If $y_v = 0$ and some G_i contains positively and negatively valuated vertices, then all the remaining components are 0-valuated.

Theorem 7.3 [F2] Let G be a connected graph with a characteristic valuation y. Two possibilities arise:

(a) There exists a block B_0 of G with both, positive and negative y-values. All other blocks have either all positive, or all negative, or only zero values.

(b) No block of G contains positive and negative values simultaneously. In this case there is a *unique* cut-vertex v with $y_v = 0$ which has a neighbour u with $y_u \neq 0$. In the case when G is a tree, the above results are strengthened in [F2] and discussed in greater details in [Me2, GM1, GM2]. In [GM1] the *trees of type I* are investigated. These are trees with a characteristic valuation y such that $y_v = 0$ for some vertex v of the tree. It is shown that every tree T of type I contains a unique vertex w such that $y_w = 0$ for every characteristic valuation y of T. This vertex is called a *characteristic vertex* of T. *Branches*, i.e. the components of $T - w$ are characterized as *active* or *passive* (every characteristic valuation is 0 on a passive branch). Their properties with respect to $\lambda_2(T)$ are investigated. In [GM2] the authors consider a more general family of matrices, $Q^{\alpha,\beta} = \alpha D(G) + \beta A(G)(\alpha, \beta \in R)$.

Characteristic valuations can be efficiently used to obtain well-behaved heuristic algorithms for various problems. Let y be a characteristic valuation of G and let $A = \{v \in V(G) | y_v \geq 0\}, B = V \backslash A$. It follows from the proof of Theorem 4.2 in [M2] that the partition $V = A \cup B$ is not too far from an optimal partition $V = A^o \cup B^o$ where the optimum means that A^o and B^o have minimal possible average outdegree. This fact can be applied to devise several *divide-and-conquer* algorithms: One solves a problem separately on A and on B and then tries to combine solutions to get an acceptable solution for the graph G.

Another type of problem where characteristic valuations can be used are *optimal labeling problems*, e.g., the bandwidth, or the min-sum problem. It is requested to arrange the vertices of a graph in a linear order $v_1, v_2, \cdots, v_n$ in such a way that the edges will not do too long jumps (if $v_i v_j$ is an edge then $|i - j|$ should be small). A reasonably good ordering is obtained by ordering the vertices of G with respect to the values of a characteristic valuation y, i.e., $v \leq u$ if $y_v \leq y_u$. The eigenvalue $\lambda_2(G)$ also gives a lower bound on the average square of jumps for *any* linear ordering $v_1, \cdots, v_n$ of $V(G)$. If $e = v_i v_j$ then *jump(e)*$:= i - j|$. Then

$$\sum_{e \in E(G)} [\text{jump}(e)]^2 \geq \lambda_2(G) \frac{n(n^2-1)}{12}. \tag{7.1}$$

The details about (7.1) with many different extensions will appear elsewhere [JM].

8. Miscellaneous

In case of regular graphs, all results about the adjacency spectrum of graphs carry over to results about the Laplacian spectrum since for a d-regular graph G

$$\mu(G,x) = (-1)^n \varphi(G, d-x) \tag{8.1}$$

where φ is the characteristic polynomial of $A(G)$. But even in the general, non- regular case, the Laplacian spectrum of G is related to the adjacency spectrum of some graph G'. Let Δ be the maximal valency of G, and let G' be the graph obtained from G by adding, at each vertex $v \in V(G) = V(G'), \Delta - d(v)$ loops. Thus G' is Δ-regular and $Q(G) = Q(G')$. Consequently,

$$\mu(G,x) = \mu(G',x) - (-1)^n \varphi(G', \Delta - x) \tag{8.2}$$

D. L. Powers has assembled a catalogue containing the eigenvalues and eigenvectors of the adjacency and the Laplacian matrices of all connected graphs with up to 6 vertices [P2] and all trees with up to 9 vertices [P1].

Kel'mans [K3] found a class of graphs which is characterized by the Laplacian spectrum. In general there are many cospectral non-isomorphic graphs. For example, it can be shown that almost all tee are cospectral.

R. Merris [Me2] derived some inequalities between the coefficients of the Laplacian polynomial $\mu(G,x)$ and the coefficients of the chromatic polynomial of G (G is a connected simple graph).

Let G be a simple graph. For a vertex $v \in V(G)$, let the *star degree* of v be defined as

$$\text{stardeg}(v) = \begin{cases} 0, & \text{if no neighbour of } v \text{ is a pendant vertex} \\ k-1, & \text{if } v \text{ has } k \geq 1 \text{ pendant neighbours} \end{cases}$$

The *star degree* of the graph G is then equal to the sum of star degrees of all its vertices. I. Faria [Fa] proved that the star degree of G is equal to the multiplicity of $x = 1$ as the root of the permanental polynomial per $(xI - B(G)), B(G) = D(G) + A(G)$. If G is bipartite then the permanental polynomial of $B(G)$ is equal to the permanental polynomial of the Laplacian matrix $Q(G)$. The author mentions (without an explicit proof) that in the case of characteristic polynomials of $B(G)$ and $Q(G)$, we can only conclude that the star degree of G is at most equal to the multiplicity of $x = 1$ as the zero of these polynomials. Some other authors studied the permanental polynomial of $Q(G)$ [BG, Me1].

C. D. Godsil [Go] proved that the polynomial $\psi(G, S; x) = \sum f_k x^k$, where $S \subseteq E(G)$ and f_k is the number of spanning trees of G with exactly k edges in S, has only real zeros. A corresponding result holds for a generalization of $\psi(G, S; x)$ to unimodular matroids and numbers of bases with specified number of elements in given subsets of the matroid. Is there a corresponding generalization of $\mu(G, x)$ to matroids where the coefficients of the polynomial relate to bases of the matroid in the same way as the coefficients of $\mu(G, x)$ related to spanning trees of G (cf. Theorem 4.3)?

REFERENCES

[1] [Ab] H. Abelson, "A note on time space tradeoffs for computing continuous functions," *Infor. Proc. Letters 8* (1979) 215-217.

[2] [AKS] M. Ajtai, J. Komlós, and E. Szemerédi, "Sorting in $c \log n$ parallel steps," *Combinatorica 3* (1983) 1-9.

[3] [A1] N. Alon, "Eigenvalues and expanders," *Combinatorica 6* (1986) 83-96.

[4] [A2] N. Alon, "Eigenvalues, geometric expanders, sorting in rounds, and Ramsey theory," *Combinatorica 6* (1986) 207-219.

[5] [AC] N. Alon, F. R. K. Chung, "Explicit construction of linear sized tolerant networks, *Proc. 1st Japan Conf. on Graph Theory and Applications.*

[6] [AGM] N. Alon, Z. Galil, and V. D. Milman, "Better expanders and superconcentrators, *J. Algorithms 8* (1987) 337-347.

[7] [AM1] N. Alon, V. D. Milman, "Concentration of measure phenomena in the discrete case and the Laplace operator of a graph," Semin. analyse fonct., Paris 1983-84, Publ. Math. Univ. Paris VII 20 (1984) 55-68.

[8] [AM2] N. Alon, V. D. Milman, "Eigenvalues, expanders and superconcentrators," *Proc. 25th FOCS*, Florida, 1984, pp. 320-322 (Extended abstract).

[9] [AM3] N. Alon, V. D. Milman, "λ_1, isoperimetric inequalities for graphs and superconcentrators," *J. Combin. Theory*, Ser. B 38 (1985) 73-88.

[10] [AnM] W. N. Anderson, T. D. Morley, "Eigenvalues of the Laplacian of a graph," *Lin. Multilin. Algebra* 18 (1985) 141-145.

[11] [Bi] N. L. Biggs, "Algebraic graph theory," Cambridge Univ. Press, Cambridge, 1974.

[12] [BMS] N. L. Biggs, B. Mohar, and J. Shawe-Taylor, "The spectral radius of infinite graphs," *Bull. London Math. Soc.* 20 (1988) 116-120.

[13] [Bo] A. G. Boshier, "Enlarging properties of graphs," Ph.D. Thesis, Royal Holloway and Bedford New College, University of London, 1987.

[14] [BG] R. A. Brualdi, J. L. Goldwasser, "Permanent of the Laplacian matrix of trees and bipartite graphs," *Discrete Math.* 48 (1984) 1-21.

[15] [Cha] I. Chavel, "Eigenvalues in Riemannian geometry," *Academic Press*, New York, 1984.

[16] [Che] J. Cheeger, "A lower bound for the smallest eigenvalue of the Laplacian, in *Problems in analysis*," (R. C. Gunnig, ed.), Princeton Univ. Press, 1970, pp. 195-199.

[17] [CFM] F. R. K. Chung, V. Faber, and T. Manteuffel, "An estimate of the diameter of a graph from the singular values and eigenvalues of its adjacency matrix," preprint, 1988.

[18] [Co] G. M. Constantine, "Schur convex functions on the spectra of graphs," *Discr. Math.* 45 (1983) 181-188.

[19] [CDGT] D. M. Cvetković, M. Doob, I. Gutman, and A. Torgasev, "Recent results in the theory of graph spectra," *Ann. Discr. Math.* 36, North-Holland, 1988.

[20] [CDS] D. M. Cvetković, M. Doob, and H. Sachs, "Spectra of graphs - Theory and applications," VEB Deutscher Verlag d. Wiss., Berlin, 1979; Acad. Press, New York, 1979.

[21] [E1] B. E. Eichinger, "An approach to distribution functions for Gaussian molecules," *Macromolecules* 10 (1977) 671-675.

[22] [E2] B. E. Eichinger, "Scattering functions for Gaussian molecules," *Macromolecules* 11 (1978) 432-433.

[23] [E3] B. E. Eichinger, "Scattering functions for Gaussian molecules. 2. Intermolecular correlation," *Macromolecules* 11 (1978) 1056-1057.

[24] [E4] B. E. Eichinger, "Configuration statistics of Gaussian molecules," *Macromolecules* 13 (1980) 1-11.

[25] [EM] B. E. Eichinger, J. E. Martin, "Distribution functions for Gaussian molecules. II. Reproduction of the Kirchhoff matrix for large molecules," *J. Chem. Phys.* 69 (10) (1978) 4595- 4599.

[26] [Fa] I. Faria, "Permanental roots and the star degree of a graph," Linear Algebra Appl. 64 (1985) 255-265.

[27] [F1] M. Fiedler, "Algebraic connectivity of graphs," *Czech. Math. J* 23 (98) (1973) 298-305.

[28] [F2] M. Fiedler, "A property of eigenvectors of nonnegative symmetric matrices and its application to graph theory," Czech. Math. J.25 (100) (1975) 619-633.

[29] [F3] M. Fiedler, "An algebraic approach to connectivity of graphs, in *Recent advances in graph theory*", Academia, Prague, 1975, pp. 193-196.

[30] [F4] M. Fiedler, "Algebraische Zusammenhangszahl der Graphen und ihre numerische Bedeutung, in *Numerische Methoden bei graphentheoretischen und kombinatorischen Problemen*," Birkhäuser, Basel, 1975, pp. 69-85.

[31] [FS] M. Fiedler, J. Sedlácek, "O w-basich orientovaných grafu," Cas. Pest. Mat. 83 (1958) 214-225 (Czech.).

[32] [Fi] M. E. Fisher, "On hearing the shape of a drum," *J. Combin. Theory 1* (1966) 105-125.

[33] [GS] H. Galina, M. Syslo, "Some applications of graph theory to the study of polymer configuration," *Discr. Appl. Math.* 19 (1988) 167-176.

[34] [Go] C. D. Godsil, "Real graph polynomials, in *Progress in graph theory*," Academic Press, 1984, pp. 281-293.

[35] [GM] M. Gromov, V. D. Milman, "A topological application of the isoperimetric inequality," American J. Math. 105 (1983) 843-854.

[36] [GM1] R. Grone, R. Merris, "Algebraic connectivity of trees," *Czech. Math. J. 37* (112) (1987) 660-670.

[37] [GM2] R. Grone, R. Merris, "Cutpoints, lobes and the spectra of graphs, preprint, 1987.

[38] [H] K. M. Hall, "r-dimension quadratic placement algorithm" *Management Sci. 17* (1970) 219-229.

[39] [JJ] J. Ja'Ja, "Time space tradeoffs for some algebraic problems," *Proc. 12th Ann. ACM Symp. on Theory of Computing*, 1980, pp. 339-350.

[40] [JM] M. Juvan, B. Mohar, "Optimal linear labelings and algebraic properties of graphs, submitted.

[41] [K1] A. K. Kel'mans, "The number of trees in a graph. I." *Automat. i Telemeh. 26* (1965) 2194-2204 (in Russian); transl. Automat. Remote Control 26 (1965) 2118-2129.

[42] [K2] A. K. Kel'mans, "The number of trees in a graph. II.," *Automat. i Telemeh. 27* (1966) 56-65 (in Russian); transl. Automat. Remote Control 27 (1966) 233-241.

[43] [K3] A. K. Kel'mans, "Properties of the characteristic polynomial of a graph," Kibernetiky - na sluzbu kommunizmu," Vol. 4, Energija, Moskva - Leningrad, 1967, pp. 27-41 (in Russian).

[44] [K4] A. K. Kel'mans, "Comparison of graphs by their number of spanning trees," *Discrete Math. 16* (1976) 241-261.

[45] [KC] A. K. Kel'mans, V. M. Chelnokov, "A certain polynomial of a graph and graphs with an extremal number of trees," *J. Combin. Theory*, Ser. B 16 (1974) 197-214.

[46] [Ki] G. Kirchhoff, "Über die Auflösung der Gleichungen, auf welche man bei der Untersuchung der linearen Verteilung galvanischer Ströme geführt wird," *Ann. Phys. Chem* 72 (1847) 497-508. Translated by J. B. O'Toole in I.R.E. Trans. Circuit Theory, CT-5 (1958) 4.

[47] [LT] T. Lengauer, R. E. Tarjan, "Asymptotically tight bounds on time space trade-offs in a pebble game," J. ACM 29 (1982) 1087-1130.

[48] [Lo] L. Lovász, "On the Shannon capacity of a graph," *IEEE Trans. Inform. Theory*, IT-25 (1979) 1-7.

[49] [Ma] C. Maas, "Transportation in graphs and the admittance spectrum," *Discr. Appl. Math.* 16 (1987) 31-49.

[50] [MM] M. Marcus, H. Minc, "A survey of matrix theory and matrix inequalities," Allyn and Bacon, Boston, Mass., 1964.

[51] [McK] B. D. McKay, private communication.

[52] [Me1] R. Merris, "The Laplacian permanental polynomial for trees," *Czech. Math. J.* 32 (107 (1982) 391-403.

[53] [Me2] R. Merris, "Characteristic vertices of trees," *Lin. Multilin. Alg.* 22 (1987) 115-131.

[54] [M1] B. Mohar, "Isoperimetric inequalities, growth, and the spectrum of graphs," *Linear Algebra Appl.* 103 (1988) 119-131.

[55] [M2] B. Mohar, "Isoperimetric numbers of graphs," *J. Combin. Theory*, Ser. B 47 (1989).

[56] [M3] B. Mohar, "Eigenvalues, diameter, and mean distance in graphs," submitted.

[57] [MP] B. Mohar, S. Poljak, "Eigenvalues and the max-cut problem," submitted.

[58] [MW] B. Mohar, W. Woess, "A survey on spectra of infinite graphs," *Bull. London Math. Soc.* 21 (1989) 209-234.

[59] [PTC] W. J. Paul, R. E. Tarjan, and J. R. Celoni, "Space bounds for a game on graphs," Math. Sys. Theory 20 (1977) 239-251.

[60] [Pi] N. Pippenger, "Advances in pebbling," *Internat. Colloq. on Autom. Lang. and Prog.* 9 (1982) 407-417.

[61] [P1] D. L. Powers, "Tree eigenvectors, preprint, 1986.

[62] [P2] D. L. Powers, "Graph eigenvectors," preprint, 1986.

[63] [Ro] A. Rouault, "Superconcentrateurs," Publ. Math. d'Orsay, No. 87-01.

[64] [Ta] R. M. Tanner, "Explicit concentrators from generalized n-gons," *SIAM J. Alg. Discr. Meth.* 5 (1984) 287-293.

[65] [To] M. Tompa, "Time space tradeoffs for computing functions, using connectivity properties of their circuits," *J. Comp. and Sys. Sci.* 20 (1980) 118-132.

[66] [T] W. T. Tutte, "How to draw a graph," *Proc. London Math. Soc.* 52 (1963) 743-767.

[67] [V] E. B. Vahovskii, "On the characteristic numbers of incidence matrices for non- singular graphs," Sibirsk. Mat. Zh. 6 (1965) 44-49 (in Russian).

[68] [Va] L. G. Valiant, "Graph theoretic properties in computational complexity," *J. Comp. and Sys. Sci.* 13 (1976) 278-285.

[69] [Wi] R. J. Wilson, "Introduction to graph theory," Longman, New York, 1972.

Partitioning Points and Graphs to Minimize the Maximum or the Sum of Diameters

Clyde Monma and Subhash Suri
BellCore
Morristown, New Jersey

ABSTRACT

Let V be a set of n points in the Euclidean plane. We consider the problem of partitioning V into two sets X and Y such that the sum of the diameters of X and Y is minimized. Our main result is an $O(n^2)$–time and $O(n)$–space algorithm for this problem. More generally, if V is the vertex–set of a weighted graph having m edges, then we can partition V into two sets with a minimum sum of diameters in time $O(nm\log n)$. The previously known best algorithm for these problems required $O(n^3\log n)$ time. We also consider a related problem where one seeks a bipartition of V into two sets so that the maximum of the two diameters is minimized. We solve this problem in time $O(n\log n)$ for the Euclidean case, and in time $O(m\log^* n)$ for an arbitrary weighted graph. Our algorithm for the Euclidean case is simpler than the one recently proposed by Asano et al., while for arbitrary graphs the best algorithm previously known required time $O(m\log n)$. The general problem of partitioning a set into $k \geq 3$ subsets under either criterion of minimization (the maximum or the sum) is known to be NP–complete.

1. Introduction

The problem of partitioning a set of entities into clusters arises frequently in many disciplines. Due to the wide range of applications, many different objective functions have been considered, thus, giving rise to many variations of this problem. Generally speaking, the basic problem of cluster analysis is to partition a set of entities into homogeneous and well-separated classes, called clusters. Separation and homogeneity are often expressed by describing dissimilarities between pairs of entities.

For optimization criteria, the notions of split and diameter have been used widely (see Delattre and Hansen [DH]).

This paper is concerned with the problem of partitioning a set of points in the Euclidean plane or the vertices of a weighted graph so as to minimize certain functions of diameters. To be precise, let V be a set of n points in the Euclidean plane. We want to partition V into two sets X and Y such that the sum of the diameters of X and Y is minimized. We present an $O(n^2)$–time and $O(n)$–space algorithm for this problem. More generally, if V is the vertex set of an arbitrary weighted graph, then we can partition V into two subsets having a minimum sum of diameters in time $O(nm\log n)$, where m is the number of edges in the graph. These results improve over the previous best algorithm, which is due to Hansen and Jaumard and requires $O(n^3\log n)$ time [HJ]. We also give an optimal solution for the following decision problem: Given an arbitrary graph G with real–valued edge weights and two real numbers r_1 and r_2, can the vertex set of G be partitioned into two subsets X and Y such that diameters of X and Y are at most r_1 and r_2, respectively? An $O(n^2)$ algorithm for this problem is presented in Hansen and Jaumard [HJ]; the Euclidean version of this problem is analogous to the "specified diameter" problem considered by Avis, also solved in $O(n^2)$ time [Av]. We give an $O(n + m)$ algorithm for this problem, where G has n vertices and m edges, which is an improvement over the previous algorithms for sparse graphs.

A related problem is to partition V into subsets X and Y such that the *larger* of the two diameters is minimized. Our results for this problem are as follows. If V is a set of points in the Euclidean plane, then we can find a minimum bipartition in time $O(n\log n)$, which is opitmal in the algebraic computation–tree model. Although an $O(n\log n)$ algorithm for this problem already exists (due to Asano et al. [ABKY]), the merit of our algorithm lies in its simplicity. If V is the vertex set of an arbitrary graph having m edges, then our algorithm finds a minimum bipartition in time $O(m\log^* n)$, where $\log^* n$ is the iterated logarithm. The latter time bound improves upon the previous best result, which runs in $O(m\log n)$ time [Av].

Partitions with either a minimum diameter or a minimum sum of diameters are of interest in many situations where an analyst's main concern is the homogeneity of the clusters. For instance, as mentioned by Hansen and Jaumard [HJ], this is the case when patients suffering from a multiform disease need to be split into clusters for treatment, or when counties within a state need to be grouped for the efficient enactment of an economic policy. Partitions that minimize the maximum diameter, however, suffer from the *dissection* effect, i.e., similar entities may be assigned to different clusters, see Cormack [Co]. This happens because clusters tend to have roughly equal diameters, which may entail the dissection of a natural cluster. Dissection effect is less damaging if the sum of the diameters is minimized, see Hansen and Jaumard [HJ]. Finally, while two clusters often are not enough for a detailed classification, one may apply our bipartitioning algorithm recursively to obtain an approximate partition into $k \geq 3$ clusters.

The paper is organized in five sections. Section 2 contains a basic lemma concerning the maximum diameter of a bipartition. In Section 3, we prove our results for minimizing the maximum diameter in a bipartition. Section 4 contains our main results, namely, the algorithms for minimizing the sum of diameters. Finally, in Section 5, we close with a few remarks concerning various other functions of diameters to which our algorithms are applicable.

2. Preliminaries

Let $G = \{V, E, d: E \rightarrow R^+\}$ be a weighted graph on the set of vertices $V = \{v_1, v_2, \ldots, v_n\}$. If two vertices v_i and v_j are joined by an edge in G, then we denote this fact by $v_iv_j \in E$. The weight of v_iv_j is denoted $d(v_i, v_j)$. In the special case where V is a set of points in the plane, the Euclidean distance function $d(\cdot)$ induces a complete weighted graph, which we call the *Euclidean graph* of V. That is, in the Euclidean graph $G = \{V, E, d: E \rightarrow R^+\}$, $v_iv_j \in E$, and $d(v_i, v_j)$ equals the Euclidean distance between the points v_i and v_j, for all $1 \leq i < j \leq n$. Throughout, we assume $|V| = n$ and $|E| = m$. For any subset $V' \subseteq V$, the *diameter* of V', denoted diam(V'), is the maximum weight of an edge in the subgraph of G induced by V', i.e.,

$$\text{diam}(V') = \max \{d(v_i, v_j) \mid v_i, v_j \in V', v_iv_j \in E\}.$$

A *maximum spanning tree* of G is a spanning tree of G having a maximum total edge weight; in general, a maximum spanning tree is not unique. In the following, we let MXST (G) denote an arbitrary but fixed maximum spanning tree of G. Consider an edge $v_iv_j \in E$ that does not belong to MXST(G). There exists an unique cycle formed by the edge v_iv_j and the path joining v_i and v_j in MXST(G). We denote this cycle by *cycle*(v_i, v_j). If cycle (v_i, v_j) has an odd (resp. even) number of edges, we call it an odd (resp. even) cycle. Th following lemma is established in Hansen and Jaumard [HJ]; for the sake of completeness, we include its proof.

Lemma 1 *Let $G = \{V, E, d: E \rightarrow R^+\}$ be an arbitrary graph and let $\{V_1, V_2\}$ be a bipartition of V, where $diam(V_1) \geq diam(V_2)$. Let v_pv_q be a maximum–weight edge in E that forms an odd cycle with MXST(G). Then the following holds.*

(1) $diam(V_1) \geq d(v_p, v_q)$.

(2) If $diam(V_1) > d(v_p, v_q)$, then there exists an edges $v_kv_l \in MXST(G)$ such that $diam(V_1) = d(v_k, v_l)$.

Proof of (1) Suppose, to the contrary, that $\text{diam}(V_1) < d(v_p, v_q)$. Clearly, v_p and v_q must lie in different components of the partition. Assume, without loss of generality, that $v_p \in V_1$ and $v_q \in V_2$. Let $v_p = x_1, x_2, \ldots, x_m = v_q$ be the ordered list of vertices on cycle(v_p, v_q), where m is odd and $x_i x_{i+1} \in$ MXST(G), for $i = 1, 2, \ldots, m-1$. By the maximality of the spanning tree $d(x_i, x_{i+1}) \geq d(v_p, v_q)$, for $i = 1, 2, \ldots, m-1$. Hence, x_i and x_{i+1} lie in different components of the partition, for $1 \leq i \leq m-1$. That is, starting with $x_1 \in V_1$, x_i must alternate between V_1 and V_2, ending in $x_m \in V_2$. That, however, is impossible since m is odd, which contradicts the assumption that $\text{diam}(V_1) < d(v_p, v_q)$.

Proof of (2) Suppose that $\text{diam}(V_1) = d(v_i, v_j) > d(v_p, v_q)$ but the edge $v_i v_j$ does not belong to MXST(G). It follows that $v_i v_j$ forms an even cycle with MXST(G). We, thus, have $v_i, v_j \in V_1$ and $d(x_i, x_{i+1}) \geq d(v_i, v_j)$, for all edges $x_i x_{i+1} \in$ cycle(v_i, v_j). Since cycle(v_i, v_j) is even, at least two other adjacent vertices on the cycle, say v_k and v_l are assigned to the same component of the partition. Since having $d(v_k, v_l) > d(v_i, v_j)$ would contradict our assumption that $\text{diam}(V_1) = d(v_i, v_j)$, the equality $d(v_i, v_j) = d(v_k, v_l)$ holds and, thus, the edge $v_k v_l$ satisfies the claim. This completes the proof. ❑

3. Algorithms for Minimizing the Maximum Diameter

Let $G = \{V, E, d{:}E \rightarrow R^+\}$ be an arbitrary weighted graph, and let $\{V_1, V_2\}$ be a bipartition of V. We say that $\{V_1, V_2\}$ is a *minimax* bipartition if $\max\{\text{diam}(V_1), \text{diam}(V_2)\}$ is a minimum over all bipartitions of V. Our first theorem states our results for minimax bipartitions.

Theorem 1 *Let $G = \{V, E, d{:}E \rightarrow R^+\}$ be an arbitrary weighted graph. There is an $O(m\log^* n)$–time and $O(m)$–space algorithm for computing a minimiax bipartition of V. If G is an Euclidean graph, then the algorithm runs in $O(n\log n)$ time and uses $O(n)$ space. The bounds for the Euclidean case are optimal in the algebraic computation–tree model.*

Proof We first describe the algorithm and then discuss its time and space complexity. Compute a maximum spanning tree of G, MXST(G). Let χ: $V \rightarrow \{1, 2\}$ be a 2–coloring of MXST(G) and let $\{V_1, V_2\}$ be the bipartition of V induced by χ, i.e., $V_k = \{v \in V \mid \chi(v) = k\}$, for $k = 1, 2$. We claim that $\{V_1, V_2\}$ is a minimax bipartition. To prove the claim, assume without loss of generality that $\text{diam}(V_1) \geq \text{diam}(V_2)$. If v_i and v_j are two vertices in V_1, then $v_i v_j$ forms an odd cycle with MXST(G). Let $v_p v_q$ be a maximum–weight edge in E that forms an odd–cycle with MXST(G). Then

$\text{diam}(V_1) \le d(v_p, v_2)$. The claim now follows easily from Lemma 1 since, for every bipartition $\{V', V''\}$ of V,

$$\max \{\text{diam}(V'), \text{diam}(V'')\} \ge d(v_p, v_q).$$

Next, we analyze the time complexity of our algorithm. Given a maximum spanning tree of G, a 2–coloring χ can be obtained in time $O(n)$. To compute a maximum spanning tree of G, we first negate all the edge weights and then use the minimum spanning tree algorithm of Fredman and Tarjan [FT], which runs in time $O(m\log^* n)$. In the Euclidean case, a maximum spanning tree can be computed in time $O(n\log n)$, using an algorithm of Monma, Paterson, Suri and Yao [MPSY]. The time complexity in the latter case is optimal since the problem of computing the diameter of a planar point set (which is known to require $\Omega(n\log n)$ time in the algebraic computation–tree model) is $O(n)$–time reducible to the minimax bipartition problem. This completes the proof of the theorem. ❑

Asano et al. [ABKY] have shown that, for the Euclidean case, there always exists a *linearly separable* minimax bipartition $\{V_1, V_2\}$, i.e., there exists a line that separates the points of V_1 and V_2. The solution obtained by using Theorem 1 may not have this property. However, Asano et al. show that, any bipartition can be transformed into a linearly separable bipartition without increasing the maximum diameter (see [ABKY]). Consequently, if desired, the solution reparted by Theorem 1 can be converted into a linearly separable minimax bipartition in additional $O(n\log n)$ time.

4. Algorithms for Minimizing the Sum of Diameters

A bipartition $\{V_1, V_2\}$ of V is called a *minimum–sum* bipartition if $\text{diam}(V_1) + \text{diam}(V_2)$ is minimum over all bipartitions of V. In this section, we develop algorithms for computing minimum–sum bipartitions. Our algorithm for arbitrary graphs runs in time $O(nm\log n)$ and the algorithm for Euclidean graphs runs in time $O(n^2)$. The two algorithms are quite similar in spirit but certain steps can be speeded up for Euclidean graphs. We first consider the problem for an arbitrary graph $G = \{V, E, d: E \rightarrow R^+\}$.

4.1 General Graphs

We start with the following decision problem: Given two real numbers r_1 and r_2, where $r_1 \ge r_2$, is there a bipartition $\{V_1, V_2\}$ of V satisfying $\text{diam}(V_1) \le r_1$ and $\text{diam}(V_2) \le r_2$? If such a bipartition exists, we call it a (r_1, r_2)–partition. We show that the problem of deciding whether a (r_1, r_2)–partition exists can be solved in time $O(m)$.

Our method is based on labeling the vertices of G with labels 1 and 2. Initially, all vertices are unlabeled. Our procedure then either labels all the vertices of G successfully, in which case the vertices labeled i are assigned to V_i, $i = 1, 2$, or it halts having discovered that no (r_1, r_2)–partition exists.

Let v_iv_j be an edge of E such that v_i is labeled but v_j is not. Our procedure uses the following two propagation rules to decide what label to use for v_j. Rule (R1) applies if $d(v_i, v_j) > r_1$, and rule (R2) applies if $r_1 \geq d(v_i, v_j) > r_2$.

(R1) [*Applies when* $d(v_i, v_j) > r_1$.] If v_i is labeled with 1 (resp. 2), then label v_j with 2 (resp. 1).

(R2) [*Applies when* $r_1 \geq d(v_i, v_j) > r_2$.] If v_i is labeled with 2, then label v_j with 1. Otherwise, leave v_j unlabeled.

We say that a *label–violation* occurs at an edge v_iv_j if either v_i and v_j both are labeled 1 and $d(v_i, v_j) > r_1$, or v_i and v_j are both labeled 2 and $d(v_i, v_j) > r_2$. Clearly, a (r_1, r_2)–partition of V exists if and only if there is a way to label all the vertices without a label–violation.

Our procedure starts by labeling an arbitrary vertex v^* with label 1. The two propagation rules (R1) and (R2) are then used to label as many vertices as possible. One of the following two events must occur:

(i) A label–violation occurs at some edge v_iv_j.

(ii) A connected subgraph of G gets labeled without any label–violations and no more vertices can be labeled using (R1) and (R2).

If event (i) occurs, we erase all the labels that propagated from v^* and restart the procedure by labeling v^* with 2. If a label–violation occurs again, we halt the procedure and report that a (r_1, r_2)–partition of V is impossible.

If event (ii) occurs and all the vertices of V have been labeled, the procedure stops and reports the (r_1, r_2)–partition given by the labels. Otherwise, we assign to V_i all the vertices labeled i, for $i = 1, 2$, and delete from G these vertices. Let $G' \subseteq G$ be the subgraph induced by the remaining (unlabeled) vertices. Start the procedure afresh on G'.

The time complexity of the procedure is clearly bounded by $O(m)$. To prove the correctness of the procedure, we reason as follows. Any (r_1, r_2)–partition of V must induce a violation–free labeling of V. All the labels propagated via rules (R1) and (R2) therefore are correct, up to the choice of the label of the start vertex v^*. Therefore, if a label–violation occurs for both the choices of labels for v^*, our procedure correctly determines that (r_1, r_2)–partition of V is infeasible. Otherwise, a (connected) subgraph of G receives a violation–free labeling. Let $V_\ell \subseteq V$ be the set of vertices that received labels and let $V_m = V - v_\ell$ be the set of remaining (unlabeled) vertices. Consider a pair of vertices $v_\ell \in V_\ell$ and $v_m \in V_m$ that are joined by an edge in G. Since both rules (R1) and (R2) fail to apply for the edge $v_\ell v_m$, one of the following two conditions must hold:

(*) $r_1 \geq d(v_\ell, v_m) > r_2$ and v_ℓ is labeled 1, or
(**) $r_2 \geq d(v_\ell, v_m)$.

In either case, v_m can be labeled either 1 or 2 without causing a label–violation at $v_\ell v_m$. In other words, the vertices of V_m can be labeled independently of labels of V_ℓ, which shows that event (ii) is handled correctly by the procedure. This completes our discussion of the decision problem, and we summarize the result as follows.

Theorem 2 *Let $G = \{V, E, d: E \rightarrow R^+\}$ be an arbitrary weighted graph, where $|V| = n$ and $|E| = m$, and let r_1 and r_2 be two real numbers. Then, there is an $O(m)$ time and space algorithm that either finds a (r_1, r_2)–partition of V, or determines that no such bipartition is possible.*

Now, we use the result of Theorem 2 to design an algorithm for finding a minimum–sum bipartition of V.

Let MXST(G) be a maximum spanning tree of G. Let $v_p v_q$ be a maximum–weight edge in E that forms an odd cycle with MXST(G). Let K be the set of edges consisting of $v_p v_q$ and the edges in MXST(G) whose weight exceeds that of $v_p v_q$, i.e.,

$$K = \{v_p v_q\} \cup \{v_i v_j \in \text{MXST}(G) \mid d(v_i, v_j) > d(v_p, v_q)\}$$

Let $\{V_1, V_2\}$ be any bipartition of V. Without loss of generality, assume $\text{diam}(V_1) \geq \text{diam}(V_2)$. We know from Lemma 1 that there always exists an edge $v_k v_l \in K$ such that $\text{diam}(V_1) = d(v_k, v_l)$. Therefore, in our search for a minimum–sum bipartition, we always can choose an edge of K to define the larger of the two diameters. Our algorithm is described as follows.

Algorithm Minimum–Sum–Partition–1

(1) Sort edges of E in non–increasing order of weight. Let E_s be the resulting list of edges.

(2) Compute a maximum spanning tree of G, MXST(G), and determine the set of edges K.

(3) For each edge–weight r_1 in K, perform a binary search on E_s to determine the smallest edge–weight r_2 for which there exists a (r_1, r_2)–partition. Report that (r_1, r_2)–partition for which $r_1 + r_2$ is minimum.

Step 1 takes $O(m\log m)$ time, by a standard sorting algorithm. Computing MXST(G) takes time $O(m\log^* n)$; use the minimum spanning tree algorithm of Fredman and Tarjan [FT] after negating all the edge weights. Given MXST(G), the set of edges K can be found in time $O(m)$, as follows. Let χ be a 2–coloring of MXST(G). Then, edges of E that form an odd cycle with MXST(G) are those edges whose both endpoints have the same color under χ. Thus, in a single scan of E, we can find the maximum–weight edge that forms an odd cycle with MXST(G). Let $v_p v_q$ be this edge. Given $v_p v_q$, the remaining edges of K can be determined from MXST(G) in $O(n)$ time. The most expensive step of the algorithm Minimum–Sum–Partition–1 is Step 3. By Theorem 2, feasibility of a (r_1, r_2)–partition can be determined in time $O(m)$ for any fixed pair of values r_1 and r_2. Since K has at most n edges and, for any fixed r_1, at most $(\log m + 1)$ pairs of r_1 and r_2 are tested, Step 3 requires $O(nm\log n)$ time. We, thus, have the following result.

Theorem 3 *Let $G = \{V, E, d{:}E \rightarrow R^+\}$ be an arbitrary weighted graph, where $|V| = n$ and $|E| = m$. There is an $O(nm\log n)$–time and $O(m)$–space algorithm for determining a minimum–sum bipartition of V.*

4.2. Euclidean Graphs

For Euclidean graphs, Theorem 3 gives $O(n^3\log n)$ time bound. In this section, we show that the geometric structure of the problem can be exploited to improve the time complexity to $O(n^2)$. We start with a few preliminaries.

Let V be a set of points in the plane in general position, i.e., no three points lie on a straight line. Following [MPSY], we adopt a general "tie–breaking" convention that guarantees distinctness of distances between all pairs of points. The convention briefly is as follows. First, assume that the points $\{v_1, v_2, \ldots, v_n\}$ are lexicographically ordered by their coordinates. Then, for any pair of points v_i and v_j, with $i < j$, define the augmented distance between them as the triple $< d(v_i, v_j), j, -i >$. The lexicographic order on this augmented distance gives the required total ordering on distances. This ordering is consistent in the following sense: for any $\varepsilon > 0$, there is another point set V' in the plane such that (1) distances between points of V' are all distinct, (2) each point of V' is within distance ε of its corresponding point in V, and (3) ordering between points of V' is exacly the same as the augmented ordering between points of V. (The reader is referred to Monma, Paterson, Suri and Yao [MPSY] for further details regarding this augmented ordering.) Therefore, in the following, we may assume that the distances between all pairs of points of V are distinct. CH(V) denotes the convex hull of V.

Lemma 2 *Let V be a set of points in the plane in general position and let $\{V_1, V_2\}$ be a bipartition of V such that $CH(V_1)$ and $CH(V_2)$ intersect. Then $diam(V_1) + diam(V_2) \geq diam(V_1 \cup V_2)$.*

Proof If either $CH(V_1) \subseteq CH(V_2)$ or vice versa, then the claim trivially holds. Otherwise, let v_p and v_q be two points such that $diam(V_1 \cup V_2) = d(v_p, v_q)$. If either $v_p, v_q \in V_1$ or $v_p, v_q \in V_2$, then the claim holds again, since $diam(V_1), diam(V_2) > 0$. So, assume without loss of generality that $v_p \in V_1$ and $v_q \in V_2$. Let x be a point where the boundaries of $CH(V_1)$ and $CH(V_2)$ intersect. The triangle inequality in the plane implies that

$$diam(V_1 \cup V_2) = d(v_p, v_q) \leq d(v_p, x) + d(x, v_q) \leq diam(V_1) + diam(V_2).$$

This finishes the proof. ❑

Recall that a bipartition $\{V_1, V_2\}$ is called linearly separable if there exists a straight line that separates $CH(V_1)$ and $CH(V_2)$. An easy consequence of Lemma 2 is that, for every set of points V in the plane, there exists a linearly separable minimum–sum bipartition.

Let v_a and v_b be two points. The common intersection of the two closed discs, each of radius $d(v_a, v_b)$, centered at v_a and v_b, is called the *lune of $v_a v_b$*, and is denoted *lune*(v_a, v_b). The following fact is proved easily.

Lemma 3 *Let v_a, v_b, v_c and v_d be four points in the plane in general position such that $d(v_c, v_d) > d(v_a, v_b)$. If v_c and v_d both lie in lune(v_a, v_b), then the line segments $v_a v_b$ and $v_c v_d$ intersect.*

Let $G = \{V, E, d: E \rightarrow R^+\}$ be the Euclidean graph of the point set V. Let MXST(G) be a maximum spanning tree of G; distinctness of all the distances among points of V ensures that MXST(G) is unique. Let v_p and v_q be the two points such that, among all edges of E that form an odd cycle with MXST(G), $d(v_p, v_q)$ is maximum. Let K be the set consisting of $v_p v_q$ and the edges of MXST(G) whose weight exceeds $d(v_p, v_q)$, i.e.,

$$K = \{v_p v_q\} \cup \{v_i v_j \in MXST(G) \mid d(v_i, v_j) > d(v_p, v_q)\}$$

By Lemma 1, for any bipartition of V, there exists an edge in K that defines the larger of the two diameters. Therefore, our problem remains to find, for each edge weight r_1 in K, a bipartition $\{V_1, V_2\}$, with $diam(V_1) = r_1$ and $diam(V_2) = r_2$, such that r_2 is as small as possible. Obviously, we are only interested in those bipartitions $\{V_1, V_2\}$ that satisfy $diam(V_1) + diam(V_2) < diam(V_1 \cup V_2)$. Our procedure, again, is based on labeling the vertices of V either 1 or 2 and is described below in more detail.

Let v_av_b be an edge in K and let r_1 be its weight, i.e., $r_1 = d(v_a, v_b)$. A *label–violation* is said to occur in a labeling of V if two vertices at distance greater than r_1 are labeled 1. The procedure starts by deleting from MXST(G) all the edges whose weights are less than or equal to r_1. Let $F(r_1)$ denote the resulting forest and let T_1, T_2, ..., T_k be the subtrees of $F(r_1)$.

Lemma 4 *Let v_i and v_j be two vertices that belong to the subtrees T_i and T_j, respectively, where $1 \le i \ne j \le k$. Then $d(v_i, v_j) \le r_1$.*

Proof Let V' be the vertex set of T_i and let V" = V – V'. Let v'v" be the maximum–weight edge that goes between V' and V". It is well known that v'v" belongs to MXST(G). (This is the familiar theorem of Prim stated in the context of maximum spanning trees, see Graham and Hell [GH].) However, V' and V" are disconnected in $F(r_1)$, implying that $d(v', v'') \le r_1$. Since $v_i \in V'$ and $v_j \in V''$, it follows that $d(v_i, v_j) \le d(v', v'') \le r_1$. ❑

It follows from Lemma 4 that if T_i consists of a single vertex v_i, then v_i can be labeled safely with 1; this cannot cause a label–violation since all other points of V are at distance no more than r_1 from v_i. We, therefore, assume from now on that all subtrees T_i have at least two vertices. Let v_iv_j be an edge of some subtree T_i. Then v_i and v_j cannot be given the same label, because $d(v_i, v_j) > r_1$. Therefore, labeling any vertex of T_i, $1 \le i \le k$, forces labels of all the vertices of T_i (recall the propagation rule (R1) from the previous section). Since each subtree in the forest can be labeled in two different ways, there still are exponentially many labelings that do not have a label–violation. We are interested in a labeling that minimizes the diameter of the points labeled 2. In the following, we show that it suffices to check only two different labelings of the points.

Without loss of generality, assume that v_av_b is placed vertically in the plane such that the endpoint v_a is above v_b, where v_av_b is the unique edge in K having the weight equal to r_1. Let T_a and T_b be the subtrees in $F(r_1)$ that contain v_a and v_b, respectively; $T_a = T_b$ is possible. Clearly, v_a and v_b both must be labeled with 1. These labels force a unique labeling of all other vertices of $T_a \cup T_b$. It remains to label the vertices of other subtrees. Consider a subtree T_i, where $i \ne a, b$, and a vertex v_i in it. By Lemma 4, $d(v_a, v_i)$ and $d(v_b, v_i)$ are both less than $d(v_a, v_b) = r_1$ and, hence, $v_i \in lune(v_a, v_b)$. Furthermore, if v_iv_j is an edge of T_i, then by Lemma 3, v_iv_j and v_av_b intersect; recall that $d(v_i, v_j) > r_1$. Thus, all the vertices of T_i, $i \neq a, b$, lie in $lune(v_a, v_b)$ and all edges of T_i intersect v_av_b.

Let V_L (resp. V_R) be the vertices of $\bigcup_{i \neq a,b} T_i$ that lie to the left (resp. right) of v_av_b; since points of V are in general position, no vertex lies on the segment v_av_b. Note that, due to the earlier exclusion of single–vertex subtrees, each T_i has vertices both in V_L and V_R.

Lemma 5 *Let $\{V_1, V_2\}$ be a bipartition such that $diam(V_1) = r_1$ and $diam(V_1) + diam(V_2) < diam(V_1 \cup V_2)$. Then, either $V_L \subseteq V_1$ and $V_R \subseteq V_2$ or vice versa.*

Proof Suppose the lemma were false. Then, there are two edges in $F(r_1)$, v_iv_i' and v_jv_j', belonging to subtrees T_i and T_j, respectively, such that (1) v_i, v_j are in V_L and v_i', v_j' are in V_R, and (2) v_i, v_j' are labeled 1 and v_i', v_j are labeled 2. Elementary geometry shows that $CH(V_1) \cap CH(V_2) \neq \varnothing$, which together with Lemma 2 contradicts the hypothesis that $diam(V_1) + diam(V_2) < diam(V_1 \cup V_2)$. Therefore, either $V_L \subseteq V_1$ and $V_R \subseteq V_2$ or vice versa. ❑

We are now ready to describe the algorithm.

Algorithm Minimum–Sum–Partition–2

(1) Compute a maximum spanning tree of G, $MXST(G)$, and determine the set of edges K.

(2) For each edge $v_av_b \in K$, perform steps (2.1)–(2.4).

(2.1) Compute $F(r_1) = \{T_1, T_2, \ldots, T_k\}$, where $r_1 = d(v_a, v_b)$.

(2.2) Label v_a, v_b with 1 and propagate labels to all other vertices of the subtrees containing v_a and v_b. Label all single–vertex subtrees with 1.

(2.3) Partition the remaining vertices into V_L and V_R, depending on whether they lie on the left or right of v_av_b.

(2.4) Label all vertices of V_L with 1 and all vertices of V_R with 2. Compute the diameter of the points of V labeled 2. Next, reverse the labels of V_L and

V_R and recomputed the diameter of points labeled 2. Keep the bipartition that gives the smaller r_2.

(3) Report the bipartition for which $r_1 + r_2$ is minimum.

Next, we analyze the time complexity of Minimum–Sum–Partition–2. In Step 1, the maximum spanning tree of G can be computed in time $O(n\log n)$ using the algorithm of Monma, Paterson, Suri and Yao [MPSY]. Computing the set of edges K takes additional $O(n\log n)$ time, as follows. Let χ be a 2–coloring of MXST(G). The edges of E that form an odd cycle with MXST(G) are exactly the ones whose both endpoints have the same color. Therefore, the maximum–weight edge in E that forms an odd cycle with MXST(G) is the diameter of either the points labeled 1 or the points labeled 2. Let v_pv_q be this edge. We can find v_pv_q in time $O(n\log n)$ because there is an $O(n\log n)$–time algorithm for computing the diameter of n planar points (see Preparata and Shamos, pp. 176 [PS]). Given v_pv_q, the remaining edges of K can be determined in $O(n)$ time from the edge set of MXST(G).

Steps (2.1), (2.2), and (2.3) are straightforward and require $O(n)$ time. Step (2.4) consists of computing diameters of certain point sets, each of which has at most n points. Since we need to perform this step repeatedly, we can improve upon the obvious $O(n\log n)$ time bound by preprocessing the points of V. In particular, after $O(n\log n)$–time and $O(n)$–space preprocessing, the diameters in Step (2.4) can be computed in time $O(n)$, as is described below. Let v' and v" be the two points that define the diameter of V, i.e., d(v', v") = diam(V). Sort the points of V angularly around v' and v". Let S'(V) and S"(V), respectively, be the two lists thus obtained. Now, in Step (2.4), let V_1 and V_2 be the sets of points labeled 1 and 2, respectively. Note that v' and v" both cannot be labeled 1 (or 2); that would imply that $\text{diam}(V_1) + \text{diam}(V_2) > \text{diam}(V_1 \cup V_2)$. Assume without loss of generality that $v' \in V_1$ and $v'' \in V_2$. Use S' and S" to arrange V_1 and V_2 in sorted order around v' and v", respectively. Connect points of V_1 and V_2 in the sorted order to obtain two (star–shaped) simple polygons, say, P_1 and P_2. Clearly, the diameter of P_i equals the diameter of V_i, $i = 1, 2$. However, the diameter of a simple polygon with n vertices can be computed in time $O(n)$ (see Preparata and Shamos, pp. 176 [PS]). Thus, each iteration of Step (2) can be performed in time $O(n)$ and there are at most n iterations. The $O(n\log n)$ cost of preprocessing is incurred once. The total time required by the algorithm Minimum–Sum–Partition–2 is, therefore, $O(n^2)$. We summarize this result.

Theorem 4 *Let V be a set of n points in the Euclidean plane. There is an $O(n^2)$–time and $O(n)$–space algorithm for finding a minimum–sum bipartition of V.*

5. Closing Remarks

We considered the problem of partitioning a set of entities V subject to minimizing the maximum or the sum of the diameters. The entities are either points in the plane with the Euclidean measure of interdistance, or they are the vertices of an arbitrary graph with real–valued weights on edges. Most of our work, however, is applicable to a wider class of functions of the diameters. We briefly describe some of the generalizations in the following.

Let $\{V_1, V_2\}$ be a bipartition of V, where V is either a set of points or the vertices of an arbitrary graph. Let $d_1 = \text{diam}(V_1)$ and $d_2 = \text{diam}(V_2)$ be the diameters of V_1 and V_2, respectively, where we assume that $d_1 \geq d_2$. Let $f(d_1, d_2)$ be a function of the two diameters such that, for any choice of parameters, f is computable in $O(1)$ time. Consider the problem of determining a bipartition of V subject to minimizing f.

All the results of Section 4 extend to the functions f that are *monotone* in d_2, i.e., if $f(d_1, d_2) \leq f(d_1, d_2')$ whenever $d_2 \leq d_2'$. If, in addition, f also is *unimodal* with respect to d_1, then a bipartition minimizing f can be determined in $O(m\log^2 n)$ time for an arbitrary n–node m–edge graph. The time complexity can be improved to $O(n\log n)$ for the Euclidean graphs. This follows because, instead of iterating Step (3) of the algorithm Minimum–Sum–Partition–1 or Step (2) of the algorithm Minimum–Sum–Partition–2 for each edge–weight in K, we now can perform a binary search on the sorted lists of weights in K. This reduces the number of iterations from n to at most $(\log n + 1)$. Observe that the function $\max\{d_1, d_2\}$ is trivially unimodal with respect to d_1. However, $f(d_1, d_2) = d_1 + d_2$ is not unimodal.

For Euclidean graphs, it is easy to realize an $O(n^3\log n)$–time algorithms under any function f that permits an optimal linearly–separable bipartition; simply compute f for all $O(n^2)$ linearly–separable bipartitions. We remind the reader that such bipartitions exist for minimizing the maximum diameter as well as for minimizing the sum of the diameters. Finally, as observed by Hansen and Jaumard [HJ], *any* function f of the diameters can be minimized in time $O(n^5)$.

The general problem of partitioning a graph into an arbitrary number of subsets is NP–complete under both the minimax or the minimum–sum diameter measure. Similarly, it is NP–complete to decide if a set of n points in the Euclidean plane can be partitioned into k subsets so that the diameter of each subset is less than some given bound d* (see Johnson [Jo]). Finally, for the minimum–sum diameter problem, one can easily obtain a polynomial–time algorithm for any fixed number of subsets, since Lemma 2 guarantees the existence of an optimal partition where every pair of subsets is linearly separable.

REFERENCES

[Av] D. Avis, Diameter Partitioning, *Discrete and Computational Geometry*, 1 (1986) 265–276.

[ABKY] T. Asano, B.K. Bhattacharya, J.M. Keil and F.F. Yao, Clustering algorithms based on minimum and maximum spanning trees, *Proc. of the Fourth Annual Symposium on Computational Geometry* (1988) 252–257.

[Co] R.M. Cormack, A review of classification, *Journal of Royal Statistical Society*, A(134), (1971) 321–367.

[DH] M. Delattre and P. Hansen, Bicriterion cluster analysis, *IEEE Transactions on Pattern Analysis and Machine Intelligence,* (2), (1980) 277–291.

[FT] M. Fredman and R.E. Tarjan, Fibonacci heaps and their uses in improved network optimization algorithms, *Journal of the Association for Computing Machinery*, 34, (1987) 596–615.

[GH] R. Graham and P. Hell, On the history of minimum spanning tree problem, *Annals of History of Computing*, 7, (1985)

[HD] P. Hansen and M. Delattre, Complete link cluster analysis by graph coloring, *Journal of the American Statistical Association*, 73, (1978) 397–403.

[HJ] P. Hansen and B. Jaumard, Minimum sum of diameters clustering, *Journal of Classification*, (1987) 215–226.

[Jo] D.S. Johnson, NP–completeness column, *Journal of Algorithms* (3) (1982) 103–195.

[MPSY] C. Monma, M. Paterson, S. Suri and F.F. Yao, Computing Euclidean Maximum Spanning Trees, *Proc. of the Fourth Annual Symposium on Computational Geometry* (1988) 241–251.

[PS] F.P. Preparata and M.I. Shamos, *Computational Geometry*, Springer Verlag, New York, NY, (1985).

On the Number of Well–Covered Trees

J. W. Moon
University of Alberta

ABSTRACT

Results are obtained on the number of well–covered trees in various families of trees.

1. Introduction

A subset I of nodes of a graph is an *independent set* if no two nodes of I are joined to each other. If, in addition, every node not in I is joined to at least one node in I, then I is a *maximal independent* set. A graph is said to be *well–covered* (cf. [15] or [19]) if all maximal independent sets have the same size. A non–trivial tree is well–covered if and only if every node that is not an end–node is joined to one and only one end–node. This result apparently first appeared implicitly in [18; p. 13]; see also [17], [2] and [3]. Our object here is to enumerate the well–covered trees in various families of trees.

We introduce some terminology and preliminary results in §2. Then we consider certain simply–generated families in §3 and some non–simply–generated families in §4; it follows from our results for these families that the probability that a tree T_{2m} with 2m nodes is well–covered is asymptotic to $\alpha \cdot \beta^m$ where α and β are constants that depend on the family being considered. We conclude in §5 with some remarks on the average number of maximal independent sets in well–covered trees in certain families.

2. Preliminaries

We recall that *plane* trees – or *ordered* trees, as they are called by some authors [7; p. 306] – are rooted trees with a specified ordering for the branches incident with each note. Let $\mathcal{F}$ denote a give family of weighted plane trees in which the tree T_n has weight $\omega(T_n)$. The *out–degree* of a node u in a rooted tree is the number of edges incident with u that lead away from the root; let $D_i(T_n)$ denote the number of nodes of out–degree i in the rooted tree T_n. We say that the family $\mathcal{F}$ is a *simple–generated* family if there exists a sequence of non-negative constants c_0 (=1), c_1, c_2, ..., such that

(2.1) $$\omega(T_n) = \Pi c_i^{D_i(T_n)}$$

for every tree T_n in $\mathcal{F}$. Let Y_n denoted the number of trees T_n in the family $\mathcal{F}$ where (here and elsewhere) the weights are taken into account, i. e.,

$$Y_n = \Sigma\omega(T_n)$$

where the sum is over all plane trees with n nodes. It is not difficult to see (cf. [8; p. 999] or [20; p. 24]) that if $\mathcal{F}$ is a simply–generated family, then its generating functions $Y = \sum_1^{\infty} Y_n x^n$ satisfies the relation

(2.2) $$Y = x\Phi(Y)$$

where

$$\Phi(Y) = 1 + c_1Y + c_2Y^2 + \dots .$$

Different sets of coefficients c_i give rise to different simply–generated families. Thus, for example, if $c_i = 1$ for $i \geq 0$ we obtain the ordinary family of plane trees; and if $c_i = 1/i!$ for $i \geq 0$ we obtain (in effect) the family of rooted labelled trees.

In the next section we shall use the following result which is equivalent to a result provided in [8; Theorem 3.1]; a closely related result was proved earlier in [14].

Lemma 1 *Suppose* $\theta(t) = a_0 + a_1t + a_2t^2 + \dots$ *is a regular function of* t *when* $|t| < R < +\infty$, *and let* $F = F(z) = F_1z + F_2z^2 + \dots$ *denote the solution of* $F(z) = z\theta(F(z))$ *in the neighborhood of* $z = 0$. *If*

(i) $a_i \geq 0$ *for* $i \geq 0$,

(ii) $a_0 > 0$ *and* $a_i > 0$ *and* $a_j > 0$ *for some distinct integers* i *and* j *such that* $gcd(i, j) = 1$, *and*

(iii) $v\theta'(v) = \theta(v)$ *for some* v, *where* $0 < v < R$, *then*

$$F_m \sim a\delta^{-m}m^{-3/2}$$

as $m \to \infty$, *where* $\delta = v/\theta(v)$ *and* $a = (\theta(v)/2\pi\theta''(v))^{1/2}$.

Suppose T is a non–trivial well–covered tree that is rooted at node r. We shall call T a *P–tree* if r has out–degree one and we shall call T a *Q–tree* if r is joined to a node of out–degree zero. Notices that every non–trivial well–covered tree T is either a P–tree or a Q–tree and that the rooted tree with two nodes is the only such tree that is both a P–tree and a Q–tree. The *principal branches* of a rooted tree T are the maximal subtrees not containing the root–node; these branches are considered as being rooted at the nodes joined to the root of T. The *secondary branches* of T are the principal branches of the principal branches of T. The following results, which we shall use in enumerating well–covered trees, are straightforward consequences of the definitions and the characterization of well–covered trees stated in the introduction.

Lemma 2 *A rooted tree T is a P–tree if and only if the root–node has out–degree one and all the secondary branches are Q–trees.*

Lemma 3 *A rooted tree T is a Q–tree if and only if one principal branch consists of a single node and all the remaining principal branches are Q–trees.*

3. Results for simply–generated families

Let $\mathcal{F}$ denote some given simply–generated family of trees. We assume that the function Φ that appears in relation (2.2) satisfies the hypothesis of Lemma 1 with $\Phi(0) = 1$ and that, in particular, $\tau\Phi'(\tau) = \Phi(\tau)$ for some τ, where $0 < \tau < R$, so that

$$(3.1) \qquad Y_n \sim c\rho^{-n} n^{-3/2}$$

as $n \to \infty$, where $\rho = \tau/\Phi(\tau)$ and $c = (\Phi(\tau)/2\pi\Phi''(\tau))^{1/2}$.

There are no non–trivial well–covered trees with an odd number of nodes; this follows immediately from the characterization of well–covered trees. Hence to enumerate the well–covered trees in $\mathcal{F}$ it will suffice to consider the generating functions $P(z) = \sum_1^\infty P_m z^m$ and $Q(z) = \sum_1^\infty Q_m z^m$ where P_m denotes the number of P–trees T_{2m} in $\mathcal{F}$ and Q_m denotes the number of Q–trees T_{2m} in $\mathcal{F}$.

Theorem 1 $P(z) = c_1 z\Phi(Q(z))$.

Theorem 2 $Q(z) = z\Phi'(Q(z))$.

Proof It follows from Lemma 2 and relation (2.1) that the number of P–trees T_{2m} with k secondary branches (all of which are Q–trees) equal the coefficient of z^m in $c_1 c_k z Q^k(z)$ for $k \geq 0$. For $Q^k(z)$ enumerates the ordered k–tuples of Q–trees, the factor z takes into account the root–node of T_{2m} and the (unique) node u joined to the root, and the factors c_1 and c_k record the weights associated with the nodes r and u. Consequently

$$P(z) = c_1 z \sum_0^\infty c_k Q^k(z) = c_1 z \Phi'(Q(z)).$$

Similarly, it follows from Lemma 2 that the number of Q–trees T_{2m} with k principal branches (one of which consists of a single node v and $k-1$ of which are Q–trees) equals the coefficient of z^m in $zkc_kQ^{k-1}(z)$ for $k \geq 1$. For, $Q^{k-1}(z)$ enumerates the ordered $(k-1)$–tuples of Q–trees, the factor k counts the number of choices for the position of the trivial branch consisting of a single node v, the factor z takes into account the root–node and the node v, and the factor ck records the weight associated with the root–node. Consequently,

$$Q(z) = z \sum_1^\infty kc_k Q^{k-1}(z) = z\Phi'(Q(z))$$

We now determine the asymptotic behavior of P_m and Q_m subject to some additional assumptions on the function Φ.

Theorem 3 *Suppose that $c_1 > 0$ and that $c_i > 0$ and $c_j > 0$ for some distinct integers i and j such that $i > 1$, $j > 1$, and $gcd(i-1, j-1) = 1$. Suppose further that $v\Phi''(v) = v\Phi'(v)$ for some v where $0 < v < R$. Then*

(3.2)
$$Q_m \sim a\delta^{-m} m^{-3/2}$$

and

(3.3)
$$P_m \sim vc_1 Q_m$$

as $m \to \infty$, where $\delta = v/\Phi'(v)$ and $a = (\Phi'(v)/2\pi\Phi''(v))^{1/2}$.

Proof Relation (3.2) follows upon applying Lemma 1 to the relation $F(z) = z\theta(F(z))$ with $F(z) = Q(z)$ and $\theta(t) = \Phi'(t)$. And it follows from Theorem 1 and a result in [9; Lemma 4] that

$$P_m \sim c_1 \Phi'(Q(\delta)) Q_{m-1}$$

as $m \to \infty$. Hence

$$P_m \sim c_1 \delta \Phi'(\upsilon) Q_m = \upsilon c_1 Q_m$$

by (3.2) and the definitions of δ and υ. (We remark that it is possible to give examples of functions $\Phi(t)$ such that the equation $\tau\Phi'(\tau) = \Phi(\tau)$ has a solution τ where $0 < \tau < R$ but the equation $\upsilon\Phi''(\upsilon) = \Phi'(\upsilon)$ does not have a solution where $0 < \upsilon < R$; so the hypothesis that υ exists cannot simply be dropped.)

We conclude this section by illustrating the preceding results for two particular simply–generated families of trees. Let $\mathcal{F}$ denoted the family of rooted labelled trees. Then (see, e. g., [11; p. 26] or [4; p. 23]) $Y = xe^Y$ and

$$Y_n = n^{n-1}/n! \sim (2\pi)^{-1/2}e^n n^{-3/2}.$$

In this case $\Phi'(t) = \Phi(t) = e^t$ so $P = Q = ze^Q$ and

$$P_m = Q_m = m^{m-1}/m! \sim (2\pi)^{-1/2}e^m m^{-3/2}.$$

These results agree with relations (3.2) and (3.3) with $\upsilon = 1$ and $\delta = e^{-1}$. (We remark that in this particular case the expression for P_m and Q_m can readily be derived from first principals upon appealing to the characterization stated in the introduction.)

Now let $\mathcal{F}$ denote the family of plane trees. Then (see,.e. g., [16; p. 197] or [4; p. 67]) $Y = x(1 - Y)^{-1}$ and

$$Y_n = \frac{1}{n}\binom{2n-2}{n-1} \sim \pi^{-1/2}4^{n-1}n^{-3/2}$$

In this case $\Phi(t) = (1 - t)^{-1}$ and $\Phi'(t) = (1 - t)^{-2}$ so $Q = z(1 - 1)^{-2}$. When we apply Lagrange's inversion formula [1;p. 148] to this relation for Q we find that

$$Q_m = \frac{1}{m}\binom{3m-2}{m-1} \sim (27\pi)^{-1/2}(27/4)^{m-1}m^{-3/2}.$$

Furthermore, it follows from Theorems 1 and 2 that

$$P = z(1 - Q)^{-1} = Q(1 - Q).$$

Appealing again to Lagrange's formula, we find that

$$P_m = \frac{1}{3m-2}\binom{3m-2}{m-1} = \frac{m}{3m-2}Q_m$$

These results agree with relations (3.2) and (3.3) with $\upsilon = 1/3$ and $\delta = 4/27$.

4. Result for some non–simply–generated families

We now consider some non–simply–generated families for which relation (2.1) does not hold. A tree T_n with n labelled nodes rooted at node 1 is a *recursive tree* if

$n = 1$ or if $n > 1$ and T_n can be constructed by joining node n to one of the $n - 1$ nodes of a recursive tree T_{n-1}; or, equivalently, T_n is a recursive tree if the labels of the nodes on any path leading away from the root form an increasing sequence (see, e. g., [12] or [8]). Let y_n denote the number of recursive trees T_n. These definitions readily imply that $y_n = (n-1)!$ and that the generating function $y = \sum_1^\infty y_n x^n/n!$ satisfies the differential equation $y' = e^y$. Let p_m denote the number of recursive P–trees T_{2m} and let q_m denote the number of recursive Q–trees T_{2m}. Lemmas 1 and 2 can be applied to obtain differential equations for the (exponential) generating functions of the numbers p_m and q_m, and these equations can then be solved to yield formulas for p_m and q_m. In this case, however, it is possible to derive formulas for p_m and q_m by appealing directly to the characterization of well–covered trees stated in the introduction.

Theorem 4 *Let $\mathcal{F}$ denote the family of recursive trees. Then*

$$q_m = (2m-1)!/2^{m-1} \tag{4.1}$$

and

$$p_m = (2m-2)!/2^{m-1}. \tag{4.2}$$

Proof There are $(2m)!/m!2^m$ ways of partitioning $2m$ labelled nodes $(1, 2, \ldots, 2m)$ into m pairs of nodes (x_i, y_i) where $x_i < y_i$ for $1 \le i \le m$; we may suppose that $x_1 = 1$. Now construct a recursive tree T_m with the nodes $x_1, \ldots, x_m$; this can be done in $(m-1)!$ ways. If we then join node y_i to node x_i for $1 \le i \le m$, it is not difficult to see that the resulting tree T_{2m} is a recursive Q–tree and that all such trees can be constructed in this way. It follows, therefore, that

$$q_m = \frac{(2m)!}{m!2^m} \cdot (m-1)! = \frac{(2m-1)!}{2^{m-1}}.$$

Before enumerating the recursive P–trees we observe that if T_{2m} is a recursive P–tree, then node 2 is the (unique) node joined to the root–node 1. There are $(2m-2)!/(m-1)!2^{m-1}$ ways of partitioning $2m$ labelled nodes $(1, 2, \ldots, 2m)$ into m pairs of nodes (x_i, y_i) where $x_1 = 1, y_1 = 2$ and $x_i < y_i$ for $2 \le i \le m$. Now construct one of the $(m-1)!$ recursive trees T_m with the nodes $2, x_2, \ldots, x_m$. Finally, join node 1 to node 2 and node y_i to node x_i for $2 \le i \le m$. As before, it is not difficult to see the the resulting tree T_{2m} is a recursive P–tree and that all such trees can be constructed in this way. Hence,

$$p_m = \frac{(2m-2)!}{(m-1)!2^{m-1}} \cdot (m-1)! = \frac{(2m-2)!}{2^{m-1}}.$$

Now let $\mathcal{F}$ denote the family of nonisomorphic rooted unlabeled trees. It is well–known ([4]; p. 52) that the generating function $Y(x) = \sum_1^\infty Y_n x^n$ for this family satisfies the relation

$$Y(x) = x \exp \sum_1^\infty T(x)^k/k.$$

Otter [13] (see also [4; p. 213] showed that $Y_n \sim c\rho^{-n}n^{-3/2}$ where $c = 4.399\ldots$ and $\rho = .3383\ldots.$ It is very easy to enumerate the Q–trees and the P–trees in this family. suppose we join each node of a tree T_m in $\mathcal{F}$ to a different new node. If the resulting tree T_{2m} is regarded as being rooted at the root–node of the original tree T_m then T_{2m} is clearly a Q–tree; and if T_{2m} is regarded as being rooted at the new node joined to the root–node of T_m then T_{2m} is a P–tree. It is not difficult to see that this implies that

$$P_m = Q_m = Y_m$$

for this particular family.

Finally, let $y(x) = \sum_1^\infty y_n x^n$ where y_n denote the number of trees T_n in the family $\mathcal{G}$ of nonisomorphic unrooted unlabeled trees. Otter [13] (see also [4; pp. 57 and 214]) showed that

$$y(x) = Y(x) + \tfrac{1}{2}(Y^2(x) - Y(x^2)),$$

where $Y(x)$ denotes the generating function for the nonisomorphic unrooted unlabeled trees, from which he deduced that $y_n \sim b\rho^{-n}n^{-5/2}$ where $b = .5349\ldots$ and, as in the rooted case, $\rho = .3833\ldots.$ Since the trees in $\mathcal{G}$ are not rooted we cannot talk about P–trees and Q–trees now. But if we let w_m denoted the number of well–covered trees T_{2m} in $\mathcal{G}$, then it follows essentially the same argument as was used in the rooted case that $w_m = y_m$.

5. Maximal Independent Sets

Let $M(T_{2m})$ denote the number of maximal independent set in the well–covered tree T_{2m} and let $N(T'_m)$ denote the number of (not necessarily maximal) independent sets in the "stripped" tree T'_m remaining after removing all the end–nodes of T_{2m}; the empty set is to be counted in $N(T'_m)$ and we assume that $m > 1$. Since every maximal independent set S of T_{2m} has a unique decomposition of the form $S = I \cup E$ where I is an independent set of T'_m and E consists of the end–nodes of T_{2m} that are not joined to any node of I, it follows that $M(T_{2m}) = N(T'_m)$. Thus to determine the total number of maximal independent sets in all well–covered trees T_{2m} in a given simply–generated family $\mathcal{F}$, it suffices to determine the total number of independent sets in the corresponding "stripped" trees T'_m. The problem of determining the total number of independent sets in certain simply–generated families of trees has been considered in [5], [6] and [10]. The approach used in theses papers can, in principle, be adapted to the present problem for simply–generated families by considering the "stripped" trees obtained from Q–trees and from P–trees separately and making use of Lemmas 2 and 3. We shall not pursue this further here other than to remark that if $e(2m)$ denotes the average number of maximal independent sets in well–covered trees T_{2m} belonging to any given family $\mathcal{F}$, then it can be shown that $(e(2m))^{1/2m}$ tends to 1.286... and to 1.293... for the labelled trees and the plane trees, respectively; by contrast, if $\mu(n)$ denotes the average number of maximal independent sets in all trees T_n in $\mathcal{F}$, then (see [10]) $(\mu(n))^{1/n}$ tends to 1.273... and to 1.239... for these same two families respectively.

6. Acknowledgements

I am indebted to Professor B. Hartnell for introducing me to well–covered trees. I am also indebted to W. Aiello for performing some calculations for me. The preparation of this paper was assisted by a grant from the Natural Sciences and Engineering Research Council of Canada.

REFERENCES

[1] L. Comtet, *Advanced Combinatorics*, Reidel, Dordrecht, 1974.

[2] O. Favaron, Very well covered graphs, *Discrete Math.* **42** (1982) 177–187.

[3] A. S. Finbow and B. L. Hartnell, A game related to covering by stars, *Ars Combinatoria* **16A** (1983) 189–198.

[4] F. Harary and E. M. Palmer, *Graphical Enumeration*, Academic Press, New York, 1973.

[5] P. Kirschenhofer, H. Prodinger and R. F. Tichy, Fibonacci numbers of graphs: II, *Fibonacci Quarterly* **21** (1983) 219–229.

[6] P. Kirschenhofer, H. Prodinger and R. F. Tichy, Fibonacci numbers of graphs: III, *Fibonacci Numbers and their Applications*, Reidel, Dordrecht, (1986) 105–120.

[7] D. E. Knuth, *The Art of Computer Programming*, vol. 1, Addison–Wesley, Reading, 1973.

[8] A. Meir and J. W. Moon, On the altitude of nodes in random trees, *Can. J. Math.* **30** (1978) 997–1015.

[9] A. Meir and J. W. Moon, Games on random trees, *Congressus Numerantium* **44** (1984) 293–303.

[10] A. Meir and J. W. Moon, On maximal independent sets of nodes in trees, *J. Graph Th.* **12** (1988) 265–283.

[11] J. W. Moon, *Counting Labelled Trees*, Canadian Mathematical Congress, Montreal, 1970.

[12] B. R. Myers and M. A. Tapia, Generation of concave node–weighted trees, *IEEE Trans. Cir Th.* CT–14 (1967) 229–330.

[13] R. Otter, The number of trees, *Ann. of Math.* **49** (1984) 583–399.

[14] R. Otter, The multiplicative process, *Ann. of Math. Statis.* **20** (1949) 206–224.

[15] M. D. Plummer, Some covering concepts in graphs, *J. Comb. Th.* **8** (1970) 91–98.

[16] G. Pólya, Kombinatorische Anzahlbestimmungen für Gruppen, Graphen und chemische Verbindungen, *Acta Math.* **68** (1937) 145–254.

[17] G. Ravindra, Well covered graphs, *J. Combin Inform System. Sci.* **2** (1977) 20–21.

[18] J. A. W. Staples, *On some subclasses of well–covered graphs*, Ph.D. Thesis, Vanderbilt University, 1975.

[19] J. A. W. Staples, On some subclasses of well–covered graphs, *J. Graph Th.* **3** (1979) 197–204.

[20] J.–M. Steyaert and P. Flajolet, Patterns and pattern–matchings in trees: an analysis, *Informations and Control* **58** (1983) 19–58.

THE UNIVERSAL GRAPHS OF FIXED FINITE DIAMETER

Lawrence S. Moss

University of Michigan
at Ann Arbor

ABSTRACT

In [3], we showed that for each natural number N *there is a graph* U_N *with the following two properties: (1)* U_N *is a countable graph of diameter* N*, and (2) for every countable graph* H *of diameter* N*, there is an isometric embedding of* H *into* U_N*. Our proof there was from first principles. In this paper, we present a new proof of the existence of these universal graphs of finite diameter. Our proof here is a little simpler than in [3], especially for even* N*, but it relies on some related results from [2]. In that paper, we constructed a countable connected graph* U_∞ *with the property that every countable connected graph is isometrically embedded in* U_∞*. These results are reviewed below, so this paper may be read on its own. Then we prove the existence of* U_N*. We also show that each* U_N *is distance homogeneous, and in fact it is the unique distance homogeneous graph with properties (1) and (2).*

The first results concerning universal graphs are due to Rado [4]. He showed that there is a countable graph which isomorphically embeds every countable graph. In our notation, Rado's graph is U_2*, the universal graph of diameter 2. (It is obvious that* U_0 *is a single point, and*

This research was supported in part by NSF grant DMS-85-01752.

U_1 is the complete graph on infinitely many vertices.) Our graphs U_N are thus higher diameter analogs of the Rado universal graph.

Our last section contains results relating the different U_N. We investigate first-order theories of the universal graphs, and our main result there is that a first order sentence is satisfied by U_∞ if and only if it is satisfied by U_N for almost all N. So the theory of U_∞ is the limit of the theories of the U_N.

1. Preliminaries

If G is any graph, then we write $d_G(x,y) = N$ to mean that the shortest path in G from x to y has length N. If N is "∞", then this means that x and y belong to different connected components of G. A map between graphs is an *isometric embedding* if it preserves all distances. All of the maps between graphs in this paper are isometric embeddings. An induced subgraph H of G is an *isometric subgraph* of G if the inclusion map is an isometric embedding. We write $H \hookrightarrow G$ in this situation. G is *distance finite* if every finite set X of vertices is contained in a finite isometric subgraph of G.

For two tuples $\overline{x}$ and $\overline{y}$ of the same length k taken from two graphs G and H, respectively, we will write $\overline{x} =_N \overline{y}$ to mean that whenever $1 \leq i,j \leq k$ and either $d_G(x_i,x_j) \leq N$ or $d_H(y_i,y_j) \leq N$, then $d_G(x_i,x_j) = d_H(y_i,y_j)$. A graph G is *distance homogeneous* if for every pair of tuples $\overline{x}$ and $\overline{y}$ of the same length such that $\overline{x} =_\infty \overline{y}$ there is some automorphism if G taking $\overline{x}$ to $\overline{y}$ pointwise.

G is *distance injective* if for all finite connected F and H, if $i : F \to G$ and $J : F \to H$, then there is an isometric embedding $k : H \to G$ such that $k \deg j = i$.

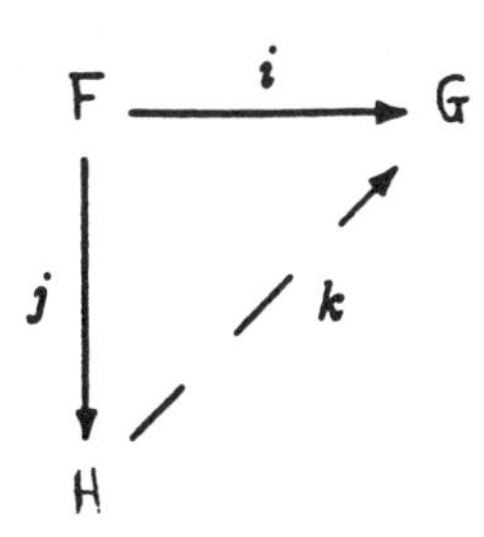

In this paper, we will need a refinement of this concept. G is *N-distance injective* if for all $i : F \to G$ and $j : F \to H$ where H is a finite graph of *diameter N*, there is some $k : H \to G$ such that $koj = i$.

Clearly, G is distance injective iff it is N-distance injective for all N, and if G is N-distance injective, then it is also M-distance injective for all $M < N$.

Finally, G is *distance full* if G is distance injective and distance finite.

In [2] we studied these concepts, and the reader is referred there for a more thorough discussion. Here is a summary of some of the results: Using the Amalgamation Lemma for distances in graphs, it is routine to show that there is a countable distance full graph. One constructs such a graph by repeatedly amalgamating all of the finite connected graphs, in all possible ways. In addition, every countable distance full graph isometrically embeds every countable connected graph. (So the disjoint union of countably many such graphs embeds isometrically every countable graph.) Every countable distance full graph is distance homogeneous. There is exactly one countable connected distance full graph, and we let U_∞ denote this graph. Our last preliminary is a general construction concerning graphs of finite diameter. Let G be any graph, and $N \in \omega$. Then we build a graph G^+, the *canonical graph of diameter N determined by G* as follows: If N is even, then we take a new point z and add chains of length $N/2$ from the z the elements of G. The chains consist of new elements, and the only neighbors that these elements have are other elements on the chains. If N is odd, then we take a new clique $C = \{c_g : g \in G\}$ in one-to-one correspondence with the elements of G. We then add a new chain of length $(N-1)/2$ from each c_g to the corresponding g.

If G has diameter N, then G is an isometric subgraph of G^+. If G did not have diameter N, then the inclusion will not preserve all distances, but it will preserve all distances of length $\leq N$. This is because all of the paths in G^+ which contain new points have length $\geq N$.

2. Saturated Cliques

The study of universal graphs of odd diameter calls for an investigation of cliques in U_∞. The idea is that to get the universal graph of diameter $2N+1$, we take a clique C, and then consider the set of all points whose distance to some element of C is N. However, it turns out that this will not work unless C has a special saturation property which we isolate below.

The main results in this section are Lemmas 1, 2 and 5. At first the reader may wish to omit all of the details of this section except the two definition paragraphs and the statements of these results.

Definitions Let C be a clique in a graph G, and let $S \subseteq G$ be finite. A *type of* S is a function $t : S \to \omega$. The type t is *consistent with* C if there is an isometric extension H of G which contains a point y such that $C \cup \{y\}$ is a clique in H, and for all $x \in S, d_H(x,y) = t(x)$. A type t is *realized in* C if there is some $y \in C$ such that for all $x \in S, d_H(x,y) = t(x)$. C is *saturated* if every type consistent with C is realized in C.

Before proving the existence of saturated cliques in U_∞, we should make a few remarks which will be used below. If t is consistent with C, then t is also consistent with every subset $D \subseteq C$. If t is realized in C, then it is consistent with C - take H to be G with a new point z whose neighbors are y and the neighbors of y in G. Also, if C is saturated, then every type consistent with it is realized by infinitely many elements of C.

Every saturated clique $C \subseteq U_\infty$ has the property that for all $x \in U_\infty$ there are exactly two distances from x to the points of C. (If not, add a new point z to U_∞, making it a neighbor only to the points of C.) We call such a clique *two-valued*, and we let $m = m_c : U_\infty \to \omega$ be such that $m(x)$ is the least distance from x to an element of C.

It is important to reformulate the notion of consistency as follows in terms of the function m: Let C be a two-valued clique in a graph G, and let S be finite. A type t of S is consistent iff (1) for all $x \in S, t(x) = m(x)$ or $t(x) = m(x) + 1$; and (2) if for all $x, y \in S, d_G(x,y) \leq t(x) + t(y)$ and $t(x) \leq t(y) + d_G(x,y)$. (1) is necessary because a point in a graph can have at most two distances to any clique, and (2) is necessary by the triangle inequality. To see that these two are sufficient, assume that t has these properties. Let $H = H(G,C,t)$ arise by adding a new point y to G with edges to all of the points in C and with a chain of length $t(x)$ to each $x \in S$. Then triangle inequalities of (2) imply that $G \hookrightarrow H$ and that for all $x \in S, d_H(x,y) = t(x)$.

Recapitulating, the consistency of a type t of S may be decided solely on the basis of the distance relations among the points of S and the action of m_C on S.

Lemma 1 There exist saturated cliques in U_∞. In fact, every finite clique D is included in a saturated clique C.

Proof Let D be any finite clique. Let $< t_i \ : \ i \in \omega >$ enumerate the types of U_∞. We construct finite cliques C_i for $i \in \omega$ as follows: Let $C_0 = D$. Given C_i, if t_i is not consistent with C_i, set $C_{i+1} = C_i$. If it is consistent, let F be a finite isometric subgraph of U_∞ containing $C_i \cup dom(t_i)$. In the notation above, let $H = H(F, C_i, t)$. So $F \hookrightarrow H$, and by distance injectivity, we can assume that $H \hookrightarrow U_\infty$. So let $C_{i+1} = C_i \cup \{y\}$; this insures that if t_i is consistent with an extension D of C_{i+1}, it is realized in D.

Now the union $\bigcup_{o \in \omega} C_i$ is saturated.

Before going on, we should state a reformulation of the notion of a saturated clique. Consider the category of cliqued graphs; i.e., the objects are structures $< G, C >$ such that G is a graph and C is a clique in G. The morphisms are the isometric embeddings which preserve membership in the clique in both directions. This category has the amalgamation property, and

$C \subseteq U_\infty$ is saturated iff $< U_\infty, C >$ has the injectivity property in the category of cliqued graphs.

Lemma 2 Let G be any countable connected graph and let D be a clique in G. Then there exists an isometric embedding $i : G \rightarrow U_\infty$ and a saturated clique $C \subseteq U_\infty$ such that $i(D) \subseteq C$.

Proof We construct C and i together, alternating the steps. The steps in constructing C are as in Lemma 1. And the steps in constructing i are as in the proof that U_∞ embeds G.

The following lemma shows that saturation is a strong form of maximality in the poset of all cliques. It will not be used in the sequel.

Lemma 3 Every saturated clique is maximal. There are maximal cliques of U_∞ which are not saturated.

Proof Let C be saturated. If $x \in U_\infty - C$, then there are points in C of different distances to x. In particular, x is not a neighbor of every element of C. So C is maximal.

For the other direction, let x and y be neighbors in U_∞. Let $< a_i : i \in \omega >$ enumerate the neighbors of x, and let $< b_i : i \in \omega >$ enumerate the neighbors of y. We construct finite cliques C_i by recursion on i. Let $C_0 = \{y\}$. Given C_{2i}, let C_{2i+1} be $C_{2i} \cup \{b_i\}$ if this is a clique and b_i is not a neighbor of x (and let it be C_{2i} otherwise). Given C_{2i+1}, let z be a neighbor of every element of C_{2i+2} but not a neighbor of a_i or x. Such a point exists by injectivity. Let $C_{2i+2} = C_{2i+1} \cup \{z\}$.

This defines the sequence, and let C be the union. Now C is a clique, and the only neighbor of x in C is y. So C is not saturated. We claim that C is maximal. If not, then for some $i, b_i \notin C$, but $C \cup \{b_i\}$ is a clique. Now in defining C_{2i+1}, we did not put b_i into C. Thus b_i is a neighbor of x. Suppose $b_i = a_j$. Then at step C_{2j+2} we added to C to some z which is not a neighbor of b_i. Thus $C \cup \{b_i\}$ is not a clique, and this is a contradiction.

Definition Let C and D be saturated cliques in U_∞. Let $\overline{x}$ and $\overline{y}$ be two n-tuples from U_∞. We write $\overline{x} \equiv_{C,D} \overline{y}$ if $\overline{x} = \infty \overline{y}$ and if $m_c(x_i) = m_D(y_i)$ for $1 \leq i \leq n$.

Lemma 4 Suppose $\overline{x} \equiv_{C,D} \overline{y}$. Then for every $x^* \in U_\infty$ there is some $y^* \in U_\infty$ such that $\overline{x}, x^* \equiv_{C,D} \overline{y}, y$.

Proof We first consider the special case when $x^* \in C$. Define t on $\overline{x}$ by $t(x_i) = d(x_i, x^*)$. This type is consistent since it is realized. So it satisfies the triangle inequalities. Now define a type u on $\overline{y}$ by $u(y_i) = t(x_i)$. Then u satisfies the same triangle inequalities, so it is consistent with D. By saturation, we may just take y^* to be a point in D realizing u.

The general case uses this observation. Consider $\overline{x}, x^*$, and let a_1 and a_2 be points from C such that $d(x^*, a_1) = m_c(x^*)$ and $d(x^*, a_2) = m_c(x^*) + 1$. By the first paragraph we can find b_1 and b_2 from D such that $\overline{x}, a_1, a_2 \equiv_{C,D} \overline{y}, b_1, b_2$. By distance homogeneity, we can find some y^* such that $\overline{x}, a_1, a_2, x^* = \infty \overline{y}, b_1, b_2, y^*$. It follows from this that $m_D(y^*) = m_C(x^*)$. Therefore $\overline{x}, x^* \equiv_{C,D} \overline{y}, y^*$.

Lemma 5 Suppose that C and D are saturated cliques in U_∞. Then there is an automorphism $\imath$ of U_∞ such that $\imath(C) = D$. In fact, given any tuples $\overline{x}$ and $\overline{y}$, such that $\overline{x} \subset_{C,D} \overline{y}$, there is an automorphism $\imath$ taking the first tuple to the second pointwise.

Proof We build the automorphism one step at a time using the last lemma.

3. Existence of Universal Graphs

Theorem 6 For each N, there is a countable graph U_N of diameter N such that for all countable G of diameter N there is an isometric embedding $i : G \to U_N$.

Proof First we consider the case of even N. Let x be any point of U_∞, and let U_{2N} be the subgraph of U_∞ induced by the set

$$S_x = \{y \in U_\infty : d_{U\infty}(x,y) = n\}.$$

Clearly U_{2N} has diameter $2N$. To show that U_{2N} is universal, let F be any countable graph of this diameter. Let z be a new point, and consider a graph H obtained by adding a new chain of length N to each point of F. This H has diameter $2N$, and F is isometrically embedded in it. Since it is countable, there is an isometric embedding $j : G \to U_\infty$. But also, there is an automorphism $\imath$ of U_∞ taking $j(z)$ to x. Let k be the restriction of $j o \imath$ to H. So k is an isometric embedding of H into U_∞ whose range is contained in U_{2N}.

We claim k is an isometric embedding when considered as a map into U_{2N}. Let $x, y \in F$. Since U_{2N} is an induced subgraph of $U_\infty, d_{U_\infty}(k(x), k(y)) \leq d_{U_N}(k(x), k(y))$. But there is a path in U_{2N} between $k(x)$ and $k(y)$ whose length is $d_F(x,y)$. Therefore, $d_{U_{2N}}(k(x), k(y)) \leq d_F(x,y) = d_{U\infty}(k(x), k(y))$.

Second, we get universal graphs for the odd diameters. Let C be a saturated clique, and let U_{2N+1} be the subgraph induced by

$$S_C = \{y \in U_\infty : (\exists x \in C) d_{U_\infty}(x,y) = N \& (\exists x \in C) d_{U_\infty}(x,y) = N+1\}.$$

We first need to see that S_C has diameter $2N+1$. Clearly the diameter is at most $2N+1$. Let x and y be any two points in C, let F be $\{x,y\}$ considered as a subgraph of U_∞, and let H be a chain $z_1, ..., z_{2N+2}$ of length $2N+1$. Then $F \hookrightarrow U_\infty$, and $F \hookrightarrow W$ via $x \mapsto z_{N+1}, y \mapsto z_{N+2}$. By distance injectivity, we can assume H is a subgraph of $U\infty$. Then z_1 and z_{2N+2} are in U_{2N+1} and their distance is $2N+1$.

Now we prove universality. Let F be a countable graph of diameter $2N+1$, and let F^+ be the canonical graph of diameter $2N+1$ determined by F. By Lemma 2, there is an isometric embedding $i : H \to U_\infty$ with the property that $i(D)$ is contained in a saturated clique. The rest of the proof is the same as in the even case, using Lemma 5.

Henceforth we will let U_N denote the graph built in the preceding proof. It is easy to see that U_{2N} does not depend on the choice of x. This is because if x and y are any two points in U_∞, then there is an automorphism taking x to y and hence S_x to S_y. Similarly, Lemma 5 implies that U_{2N+1} does not depend on the choice of C.

4. Distance Homogeneity and Uniqueness of the Universal Graphs

In this section, we characterize the universal graph $_N$ as the unique countable distance homogenous graphs of diameter N which isometrically embeds every countable (or even every finite) graph of diameter N. In so doing, we need to prove a few other properties of these graphs. These are all consequences of the following technical lemma which is a refinement of Theorem 6.

Lemma 7 Let F be a finite isometric subgraph of U_∞ of diameter $\leq N$. There is an automorphism $\imath$ of U_∞ such that $\imath(F) \subseteq U_N$ and $\imath$ is the identity on $F \cap U_N$.

Proof We need to consider the cases of even and odd diameter separately. We omit the even case, because it is easier and because it is similar to the proof of Theorem 6.

For the odd case, let C be such that in the notation from the proof of Theorem 6, $U_{2N+1} = S_C$. Given F, let F^+ be the canonical graph of diameter N determined by F, and let D be the new clique in F^+. Now F^+ is isometrically embedded in U_∞, and by distance homogeneity, we may assume that the embedding is the identity on F. Let E be a saturated clique extending D by Lemma 1. Consider the set $F \cap U_{2N+1}$ and recall the function m which gives the minimum distance to a clique. By the definition of U_{2N+1} and the construction of D we see that $m_E(x) = m_E(x) = N$ for $x \in F \cap U_{2N+1}$. Thus $\overline{x} \equiv_{C,E} \overline{x}$. Now by Lemma 5, let $\imath$ be an automorphism of U_∞ taking E to C and fixing $\{x_1, ..., x_n\}$ pointwise. This $\imath$ takes F to a subset of U_{2N+1}.

Corollary 8 For all N, U_N is an isometric subgraph of U_∞.

Proof Given x and y, let F be a path in U_∞ between them.

Henceforth we shall use Corollary 8 without mentioning it. For example to show that a subgraph F of U_N is an isometric subgraph, it is sufficient to show that $F \hookrightarrow U_\infty$.

Corollary 9 U_N is distance finite and distance homogeneous for all N.

Proof Let S be a finite subset of U_∞. Let F be S together with paths between the elements of S, and let F^+ be the canonical graph of diameter N determined by F. The distances between the points in S are the same when measured in F^+ as in F, and F^+ is isometrically embedded in U_∞. By Lemma 7, we can assume that the embedding maps into U_N and that it is the identity on S. Thus U_∞ is distance finite.

For distance homogeneity, suppose $x_1, \cdots, x_n = \infty y_1, \cdots, y_n$ are tuples from U_N, and let x^* also be a point in U_N. We know that there is some $y^* \in U_\infty$ such that $x_1, \cdots, x_n x^* =_\infty y_1, \cdots, y_n, y^*$. Let F be a finite isometric subgraph of U_∞ containing $y_1, ..., y_n$ and y^*. The diameter of F may be taken to be at most N by the argument of the last paragraph, since the distances between the points in the tuples are at most N. Apply Lemma 7. The point we want is $\imath(y^*)$.

For the next result, recall the definition of N-distance injectivity from Section 1.

Corollary 10 U_N is N-distance injective for all N.

Proof Any $i : F \to U_N$. may be regarded as an isometric embedding into U_∞. So there is some k such that $koj = i$. This k is a map into U_∞, but since its image $k(H)$ has diameter N, we may assume by Lemma 7 that it maps into U_N.

Lemma 11 Fix N. U_N is (up to isomorphism) the only countable distance finite N-distance injective graph of diameter N.

Proof This is a routine back and forth argument.

The following is our main result in this section:

Theorem 12 For all N, U_N is (up to isomorphism) the only countable distance homogeneous graph of diameter N which isometrically embeds each finite graph of diameter N.

Proof We know by Corollaries 9 and 10 that U_N has the properties mentioned above. For uniqueness, suppose that G has them. We show that G is distance finite and N-distance injective. The arguments are virtually the same as those for Corollaries 9 and 10, and we omit them. Then by Lemma 11, $G = U_N$.

5. Additional Characterizations

One of the basic characterizations of the Rado universal graph U_2 is that it is the only countable graph G with the property P_N for all N. P_N is the following condition:

(P_N) For every two disjoint sets U and V of vertices of size at most N, there is a point $x \in G$ which is a neighbor of every element of U and of no element of V.

In this section we present two properties of a similar flavor that capture the other universal graphs. Consider first the following property of a graph G:

(Q_N) Let $V_1, V_2, ... V_{N-1}$ and W be finite subsets of G. Suppose that whenever $1 \leq i \leq j \leq N-1, x \in V_i$, and $y \in V_j$, we have $j - i \leq d_G(x,y) \leq j + i$. Finally, suppose that whenever $1 \leq i \leq N-1, x \in V_i$ and $y \in W$, we have $N - i \leq d_G(x,y)$. Then there is some $z \in G$ such that (a) for all $i \leq N$ and all $x \in V_i, d_G(x,z) = i$; and (b) for all $y \in W, d_G(y,z) \geq N$. The motivation for the conditions is as follows: Let z be a point in any graph, and for $1 \leq i \leq N-1$, let V_i be the set of points whose distance from z is i. Let W be the set of points whose distance is at least N. Then $V_1, \cdots V_{N-1}, W$ have the properties above. The condition Q_N says that whenever such a point z could possibly exist, it must exist.

Lemma 13 For all N, if $N \leq M \leq \infty$, then U_M has Q_N.

Proof Let F be a finite isometric subgraph of U_N containing $V_1 \cup \cdots \cup V_{N-1} \cup W$. Let H_0 be obtained by adding a new point w together with chains of length i to each element of V_i. It should be checked that $F \hookrightarrow H_0$, and that w satisfies the conclusion of Q_N. If M is finite, we cannot apply injectivity unless we know that the diameter of H_0 is at most M. This is not in general true, but we can consider instead the canonical graph of diameter M determined by H_0. Call this graph H. H_0 need not be isometric subgraph of H, but F still is. Also, the distances between w and the points in F are not shortened as we pass to H. Then by M-distance injectivity, we get a map $k : H \to U_N$, and we let $z = k(w)$.

It follows from results below that U_N is the only countable graph of diameter N with Q_N. Our proof involves another property, R_N given as follows:

(R_N) Let H be any graph, and let $m \in \omega$ be arbitrary. Suppose $\overline{x}$ and $\overline{y}$ are tuples of length m from G and H respectively, and suppose that $\overline{x} =_{2(N-1)} \overline{y}$. Then for any $y^* \in H$ there is some $x^* \in G$ such that $\overline{x}, x^* =_{N-1} \overline{y}, y^*$.

Lemma 14 For all G, Q_N iff R_N.

Proof Let G be any graph with Q_N. Fix $\overline{x}, \overline{y}$, and y^*. For $1 \leq i \leq N-1$, let $V_i = \{x_a : 1 \leq a \leq n \& d_H(y^*, y_a) = i\}$. Let $W = \{x_a : d_H(y^*, y_a) \geq N\}$. We verify the hypothesis of Q_N for these sets. Suppose $i \leq j \leq N-1, x_a \in V_i$, and $x_b \in V_j$. Then the triangle inequality tells us that $d_H(y_a, y_b) \leq 2(N-1)$. Hence $d_H(y_a, gy_b) = d_G(x_a, x_b)$. And this distance is between $j - i$ and $j + i$.

Next, suppose that $x_a \in V_i$ and $x_b \in W$. We need to see that $d_G(x_a, x_b) \geq N - i$. Suppose not. Then $d_H(y_a, y_b)$ is the same, and so

$$d_H(y^*, y_b) \leq d_H(y^*, y_a) + d_H(y_a, y_b) < i + (N - i) = N.$$

This contradicts the assumption that $x_b \in W$.

Thus all of the hypotheses for Q_N are satisfied. Let x^* be the point z given by Q_N. This proves R_N. The argument that R_N implies Q_N is similar to the proof of Lemma 13.

Theorem 15 For all N, U_N is the only countable graph of diameter N with property R_N.

Proof By Lemmas 13 and 14, U_N has R_N. To prove uniqueness, we use a back and forth argument to show that any two graphs G and H of diameter N with R_N must be isomorphic. Suppose that $\overline{x}$ and $\overline{y}$ are two n-tuples from these graphs, and suppose that $\overline{x} =_\infty \overline{y}$. A fortiori, $\overline{x} =_{2(N-1)} \overline{y}$. Let $x^* \in G$. By R_N, we get some y^* such that $\overline{x}, x^* =_{N-1} \overline{y}, y^*$. The important point is that since both G and H have diameter N, we actually have "$=_\infty$" here.

We do not know whether there is a uniqueness result for the graphs of diameter N which have R_M for $M < N$. (We suspect that there are many such graphs.) We also do not know the relation of R_N to the other properties studied in this paper, such as distance finiteness and distance homogeneity.

6. The First Order Theories of the Universal Graphs

The goal of this section is to prove a result which illustrates a sense in which U_∞ is a kind of limit of the graphs U_N. Since the U_N do not sit inside each other in any obvious way, we cannot simply take a union. Our result is based on the first order logic. We have in mind the very simple language containing exactly one nonlogical symbol, a binary relation symbol R. All of the results in this paper concern this language. We show that the first order theory of U_∞ is the limit of the first order theories of its approximations U_N.

Consider a game $\mathcal{G}_{m,r}(G, H)$ played by two players called I and II. This game has m rounds, and in the course of these rounds, the players pick m-tuples $\overline{x}$ from G and $\overline{y}$ from H. More concretely, a play of the game is as follows: player I picks a point in either G or H, and then player II responds with a point in the other graph. They repeat this m times, and at in each of the m rounds, I may pick in either structure. In this way, they select tuples $\overline{x}$ from G and $\overline{y}$ from H. Note that $\overline{x}$ consists of the points chosen from G by either player; the order of points in this tuple is the order in which the points were selected. Given such a round of the game, we say that II wins iff $\overline{x} =_r \overline{y}$. We are interested in the question of which player has a winning strategy in this game.

Theorem 16 Let $m \geq 2$, and suppose that $r \cdot 2^{m-2} < N, M \leq \infty$. Then player II has a winning strategy in $\mathcal{G}_{m,r}(U_N, U_M)$.

Proof By induction on m. When $m = 2$ this follows from the $(r+1)$-distance injectivity of U_N and U_M. Assume it true for m. Fix r, and suppose $r \cdot 2^{m-1} < N, M$. Therefore $(2r) \cdot 2^{m-2} < N, M$, so we have a winning strategy σ for II in $\mathcal{G}_{m,2r}(U_N, U_M)$. We get a winning strategy for II in $\mathcal{G}_{m+1,r}(U_N, U_M)$ by playing the first m rounds according to σ. This gives m-tuples $\overline{x} =_{2r} \overline{y}$. For the last move, note that since $r \cdot 2^{m-1} \geq r+1, U_N$ and U_M both have property R_{r+1}. So II can always play to insure that the $(m+1)$-tuples will be r-equivalent.

The well-known Ehrenfeucht-Fraïssé game of length m is $\mathcal{G}_{m,1}(G, H)$. It has an important connection to first order logic by a theorem which we will need. To state it, we need to recall the inductive definition of the *quantifier rank* of a formula ϕ of first order logic, $qr(\phi)$:

$$qr(\phi) := 0 \text{ if } \phi \text{ is atomic}$$

$$qr(\neg\phi) := qr(\phi)$$

$$qr(\phi \wedge \psi) = qr(\phi \vee \psi) := \max(qr(\phi), qr(\psi))$$

$$qr(\exists x)\phi) = qr((\forall x)\phi) := 1 + qr(\phi).$$

For example, consider the formulas $\psi_m(x,y)$ defined by recursion on m: $\psi_0(x,y)$ is $(x = y) \vee R(x,y)$, and given $\psi_m, \psi_{m+1}(x,y)$ is $(\exists z)(\psi_m(x,z) \wedge \psi_m(z,y))$. Then each $\psi_m(x,y)$ has quantifier rank m. And for all graphs

$G, G \models \psi_m[x, y]$ iff $d_G(x, y) \leq 2^m$.

Recall the Ehrenfeucht-Fraïssé Theorem (cf. [1]): Suppose that II has a winning strategy in $\mathcal{G}_{m,1}(G, H)$, and let ϕ be a sentence of quantifier rank m. Then $G \models \phi$ iff $H \models \phi$.

From this and Theorem 16, we see that if ϕ is a sentence of quantifier rank $m \geq 2$ and $N > 2^{m-2}$, then $U_\infty \models \phi$ iff $U_N \models \phi$. (This lower bound on N is sharp. Consider the sentence $(\exists x)(\exists y)\neg\psi_{m-2}(x, y)$. This sentence has quantifier rank m and it is true in U_∞ but false in $U_{2^{m-2}}$.)

Thus with the natural meaning of limit,

$$\text{Theory}(U_\infty) = \lim_{N\to\infty} \text{Theory}(U_N).$$

We might also note that the results of this section show that the first order theories of the U_N are all complete. (A set of axioms for Theory (U_N) may be obtained from Theorem 15.) If we extend the language of graphs to include additional predicates for finite distances, then Theory(U_N) is the model completion of the theory of graphs of diameter N. We conjecture that every finite subtheory of it has a finite model. One way to express this conjecture is as an injectivity principle: For all N, there is a finite G such that for all F and H of size at most N and all $i : F \to G$ and $j : F \to H$ there is some $k : H \to G$ such that $koj = i$.

REFERENCES

[1] Ehrenfeucht, A., "An Application of Games to the Completeness Problem for Formalized Theories," *Fundamenta Mathematicae* 49 (1961), 129-141.

[2] Moss, L S., "Distanced Graphs," to appear.

[3] Moss, L. S. "Existence and Nonexistence of Universal Graphs," *Fundamenta Mathematicae*, to appear.

[4] Rado, R., "Universal Graphs and Universal Functions," *Acta Arithmetica* 9 (1964), 331-340.

On the Difference Between the Domination and Independent Domination Numbers of Cubic Graphs

Christine Mynhardt

University of South Africa

ABSTRACT

Barefoot, Harary and Jones [2] conjectured that $K(3,3)$ and $C_5 \times K_2$ are the only 3–connected cubic graphs for which the domination and independent domination numbers, γ and i respectively, differ. They also conjectured the existence of an infinite class of cubic graphs with connectivity one for which $i - \gamma$ becomes unbounded. We disprove the former conjecture by describing an infinite class of 3–connected cubic graphs for which $\gamma \neq i$, and prove the latter conjecture by constructing a class of graphs satisfying the given requirements.

1. Introduction

Unless stated otherwise we use the notation and terminology of [3]. A set D of vertices of a graph $G = (V, E)$ is a *dominating set* if each vertex of G not in D is adjacent to at least one vertex in D. A set I of vertices is *independent* in G if no two vertices in I are adjacent; if, in addition I is also a dominating set then I is called an *independent dominating set* of G. Note that the independent dominating sets of G are exactly the maximal independent sets of G (see [4, p. 309]). A vertex x of a subset X of V is *redundant* in X if its closed neighbourhood is contained in the union of the closed neighbourhoods of the vertices in $X - \{x\}$, and X is *irredundant* if it contains no redundant vertices.

Two parameters are associated with each of the above concepts in the following way. The *domination number* $\gamma(G)$ *(upper domination number* $\Gamma(G)$) of G is the smallest (largest) number of vertices in a minimal dominating set of G. (In [3], $\sigma(G)$ is used instead of $\gamma(G)$.) The *independent domination number* $i(G)$ (*independence number* $\beta(G)$) of G is the smallest (largest) number of vertices in a maximal independent set of G, and finally, the *irredundance number* $ir(G)$ (*upper irredundance number* $IR(G)$) of G is the smallest (largest) number of vertices in a maximal irredundant set in G. Since every maximal independent set is a minimal dominating set (see [4, p. 309]), and every minimal dominating set is a maximal irredundant set (see [7]), it follows that for any graph G,

$$ir(G) \le \gamma(G) \le i(G) \le \beta(G) \le \Gamma(G) \le IR(G).$$

Various authors have found sufficient conditions for two or more of these parameters to be equal. These include [1], [5], [8] and [13] for the lower parameters and [6], [8], [11] and [12] for the upper parameters. Of specific importance to the present paper are conditions under which equality of the domination and independent domination numbers occurs. In this respect Allan and Laskar proved in [1] that if G is $K(1, 3)$–free, then $i(G) = \gamma(G)$. This extended an earlier result by Mitchell and Hedetniemi [13] that if G is the line graph of a tree, then $i(G) = \gamma(G)$. Harary and Livingston characterised the trees and the caterpillars for which these two parameters are equal in [9] and [10] respectively, and in [2], Barefoot, Harary and Jones conjectured that the only 3–connected cubic graphs for which $i \neq \gamma$ are the graphs $K(3, 3)$ and $C_5 \times K_2$. We disprove this conjecture by describing a known infinite class of 3–connected cubic graphs containing $C_5 \times K_2$ for which $i = \gamma + 1$.

Barefoot, Harary and Jones [2] further constructed an infinite class of cubic graphs of connectivity 2 for which the difference $i - \gamma$ becomes unbounded, and conjectured that an infinite class of cubic graphs of connectivity 1 with the same property existed. We prove this conjecture by constructing a class of graphs with the desired properties.

2. A Class of 3–Connected Cubic Graphs with $i \neq \gamma$

For any positive integer k, let $G_k = C_k \times K_2$ – note that G_k is 3–connected and 3–regular. Suppose that $V(G_k) = U \cup W$ with $U \cap W = \phi$ and $G_k\langle U\rangle = G_k\langle W\rangle = C_k$. Say $U = \{u_0, ..., u_{k-1}\}$, $W = \{w_0, ..., w_{k-1}\}$ and assume that u_j and w_j are adjacent for each $j = 0, ..., k-1$.

Theorem 1 *If $k \equiv 5 (mod\ 12)$, then*

$$\gamma(G_k) = \lceil k/2 \rceil \text{ and } i(G_k) = \lceil k/2 \rceil + 1.$$

Proof For any graph G,

$$\gamma(G) \ge \lceil |V(G)| / (\Delta(G) + 1) \rceil.$$

Hence to show that $\gamma(G_k) = \lceil k/2 \rceil$ we only need to find a dominating set of G_k of cardinality $\lceil k/2 \rceil$. Let $k = 12m + 5$ for some integer m and let $D = \{ u_{4j} \mid j = 0, ..., 3m\} \cup \{w_{4j+2} \mid j = 0, ..., 3m\} \cup \{w_{12m+3}\}$. Then D dominates G_k (the routine details are omitted) and $|D| = 6m + 3 = \lceil k/2 \rceil$. Since w_{12m+2} and w_{12m+3} are adjacent, D is not independent.

Let $I = (D - \{w_{12m+2}\}) \cup \{w_{12m+1}, u_{12m+2}\}$. Then I is an independent dominating set of G_k of cardinality $\lceil k/2 \rceil + 1$. Suppose that S is an independent dominating set of G_k of cardinality $\lceil k/2 \rceil = 6m + 3$. Since G_k is 3–regular and S is independent, each vertex of S dominates exactly three vertices of $G_k - S$. But $G - S$ has $18m + 7$ vertices whereas $3|S| = 18m + 9$, which by the pigeon hole principle implies that $G_k - S$ either contains two vertices, each of which is dominated by two vertices in S, or one vertex which is dominated by three vertices in S, while all remaining vertices of $G_k - S$ are dominated by exactly one vertex in S.

Suppose firstly that u_0 (say) is dominated by the vertices w_0, u_1 and u_{k-1} in S. Then w_{k-1} and w_1 are dominated by u_{k-1}, w_0 and u_1, w_0 respectively, which is impossible. Therefore $G - S$ has two vertices which are each dominated by two vertices in S.

Say u_0 is dominated by two vertices in S – without losing generality we may assume that u_1 is one such vertex. If $u_{k-1} \in S$, then in order for w_0 to be dominated, w_0 is in S since S is independent. But we have already shown this to be impossible, hence $u_{k-1} \notin S$ and $w_0 \in S$, so that w_1 is the remaining vertex of $G_k - S$ dominated by two vertices in S. Since w_2 is dominated and S is independent, it follows that $w_3 \in S$. (Note that $w_2 \notin S$ for otherwise w_1 is dominated by three vertices in S.) Similarly, u_4 is dominated and $u_4 \notin S$, consequently $u_5 \in S$. Continuing in this way we deduce that $S = \{w_0\} \cup \{w_{4\ell+3} \mid \ell = 0, ..., 3m\} \cup \{u_{4\ell+1} \mid \ell = 0, ..., 3m\}$ – however, this is impossible since u_{k-1} is not dominated while w_{k-1} is dominated by w_0 and w_{k-2}.

This proves that $i(G_k) = \lceil k/2 \rceil + 1$. ❑

The graphs $C_k \times C_2$ for $k \equiv 5 (\text{mod } 12)$ therefore form an infinite class of 3–connected cubic graphs for which $i = \gamma + 1$.

3. A Class of Cubic Graphs with Connectivity One for which $i - \gamma$ becomes Unbounded

Let F, G and J with $V(F) = \{v_1, ..., v_5\}$ and $V(G) = \{w_1, ..., w_6\}$ be the graphs depicted in Figure 1. We construct the graph H_k as follows: Let $V(H_k) = U \cup V \cup W$ (disjoint union), where

$$U = \bigcup_{i=1}^{2k-1} U_i \qquad \text{for } U_i = \{u_{i1}, u_{i2}\},$$

$$V = \bigcup_{i=0}^{4k-1} V_i \quad \text{for } V_i = \{v_{i1}, \ldots, v_{i5}\} \text{ and}$$

$$W = \bigcup_{i=1}^{2k-2} W_i \quad \text{for } W_i = \{w_{i1}, \ldots, w_{i6}\}.$$

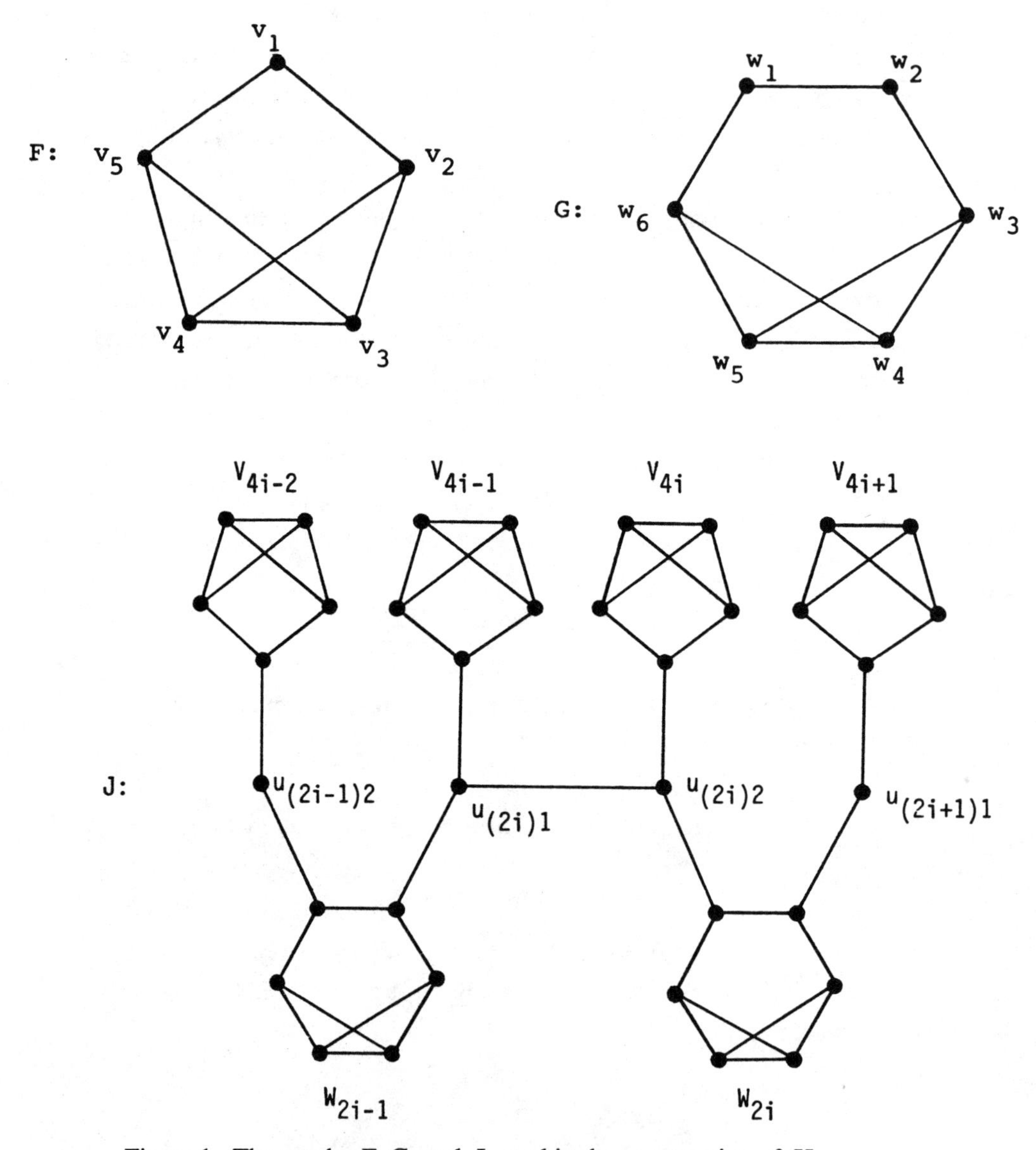

Figure 1. The graphs F, G and J used in the construction of H_k.

Add edges in such a way that

$H_k\langle U_i\rangle = K_2$ for each $i = 1, ..., 2k-1$,

$H_k\langle V_i\rangle = F$ such that v_{ij} corresponds to the vertex v_j of F

for every $j = 1, ..., 5$ and $i = 0, ..., 4k-1$, and

$H_k\langle W_i\rangle = G$ such that w_{ij} corresponds to the vertex w_j of G

for every $j = 1, ..., 6$ and $i = 1, ..., 2k-2$. Add further edges such that for

$$X_i = (V_{4i-2} \cup V_{4i-1} \cup V_{4i} \cup V_{4i+1}) \cup (W_{2i-1} \cup W_{2i}) \cup \{u_{(2i-1)2}, u_{(2i)1}, u_{(2i)2}, u_{(2i+1)1}\},$$

where $i = 1, \ldots, k-1$,

$$H_k\langle X_i\rangle = J_i = J.$$

Finally add the edges $\{v_{01}u_{11}, v_{11}u_{11}, v_{(4k-2)1}u_{(2k-1)2}, v_{(4k-1)1}u_{(2k-1)2}\}$. The sresulting graph is H_k. Note that H_k contains a string of $k-1$ copies $J_1, \ldots, J_{k-1}$ of J, where for each $i = 1, \ldots, k-2$, the copy J_i is joined to J_{i+1} by the edge $u_{(2i+1)1}u_{(2i+1)2}$. We illustrate H_1 and H_2 in Figures 2 and 3 respectively. Note that H_k is 3–regular and has connectivity one.

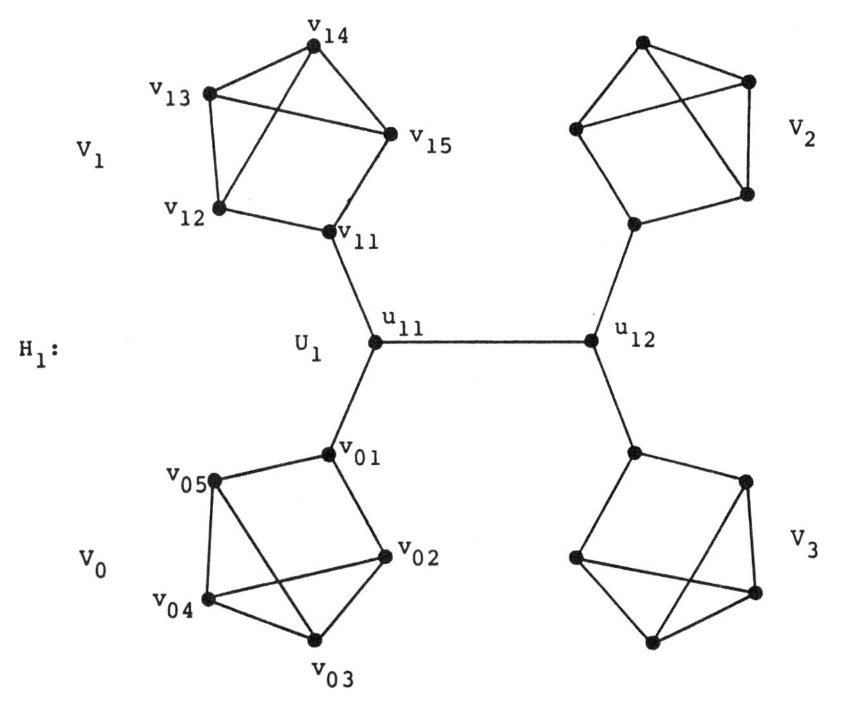

Figure 2. The graph H_1

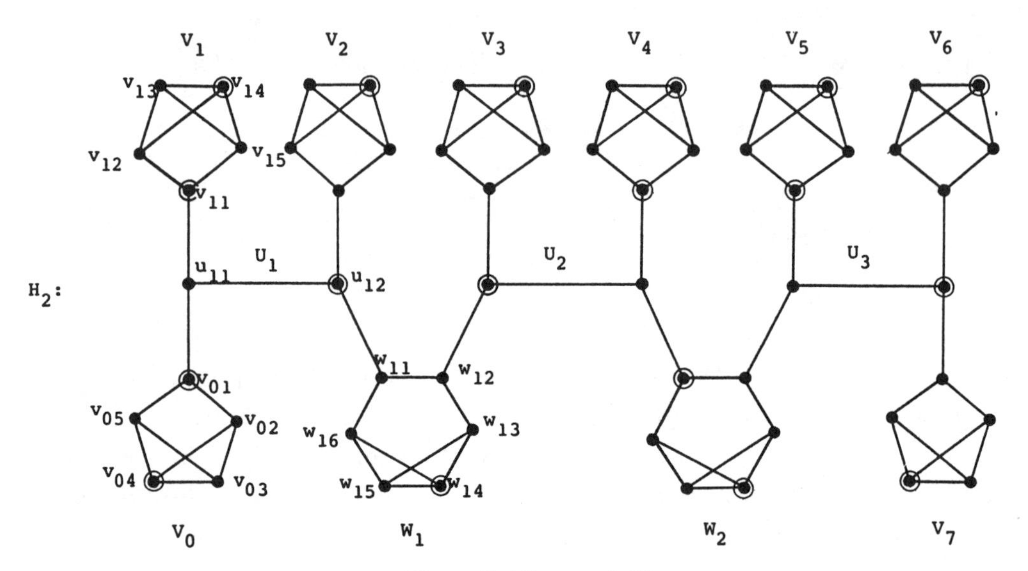

Figure 3. The graph H_2

We claim that $\gamma(H_k) = 10k - 4$ while $i(H_k) = 11k - 4$. In order to prove this, we first prove that it holds for $k = 1$.

Lemma 1 *$\gamma(H_1) = 6$ and $i(H_1) = 7$.*

Proof A dominating set of H_1 of cardinality six is given by $\{u_{11}, u_{12}, v_{04}, v_{14}, v_{24}, v_{34}\}$ and since

$$\lceil |V(H_1)| / (\Delta(H_1) + 1) \rceil = 6,$$

we have that $\gamma(H_1) = 6$. An independent dominating set of H_1 of cardinality seven is given by $\{u_{11}, v_{04}, v_{14}, v_{22}, v_{25}, v_{32}, v_{35}\}$. Further, if D is any independent dominating set of H_1, then at most one of u_{11} and u_{12} is in D. If (say) $u_{12} \notin D$, then at least two vertices in each of V_2 and V_3 are in D in order to dominate $V_2 \cup V_3$. No vertex in $V_2 \cup V_3$ dominates any vertex in $V_0 \cup V_1 \cup \{u_{11}\}$ and no subset of $V_0 \cup V_1 \cup \{u_{11}\}$ containing fewer than three vertices dominates $V_0 \cup V_1 \cup \{u_{11}\}$; hence $|D| \geq 7$. We have thus proved that $i(H_1) = 7$. ❑

Now let $J = J_1$, $L = J_1 - \{u_{12}\}$ and $M = J_1 - \{u_{12}, u_{31}\}$. In our next result we determine the domination and independent domination numbers of J, L and M.

Lemma 2 $\gamma(J) = 10$,
$\gamma(L) = i(J) = i(L) = 11$ *and*
$\gamma(M) = i(M) = 12$.

Proof Throughout the proof we use the fact that if D is a set which dominates $\langle V_i \rangle$, and v_{i1} is not dominated by $D - V_i$, then $|D \cap V_i| \geq 2$, and if D dominates $\langle W_i \rangle$ and at least one of w_{i1} and w_{i2} is not dominated by $D - W_i$, then $|D \cap W_i| \geq 2$.

Let $N = H_k\langle V_3 \cup V_4 \cup U_2 \rangle$. By the above observation it is clear that $\gamma(N) = i(N) = 4$. Also by the above observation, if D is any dominating set of M, then D contains at least two vertices of each of V_2, V_5, W_1 and W_2, and it follows that $|D| \geq 12$. The set

$$D = \left(\bigcup_{i=2}^{5} \{v_{i2}, v_{i5}\} \right) \cup \{w_{12}, w_{14}, w_{21}, w_{24}\}$$

is an independent dominating set of M of cardinality 12 and hence $\gamma(M) = i(M) = 12$. Similarly, $\gamma(L) = i(L) = 11$ and $\gamma(J) = 10$.

To prove that $i(J) \geq 11$, note that if S is an independent dominating set of J, then at most one of u_{21} and u_{22} is in S. If (say) $u_{22} \notin S$, then S contains at least two vertices of each of V_4 and W_2 and the result follows. An independent dominating set of J of cardinality 11 is given by

$$S = \{u_{12}, v_{24}, w_{14}, u_{21}, v_{34}, v_{42}, v_{45}, w_{21}, w_{24}, u_{31}, v_{54}\},$$

consequently $i(J) = 11$. ❑

Note that D dominates H_k, then for any $i = 1, ..., k-1$, $D \cap V(J_i)$ corresponds to a dominating set of J, L or M, depending on whether two, one or none of the vertices $u_{(2i-1)2}$ and $u_{(2i+1)1}$ of J_i are contained in D.

Theorem 2 *For any positive integer k,*

$$\gamma(H_k) = 10k - 4 \text{ and } i(H_k) = 11k - 4.$$

Proof A dominating set of H_k of cardinality $10k - 4$ is given by

$$U \cup \left(\bigcup_{i=0}^{4k-1} \{v_{i4}\} \right) \cup \left(\bigcup_{i=1}^{2k-2} \{w_{i4}\} \right).$$

We also claim that

$$\left(\bigcup_{i=0}^{4k-1}\{v_{i4}\}\right)\cup\left(\bigcup_{i=1}^{2k-2}\{w_{i4}\}\right)\cup\left(\bigcup_{i=1}^{k-1}\{u_{(2i-1)2},u_{(2i)1},w_{(2i)1}\}\right)\cup\left(\bigcup_{i=0}^{k-1}\{v_{(4i)1},v_{(4i+1)1}\}\right)\cup\{u_{(2k-1)2}\}$$

is an independent dominating set of H_k of cardinality $11k-4$. That this is indeed so, can most easily be seen by noticing that the relevant vertices are the circled vertices of H_2 in Figure 3; the genaral case is similar.

If, for $k \geq 2$, H_k has a dominating set D with $|D| < 10k-4 = 6+10(k-1)$, then either D contains at most five vertices of $Y = V_0 \cup V_1 \cup V_{4k-2} \cup V_{4k-1} \cup \{u_{11}, u_{(2k-1)2}\}$, or D contains at most nine vertices of a subgraph J_i of H_k. Since neither case is possible, $\gamma(H_k) = 10k-4$.

If H_k has an independent dominating set S with $|S| < 11k-4 = 7+11(k-1)$, then either $|S \cap Y| \leq 6$ (in which case $|S \cap Y| = 6$), or S contains at most ten vertices of J_i for some i. By Lemma 2 the latter possibility does not occur, hence suppose $|S \cap Y| = 6$. Then $\{u_{11}, u_{(2k-1)2}\} \subseteq S$ and $\{u_{12}, u_{(2k-1)1}\} \cap S = \phi$. But then for at least one i, say $i = 1$, $S \cap V(J_1)$ corresponds to an independent dominating set of M and hence $|S \cap V(J_1)| \geq 12$. Since Lemma 2 implies that $|S \cap V(J_i)| \geq 11$, it now follows that

$$|S| \geq 6+12+11(k-2) = 11k-4,$$

a contradiction. Therefore $i(H_k) = 11k-4$. ❑

REFERENCES

[1] R.B. Allan and R. Laskar, On domination and independent domination numbers of a graph, *Discrete Math.* **23** (1978), 73 – 76.

[2] C. Barefoot, F. Harary and K.F. Jones, What is the difference between α and α' of a cubic graph? Preprint.

[3] M. Behzad, G. Chartrand and L.Lesniak–Foster, *Graphs and Digraphs*, Prindle, Weber and Schmidt, Boston, 1979.

[4] C. Berge, *Graphs and Hypergraphs*, North–Holland, Amsterdam, 1973.

[5] B. Bollobás and E.J. Cockayne, Graph–theoretic parameters concerning domination, independence and irredundance, *J. Graph Theory* **3** (1979), 241 – 249.

[6] E.J. Cockayne, O. Favaron, C. Payan and A.G. Thomason, Contributions to the theory of domination, independence and irredundance in graphs, *Discrete Math.* **33** (1981), 249 – 258.

[7] E.J. Cockayne, S.T. Hedetniemi and D.J. Miller, Properties of hereditary hypergraphs and middle graphs, *Canad. Math. Bull.* **21** (4) (1978), 461 – 468.

[8] O. Favaron, Stability, domination and irredundance in graphs, *J. Graph Theory* **10** (1986), 429 – 438.

[9] F. Harary and M. Livingston, Characterisation of trees with equal domination numbers, *Congressus Numerantium* (to appear).

[10] F. Harary and M. Livingston, The caterpillars with equal domination and independent domination numbers, *Annals of Discrete Math* (to appear).

[11] M.S. Jacobson and K. Peters, Chordal graphs and upper irredundance, upper domination and independence. Preprint.

[12] M.S. Jacobson and K. Peters, Upper domination, independence and irredundance numbers for peripheral graphs.Private communication.

[13] S. Mitchell and S.T. Hedetniemi, Independent domination in trees, *Proc. of Eight S.E. Conference on Combinatorics, Graph Theory and Computing* (1977), 489 – 509.

A REPORT ON THE ALLY RECONSTRUCTION PROBLEM

Wendy Myrvold

University of Victoria

British Columbia

ABSTRACT

The ally-reconstruction number of a graph G, ally-rn(G), is the minimum number of vertex-deleted subgraphs required in order to identify G up to isomorphism. In this paper, we report on some recent results for this parameter.

1. Introduction

In this paper, we report on some new results for the reconstruction problem. Readers unfamiliar with this problem should refer to Bondy and Hemminger [4] which is a good survey of papers up to 1976 with a comprehensive bibliography. All graphs considered in this paper are simple, finite and undirected. The *order* of a graph is the number of vertices. Any further graph theoretic terminology can be found in [3].

The ally-reconstruction number of a graph G, *ally-rn(G)*, is the minimum number of vertex- deleted subgraphs of G required in order to identify G up to isomorphism. This parameter was introduced as the reconstruction number of a graph by Harary and Plantholt in 1985 [5]. In this paper, we survey what is known about the ally-reconstruction number of a graph and explain how this problem relates to the reconstruction problem.

2. Graphs with Small Ally-Reconstruction Number

One simple observation that was made by Harary and Plantholt [5, p. 451, Observation 1] is that the ally- reconstruction number of any graph is always at least three. We have included a proof here since it was omitted in [5].

Observation 2.1 For any graph G, ally-rn$G \geq 3$.

Proof Suppose ally-rn$(G) = 2$. Then there are two cards, $G - u$ and $G - v$, which uniquely identify G. Assume without loss of generality that G does not have the edge (u, v). Then $H = G + (u, v)$ also has these two cards. The graph H is not isomorphic to G because it has a different number of edges.

We now show that almost all graphs can be reconstructed from any three cards of their deck and as a result, almost all graphs have ally-reconstruction number three. By 'almost all', we mean that the percentage of graphs on n vertices for which the statement is true approaches one hundred percent as n approaches infinity. This result was conjectured by Harary and Plantholt [5, p. 454, Conjecture 4] and proved by Bollobás [2]. We give an alternative proof which is a simple extension of the proof by Müller [13] that almost all graphs are reconstructible.

Lemma 2.2 Almost all graphs are reconstructible from any three cards.

Proof In [13], Müller proves that given $\epsilon > 0$, the k-vertex- deleted subgraphs of almost all graphs are pairwise nonisomorphic if $k \leq \frac{n}{2}(1 - \epsilon)$. All that is needed for this proof is that almost all graphs have mutually nonisomorphic one, two, and three-vertex-deleted subgraphs. Because this is true, the following reconstruction algorithm will reconstruct almost all graphs from any three cards.

Reconstruction Algorithm for a Random Graph **Input**: Any three cards, say $G - a, G - b$, and $G - c$ from a graph G.

Output: If the output of this algorithm is a graph H, then we are guaranteed that any graph with the three cards input is isomorphic to H. Otherwise, the output is the message "Algorithm failed to reconstruct G". In this case, there still may only be one reconstruction from the cards given or there may be several.

1. Find all vertices b' of $G-a$ so that $G-a-b'$ is a card of $G-b$. If there is more than one such vertex b', report "Algorithm failed to reconstruct G".
2. Find all vertices a' of $G-b$ so that $G-b-a'$ is a card of $G-a$. If there is more than one such vertex a', report "Algorithm failed to reconstruct G".
3. If there is more than one isomorphism f from the vertices of $G-a-b'$ to those of $G-b-a'$, then report "Algorithm failed to reconstruct G". Otherwise, add a vertex b to $G-b$ which is incident to vertex w on $G-b$ if and only if vertex b' is adjacent to w' on $G-a$, and $f(w')=w$.
4. At this point, the structure of G has almost been determined. The only thing not known is whether or not G has edge (a,b). This is determined from $G-c$ by labeling vertices a and b on this card as in step one.

Since for almost all graphs, the 2-vertex-deleted subgraphs are mutually nonisomorphic, steps one, two and four are almost always carried out successfully. Since almost all graphs have mutually nonisomorphic 3-vertex-deleted subgraphs, there is almost always only one isomorphism from $G-a-b'$ to $G-b-a'$ in step three. Hence, given any three cards from a random graph, the preceding algorithm almost always succeeds in creating a unique reconstruction.

3. Graphs With Large Ally-Reconstruction Number

What is the largest ally-reconstruction number that a graph on n vertices can have? It has been conjectured by Harary and Plantholt [5, p. 453, Conjecture 1] that the answer is $\frac{n}{2}+2$. A proof would be remarkable since this conjecture implies the reconstruction conjecture. This conjecture is true for the families of graphs considered in the following section. We present here the only known family of graphs with ally-reconstruction number equal to this value.

Lemma 3.1 [5] There exists an infinite family of r-regular graphs which have ally-reconstruction number equal to

$$\lfloor \frac{n}{2} \rfloor + 2.$$

Proof Let G be the r-regular graph consisting of two copies of K_{r+1}. Let H consist of K_{r+2} and K_r. This graph pair is shown in Figure 1 for $r = 3$. All cards of G are isomorphic to the cards created from H be deleting any of the $r+2$ vertices in the K_{r+2}. Hence, any subset of the deck of G which the ally chooses has at least $r+3$ cards.

Finally, we would like to quickly mention a counterexample to a remark made by Harary and Plantholt [5, p. 454, Conjecture 4]. They suggested that if G is of odd order, then perhaps *ally-rn(G)* $= 3$. The graph family consisting of $G_p = 3K_p$ is a family of graphs with *ally-rn*$(G_p) = p+1$ (the proof of this is similar to that of Lemma 3.1). These graphs have odd order when p is odd (the number of vertices is 3 p).

Figure 1. Regular graph G with ally-rn 6

4. Results for Specific Families of Graphs

As is customary for various reconstruction problems, the ally-reconstruction problem has been studied for various classes of graphs. The *class-reconstruction number* of a graph with respect to some particular class C of graphs is the number of cards needed to reconstruct a graph G assuming that we already know that G is a member of the class C. The class-reconstruction number has been studied for trees [1], unicyclic graphs [6], and maximal planar graphs [7]. Since these results are surveyed fairly thoroughly by Lauri [8], we will not cover them here. Instead, we will concern ourselves with what is known about the ally-reconstruction number of graphs in particular classes. The difference between these two problems is that in the latter case, we must also be able to determine that a graph is a member of the class being studied.

Harary and Plantholt [5, p. 454, Conjecture 3] conjectured that trees on five or more vertices have ally- reconstruction number three. Their conjecture is restricted to trees on five or more vertices since the ally- reconstruction number of a path on four vertices is four. This conjecture is proved in [11].

Theorem 4.1 For any tree T with $n \geq 5$ vertices, *ally-rn(T)* = *3*.

In the proof of this result, we first describe how to select three cards for any tree T from which T is reconstructible. Next, we show that any graph having these cards must be a tree. Finally, we prove that any tree with the three cards selected is isomorphic to T. The main idea used in this proof is different from that used by Harary and Lauri for the class-reconstruction number of a tree in [1].

The ally-reconstruction problem has also been solved for disconnected graphs. In [10], it is proved that unless a disconnected graph G has all its components isomorphic, then the ally-reconstruction number of G is three. If G has isomorphic components of order c, then it has ally-reconstruction number at most $c+2$. The family of graphs described in Lemma 3.1 can be used to construct families of graphs which show that this result is tight.

Theorem 4.2 [10, pp. 59-65] Let G be a disconnected graphs. If all components of G are isomorphic with order c, then *ally-rn(G)* $\leq c+2$. Otherwise, *ally-rn(G)* = *3*.

If a graph G is reconstructible from some set C of its deck the complement $\overline{G}$ of G is reconstructible from the corresponding cards of $\overline{G}$. This is true because the cards of $\overline{G}$ are the complements of the cards of G. One result of this observation is that Theorem 4.2 also applies to graphs which are the complements of disconnected graphs (and less interesting perhaps, Theorem 4.1 applies to graphs whose complements are trees).

The only other class of graphs for which this problem has been solved is for regular graphs. To understand where the following theorem comes from, observe that the complement of an r-regular graph is an $(n-1-r)$-regular graph.

Theorem 4.3 [12, pp. 18-24] An r-regular graph G has

$$\text{ally-rn}(G) \leq \min\{r+3, [(n-1)-r]+3\}.$$

This result is proved by first showing that the degree sequence of a regular graph is reconstructible from the number of cards mentioned above. The regular graph can then easily be reconstructed from any one card by adding a new vertex to the card and placing edges from the new vertex to all the vertices on the card of degree $r-1$. With little extra work, the following corollary can be obtained.

Corollary 4.4 [10, pp. 22-23] If G is an r-regular graph on n vertices, then *ally-rn*$(G) \leq \lfloor \frac{n}{2} + 2 \rfloor$.

For all the classes mentioned here, we have been able to prove that the ally-reconstruction number is at most $\lfloor \frac{n}{2} \rfloor + 2$.

5. Open problems

In this final section, we make some remarks concerning why we feel the ally-reconstruction problem is an important new area for reconstruction fanatics. No attempt has been made to screen out statements that may be controversial.

Nobody has been able to find a family of graphs or even a single graph having ally-reconstruction number greater than $\frac{n}{2}+2$. If we define the ally-reconstruction number of a non-reconstructible graph on n vertices to be $n+1$, even the graphs in the non-reconstructible pair of graphs on two vertices satisfy the condition that the ally-reconstruction number is at most $\frac{n}{2}+2$. If there does exist a graph with ally-reconstruction number $n+1$ (i.e. a non-reconstructible graph), then one might also expect to find reconstructible graphs with reconstruction number greater than $\frac{n}{2}+2$. The apparent absence of such graphs suggests strongly that perhaps the reconstruction conjecture is true.

The ally-reconstruction number could be thought of as a measure of how hard it is to reconstruct a graph. Regular graphs are one of the easiest cases for the general reconstruction problem, in spite of their incredible symmetry. Using the ally-reconstruction number as a measure, they appear to be one of the hardest cases. This is perhaps more appropriate.

The results mentioned in the previous section were generally proved by first explaining how to select the ally's cards from the deck of the graph. The next step was to explain how to reconstruct the graph from these cards together with a proof that there was only one way to do this (i.e. a proof that all reconstructions of the cards selected are isomorphic). This type of argument gives much more information concerning why a graph is reconstructible than previous results which were proved by contradiction (see for example [9]). Hopefully, someone will be able to use this additional insight to make further advances on the reconstruction conjecture.

At a time when most work on the reconstruction conjecture has come to a standstill, ally-reconstruction is flourishing. It is hoped that the ideas presented here will spark new interest in the reconstruction problem and that someday this (in)famous conjecture will be resolved.

REFERENCES

[1] F. Harary and J. Lauri, "On the class-reconstruction number of trees," *Oxford Quarterly Journal of Mathematics*, 1987. To appear.

[2] B. Bollobás, "Almost every graph has reconstruction number three," *Journal of Graph Theory*, 1987. To appear.

[3] J. A. Bondy and U. S. R. Murty, *Graph Theory with Applications*, Elsevier North Holland, Inc., New York, 1976.

[4] J. A. Bondy and R. L. Hemminger, "Graph reconstruction - A survey," *Journal of Graph Theory*, vol. 1, no. 3, pp. 227-268, 1977.

[5] F. Harary and M. Plantholt, "The graph reconstruction number," *Journal of Graph Theory*, vol. 9, pp. 451-454, 1985.

[6] F. Harary and M. Plantholt, "The class-reconstruction numbers of unicyclic graphs," 1987. Submitted.

[7] F. Harary and J. Lauri, "The class-reconstruction number of maximal planar graphs," *Graphs and Combinatorics*, vol. 3, no. 1, pp. 45-53, 1987.

[8] J. Lauri, "Graph reconstruction - Some techniques and new problems," *Ars Combinatoria*, 1987. Submitted.

[9] W. J. Myrvold, M. N. Ellingham, and D. G. Hoffman, "Bidegreed graphs are Edge Reconstructible," *Journal of Graph Theory*, vol. 11, pp. 281-302, 1987.

[10] W. J. Myrvold, "*Ally and Adversary Reconstruction Problems*, Ph.D. thesis, University of Waterloo, 1988.

[11] W. J. Myrvold, "The ally-reconstruction number of a tree with five or more vertices is three, " *Journal of Graph Theory*, 1989. To appear.

[12] W. J. Myrvold, "The ally-reconstruction number of a disconnected graph," *Ars Combinatoria*, 1989. To appear.

[13] V. Müller, "Probabilistic reconstruction from subgraphs," *Commentationes Mathematicae Universitatis Carolinae*, vol. 17, pp. 709-719, 1976.

Random Superposition: Multigraphs

E.M. Palmer
Michigan State University

Dedicated to J. Howard Redfield, 1879 – 1944

ABSTRACT

A new probability model for graphs based on the superposition principle of J.H. Redfield is introduced, thereby revealing a great array of problems involving certain types of multigraphs. We focus on the simplest possible special case in which each superposition is the union of a collection of isolated edges. Some fundamental results are derived for these random multigraphs. In particular, the distribution of multiple edges is sketched and a sharp threshold for connectivity is determined. Numerical evidence to support the theory is provided for small multigraphs with less than 7000 vertices.

1. Introduction

Let H be any graph of order n and let $s(H)$ denote the order of the automorphism group of H. Hence, the number of ways to label the vertices of H is $n! / s(H)$. Now select a random list of r labeled copies

$$H_1, H_2, ..., H_r$$

of H, with replacement. So we have an r–permutation, with repetition, of a k–set, where $k = n! / s(H)$. Now form the random superposition graph, denoted $G(H)$, from the list by identifying vertices with the same label. Note that $G(H)$ is a multigraph of order n in which the r isomorphic factors retain their identity. Thus we have a probability model in shich the sample space has

$$(1.1) \qquad (n! / S(H))^r$$

equally likely elements.

Suppose we are given a particular superposition graph $G(H)$ of order n and the identities of the r copies of H are revealed by attaching the label 1 to each edge of the first copy of H, 2 to the second copy and so on. Then the contribution to (1.1) of all superpositions isomorphic to G(H) is

(1.2) $$(r!/a)(n!/b)$$

where a is the number of permutations of the edge labels that leave G invariant, and b is the number of permutations of the vertex labels that leave G invariant, i.e. preserve edge labels. Otherwise put, (1.2) is the number of ways to label a superposition G(H).

Redfield introduced the notion of superposition in his famous 1927 paper [R27], in which he developed a very broad method for counting unlabeled superpositions. For background material and the development of this topic there are the important articles [H60], [HR84], [Re59] and [Re60] by Harary, Read and Robinson mentioned in the references. See especially the 1984 issue vol.8(2) of the *Journal of Graph Theory* that is dedicated to the memory of J.H. Redfield and contains a beautiful biographical sketch by Keith Lloyd [L84].

Many questions could be raised about the nature of the random superposition graph G(H) that results from various choices of the basic factor H and the number of copies in the superposition. In this article we focus on the simple special case in which H is the graph of order n with exactly one edge, i.e. the disjoint union $H = K_2 \cup \overline{K}_{n-2}$. There are $\binom{n}{2}$ ways to label H and we form the sample space of superpositions with q copies of H. Hence there are $(n(n-1)/2)^q$ superpositions G(H) and each one is a multigraph with q edges. We call q the *size* of G(H). Recall that the copies of H retain their identity in the superposition so each of the edges may be regarded as having a different label from 1 to q.

We use C(n, k) to denote the binomial coefficient $\binom{n}{k}$ and illustrate the sample space with $n = 4$ and $q = 3$. There are $C(4, 2)^3 = 216$ superpositions in the sample space. There are 6 basic types described as follows:

G_1 has one edge of multiplicity 3 and two isolated vertices

G_2 has one edge of multiplicity 2 and one isolated vertex

G_3 has one edge of multiplicity 2 and no isolated vertices

G_4 is a path

G_5 is the bipartite tree $K_{1,3}$

G_6 consists of a triangle and an isolated vertex.

The next equation illustrates the application of formula (1.2) to the graphs in the order listed above, thereby counting the number of elements in the sample space in another way:

(1.3) $$(3!/3!)(4!/4) + (3!/2)(4!/4) + (3!/2)(4!/1) + (3!/2)(4!/1) + (3!/3!)(4!/1) + (3!/3!)(4!/1) = 216.$$

2. Distribution of Multiple Edges

It is convenient to denote the binomial coefficient $C(n, 2)$ by the single letter N. Then our sample space of superpositions of q copies of H has N^q elements. Before any investigation of the structure of a random superposition, some basic facts about multiple edges must be determined. Let $X_m(G)$ denote the number of edges of multiplicity m in the superposition graph $G = G(H)$. Then the expected value of this random variable X_m for $m \geq 0$ is

(2.1) $$E(X_m) = N\, C(q, m)\, (1/N)^m\, (1 - 1/N)^{q-m}.$$

The justification of this formula proceeds as follows. The first factor N is the number of ways to select two vertices, say u and v. The second factor $C(q, m)$ counts the number of ways to select labels (or different copies of H) for the m edges joining u and v. Then $(1/N)^m$ is the probability that the m selected edges actually join u and v, and finally $(1 - 1/N)^{q-m}$ is the probability that the other $q - m$ edges do not join u and v. We want to know about the likelihood of edges of fixed multiplicity m in a random superposition graph when n and q are large. For example, suppose $m = 3$ and n is very large. If q is too small, it is quite likely that the random superposition will not have any edges of multiplicity 3. But if we increase q suffciently, edges of multiplicity 3 will appear with high probability. But if q is increased by too great an amount, every pair of vertices may be adjacent and even with multiplicity greater then 3. The critical values of q can be determined from the expectation $E(X_m)$ as can be seen in the next theorem. All limits in this article are taken with respect to $n \to \infty$.

Theorem 2.1 *With fixed $m \geq 1$, let*

(2.2) $$q = n^{2-2/m}\, \omega_n \text{ where } \omega_n \to \infty$$

but

(2.3) $$\omega_n = o(\, n^{2/(m(m+1))}\,).$$

Then $E(X_m) \to \infty$ but $E(X_{m+1}) \to 0$.

Proof First observe that the hypothesis (2.2) implies that

$$(1 - 1/N)^q \sim 1.$$

Then we observe that since m is fixed,

(2.4) $$E(X_m) \sim (2^{m-1}/m!)\,(q/(n^{2-2/m}))^m,$$

and so $E(X_m) \to \infty$.

On the other hand,

(2.5) $$E(X_{m+1}) \sim (2^m/(m+1)!)\,(q/(n^{2-2/(m+1)}))^{m+1},$$

and now it follows from (2.3) that $E(X_{m+1}) \to 0$. ❑

Let $\mathcal{A}_m$ denote all the superpositions of order n that have edges of multiplicity exactly m. As usual $P(\mathcal{A}_m)$ denotes the probability of this event. The second moment method (see [Pa85] section 3.1 or [Sp87] lecture 3) can be used to get the first part of the next corollary. The second part follows from the fact that $E(X_{m+1}) \to 0$.

Corollary 2.1 *Under the same hypothesis as the theorem above, $P(\mathcal{A}_m) \to 1$ but $P(\mathcal{A}_m) \to 0$, i.e. almost all superpositions have edges of multiplicity m but have none of multiplicity $m + 1$.*

If $q = n^{2-2/m} c$, with constant $c > 0$, then

(2.6) $$E(X_m) \sim (1/2)\,(2c)^m / m! = \mu$$

and with a bit of work it can be shown that the edges of multiplicity m are distributed according to Poisson's Law with mean μ. Here is an application. Let $\langle x \rangle$ denote the nearest integer to x and suppose

(2.7) $$q = \langle c\, n \log n \rangle$$

Then it is easy to see from Theroem 2.1 that

$$E(X_2) \to \infty \text{ but } E(X_3) \to 0,$$

so almost all superpositions have edges of multiplicity 1 and 2 but none of multiplicity 3. In fact, with q defined by (2.7), it follows fron (2.1) that

$$E(X_1) \sim c\, n \log n \qquad \text{and} \qquad E(X_2) \sim (c \log n)^2$$

so the edges of multiplicity 1 still dominate. Recall that in the Erdös–Rényi probability model for graphs [ErR59], almost all are connected if q has the form (2.7) and $c > 1/2$. The fact that the edges of multiplicity 1 dominate in our model indicates that the threshold for connectivity may be the same in both models.

If the number of edges is a bit larger than N, then edges of every (fixed) multiplicity appear. The next theorem describes the range of values of q for which this phenomenon occurs.

Theorem 2.2 *For constant $c > 0$ and $\omega_n \to \infty$ with $\omega_n = o(N)$, if q is in the interval*

(2.8) $$cN < q < N \log (N/\omega_n),$$

then $E(X_m) \to \infty$ for every fixed $m \geq 0$.

Proof Since $q = o(N^2)$, we have

$$(1 - 1/N)^q \sim e^{-q/N}$$

and hence from (2.1)

(2.9) $$E(X_m) \sim (1/m!)\, q^m e^{-q/N} / N^{m-1}.$$

It follows from (2.8) that the right side of (2.9) is bounded below by $(c^m/m!)\omega_n$. ❑

An application of the second moment method yields the next corollary.

Corollary 2.2 *Under the same hypothesis as the theorem above*

$$P(\mathcal{A}_m) \to 1,$$

i.e. almost every superposition has edges of every (fixed) multiplicity.

As the edges accumulate in the random superpostion graph the edge multiplicities increase until every pair of vertices is joined by at least one edge. That is, the minimum multiplicity is at least 1 and we say that the graph is *complete* or *full*. Then the edges of multiplicity 1 diminish in number and eventually disappear, and so on. Next we determine when these events take place.

Theorem 2.3 *With fixed* $m \geq 0$, *let*

(2.10) $\quad q = N(\log N + m \log\log N + \log \omega_n)$

where $\omega_n \to \infty$ *but*

(2.11) $\quad \omega_n = o(\log N).$

Then $E(X_m) \to 0$ *but* $E(X_{m+1}) \to \infty$.

Proof Since $q = o(N^2)$, we have (2.9). On substitution of q from (2.10) in (2.9), we find

(2.12) $\quad E(X_m) \sim (1/m!) / \omega_n \to 0$

and similarly

(2.13) $\quad E(X_{m+1}) \sim (1 / (m+1)!)(\log n) / \omega_n \to \infty.$ ❑

As before, there is an accompanying result that can be established with a bit more effort.

Corollary 2.2 *Under the same hypothesis as the theorem above*

$$P(\mathcal{A}_m) \to 0 \quad \text{but} \quad P(\mathcal{A}_{m+1}) \to 1,$$

i.e. almost all superpostion graphs have no edges of multiplicity m but edges of multiplicity m + 1 persist.

Note in particular that if $m = 0$ in (2.10) then for

(2.14) $\quad q = N(\log N + \log \omega_n)$

almost all superpositions are complete but there are still edges of multiplicity 1. This observation should be contrasted with the Erdős–Rényi Model where X_0 also stands for the number of pairs of vertices of multiplicity 0. In this model the sample space consists of $C(N, q)$ equiprobable labeled graphs of order n and size q. Then obviously $E(X_0) = N - q$ and so

(2.15) $E(X_0) \to 0$ if $q = N + o(1)$

but

(2.16) $E(X_0) \to \infty$ if $q = N - \omega_n$ where $\omega_n \to \infty$.

Thus the duplication of edges in the superposition model causes a substantial delay in achieving a complete graph. This delay is also apparent in the variation of the Erdős–Rényi Model in which the q edges are selected with repetition allowed. Then the sample space consists of $C(N + q - 1, q)$ equally likely labeled multigraphs of order n with q (unlabeled) edges. And we have

(2.17) $E(X_0) = N\, C(N + q - 2, q) / C(N + q - 1, q)$,

from which the following facts emerge. If $\omega_n \to \infty$, then

(2.18) $E(X_0) \to 0$ if $q = N^2\omega_n$

but

(2.19) $E(X_0) \to \infty$ if $q = N^2/\omega_n$.

The effect of the duplication of edges is really apparent here where far more edges are necessary to gaurantee a complete multigraph. For edges of any multiplicity $m \geq 0$ in this model we have the formula

(2.20) $E(X_m) = N\, C(N + q - m - 2, q - m) / C(N + q - 1, q)$.

A little algebra shows that if $q = o(N)$, the behavior of $E(X_m)$ is exactly the same as in Theorem 2.1. But if $N = o(q)$, we find the following:

(2.21) $E(X_m) \to 0$ if $q = N^2\omega_n$, $\omega_n \to \infty$

and

(2.22) $E(X_m) \to \infty$ if $q = N^2/\omega_n$, $\omega_n = o(N)$.

Note that this result is independent of m. Evidently for q roughly between N and N^2, edges of all finite multiplicities appear but when q passes N^2, they vanish. Hence in this model they last longer and their termination is sudden compared to the random superposition model in which their endurance is brief and extinction comes in stages.

We conclude this section with some remarks about the comparison of the number q of edges in each superposition G and the number of edges in the underlying graph, namely $N - X_0(G)$. From formula (2.1) it follows that $E(X_0)^2 \sim E(X_0{}^2)$ whenever $E(X_0) \to \infty$. So almost all superpositions will have X_0 fairly close to the mean $E(X_0)$. We also find that if $q = N/\omega_n$, where $\omega_n \to \infty$ but $\omega_n = o(N)$, then

(2.22) $N - E(X_0) \sim q$,

i.e., the expected number of edges in the underlying graph of the random superposition and the total number of edges in the superposition are asymptotically equivalent. This result covers the range of values of q in Theorem 2.1. Hence some properties of random superpositions may be similar to those of random graphs in the Erdös–Rényi model. Of course, the arithmetic difference between q and $N - E(X_0)$ is $o(1)$ only if almost all superpositions have no edges of multiplicity 2. But this requires $q/n = o(1)$.

When q lies in the gap specified by (2.8), the expected number of edges in the underlying graph is substantially less then q. For example if $q = N$,

(2.23) $$N - E(X_0) \sim (1 - 1/e)q.$$

And if $q = N \log n$,

(2.24) $$N - E(X_0) \sim (1 / \log n)q.$$

There are many other comments that could be made about the distribution of multiple edges in these models. But the theorems above are sufficient for our investigation which concentrates on the range of q for which there are still edges of multiplicity zero. Of course the results of this section are of a fundamental combinatorial nature. They are just straightforward solutions of standard occupancy problems for several probability models. They have been recast in the terminology of random graphs for the convenience of the reader. The corresponding results for superposition of graphs with more than one edge will be substantially more difficult.

For another probability model for random multigraphs see [Go79] which is briefly mentioned in [Bo85].

3. Connectivity

How many edges should a random superposition have so that it is almost surely connected? Erdös and Rényi [ErR59] found that about n log n edges are needed when multiple edges are forbidden. In the superposition model when q has the form of (2.7), we find there are sure to be edges of multiplicity 2. But the next theorem shows that despite these redundancies, the threshold for connectivity is exactly the same as that found by Erdös and Rényi. In fact, the theorem also holds for the Erdös–Rényi model with repetition of edges allowed. The proof generally follows the pattern set by the founding fathers [ErR59] but there are interesting differences here and there. For example, we need the following lemmas because the edge probability for one pair of vertices is not independent of the edge probability of another pair. Let $\{(u_i, w_i) \mid i = 1$ to $k\}$ be a set of k different pairs of vertices and let $\mathcal{A}_i$ be the event that u_i and w_i are joined by an edge of multiplicity at least 1.

Lemma 1 *If $q < N$, then the probability that k specified pairs of vertices are adjacent has the bound:*

(3.1) $$P\Big(\bigcap_{i=1}^{k} \mathcal{A}_i \Big) < (q/N)^k.$$

Proof From the definition of conditional probability

(3.2) $$P\left(\bigcap_{i=1}^{k} \mathcal{A}_i\right) = P\left(\bigcap_{i=1}^{k-1} \mathcal{A}_i \mid \mathcal{A}_k\right) P(\mathcal{A}_k),$$

and from the definition of the superposition model,

(3.3) $$P\left(\bigcap_{i=1}^{k-1} \mathcal{A}_i \mid \mathcal{A}_k\right) < P\left(\bigcap_{i=1}^{k-1} \mathcal{A}_i\right).$$

Hence we have

(3.4) $$P\left(\bigcap_{i=1}^{k} \mathcal{A}_i\right) < \prod_{i=1}^{k} P(\mathcal{A}_i).$$

The rules of complementary probabilities tell us that

(3.5) $$P(\mathcal{A}_i) = 1 - (1 - 1/N)^q = \sum_{j=1}^{q} (-1)^{j+1} C(q,j) (1/N)^j$$

Now an application of the ratio test shows that the terms of the sum are decreasing, i.e.

(3.6) $$C(q, j)(1/N)^j > C(q, j+1)(1/N)^{j+1}$$

provided that $q < N$. Therfore for each $i = 1$ to k

(3.7) $$P(\mathcal{A}_i) < q/N,$$

and combining this inequality with (3.4) finishes the proof. ❑

We also need to estimate thr probability of the event that some pairs of vertices are adjacent and others are not. Suppose

(3.8) $$\mathcal{A} = \bigcap_{i=1}^{r} \mathcal{A}_i \quad \text{and} \quad \mathcal{B} = \bigcap_{i=r+1}^{r+s} \bar{\mathcal{A}}_i.$$

Note that Lemma 1 can be applied when the number N of pairs of vertices that are allowed to have edges is reduced to $N - s$. Then we have the conditional probability of the next lemma.

Lemma 2 *If $q < N - s$, the probability that r pairs of vertices are adjacent given that s other pairs are not adjacent is bounded by*

(3.9) $$P(\mathcal{A} \mid \mathcal{B}) < (q/(N-s))^r.$$

But we have the simple exact formula for $P(\mathcal{B})$ of the next lemma.

Lemma 3 *The probability that s specified pairs of vertices are not adjacent is*

(3.10) $$P(\mathcal{B}) = (1 - s/N)^q.$$

Hence Lemmas 2 and 3 can be combined to obtain an upper bound for the probability of the intersection of $\mathcal{A}$ and $\mathcal{B}$.

Lemma 4 *If $q < N - s$, the probability that r specified pairs of vertices are adjacent and s others are not is bounded by*

$$(3.11) \qquad P(\mathcal{A} \cap \mathcal{B}) < (q/(N-s))^r (1 - s/N)^q.$$

It is interesting to observe that an exact formula for the probability $P(\mathcal{A} \cap \mathcal{B})$ can be worked out from formula (3.10) and the following recurrenece relation in which we have used $P(r, s)$ to denote $P(\mathcal{A} \cap \mathcal{B})$:

$$(3.12) \qquad P(r, s) = P(r-1, s) - P(r-1, s+1).$$

On solving the recurrence relation we find

$$(3.13) \qquad P(\mathcal{A} \cap \mathcal{B}) = \sum_{k=0}^{r} C(r, k)\,(-1)^k\,(1 - (s+k)/N)^q.$$

On dividing the right side of (3.13) by the right side of (3.10), after some simplification we arrive at an exact formula for the conditional probability:

$$(3.14) \qquad P(\mathcal{A} \mid \mathcal{B}) = \sum_{i=0}^{q} C(q, i)\,(-1/(N-s))^i \sum_{k=0}^{r} C(r, k)\,(-1)^k\,k^i.$$

Next we find the neat identity (see formula 1.13 in Gould's Tables [Go72]) for the second sum in (3.14):

$$(3.15) \qquad \sum_{k=0}^{r} C(r, k)\,(-1)^k\,k^i = \begin{cases} 0 & \text{if } 0 \le i < r \\ (-1)^r\, r! & \text{if } i = r \end{cases}.$$

So the first sum in (3.14) can begin at $i = r$ and the contribution to $P(\mathcal{A} \mid \mathcal{B})$ of this term is $C(q, r)/(N-s)^r$ from which (3.9) can be derived after more work on the remaining terms of the sum. Hence it seems that the exact formulas are harder to deal with than the more conbinatorial approach that was taken in the lemmas. At any rate, we are now ready for the main result of this section.

Theorem 3.1 *Let x be any (fixed) real number and set*

$$(3.16) \qquad q = \langle (n/2)(\log n + x) \rangle$$

and denote by C_n the set of connected superposition graphs of order n and size q. Then

$$(3.17) \qquad P(C_n) \to e^{-e^{-x}}.$$

Proof For $i = 1, 2, \ldots, \lfloor n/2 \rfloor$, let $Y_i(G)$ be the number of components of order i in the superposition graph G. Now define

$$X = Y_1$$

and

$$Y = Y_2 + \ldots + Y_{\lfloor n/2 \rfloor}.$$

Then we have the upper and lower bounds for $P(C_n)$:

$$(3.18) \qquad P(X = 0) - P(Y \ge 1) \le P(C_n) \le P(X = 0).$$

Since $P(Y \geq 1) \leq E(Y)$, we have

(3.19) $$P(X = 0) - E(Y) \leq P(\mathcal{C}_n) \leq P(X = 0).$$

Now we estimate the upper bound $P(X = 0)$ by using the binomial moments. First observe that

(3.20) $$E(X) = n(C(n-1, 2) / N)^q = n(1 - 2/n)^q$$

and a few simple steps show that

(3.21) $$\log E(X) = -x + o(1).$$

Hence the first binomial moment $S_1 = E(X)$ is asymptotic to e^{-x} and just as observed by Erdös and Rényi [ErR59] in the case of random graphs, it can be shown that the binomial moments for $k = 0, 1, 2, \ldots$ are given by

(3.22) $$S_k \sim (e^{-x})^k/k!.$$

And so the isolated vertices are distributed according to Poisson's law with mean $\mu = e^{-x}$ and we have

(3.23) $$P(X = 0) \to e^{-e^{-x}}.$$

It remains to show that $E(Y) \to 0$, i.e. almost all random graphs have one large component and perhaps some isolated vertices. We begin with the next equality which holds for $k \geq 1$ and follows from the definition of expectation:

(3.24) $$E(Y_k) = C(n, k) P(\mathcal{D} \cap \mathcal{B}),$$

where the binomial coefficient $C(n, k)$ counts the number of ways to select k labels for the vertices of the component of order k, $\mathcal{D}$ is the event that the superposition graph G has a connected subgraph spanning the specified vertices with labels 1 to k and $\mathcal{B}$ is the event that it has no edges joining these k vertices with the other $n - k$ vertices. Recall that Cayley's theorem says that there are k^{k-2} labeled trees of order k. Since every connected graph of order k contains a spanning tree of order k, we have

(3.25) $$P(\mathcal{D} \cap \mathcal{B}) \leq k^{k-2} P(\mathcal{A} \cap \mathcal{B}),$$

where $\mathcal{A}$ is the event that $k - 1$ specified edges are present in the subgraph on the vertices with labels 1 to k. Now we apply lemma 4 to (3.25), combine (3.24) and arrive at the next inequality, which holds for $k \geq 1$:

(3.26) $$E(Y_k) \leq C(n, k) k^{k-2} (q / (N - k(n-k)))^{k-1} (1 - k(n-k) / N)^q.$$

To estimate the right side of (3.26) we begin by applying Stirling's formula for $k!$ and find that

(3.27) $$C(n, k) k^{k-2} \leq O(1) n^k e^k.$$

Since $k(n-k)$ is largest when $k = n/2$, we have for $n \geq 3$:

(3.28) $$1 - k(n-k) / N \geq 1/4,$$

and this can be used to show that

(3.29) $$(q / (N - k(n-k)))^{k-1} \leq (4q/N)^{k-1}.$$

Since $1 \le k \le n/2$,

(3.30) $$k/(n-1) \le k(n-k)/N,$$

and so

(3.31) $$(1 - k(n-k)/N)^q \le (1 - k/(n-1))^q < e^{-qk/n}.$$

Now we can use the formula (3.16) for q together with (3.27), (3.29) and (3.31) to obtain the bound

(3.32) $$E(Y_k) = \mathcal{O}(n/\log n)\,(\mathcal{O}(1)\log n/\sqrt{n}\,)^k.$$

Hence if $k \ge 3$, then $E(Y_k) \to 0$ but if $k = 2$, formula (3.32) tells us that $E(Y_2) = \mathcal{O}(\log n)$. So when $k = 2$, we can improve the bound from (3.31) by using

(3.33) $$(1 - 2(n-2)/N)^q \le \mathcal{O}(n^{-2}).$$

Then $E(Y_2) \to 0$ and for $k \ge 3$

(3.34) $$\sum_{k=3}^{\infty} E(Y_k) = \mathcal{O}(1)\,(\log n)^2/\sqrt{n} \to 0.$$

Hence (3.17) is established. ❑

The theorem also holds for the Erdös–Rényi model with replacement. In this case, an explicit formula for $P(\mathcal{A} \cap \mathcal{B})$ is immediate:

(3.35) $$P(\mathcal{A} \cap \mathcal{B}) = C(N - s + q - r - 1, q - r) / C(N + q - 1, q),$$

and from this follows the approproate inequality corresponding to formula (3.11) of Lemma 4:

(3.36) $$P(\mathcal{A} \cap \mathcal{B}) < (q/(N - s - 1))^r (1 + q/(N-1))^{-s}.$$

Now (3.36) can be used used to show that $E(Y) \to 0$ just as in the superposition model.

4. Numerical Results

For small values of n and q, we can calculate the exact number $F_{n,q}$ of connected superpositions of order n and size q from the following recurrence relation. For $n \ge 2$ and $q \ge n - 1$:

(4.1) $$N^q = F_{n,q} + (n-1)F_{n-1,q} + C(n-1, 2)^q + \sum_{k=2}^{n-2} C(n-1, k-1) \sum_{i=k-1}^{q} C(q, i)\, F_{k,i}\, C(n-k, 2)^{q-i}.$$

This relation is derived by expressing the total number N^q of superpositions as the sum for $k = 1$ to n of the number for which the component that contains the vertex with label number 1 has order k. The first three terms on the right side of (4.1) arise when $k = n, n-1$ and 1 respectively. In the double sum $C(n-1, k-1)$ is the number of ways to select the other $k - 1$ labels for the component containg vertex # 1. Next C(q, i) is the number of ways to choose i edges for this component. Then $F_{k,i}$ is the number of components with these specified vertex and edge labels. Finally $C(n-k, 2)^{q-i}$ is the number of ways to fill in the remaining $n-k$ vertices with the other $q-i$ edges.

Now we introduce notation for the probability of connectivity for $n \geq 1$ and $q \geq n-1$:

(4.2) $$P_{n,q} = F_{n,q} / N^q.$$

Then (4.1) can be transformed into a recurrence relation for probabilities by dividing both sides by N^q and then using (4.2) to eliminate each occurence of F. The result for $n \geq 2$ and $q \geq n-1$ is a recurrence relation for the probability $P_{n,q}$ that a random superposition of order n and size q is connected:

(4.3) $$P_{n,q} = 1-(n-1)((n-2)/n)^q P_{n-1,q} - ((n-2)/n)^q + \sum_{k=2}^{n-2} C(n-1, k-1) \sum_{i=k-1}^{q} C(q, i) P_{k,i} C(k, 2)^i C(n-k, 2)^{q-i} / N^q.$$

A double precision Fortran program used this relation on a Zenith–158 PC to calculate the numbers in the Table. If $x = 0$ in formula (3.16) of Theorem 3.1, then $q = \langle (n \log n) / 2 \rangle$ as in the table and according to the theorem, $P_{n,q}$ should approach $1/e = .367879...$. For $n > 40$ the computations required by the recurrence relation were beyond the range of the program. But $P_{n,q}$ can then be calculated using the bounds in equation (3.19). However, as in the case of the Erdös–Rényi model (see [BoT85] for extensive data and relevant equations for computing the probability of connectivity for ordinary random graphs), to achieve better accuracy it was found necessary to redefine X and Y as follows:

$$X = Y_1 + Y_2$$

and

$$Y = Y_3 + \ldots + Y_{\lfloor n/2 \rfloor}.$$

The formula for the binomial moments is rather more complicated now. We need S_r, the expected number of r–sets of components of order one or two. And now we must take into consideration the multiplicity of the edges in the components of order two. If $n = 100$ and $q = 230$, then the expected number of components of order two with edges of multiplicity 3 is less than .0000085 and this is also an upper bound on the probability of such a component. Hence in the next formula we only include edges of multiplicity 2. For $r \geq 1$, we find

(4.4) $$S_r = (1/r!) \sum_{s=0}^{r} (n)_{r+s} C(r,s) (1/2)^s (1/N)^s (C(n-r-s, 2)/N)^{q-s} \sum_{i=0}^{s} C(s, i) (q)_{s+i} (1/2)^i (1/N)^i.$$

In this formula, r is the number of components of order one or two, s is the number of order two. So $r-s$ is the number of isolated vertices. Then in the second sum, i is the number of components of order two with edges of multiplicity 2, while s $-i$ is the number of multiplicity 1. In our interactive program, that could only handle n up to about 7000, only about ten binomial moments (i.e. $r \le 10$) were needed to obtain the first 5 digits of $P(X = 0)$. Estimates of $E(Y_k)$ were made from (3.26) but these declined so rapidly that the contributions for $k \le 10$ were always sufficient to obtain good results. It was found that for $n = 500$, 1000, 2000, 3000, 5000, and 7000, the upper and lower bounds for $P_{n,q}$ when rounded to three digits, all resulted in the same number, namely 0.368.

n	q	Pn,q
5	4	.30000
10	12	.37487
15	20	.32261
20	30	.35469
25	40	.34851
30	51	.35874
35	62	.35553
40	74	.36649

Table 1 Probability of connectivity with $q = \langle (n \log n) / 2 \rangle$.

REFERENCES

[Bo85] B. Bollobás, *Random Graphs*, Academic, New York (1985).

[BoT85] B. Bollobás and A. Thomason, Random graphs of small order, *Annals of Discrete Math.* **28** (1985) 47 – 97.

[ErR59] P. Erdös and A. Rényi, On random graphs I, *Publ. Math. Debrecen* **6** (1959) 290 – 297.

[Go79] E. Godehart, An extension of the theorems of Erdös–Rényi–type to random multigraphs, *Proc. Sixth Conf. Probability Theory,* (B. Bereanu *et al.*, eds.) Ed. Acad. R.S. România, Burcharest (1981) 417 – 425.

[Go72] H.W. Gould, *Combinatorial Identies,* H.W. Gould, Morgantown (1972).

[H60] F. Harary, Unsolved problems in the enumeration of graphs, *Magyar Tud. Akad. Mat. Kutató Int. Közl* **5** (1960) 63 – 95.

[HR84] F. Harary and R.W. Robinson, The rediscovery of Redfield's papers, *J. Graph Theory* **8** (1984) 191 – 193.

[L84] E.K. Lloyd, J. Howard Redfield 1879 – 1944, *J. Graph Theory* **8** (1984) 195 – 203.

[Pa85] E.M. Palmer, *Graphical Evolution: an introduction to the theory of random graphs,* Wiley–Interscience Series in Discrete Mathematics, New York (1985).

[Re59] R.C. Read, The enumeration of locally restricted graphs I, *J. London Math. Soc.* **34** (1959) 417 – 436.

[Re60] R.C. Read, The enumeration of locally restricted graphs II, *J. London Math. Soc.* **35** (1960) 334 – 351.

[R27] J.H. Redfield, The theory of group–reduced distributions, *Amer. J. Math.* **49** (1927) 433 – 455.

[Sp87] J. Spencer, *Ten lectures on the probabilistic method,* SIAM, Philadelphia (1987).

Graphical Designs[1]

T.D. Parsons[2]
the late of the California State University at Chico, CA

Tomaž Pisanski
University of Ljubljana, Yugoslavia

ABSTRACT

Let G *be a graph on* n *vertices* $1, 2, ..., n$. *Let*

$$G = G_1 \oplus G_2 \oplus ... \oplus G_p$$

be an arbitrary factorization [= edge decomposition] of G. *Then there exist an integer* $d > 0$, *and* n *vectors* $x_1, x_2, ..., x_n$ *from* $\{-1, +1\}^d$, *and a set of distinct integers* $\{a_0, a_1, ..., a_p\}$ *such that* $a_0 = 0$, $a_k < 0$, *for* $k \neq 0$ *and that vertex* i *is adjacent to vertex* j *in* G_k *if and only if* $\langle x_i, x_j \rangle = a_k$, *and* $\langle x_i, x_j \rangle = a_0$ *if* i *is not adjacent to* j *in* G, *where* $\langle x_i, x_j \rangle$ *is the inner product of* x_i *and* x_j.

In previous work [PP1, PP2, PP3] we have investigated vector representation of graphs. Vertices are represented as vectors in some vector space with a given bilinear form. Two vertices are adjacent if and only if the form evaluated for the corresponding two vectors attains a value from a specified set of values. The purpose of this work is to extend this notion to an edge–disjoint decomposition of a graph.

Motivation for our work can be found, for instance, in [PP3]. For the sake of self–sufficiency we rephrase here the definition from [PP3] of a good Hadamard representation. For an arbitrary graph G on the vertex set $\{1, 2, ..., n\}$ let $M(G)$ be an n by d matrix with entries $+1$ or -1 such that $M(G)M(G)^t = dI_n + t\,A(G)$, where t is a negative integer and $A(G)$ is the adjacency matrix of G. If such a matrix exists we call it a *representation matrix* of a *good Hadamard representation* of the graph G. Moreover, t is called the *symbol* and d is the *dimension* of the representation. As usal $M(G)^t$ denotes the transpose of $M(G)$. If the rows of $M(G)$ are considered as vectors $x_1, x_2, ..., x_n$, from a real d–space with cordinates $+1, -1$, then any two

1. Work supported in part by a grant from the California State University at Chico and by the NSF (Grant DMS – 8717441)
2. Tory suddenly died on April 2, 1987, a few days after we obtained this result.

vectors x_i and x_j are orthogonal if and only if the corresponding vertices i and j are not adjacent in G. Let $d_1^*(G)$ denote the minimum dimension of any Hadamard representation of graph G.

By Theorem 12 of [PP3] we have $d_1^*(G) \leq 2n^2 - 2n$; that is $2n^2 - 2n$ is an upper bound for the minimum dimension of a good Hadamard representation. If M is the matrix of a representation of G and t is the symbol of the representation we may easily produce infinitely many representations with infinitely many different symbols. Let M_k denote the matrix $M_k = [M|M|...|M]$ which is obtained by putting k copies of M next to each other. Clearly M_k is matrix of a good (kd)–dimensional Hadamard representation of G with the symbol kt.

Lemma *Given an arbitrary graph H and an arbitrary finite set X of nonpositive integers there is a negative integer c such that c does not belong to X and that H has a representation with the symbol c.*

Proof By Theorem 12 of [PP3] H admits a good Hadamard representation M with symbol t, for some negative integer t. By the above argument $M_k = [M|M|...|M]$ (k-times) is a Hadamard representation with the symbol kt. For k = 1, 2, 3, ..., we get infinitely many distinct values for c where c = kt. Therefore there exists a positive integer k such that $c \notin X$ and that H admits a good Hadamard representation with the symbol c. ❑

Theorem *Let G be a graph on n vertices 1, 2, ..., n. Let*

$$G = G_1 \oplus G_2 \oplus ... \oplus G_p$$

be an arbitrary factorization [= edge decomposition] of G. Then there exists an integer $d > 0$, and n vectors $x_1, x_2, ..., x_n$ from $\{-1, +1\}^d$, and a set of distinct integers $\{a_0, a_1, ..., a_p\}$ such that $a_0 = 0$, $a_k < 0$, for $k \neq 0$ and that vertex i is adjacent to vertex j in G_k if and only if $\langle x_i, x_j \rangle = a_k$, and $\langle x_i, x_j \rangle = a_0$ if i is not adjacent to j in G, where $\langle x_i, x_j \rangle$ is the inner product of x_i and x_j.

Proof By induction on p. For p = 1 we simply use Theorem 12 of [PP3]. Let $G = G' \oplus G_{p+1}$ and $G' = G_1 \oplus G_2 \oplus ... \oplus G_p$. By the induction hypothesis the Theorem holds true for any factorization of G' with p factors. Let us denote by d', $a_0', a_1', ..., a_p', x_1', x_2', ..., x_n'$, the parameters corresponding to the above factorization of G' with representation M'. Let $X = \{a_0', a_1', ..., a_p'\}$ and let $H = G_{p+1}$. If we apply our Lemma to these values we get a representation, say with vectors y_i, a representation matrix M_0 and with the symbol c for G_{p+1}. By combining $M = [M'|M_0]$ we get the required representation with the parameters $d = d' + d_0$, $x_i = [x_i'|y_i]$ and $a_k = a_k'$ for k = 0, 1, ..., p and $a_{p+1} = a_0' + c$. ❑

There are at least two other interpretations of our Theorem.

Corollary A *Let us select an arbitrary (not necessarily proper) edge–coloring of the complete graph* K_n. *Then there exists a d–cube graph* Q_d, *and a set of vertices* $v_1, v_2, \ldots, v_n$ *of* Q_d, *such that two edges* ab *and* cd *of* K_n *have the same color if and only if the distance from* v_a *to* v_b *in* Q_d *is the same as the distance from* v_c *to* v_d.

Proof By replacing each occurrence of -1 by 0 in x_i in our Theorem we transform x_i into v_i. Thus v_i is a d–dimensional 0–1 vector which naturally represents a vertex of Q_d. If $\langle x_i, x_j \rangle = s$, then the distance between v_i and v_j is $(d - s)/2$. ❑

Corollary B *Let us select an arbitrary (not necessarily proper) edge coloring of the complete graph* K_n. *Then there exists a family of sets* $\{V_1, V_2, \ldots, V_n\}$ *such that two edges* ab *and* cd *of* K_n *have the same color if and only if the symmetric difference of* V_a *and* V_b *has the same number of elements as the symmetric difference of* V_c *and* V_d.

Proof If 0–1 vectors are considered as characteristic vectors of subsets of some d–set then the Corollary A translates directly into Corollary B. ❑

During the conference in Kalamazoo Tom Tucker pointed out that Corollary B remains true if we replace the symmetric difference by intersection. A simple modification of the well–known proof that each graph is an intersection graph applies.

Proposition *Let us select an arbitrary (not necessarily proper) edge–coloring of the complete graph* K_n. *Then there exists a family of sets* $\{U_1, U_2, \ldots, U_n\}$ *such that any two edges* ab *and* cd *of* K_n *have the same color if and only if the intersection of* U_a *and* U_b *has the same number of elements as the intersection of* U_c *and* U_d.

Proof Assume that the edges of K_n are being colored by positive integers. For each vertex i we construct the set U_i as follows. If an edge $ij = ji = e$ is colored C then let U_i contain the C pairs $(e, 1), (e, 2), \ldots, (e, C)$. If ab and cd are two edges colored with the same color, say C, then the intersection of U_a with U_b has cardinality C and so does the intersection of U_c and U_d. ❑

REFERENCES

[PP1] T.D. Parsons and T. Pisanski, Inner product representation of graphs, Proceedings of the Sixth Yugoslav Seminar on Graph Theory, Dubrovnik 1985, 151 – 157.

[PP2] T.D. Parsons and T. Pisanski, Exotic n–universal graphs, J. Graph Theory, 12 (1988) 155–158.

[PP3] T.D. Parsons and T. Pisanski, Vector representations of graphs, Discrete Math., to appear.

Properties of Non–Minimum Crossings for Some Classes of Graphs

B. L. Piazza
The University of Southeren Mississippi

R. D. Ringeisen*
Clemson University

S. K. Stueckle
The University of Idaho

ABSTRACT

In this paper we investigate drawings of some graphs which are not minimum crossing drawings. Interest in the study of non-minimum drawings has arisen in many papers and has been of growing interest in recent years. Our goal here will be to present an upper bound for the number of crossings in a good drawing of a graph, modify this bound somewhat and display graphs which meet this new bound. In some cases we examine the possibility of interpolation results for numbers of crossings which lie between the usual and the maximum crossing numbers. After looking at graphs which do meet this new upper bound, we briefly look at the unsolved problem of determining the maximum crossing number of the n–cube.

1. Introduction

In this paper we investigate drawings of some graphs which are not minimum crossing drawings. Interest in the study of non-minimum drawings has arisen in many papers and has been of growing interest in recent years. Among those papers which have examined this phenomenon have been those by Harborth [4,5], Kleitman [6], Eggleton [2], Woodall [10], Ringel [9], and several authors in [3]. Our goal here will be to present an upper bound for the number of crossings in a good drawing of a graph,

* Partial support from the U. S. Office of Naval Research.

modify this bound somewhat and display graphs which meet this new bound. In some cases we examine the possibility of interpolation results for numbers of crossings which lie between the usual and the maximum crossing numbers. After looking at graphs which do meet this new upper bound, we briefly look at the unsolved problem of determining the maximum crossing number of the n–cube.

Throughout the paper we will only consider the so called "good" drawing of graphs in the plane. Informally, a drawing is a *good drawing* if no edge crosses itself, no two edges cross more than once, no two edges incident with the same vertex cross, and no more than two edges cross at a point. The *crossing number* of a graph, which is well studied but known for very few classes of graphs (see Kleitman [7], Beineke and Ringeisen [1], and Ringeisen and Beineke [8]), is the minimum number of crossings among all such drawings. The maximum crossing number is the maximum number of crossings among all good drawings in the plane. We will not study the crossing number here and will denote the maximum crossing number as $cr^M(G)$.

The first graph for which the maximum crossing number was studied was the complete graph. Ringel [9] showed that $cr^M(K_n)$ is n choose four with the key to such a determination being that the maximum crossing number of the complete four graph is one. An immediate corollary to this result is the following, which we single out as a remark because of its importance later in this paper.

Remark 1 *The maximum crossing number of the four cycle is one.*

Of course the result on complete graphs gives an upper bound for the maximum crossing number of any graph. In this paper we look at some other upper bounds and distinguish some graphs which meet them.

2. Upper Bounds and Some Previous Results

Given any edge in a good drawing of a graph, it is clear that it many not cross itself or any edge incident with it. The following lemma states this fact formally.

Lemma 1 *For an edge $e = uv$ in a graph G the number of edges which it may cross is bounded above by $E - (\text{degree } u + \text{degree } v - 1)$, where E is the number of edges of G.*

If we then tally up this bound over all such edges of the graph we obtain an upper bound for the maximum crossing number, which we give as the following theorem because of the significance of the result, although the proof is immediate.

Theorem 1 *For any graph G with E edges, we have*

$cr^M(G) \leq (\Sigma(E - \deg u - \deg v + 1))/2$, *where the sum is taken over all edges* $e = uv$
$= (E^2 + E - \Sigma \deg^2 u)/2$, *where the sum is all over vertices* u.

For convenience, we will denote the number on the right side of the inequality in Theorem 1 as $\mathcal{D}$(G). We note that for the complete graphs this number is larger than the actual maximum crossing number. However, there are graphs which achieve the bound. The following theorem is due to Harborth [5] in a upcoming paper.

Theorem 2 *For a cycle C_n, $n \neq 4$, we have that* $cr^M(C_n) = n(n-3)/2 = \mathcal{D}(Cn)$

In order that the reader may get a feeling for this work we display in figures 1 and 2 the drawings which realize the upper bound. Throughout the remainder of the paper, if a drawing has more than two edges crossing at a point then it is presumed that the edges may be slightly perturbed to give a proper drawing.

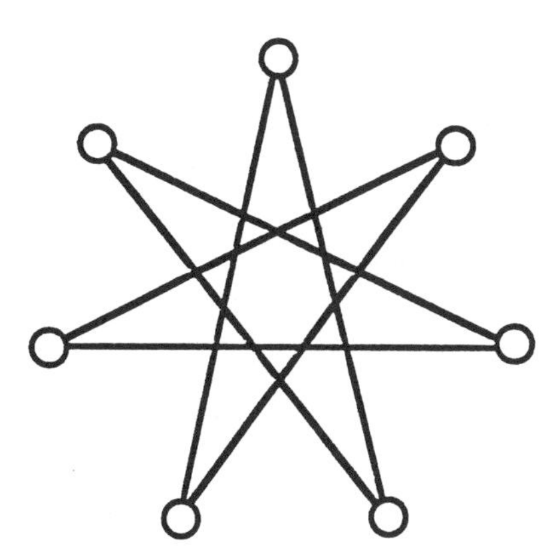

Figure 1. Drawing the odd cycles

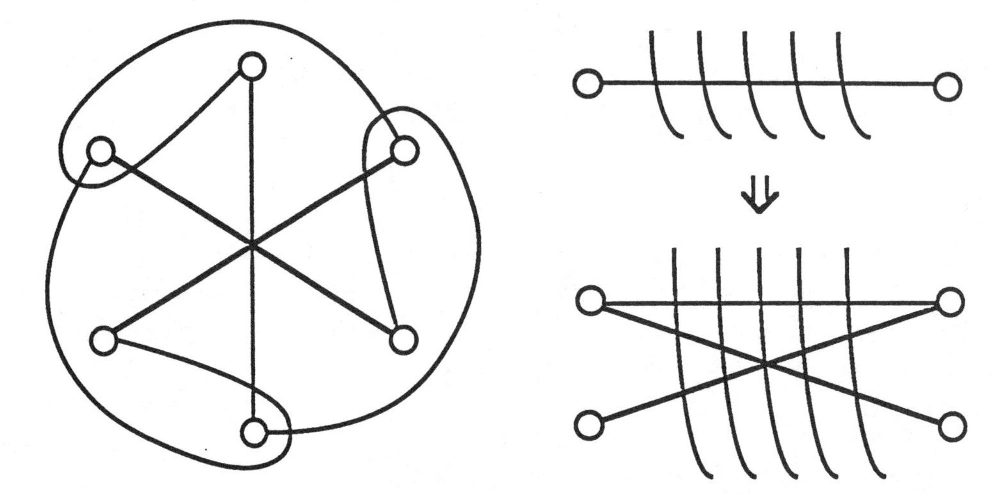

Figure 2. Drawing the even cycles

Note that the odd cycles meet the upper bound rather "naturally," while a bit more sophistication is needed in the even case. What is illustrated in figure 2 is a drawing for C_6 and then a method to go from one even number to the next by deleting an edge in the given drawing and replacing it with three edges as shown. The curved lines show that each edge would still cross all edges it may "legally" cross according to Theorem 1.

The drawings for cycles also illustrate another property of drawings which a given class of graphs may or may not possess. We will use the words *interpolation property* to mean that the graph can attain any number of crossing between its maximum crossing number and its crossing number. Notice that by modifying the drawings for cycles, one can drop the numbers of crossings by exactly one, thereby showing that cycles have the interpolation property. This property for the complete graphs is studied by Eggleton, Guy, Harborth, and Ringeisen in [3]. Kleitman [5] had earlier show that the parity of the number of crossings for any two drawings of odd complete graphs must be the same, thereby showing that these graphs do not have the interpolation property. In the four author paper [3], it is shown that there are "gaps" considerably larger than one when trying to obtain all possible number of crossings for complete graphs. These graphs then are graphs which are greatly lacking relative to the interpolation property.

Main Results

It seems natural to examine drawings of other simple planar graphs. We look next at trees and then at unicyclic graphs. Throughout this section we will use the phrase "all eligible edges" to mean those edges which a given edge may cross relative to Theorem 1 in a good drawing.

The fact that trees and connected unicyclic graphs meet the upper bound is proved using the language of "thrackles" in [10]. However, we give different proofs here which allow us to also conclude that the graphs have interpolation property.

Theorem 3 *Let T be a tree with p vertices. Then T has a drawing on the plane with r crossings, for all r,* $0 \le r \le \mathcal{D}(T) = (p(p-1) - \Sigma \deg^2 u)/2$*, where the sum is taken over all vertices u.*

Proof We first prove that any tree can be drawn with $\mathcal{D}(T)$ crossings and then examine those drawings in order to show the interpolation property. We proceed by induction on the number of vertices p. When $p \le 2$, it is clear that T can only be drawn with zero crossings. Assume that any tree on $p \ge 2$ vertices can be drawn with the appropriate number of crossings. Let T be a tree with $p + 1$ vertices with an endvertex v adjacent to a vertex u and let $T' = T - v$. By the induction hypothesis, we find a drawing D of T' with $\mathcal{D}(T')$ crossings.

Let uw, $w \neq v$, be an edge of T and thus an edge of T'. Then uw crosses all eligible edges of T' in D. We create a drawing of T by adding vertex v of T' in such a way that the edge uv is drawn "parallel" to uw, crossing all edges which uw crosses and then looping around vertex w so that it crosses all edges incident with w except uw. (See figure 3 for an illustration of the path with 6 vertices.) Note that none of the edges which uv crosses are incident with u and that uv crosses no edge more than once. In the newly created drawing uv rosses all of its eligible edges and such remains true of all of the other edges of T. Hence we have created a drawing of T in which every edge crosses all of its eligible edges and hence a drawing which has $\mathcal{D}(T)$ crossings.

To show that interpolation holds for any tree we only need note that a similar induction argument can be made. In this case we move from the drawings for T' to the drawings for T in a similar manner. However, when placing the "parallel" edge uv into a drawing we can create the various drawings necessary for T by placing the vertex v so as to cross one of its eligible edges at a time. The proof is then complete.

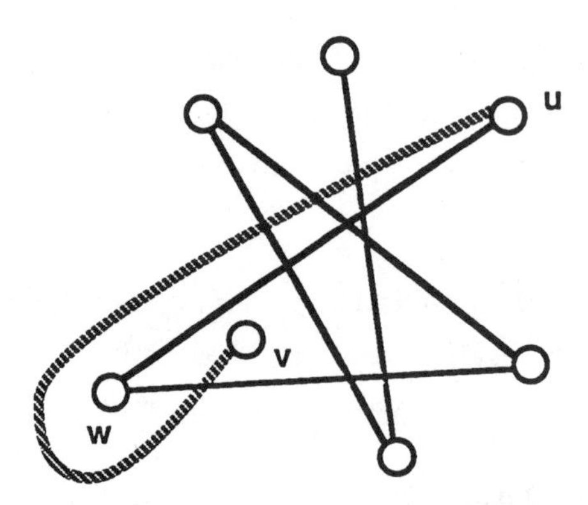

Figure 3. A tree (path) with max crossings.

The above proof can be modified so as to give a nice corollary.

Corollary 1 *If G is a connected unicyclic graph, where the cycle does not have four edges, then $cr^M(G) = \mathcal{D}(G)$, and G has the interpolation property.*

Proof If G is a cycle, then the aforementioned result of Harborth applies. Otherwise, G must have a vertex of degree one and again such may be deleted and the induction hypothesis invoked on the remaining graph. The deleted edge may then be replaced by using the same argument as in the previous proof. The interpolation property follows in exactly the same manner and the proof is complete. ❑

Note that if the cycle has 4 edges then a similar induction may be used giving the $cr^M(G) = \mathcal{D}(G) - 1$.

When one begins to further examine maximum crossing number, it looks as if no other graphs G are likely to reach $\mathcal{D}(G)$. In fact the following conjecture which we make here is similar to the "thrackel conjecture," although the terminology is different. (See Woodall [10]).

Conjecture *A connected graph G has $cr^M(G) = \mathcal{D}(G)$ if and only if G has at most one cycle C and $C \neq C_4$.*

An area of current interest, albeit of little progress, is in determining the maximum crossing number of the n–cubes. Such is not known even for $n = 3$. One notes that the n–cube has many four cycles, and that such are likely to play an important role in an upper found for any graph since a four cycle itself does not meet the previous upper bound. Hence, one needs to subtract the number of nonidentical C_4's from the upper bound. Also, since $\mathcal{D}(K_4)$ is three while at the same time K_4 has exactly three four cycles, the number of nonidentical K_4's must be added back in. Consequently, the following modified upper bound seems appropriate.

Remark 2 *For any graph G,* $cr^M(G) \leq \mathcal{D}(G) - \mathcal{N} + \mathcal{M}$, *where* $\mathcal{N}$ *and* $\mathcal{M}$ *are the number of nonidentical* C_4*'s and* K_4*'s in G, respectively.*

In this sequel we will denote $\mathcal{D}(G) - \mathcal{N} + \mathcal{M}$ by $\mathcal{D}'(G)$. Notice that if G is a tree or a unicyclic graph, where the cycle is not a four cycle, then $\mathcal{D}'(G) = \mathcal{D}(G)$ and if G is unicyclic where the cycle is a four cycle, then $\mathcal{D}'(G) = \mathcal{D}(G) - 1$, the bound which was met earlier. Furthermore, if G is a complete graph, then a simple counting argument shows that $\mathcal{D}(G) = \mathcal{N}$, giving the result that $\mathcal{D}'(G) = \mathcal{M}$, which is the result in [9].

An interesting question is whether any collection of graphs other than these immediate examples meet the modified bound $\mathcal{D}'$. The following results came about as a direct consequence of attempts to solve this problem for the n–cubes. Of course, the so called "ladder" graphs and "prism" graphs are important classes of graphs in their own right.

Theorem 4 *For the ladder graphs* $L_{2n} = P_n \times K_2$, *where* P_n *is the path with n vertices, we have* $cr^M(L_{2n}) = \mathcal{D}'(L_n)$.

Proof We need to display drawings of the L_{2n} which have the proper number of crossings. The drawings are illustrated in figure 4 for L_{12}, where 4a is a modification of the drawing with the proper number of crossings, while 4b shows the labeling of vertices when the graph is drawn in what we refer to as "standard form".

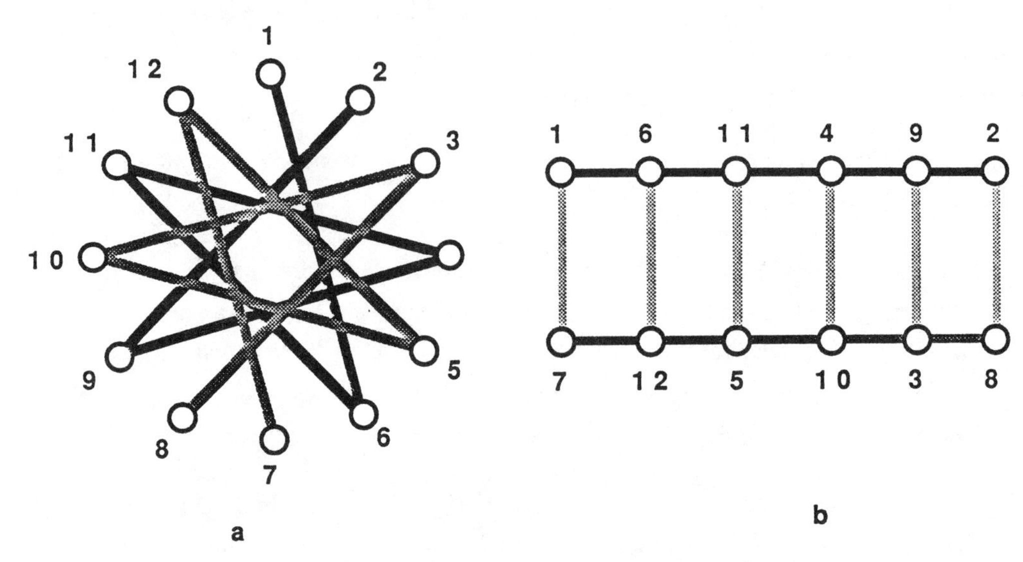

Figure 4. The ladder graph

The only four cycles in the ladder graph are the obvious ones and their are no K_4 subgraphs. To see that the desired drawing may be created for any ladder graph notice that the path P_n can be drawn with the maximum number of crossings by placing the vertices labeled 1 through $2n$ in a circular fashion, then joining vertex 1 to vertex $1 + (n - 1)$, then this vertex n to $n + (n - 1)$, etc. Now referring to figure 4a, notice that another copy of P_n can be drawn by beginning at vertex $n + 1$ and again adding $(n - 1)$ to reach vertex $2n$. Continue to create this second path by adding $(n - 1)$ to each successive vertex until the path is completed. One can use modular arithmetic to show that such a drawing always results in any "upper horizontal" edge in the standard drawing of L_{2n} (figure 4b) parallel to the corresponding "lower horizontal" edge in its four cycle. Notice that the "vertical" edges in the standard drawing each have endpoints which are diametrically opposite in figure 4a. When these edges are added directly across the interior of the drawing one obtains a drawing of L_{2n} which has $\mathcal{D}(L_{2n})$ crossings. The proof is then complete. ❑

Although the prism graphs are only a slight modification of the ladder graphs, the fact that they also meet the $\mathcal{D}$ bound is much more difficult to see. Nevertheless, we do have the following result.

Theorem 5 *The prism graphs $R_{2n} = C_n X K_2$, for $n \neq 4$ and C_n is the cycle graph on n vertices, have $cr^M(R_{2n}) = D'(R_{2n})$.*

Proof Again, drawings need to be displayed which meet the proper criteria. Since drawings which maximize the number of crossings for the cycle graphs are more complicated for even cycles than for the odd, one might expect that the prism formed when the base cycle is even might be more complicated than that for odd cycles. Such is certainly the case. We first display a drawing for the odd cycle case and then use a procedure similar to that used by Harborth [5] to obtain the even cycle case. We note that the only four cycles in R_{2n} are the obvious ones and that there are no K_4 subgraphs.

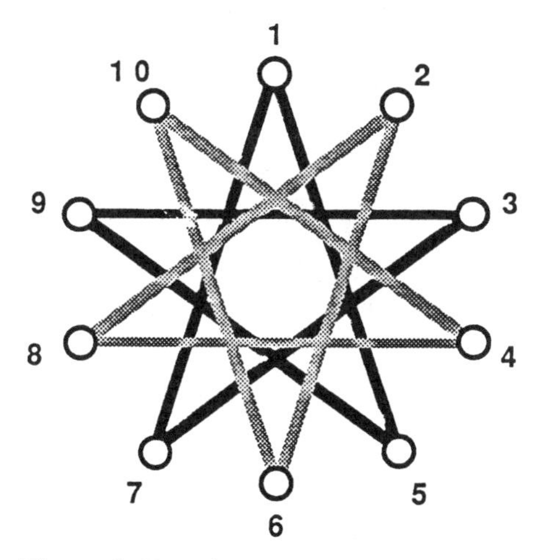

Figure 5. Drawing the Odd Cycle Prisms

Suppose n is odd. Then a drawing of R_{2n} with D' crossings can be obtained in a manner similar to that used for the ladder graphs. (Refer to figure 5.) The cycle C_n can be drawn with the maximum number of crossings by placing vertices labeled 1 through 2n in a circular fashion, then joining vertex 1 to vertex 1 + (n – 1), then this vertex n to n + (n – 1) and so on until vertex 1 is reached again. Notice that another copy of C_n can be drawn beginning at vertex 2 and again adding (n – 1) to each new vertex until vertex 2 is reach. (In what follows we consider the "standard drawing" of the prism to be that illustrated for R_{12} i n figure 6b.) As in the proof to Theorem 4, such a drawing results in any "upper horizontal " edge in the standard drawing parallel to the "lower horizontal" edge in its four cycle. Again, the "vertical" edges in the standard drawing each have endpoints which are diametrically opposite in figure 5. Adding these edges directly across the interior results in a drawing of R_{2n} with the proper number of crossings.

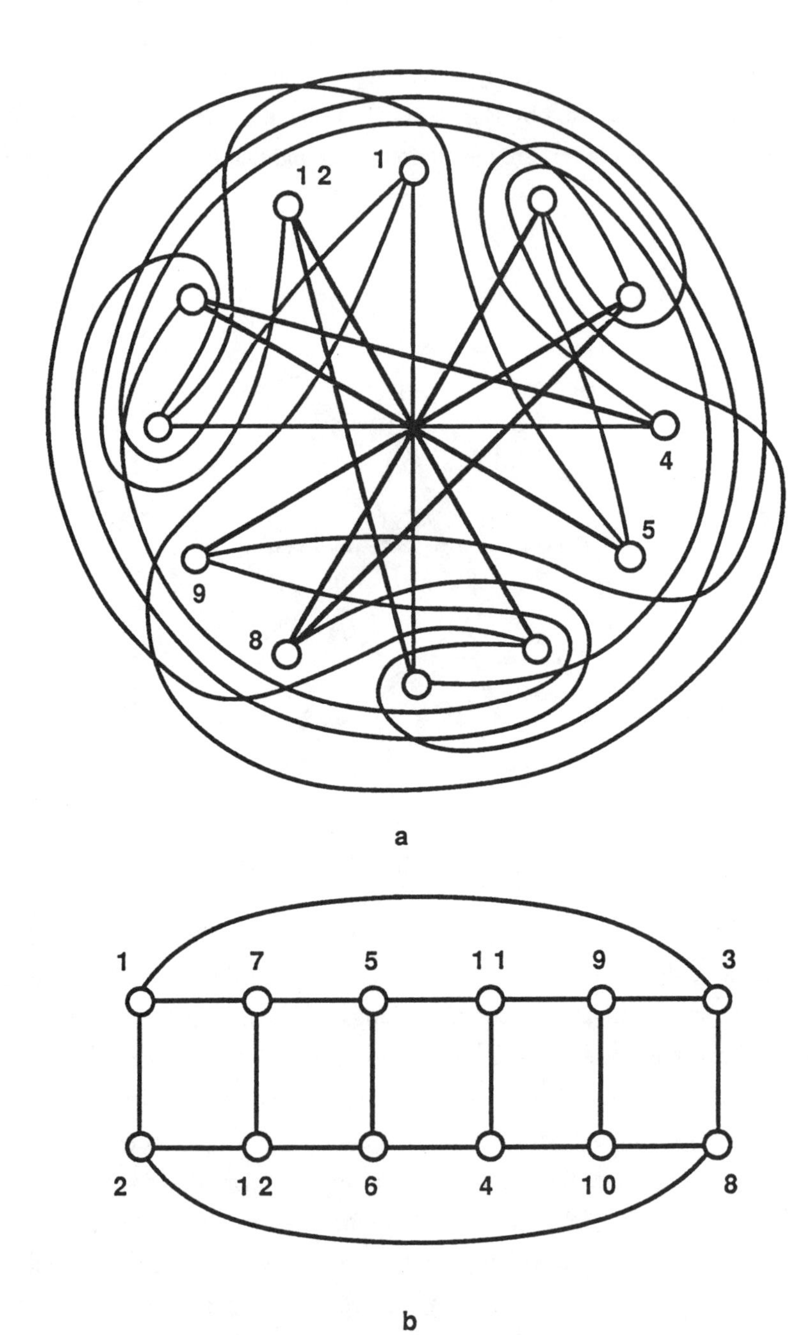

Figure 6. Drawing R_{12}

The drawing for R_{12} with $\mathcal{D}'$ crossings is given in figure 6, where the vertices are assumed to be labeled clockwise. Notice that the drawing induced by the odd numbered vertices and that induced by even number vertices are both essentially the even cycle drawing of figure 2. So we now consider R_{2n} where n i s even and $n \geq 8$.

Henceforth we consider the two n cycles which comprise the vertices of R_{2n} to be labeled 1, 2, . . ., n and 1*, 2*, . . ., n*, where we assume that the edges are i, i*, i (i + 1), and i* (i+1)*, for i = 1, . . ., n and all arithmetic is modulo n. $\mathcal{D}'(R_{16}) = 220$ and a drawing with 220 crossings is obtained from figure 7 by taking the union of the drawings in 7a and 7b, where the vertices in part a are superimposed on the vertices in part b. The proof now proceeds by induction, where the just illustrated case $n = 8$ is the basis and the induction is in increments of two.

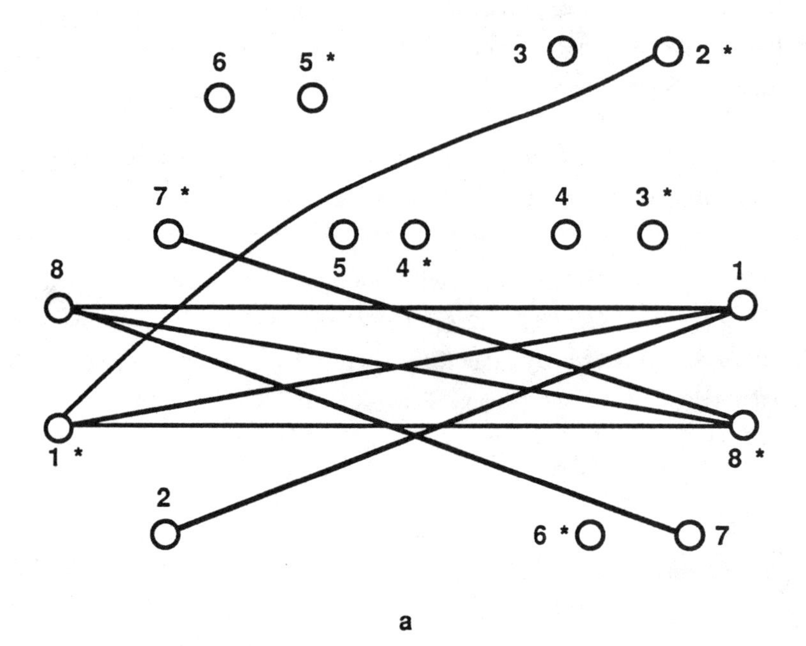

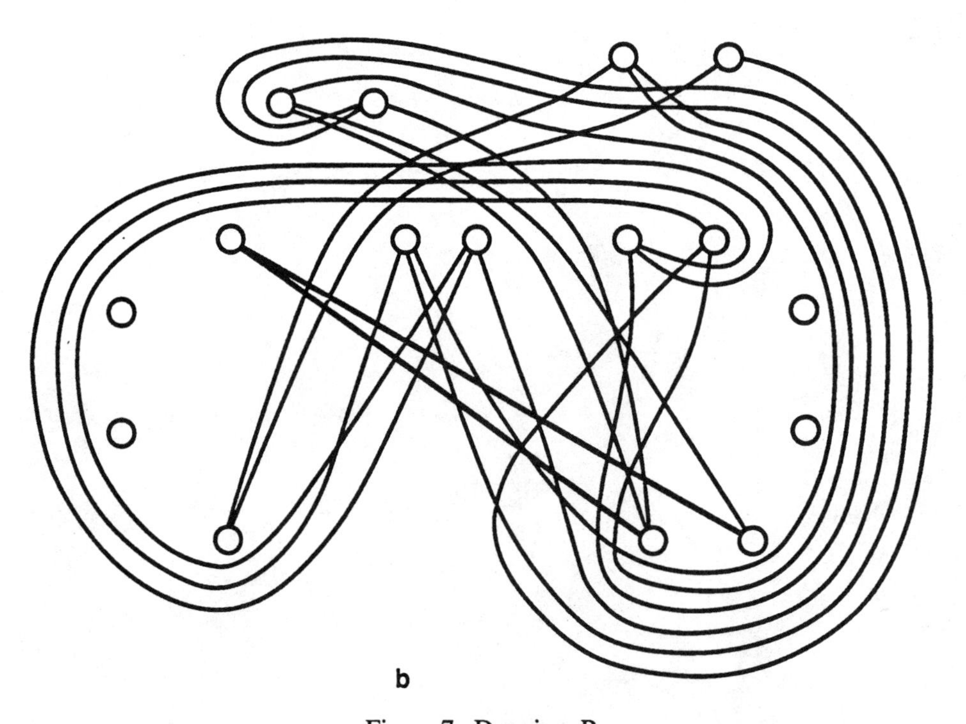

Figure 7. Drawing R_{16}

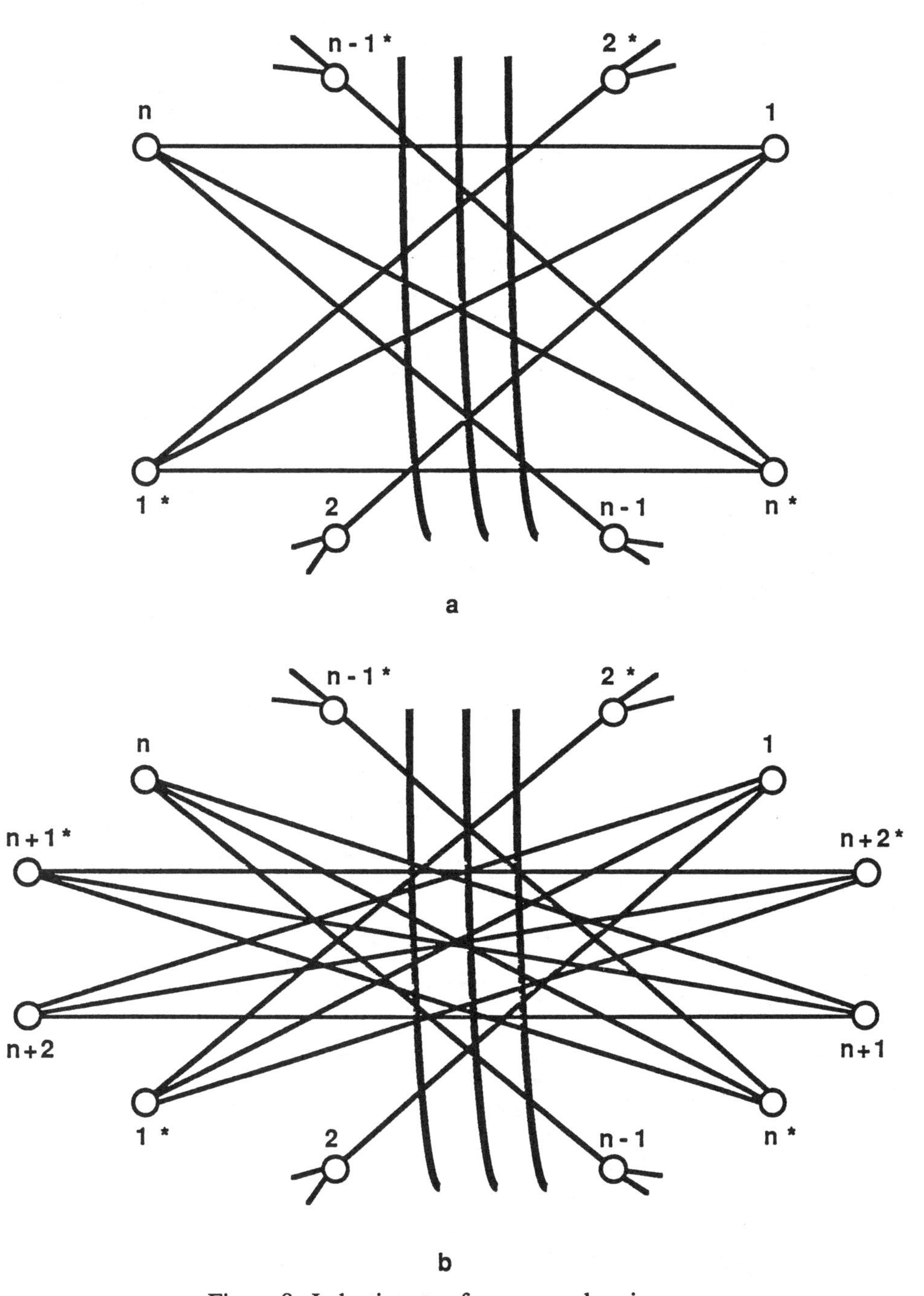

Figure 8. Induction step for even cycle prisms

The induction step is illustrated in figure 8. Figure 8a shows part of a special drawing of R_{2n} where the vertices labeled 1, 2, n – 1, n, 1*, 2*, (n – 1)*, and n* are drawn as indicated. Notice that all the legal crossings for edges among these vertices do occur. The curved lines indicate that all the other edges of the graph pass through the drawing in that way. Notice that the drawing for R_{16} has the appropriate vertices drawn in this manner. Hence, we may assume that R_{2n} has such a drawing.

To obtain the desired drawing of $R_{2(n+1)}$, delete the edges 1 n and 1* n*. Then place the vertices (n + 1), (n + 2), (n + 1)*, and (n + 2)* along with the edges n (n + 1), (n + 1) (n + 2), (n + 2) 1, n* (n + 1)*, (n + 1)* (n + 2)*, (n + 2)* 1* (n+1)(n+1)*, and (n+2)(n+2)* on the drawing as in figure 8b. Notice that all the other edges of $R_{2(n+1)}$ still cross through the structure as desired. Furthermore, in the new drawing, the vertices 1, n, n + 1, n + 2, 1*, n*, (n + 1)*, and (n + 2)* along with their appropriate edges form the structure determined by figure 8a. Hence the induction is complete and the theorem follows.

We finish this paper with a look at a drawing of the three cube. Figure 9 displays a drawing of Q_3, which is isomorphic to the excluded case R_8 from the last theorem, with 34 crossings. Of course $\mathcal{D}'(Q_3) = 36$ and thus we do not know if this drawing gives the maximum crossing number for the graph. In fact, we are at this point uncertains as to whether to conjecture that the cube graphs will meet the upper bound $\mathcal{D}$ or whether they will not. The problem is certainly and interesting one. So we close with the

Question Do the cube graphs meet the upper bound $\mathcal{D}$?

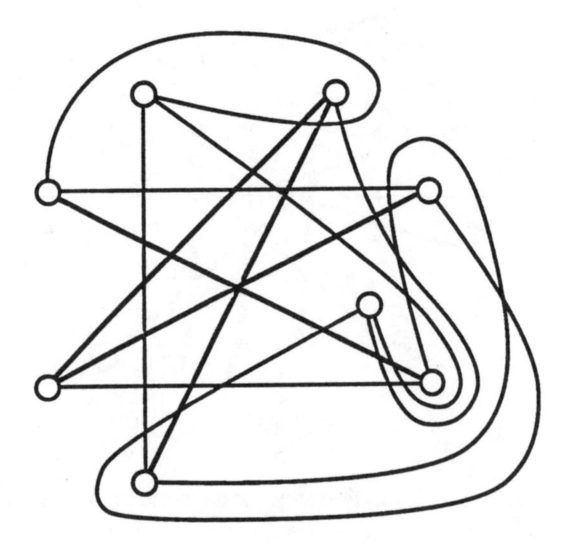

Figure 9. Drawing Q_3

REFERENCES

[1] L. W. Beineke and R.D. Ringeisen, On the Crossing Number of Products of Cycles and Graphs of Order of Four, *J. of Graph Theory* **4** (1980), 145–155.

[2] R. B. Eggleton, *Crossing Number of Graphs*, Ph.D. Dissertation, University of Calgary 1973.

[3] R. B. Eggleton, R. Guy, H. Harborth, and R. D. Ringeisen, Non-Minimal Crossings for the Complete Graphs, in progress.

[4] H. Harborth, Drawings of the Cycle Graph, *Congressus Numerantium*, to appear.

[5] H. Harborth, Parity of Number of Crossings for Complete n-partite Graphs, *Math.Slov.* **26** (1976), 77-95.

[6] D. J. Kleitman, A note on the Parity of Numbers of Crossings of a Graph, *Journal of Combinatorial Theory (B)* **21** (1976), 88-89.

[7] D. J. Kleitman, The Crossing Number of $K_{5,n}$, *J. Combinatorial Theory* **9** (1970), 315-323.

[8] R. D. Ringeisen and L. W. Beineke, The Crossing Number of C_3 X C_n, *J. Combinatorial Theory* **24** (1978), 134-136.

[9] G. Ringel, Extremal Problems in the Theory of Graphs, in *Theory of Graphs and its Applications*, Proceedings of the Symposium in Smolenice, June 1963 (M. Fiedler, Ed.), 85-90.

[10] R. D. Woodall, Thrackles and Deadlock, in *Combinatorial Mathematics and its Applications*, Proceedings of a Conference held at the Mathematical Institute, Oxford, 1969 (D. A. Welsh, ed), 335-348.

Partially Distance–Regular Graphs

David L. Powers
Clarkson University

ABSTRACT

A graph G is distance(s)–regular if, for each vertex u, there is a partition of the vertices into cells $V_0(u)$, $V_1(u)$, ..., $V_t(u)$, such that: $V_k(u) = \{v : dist(v, u) = k\}$ for $k = 0, 1, ..., s$; and for all j, k and u, there are b_{jk} edges from each vertex of $V_j(u)$ to vertices of $V_k(u)$ (independent of u). The eigenvalues and eigenvectors of the adjacency matrix A of G are determined by those of the matrix $B = [b_{jk}]$. If the columns of a matrix Z are orthogonal basis for an eigenspace of A, then the mapping that takes a vertex to the corresponding row of Z allows a geometric interpretation of G. For a distance(1)–regular graph, there are conditions that guarantee that the mapping is 1–1, adjacency in G becomes proximity in euclidean space, and the automorphism group of G is isomorphic to the orthogonal symmetry group of the rows of Z.

1. Introduction and Definitions

The objective of this paper is to show that certain results that are known for distance–regular graphs can be generalized to less regular graphs, including vertex– or arc–transitive graphs. The book of Biggs [1] contains theorems showing that the eigenvalues of a distance–regular graph (i.e., of the adjacency matrix), their multiplicities, and the associated eigenvectors are determined by the eigenvalues and eigenvectors of a related, but usually much smaller, matrix (intersection array) that summarizes some graph properties. Terwilliger [8] and Godsil [4] used these ideas to relate eigenvalue multiplicites to structure of distance–regular graphs. In [7] connections were found between eigenvectors and the automorphism group of a distance–regular graph.

This work was supported by the Office of Naval Research under grant N00014–85–04097.

Throughout, G is a connected graph of diameter d, with vertex set $V = \{1, 2, ..., n\}$ and adjacency matrix A; e is a column matrix of 1's, and e_i is column i of the identity matrix (dimensions determined by context).

A *coloration* or *equitable partition* of G is a partition of the vertex set into cells $V_0, V_1, ..., V_t$, such that each vertex in V_i is adjacent to b_{ij} vertices in V_j. Let X be the *indicator* of the partition: the v, i–entry of X is 1 if vertex v is in V_i or 0 if not. A partition with indicator matrix X is a coloration if and only if $AX = XB$, where $B = [b_{ij}]$ is called the *coloration matrix.*

We begin by defining two parametrized classes of graphs, suggested by the work of Terwilliger [8]. We will refer to these sets of vertices: $D_k(u) = \{v : \text{dist}(u, v) = k\}$, $k = 0, 1, ..., d$.

Definition A graph G is *distance(s)–transitive* if G is vertex–transitive and, for each vertex u, $D_k(u)$ is an orbit of the stabilizer of u for $k = 0, 1, ..., s$. A graph G is *distance(s)–regular* if, for each vertex u there is a coloration with cells $V_0(u), V_1(u), ..., V_t(u)$, such that $V_k(u) = D_k(u)$ for $k = 0, 1, ..., s$, and the coloration matrix is independent of u. We say that such a coloration with $V_0(u) = \{u\}$ is *centered at* u.

It is easy to see that distance(s)–transitive is just vertex–transitive if $s = 0$, arc–transitive if $s = 1$ and distance–transitive if $s = d$; similarly, distance(d)–regular is distance–regular. Distance(s)–regular implies that the leading principal submatrix of B of order $s + 1$ is tridiagonal. A distance(0)–regular graph is degree–regular; we denote its degree by r.

Lemma 1 *If G is distance(s)–transitive, then G is distance(s)–regular.*

Proof Let u be a fixed vertex, and partition the vertices of G into the orbits of the stabilizer of u, $V_0(u), V_1(u), ..., V_t(u)$. We may assume that the cells of the partition are labeled so that $V_k(u) = D_k(u)$ for $k = 0, 1, ..., s$. It is well known that the orbits of any subgroup of the automorphism group of G is a coloration [2, p. 119]; thus we have a partition centered at u that is a coloration. If v is any vertex, let g be an automorphism of G that takes u to v. Then $V_k(v) = g(V_k(u))$ for $k = 0, 1, ..., t$ defines a partition cenetered at v that is a coloration. It is easy to show that the partitions thus constructed are independent of the initial vertex u and of the automorphisms used to transfer the initial coloration to other vertices, and that all have the same coloration matrix. ❑

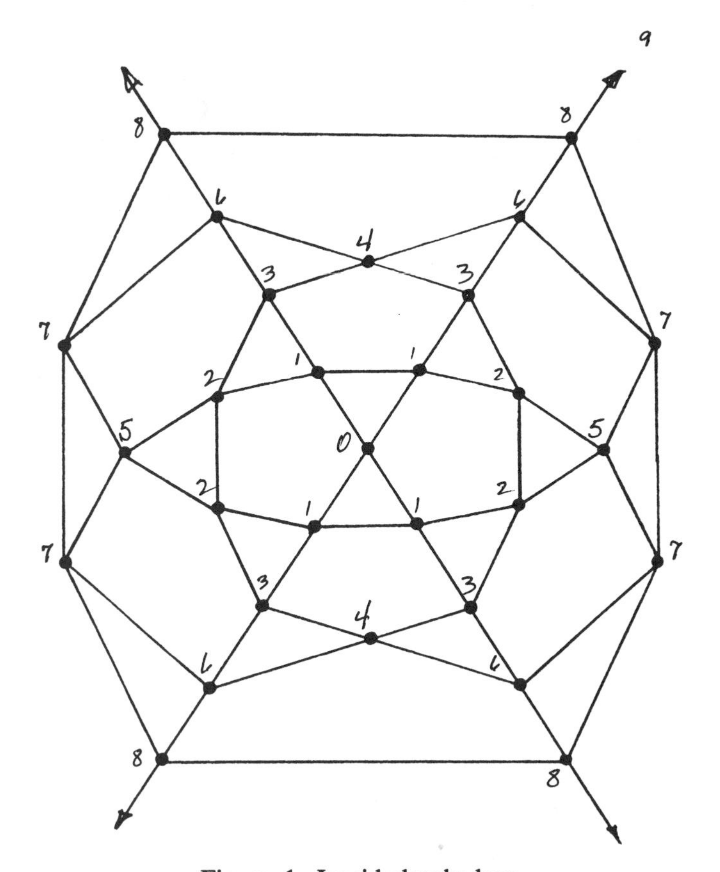

Figure 1. Icosidodecahedron

Example Figure 1 shows the skeleton of an icosidodecahedron, which is an arc–transitive graph. The number next to a vertex shows the orbit it is in.

The connection of the distance(s)–regular graph to linear algebra comes through the coloration. For each vertex u we define the indicator matrix X_u for the partition centered at u. It is easy to see that $X_u^T X_u = \text{diag}\{r_0, r_1, \ldots, r_t\} = R$, where $r_k = |V_k(u)|$. The fact that the partition is a coloration is expressed by the equation $AX_u = X_u B$, where B is the coloration matrix of order $t + 1$. We also define the *relation matrices* $A_0, A_1, \ldots, A_t$ by requiring the v, u–entry of A_k to be 1 if vertex v is in $V_k(u)$ or 0 if not. The following lemma, although fundamental, is trivial to prove.

Lemma 2 *For any vertex u and any integer $k = 0, 1, \ldots, t$, column u of A_k is column k of X_u : $A_k e_u = X_u e_k$.*

Thus the information contained in the indicator matrices is the same as that in the relation matrices, in a different arrangement. Let us also note that in a distance(0)–regular graph, $A_0 = I$, the sum of all the relation matrices is ee^T, and the adjacency matrix A of G is a sum of some relation matrices. (If G is distance(1)–regular then A is the relation matrix A_1.) In general, the relation matrices are not symmetric. For an example, note that in Figure 1, if u is the central vertex and v is in $V_5(u)$, then u is in $V_4(v)$.

2. Matrices

A key result in the theory of distance–regular graphs concerns the algebra of polynomials in A. Here we consider instead the subspace $\mathcal{V}$ of $n \times n$ matrices that is spanned by the relation matrices.

Theorem 1 *Let G be a distance(0)–regular graph with adjacency matrix A, relation matrices $A_0, A_1, ..., A_t$, and coloration matrix B.*

(1) $A_0, A_1, ..., A_t$ form a basis of $\mathcal{V}$.

(2) $\mathcal{V}$ is invariant under left multiplication by A.

(3) For $k = 0, 1, ..., t$,

$$A_k = \sum_{j=0}^{t} b_{jk} A_j.$$

That is, left multiplication by A is represented by B, relative to the basis $A_0, A_1, ..., A_t$.

(4) The space $\mathcal{V}$ contains the algebra of polynomials in A.

Proof (1) Since each of the A's has 0's where another has 1's, they are linearly independent. (2) & (3) For any vertex u and any $k = 0, 1, ..., t$,

$$\begin{aligned} AA_k e_u &= AX_u e_k = X_u B e_k \\ &= X_u \sum_{j=0}^{t} b_{jk} e_j = \sum_{j=0}^{t} b_{jk} A_j e_u \end{aligned}$$

This equality holds for each $u = 1, 2, ..., n$; thus both statments are proved. (4) Since $A_0 = I$, it follows from (3) that every power of A, and hence every polynomial in A, is in $\mathcal{V}$. ❑

Corollary 1.1 *Let G be a vertex–transitive graph. The number of orbits of a vertex stabilizer is greater than or equal to the number of distinct eigenvalues of A.*

Proof The number of distinct eigenvalues of A is the number of independent powers of A, which is not more than the dimension of $\mathcal{V}$, which is $t + 1$, the number of orbits of a vertex stabilizer. ❑

In the case of distance–regular graphs, the number of distinct eigenvalues of A is exactly $1 + t = 1 +$ the diameter of G. For the graph of Figure 1, the number of orbits of a vertex stabilizer is 10, while A has only 7 distinct eigenvalues. The difference can be partly explained by the fact that the powers of A span a space of symmetric matrices, which is a proper subspace of $\mathcal{V}$ if any of the A_k are not symmetric.

Theorem 2 *Let G be a distance(0)–regular graph with adjacency matrix A, relation matrices $A_0, A_1, \ldots, A_t$, and coloration matrix B. Let L be the projector of A associated with an eigenvalue α. Then there is a column $y = [y_0, y_1, \ldots, y_t]^T$, such that $L = y_0A_0 + y_1A_1 + \ldots + y_tA_t$, and $By = \alpha y$.*

Proof The existence of y follows directly from Theorem 1, since L is a polynomial in A. Now

$$Le_u = y_0A_0e_u + \ldots + y_tA_te_u$$
$$= y_0X_ue_0 + \ldots + y_tX_ue_t = X_uy.$$

With two applications of this equality, the equation $AL = \alpha L$ gives

$$ALe_u = AX_uy = X_uBy = \alpha Le_u = \alpha X_uy.$$

Since X_u has independent columns, it follows that $By = \alpha y$. Clearly $y \neq 0$. ❑

Corollary 2.1 *Every eigenvalue of A is an eigenvalue of B.*

This is also a corollary of a theorem of Godsil and McKay [5] that applies to the more general class of walk–regular graphs: those for which each power A^k has uniform diagonal.

It remains to identify the eigenvector y, especially in the case where α is a multiple eigenvalue of B. Suppose that p is the Lagrange interpolating polynomial that takes value 1 at α and value 0 at the other eigenvalues of A. Then the projector for A associated with α is $L = p(A)$. Moreover, since A and B have the same spectrum, $M = p(B)$ is the projector for B associated with α, and $LX_u = X_uM$. We have seen that $Le_u = X_uy$, but also $Le_u = LX_ue_0 = X_uMe_0$. Since X_u has independent columns, we may conclude that $y = Me_0$.

The multiplicity m of the eigenvalue α in the spectrum of A can be found very simply from y, for $m = tr(L) = ny_0$, since $A_0 = I$ is the only relation matrix with nonzero diagonal. However, if y can be found by means other that the equation $y = Me_0$ (e.g. if α is a simple eigenvalue of B), it will be useful to have another relation for the multiplicity. We start from an elementary property of projectors.

$$m = tr(L) = tr(L^2) = tr(L^TL)$$
$$= \sum_{u=1}^{n} \| Le_u \|^2 = n \| Le_u \|^2 = n \| X_uy \|^2 = n \sum_{i=0}^{t} r_iy_i^2.$$

Now let x be defined by $y = y_0 x$. By comparing the expression above for m with $m = ny_0$, we determine that the multiplicity of is [1, p. 143]

$$m = n / \sum_{i=0}^{t} r_i x_i^2.$$

and $y_0 = m/n$.

For future convenience, define x to be a *leading eigenvector* of B if x is any positive multiple of Me_0, where M is the projector of B associated with the same eigenvalue. The x above is a leading eigenvector with $x_0 = 1$.

3. Geometry

Now we investigate properties of eigenvectors of distance(0)–regular graph G. If α is an eigenvalue of A with multiplicity m, a *complete eigenmatrix* assocaited with α is an $n \times m$ matrix Z satisfying $AZ = \alpha Z$ and $Z^T Z = I$. Row u of Z is denoted by $w_u = e_u{}^T Z$. Thus we have a mapping from the vertices of the graph to a set of points in m–dimensional space. The algebraic investigation of this mapping depends for the most part on viewing the projector L of A as the Gram matrix of the rows of $Z : L = ZZ^T = [(w_u, w_v)]$, where $(w_u, w_v) = w_u w_v{}^T$.

Lemma 3 *Let G be a distance(0)–regular graph, Z a complete eigenmatrix associated with eigenvalue α of A, x a leading eigenvector of B associated with α.*

(1) All rows of Z have the same euclidean length, and there is a positive constant b such that $(w_u, w_u) = bx_0$;

(2) the inner product of two rows of Z is given by $(w_u, w_v) = bx_k$ if vertex u is in $V_k(v)$.

Proof By Theorem 2, the projector has the representation

$$L = b(x_0 A_0 + x_1 A_1 + \ldots + x_t A_t) \quad (1)$$

where x is a leading eigenvector and b is a positive constant. For any u, the product $(w_u, w_u) = (e_u, Le_u) = bx_0$, since $A_0 = I$ is the only relation matrix with a nonzero diagonal. (In fact, it is clear that $bx_0 = m/n$.) Similarly, (w_u, w_v) is the u, v–entry of L and hence equals bx_k if A_k is the relation matrix that has a 1 in the u, v–position. ❑

Of course, it is desirable that the mapping from vertices of G to the points whose coordinates are the rows of Z be one–to–one; that is, that the rows of Z be distinct. This will be true if a simple condition is met by a leading eigenvector of B.

Theorem 3 *Let G be a distance(0)–regular graph, Z a complete eigenmatrix associated with an eigenvalue α of A, x a leading eigenvector of B associated with α. The rows of Z are distinct if and only if $x_0 > x_k$ for all $k > 0$.*

Proof Since the rows of Z all have the same euclidean length, rows u and v ($v \neq u$) are distinct if and only if their inner product, (w_u, w_v), is strictly less than (w_u, w_u). That is, if and only if the u, u–entry of L is larger than any other entry in row u. Form Eq. (1) this is equivalent to the condition that $x_0 > x_k$ for all $k > 0$ in any leading eigenvector x. ❑

To see that the theorem is not trivial, consider the arc–transitive graph G with 2m vertices ($m \geq 5$) in which vertex u is adjacent to vertices $u + 1, u - 1, u + m - 1, u - m + 1$ (addition modulo 2m). If Z is a complete eigenmatrix associated with any eigenvalue $\alpha \neq 0$, then rows u and $u + m$ of Z are identical.

From here on, we consider distance(1)–regular graphs. In order to avoid tedious repetitions, we will assume that the following notational hypotheses apply: G is a connected, distance(1)–regular graph with adjacency matrix A, α is an eigenvalue of A with multiplicity m, Z is a complete eigenmatrix associated with α, $L = ZZ^T$, and x is a leading eigenvector of B associated with α. Recall that the relation matrix A_1 is the adjacency matrix A of G. The first row of the coloration matrix B is

$$[0 \quad r \quad 0 \ldots 0]$$

Hence, it is easy to see that in a leading eignevector x, $x_1/x_0 = \alpha/r$.

We will refer to the group of permutations that commute with a given matrix, defined by

$$\text{perm}(M) = \{P : P \text{ a permutation and } PM = MP\}.$$

Theorem 4 *Let G be distance(1)–regular. If $x_0 > x_k$ for all $k \neq 0$ and $x_1 \neq x_k$ for all $k \neq 1$, then the group*

$$orth(Z) = \{S : ZS = PZ \text{ for some permutation } P\}$$

is isomorphic to the group of automorphisms of G.

Proof It is easy to show that orth(Z) is a group of orthogonal matrices, and that the mapping of perm(ZZ^T) onto orth(Z) given by $R = Z^TPZ$ is a group homomorphism with kernel $\{P : PZ = Z\}$ [3], [7]. However, since the rows of Z are distinct by Theorem 3, the kernel is trivial. Thus, the isomorphism of orth(Z) and perm(L) is established.

Next consider the representation of L in the proof of Lemma 3. Since $x_1 \neq x_k$ for all $k \neq 1$, the number bx_1 appears as an entry of L exactly where 1 appears as an entry of $A_1 = A$. Thus, a permutation matrix P that commutes with L must also commute with A. On the other hand, L is a polynomial in A, so any matrix that commutes with A commutes with L. Therefore, perm(L) = perm(A), and the latter is well known to be isomorphic to the group of automorphisms of G. ❑

Theorem 5 *Let G be distance(1)–regular. If $x_0 > x_1$ and $x_1 > x_k$ for $k > 1$, then u and v are adjacent vertices of G if and only if w_u and w_v are nearest neighbors among the rows of Z.*

Proof Note that the rows of Z are distinct by Theorem 3, since $x_0 > x_k$ for $k > 0$. Next, $x_1 > x_k$ for all $k > 1$ if and only if the inner product (w_u, w_v) for v adjacent to u (v in $V_1(u)$) is greater than the inner product (w_u, w_v) for v not adjacent to and not equal to u (v in $V_k(u)$, $k > 1$). ❑

Note that under the hypotheses of this theorem, the mapping from vertices to points of m–space is one–to–one, preserves the automorphism group, and carries adjacency into proximity. The conditions of the theorem are always fulfilled for the second eigenvalue of a distance–regular graph [4], [7].

Godsil [3] introduced the idea of the polytope associated with a graph eigenvalue: if Z is a complete eigenmatrix associated with α, then the convex hull of the points whose coordinates are the rows of Z is the polytope $C(\alpha)$. If the hypotheses of Theorem 3 are satisfied, $C(\alpha)$ has n extreme points; the group orth(Z) referred to in Theorem 4 is the symmetry group of $C(\alpha)$, including reflections. This polytope inherits some properties of the graph.

Theorem 6 *Let G be distance(1)–regular graph, let $0 \leq \alpha < r$, and let $x_0 > x_1$ and $x_1 > x_k$ for $k > 1$. If U is a set of vertices of G that induce a complete graph K_q in G, then the corresponding points, w_u, are extreme points of a face F of $C(\alpha)$, and F is a simplex.*

Proof Let K be the $q \times q$ submatrix of L formed from entries whose indices are in U. (See Eq. 1.) Since all the vertices of U are mutually adjacent, $K = b(x_1 ee^T + (x_0 - x_1)I)$. The eigenvalues of K are $b(x_0 - x_1) = bx_0(r - \alpha)$ and $b(x_0 + (q-1)x_1) = bx_0(r + (q-1)\alpha)$, both positive. Thus the rank of K is q, and the rows of Z whose indices are in U are independent.

Next, let g be the sum of the rows w_u for which u is in U. then $(g, w_v) = b(x_0 + (q-1)x_1) = \gamma$, say, if v is in U, but $(g, w_v) \leq qbx_1 < \gamma$ if not. Then the hyperplane $H = \{w : (g, w) = \gamma\}$ supports $C(\alpha)$ and contains only those extreme points of $C(\alpha)$ corresponding to vertices in U. These must be extreme points of a $(q-1)$–face F of $C(\alpha)$, and, since they are equidistant, must form a simplex. ❑

Corollary 6.1 *If u and v are adjacent vertices in G, then w_u and w_v are adjacent in the skeleton of $C(\alpha)$.*

Proof Vertices u and v induce a K_2. By the theorem, the corresponding points form a 1-face of $C(\alpha)$. ❑

Corollary 6.2 *If q is the clique number of G, then the multiplicity of α is at least q.*

Corollary 6.3 *If the multiplicity of α in the spectrum of A is 3, then G is planar. If the multiplicity is 2, then G is an n-cycle.*

Proof By Corollary 6.1, a copy of G is an edge subgraph of the skeleton of $C(\alpha)$. If the multiplicity of α is 3, then $C(\alpha)$ is a polyhedron. If the multiplicity is 2, the same argument applies, but $C(\alpha)$ must be a regular n-gon. The copy of G that is an edge subgraph of skel($C(\alpha)$) is connected and vertex-transitive, so it must be an n-cycle. ❑

In [3] Godsil mentions that in an edge- and vertex-transitive graph G, the value of (w_u, w_v) must be the same for all pairs u, v of adjacent vertices. Thus, in such a graph, assume that the cells of the coloration centered at u are numbered so that $V_1(u), \ldots, V_q(u)$ contain all the vertices adjacent to u. Let ξ be the common value of $x_1, \ldots, x_q$ in a leading eignevector of B. The results if Theorems 4, 5, and 6 may be extended to vertex- and edge-transitive graphs by substituting the following hypotheses. (4') Let $x_0 > \xi$ and $\xi \neq x_k$ for $k > q$. (5') and (6') Let $x_0 > \xi$ and $\xi > x_k$ for $k > q$. D.F. Holt [6] gives an example of a vertex- and edge-transitive graph that is not distance(1)-regular, and for which these special hypotheses are satisfied.

REFERENCES

[1] N. Biggs, *Algebraic Graph Theory*. Cambridge University Press, London (1974).

[2] D.M. Cvetković, M. Doob and H. Sachs, *Spectra of Graphs*. VEB, Berlin/Academic Press, New York (1980).

[3] C.D. Godsil, Graphs, groups and polytopes. In *Combinatorial Mathematics VI (Canberra 1977),* D.A. Holton and J. Seberry, eds. Springer, Berlin/New York (1978), 157 – 164.

[4] C.D. Godsil, Bounding the diameter of distance–regular graphs. Submitted (1987).

[5] C. Godsil and B.D. McKay, Feasibility conditions for the existence of walk–regular graphs, Linear Algebra and Applications 30 (1980), 51 – 61.

[6] D.F. Holt, A graph which is edge transitive but not arc transitive. Journal of Graph Theory, 5 (1981) 201 – 204.

[7] D.L. Powers, Eignevectors of distance–regular graphs. SIAM Journal on Matrix Analysis and Appl., 9 (1988), 399 – 407.

[8] P. Terwilliger, Eigenvalue multiplicities of highly symmetric graphs, Discrete Math., 41 (1982), 295 – 302.

On Enumeration of Complete Matchings in Hexagonal Lattices

M. Randić
Drake University

ABSTRACT

We consider a class of polyhex graphs which are lattices, i.e., graph which when suitably oriented allow identifiaction of the "master" and the "slave" vertices. An elegant algorithm for enumeration of the perfect matchings in such lattices is outlined. The algorithm combines elements of an earlier algorithm valid for chain-like fused polyhex graphs with an addition rule which counts "interior" paths from the "master" to the slave "vertex".

1. Introduction

An important part of organic chemistry is concerned with the properties of the so called polycyclic conjugated hydrocarbons, the essential structural element of which is a six-membered ring of carbon atoms. If we ignore the peripheral hydrogen atoms such molecules can be viewed as fused polyhex units. Of interest in mathematical chemistry is to see to what extent the properties of these compounds are reflected in the combinatorial and the topological variations associated with different mode of fusion of polyhex units.

The simplest polyhex corresponds to molecule C_6H_6, benzene, which is customarily taken as a representative compound of the so called aromatic hydrocarbons to which other compounds are compared. Benzene is very stable, the "extra" stability is attributed to the fact that for this molecule one can write two so called valence structures:

August Kekulé in 1876 proposed the above valence structures of benzene. He thus indicated that there are molecules for which theoretical models require two or more molecular structural formula. Such structural formulas are refered to in chemical litereature as Kekulé valence structures. It soon became apparent that molecules showing a greater "extra" stability are those having a larger number of Kekulé valence forms. Hence, it has been of a considerable interest in theoretical organic chemistry to find for a given molecule the number of possible valence structures, or in the language of mathematics, to find the number of perfect matchings. As has been customary in chemistry for long time, we will refer to the number of perfect matchings as K, to honor August Kekulé.

2. Catacondensed polyhexes – Algorithm of Gordon and Davison

If each polyhex ring in a structure, with the exception of the terminal rings, has at most two neighbors we obtain so called cata-condensed system. In Fig. 1 we illustrate one such case. Gordon and Davison (1) already in 1952 arrived at an unusually elegant algorithm for enumeration of perfect matchings in such polyhex cata-condensed systems. The algorithm is illustrated in the upper part of Fig. 1, where numbers are inserted in hexagons successively, from the left to the right. The numbers represent the values of K for the structures obtained by truncation at any particular ring. According to Gordon and Davison one inscribes 2 in the first ring, which signifies the fact that benzene (a single hexagon system) has two Kekulé structures, or two perfect matchings. For each linearly added ring one increases the number by one, thus obtaining 3, 4. However, at the site of "change of direction" of fusion, instead of increasing the number by "one" we add the number assigned to the previous ring. Hence 3 + 4 gives 7. One continues adding this "previous" number (3 in our example) until one again comes to a ring with a "change" of direction of fusion. Hence we have, 7, 10, 13. The process of assignment of numbers to rings continues until all rings are assigned. In fully linear systems the number of perfect matching is therefore simply given by the number of fused rings + 1. If at each ring there is a "change of direction" of fusion we obtain as result Fibonacci numbers 2, 3, 5, 8, 13, 21, ... Recent revived interest in simple and elegant algorithms for finding K in general polyhex systems has been for the most part stimulated by the pioneering work of Gordon and Davison.

Although the algorithm of Gordon and Davison is remarkably simple, it has been recently pointed out that it could be even further simplified (2). As illustrated in the lower part of Fig. 1, instead of constructing successive terms from the left to the right we selected a single "kink" ring in the center of the structure, and applied the algorithm of Gordon and Davison to the two disjoint fragments obtained by erasing the selected ring (shown as shaded). The K for the molecule is obtained by multiplying the results for the two fragments to which one adds a similar result obtained from the even smaller fragments derived by erasing also all "linearly" fused rings to the ring previously selected.

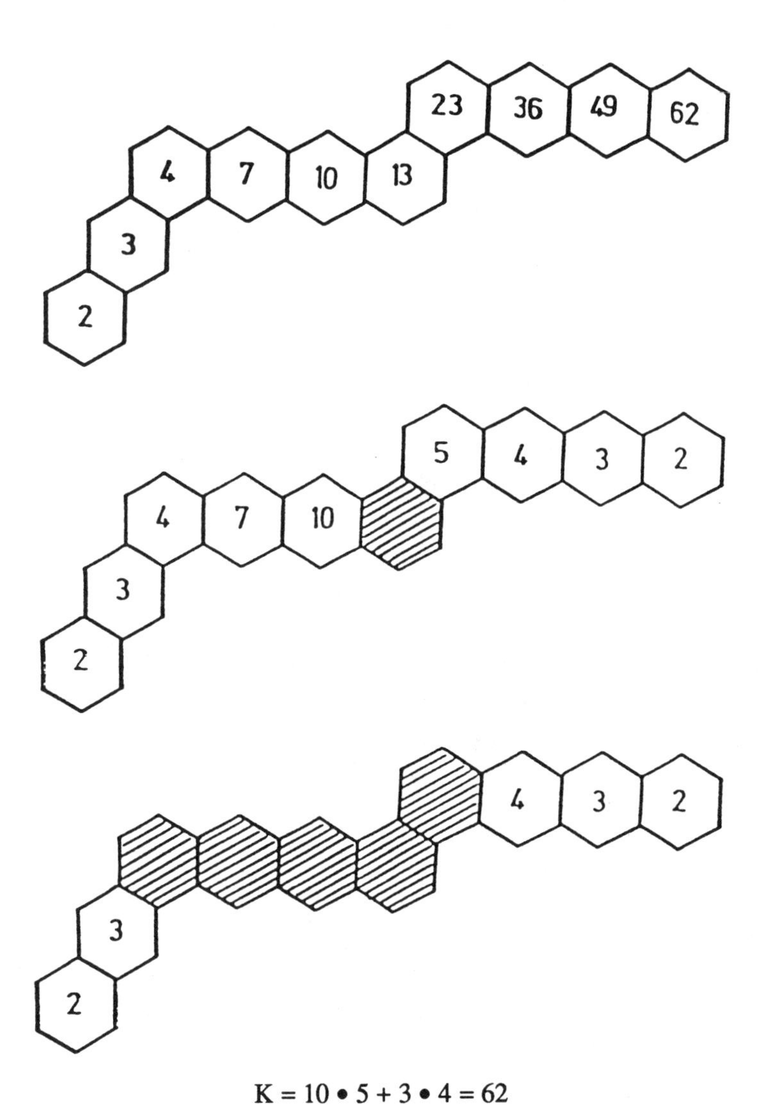

$$K = 10 \bullet 5 + 3 \bullet 4 = 62$$

Figure 1

As discussed by Gordon and Davison their algorithm can be extended to branched cata-condensed systems. Their particular generalization, however, involves some non-unique "surgery" at the branching site in the structure and thus apparently much of its original elegance is lost. However, this is not the case with extension of the above outlined "erasure" procedure. As illustrated in Fig. 2, one obtains K in branched systems immediately by applying the same "erasure" algorithm to the ring with three neighbors. The only difference is that now we have to combine the results from three disjoint fragments, rather than two, as was the case before.

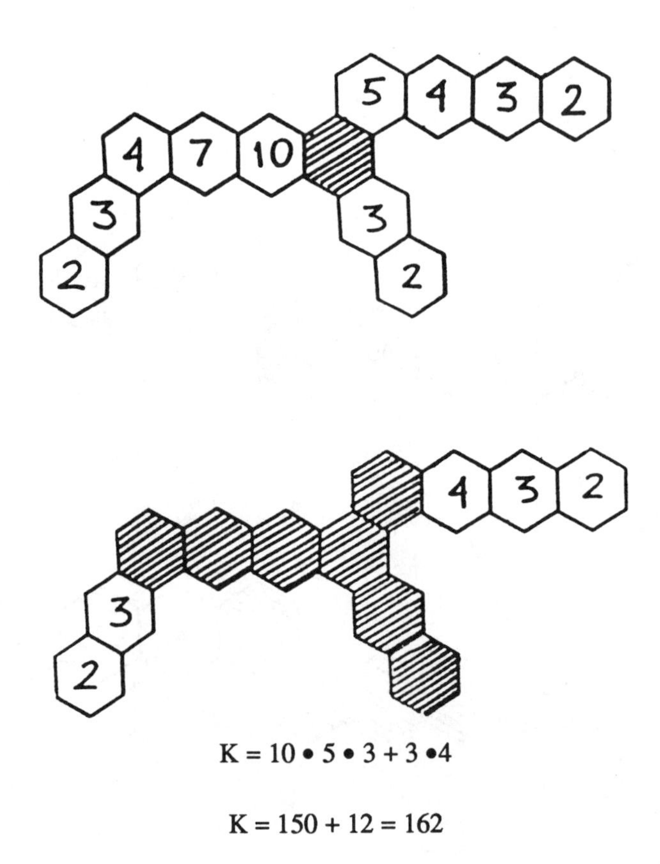

Figure 2

3. Pericondensed Systems

There have been some attempts to extend the Gordon and Davison type of approach to pericondensed benzenoid systems, that is, to polyhex structures obtain by unrestricted fusion of hexagonal rings. There are no analogous simple algorithms that allow one to determine K in these more general cases. Some progress was reported for construction of polyhex systems for which K = 0 (3), but these cases, in view of a postulate of E. Clar, a doyen of chemistry of benzenoid hydrocarbons, that such systems cannot exist (4), are of limited chemical relevance. More pertinent is the recently outlined algorithm which allows one to construct all "augmented" polyhex systems of prescribed K, if the number of Kekulé structures of the smaller components and their fragments is known (5).

Gutman and Cyvin in several papers (6) considered selected subclasses of polyhex molecules for which they were apparently able to extend Gordon and Davison type algorithm. Their work has shown that simple algorithms are possible in some special cases of pericondensed polyhexes. It is unclear, however, from their work if there exist additional structures for which application of a simple algorithm will work, and how one finds such structures.

4. Polyhex Lattices Algorithm for K

We would like to report here that in the case of polyhex (benzenoid) lattices we can derive K, the number of Kekulé structures or perfect matchings, by a very simple and elegant manner, which can be viewed as an extension of the Gordon and Davison algorithm. Polihex lattice is a lattice composed of fused hexagonal rings, i.e., the vertices of fused polyhex rings form a lattice. Such graphs can be embedded in a plane and oriented so that one vertex (ring) is at the top (refered to as "master" vertex) and one vertex (ring) is at the bottom (refered to as "slave" vertex). The shape of the lattice is not important, and it may be "triangular", convex or not. In such lattices we can differentiate the "peripheral" rings and "inside" rings. However, for the algorithm to be outlined it is more useful to differentiate between rings with two predecessor rings and the rings with only one predecessor. The former class allows immediately the assignment of the partial K-values by simply adding the known predecessor values. The assignment of the value for the other class follows Gordon and Davison algorithm. The algorithm for finding the number of perfect matchngs is as follows:

Start with the "slave ring" at the bottom and assign to it value 2 (same as in benzene and in the initialization of Gordon and Davison algorithm). Consider now separately each of the two sides of peripheral rings and assign to them the corresponding numbers as dictated by the Gordon and Davison procedure. These are the rings with one "predecessor". Consider the assignment of partial K–values for rings which have both lower rings assignment completed, by adding the values of the two predecessor rings. By exhausting the assignment in this step one continue to apply Gordon and Davison algorithm to additional peripheral rings for which this was not possible previously. In

Fig. 3 we have illustrated the process, step by step, on a relatively large polyhex lattice. Observe that at each step we enlarge the sublattice of labeled rings of the lattice initially considered.

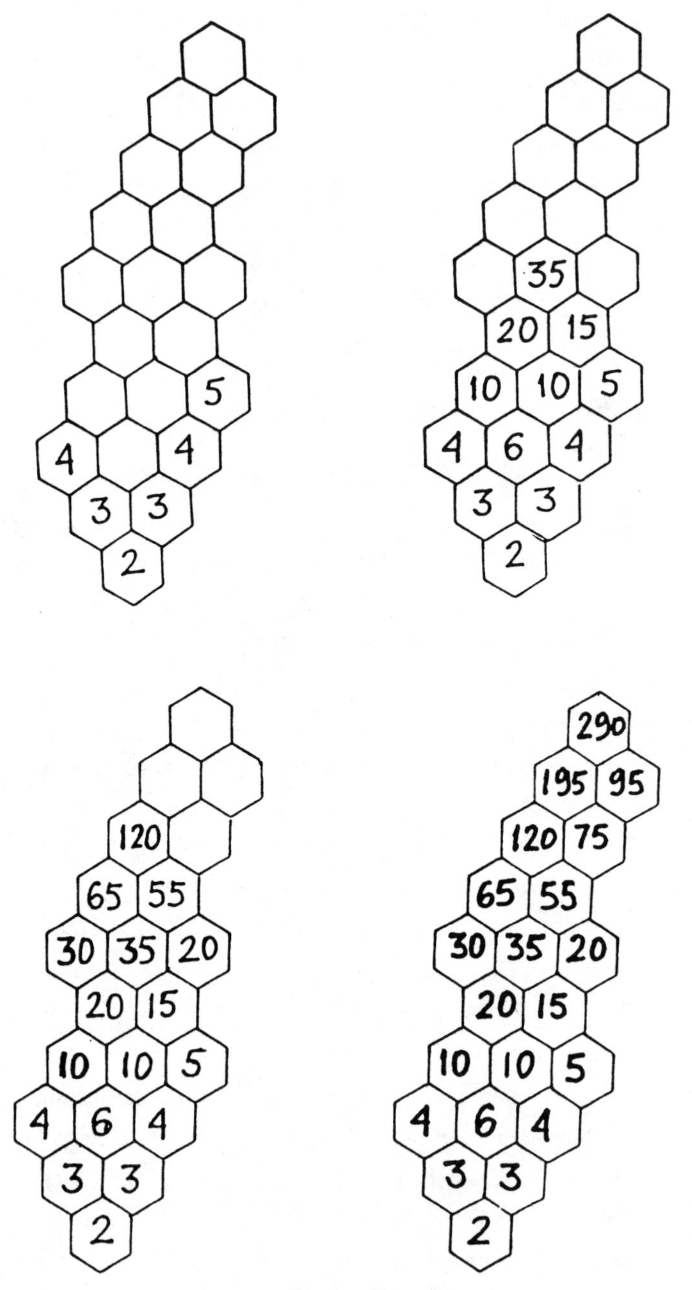

Figure 3

The basis for the outlined algorithm follows from the fact that counting paths from the master to the slave vertex, as has been discussed in the literature (7), is equivalent to counting perfect matching. The "seed" of this observation is already visible in work of Gordon and Davison, but use of count of paths has been fully recognized by a number of more recent investigators.

In concluding the brief outline of the algorithm we would like to point out that the assigned partial K-values hold for any proper sublattice of the original lattice as is illustrated in Fig. 4. By "proper" we mean any sublattice which defines a "catchment" domain for the "slave" ring.

Figure 4

REFERENCES

[1] M. Gordon and W. H. T. Davison, Theory of Resonance Topology of Fully Aromatic Hydrocarbons. I., *J. Chem. Phys.* 20 (1952), 428-435.

[2] M. Randić, On the Enumeration of Kekulé Structures in Cata-Condensed Benzenoids, *Chem. Phys. Lett.* (submitted)

[3] H. Hosoya, How to Design Non-Kekulé Polyhex Graphs ?, *Croat. Chem. Acta*, 59 (1986), 583-590.

[4] E. Clar, *Aromatic Sextet*, J. Wiley & Sons, New York (1972)

[5] M. Randić, On Construction of Benzenoids Having a Same Number of Kekulé Valence Structures, in *Valence bond Theory and Chemical Structure* (D. J. Klein and N. Trinajstić, editors), Elsevier Sci. Publ., Amsterdam (in press)

[6] S. J. Cyvin and I. Gutman, Topological Properties of Benzenoid Systems. Part XXXVI. Algortihm for the Number of Kekulé Structures in Some Peri-Condensed Benzenoids, *MATCH* (Comm. Math. Chem.), 19 (1986), 229 – 242.

S.J. Cyvin, Extended Application of and Algorithm for the Number of Kekulé Structures, *MATCH* (Comm. Math. Chem.), 22 (1987), 101-103.

[7] N. Trinajstić and P. Krivka, On the Reduced Graph Model, in *Matheamtics and Computational Concepts in Chemistry*, (N. Trinajstić, ed.)

W. He and W. He, On Kekulé Structure and P-V Path Method, in *Graph Theory and Topology in Chemistry*, (R.B. King and D.H. Rouvray, eds.), Elsevier Sci. Publ., Amsterdam (1987), 476-483.

Drake University, Des Moines, Iowa 50311 and Ames Laboratory, DOE, Iowa State University, Ames, Iowa 50011

Chromatic Roots of Families of Graphs

Ronald C. Read
University of Waterloo

Gordon F. Royle
University of Waterloo

ABSTRACT

It has been known for a long time that when the zeros of the chromatic polynomials of certain families of graphs are plotted in the complex plane, the points display a clearly non–random pattern, seeming to lie on or near to certain curves in the plane.

A recent theorem of Beraha, Kahane and Weiss enables some of this behaviour to be explained. Specifically, the limit points of the chromatic roots for recursive families of graphs can now be recognized, and the curves on which they lie can be identified. We present a number of examples for which the appropriate computations have been carried out.

This is not the whole story, however. In addition to the roots lying near the limit points of zeros, some recursive families of polynomials have isolated roots, which themselves exhibit regular patterns, as yet unexplained. Moreover, the chromatic roots of other, non–recursive, families of graphs, such as cubic graphs, show further kinds of regularity which also call for explanation.

1. Introduction

We shall assume that the reader is familiar with the basic facts and theorems about chromatic polynomials as given in, for example, [6]. In the usual definition of a chromatic polynomial as the number of ways of colouring a graph in a given number of colours, the variable is, of course, an integer; once the polynomial is defined however, the variable can be regarded as a complex number, and this we shall do in the present paper. For this reason we shall use the letter z in place of the traditional λ. We denote the chromatic polynomial of G by $P_G(z)$.

* The research for this paper was supported by grant A8142 from the Natural Sciences and Engineering Research Council, Canada.

The chromatic roots of a graph G are the roots of the polynomial equation

$$P_G(z) = 0.$$

Consider a family of graphs defined by some common property. If the chromatic roots of the graphs in this family are plotted in the complex plane it is usually found that the points representing these roots display a markedly nonrandom pattern. The problem that we address in this paper is that of explaining some of these patterns.

2. Recursive families of graphs

A recursive family of graphs is a sequence of graphs – one for each integer n – such that the Tutte polynomials satisfy a linear recurrence. See [2] or [10] for further details. It follows that the chromatic polynomials for such a family also satisfy a linear recurrence, which will be of the form

$$P_n(z) = \sum_{r=1}^{k} a_r(z)\, P_{n-r}(z).$$

In general, such a recurrence will have a solution of the form

$$P_n(z) = \sum_{i=1}^{k} \alpha_i(z)\, [\lambda_i(z)]^n$$

where the functions $\lambda_i(z)$ are the roots of the auxiliary equation

$$t^k - a_1(z)\, t^{k-1} - a_2(z)\, t^{k-2} - \dots - a_k(z) = 0.$$

As an example consider the family of graphs called ladders or prisms as illustrated in Figure 1. Let $L_n(z)$ be the chromatic polynomial of the ladder on 2n vertices. It was shown in [2] that $L_n(z)$ satisfies a linear recurrence, the solution to which is

$$L_n(z) = (z^2 - 3z + 3)^n + (z-1)(3-z)^n + (z-1)(1-z)^n + (z^2 - 3z + 1).$$

Thus these chromatic polynomials can be easily found and their complex roots determined, provided n is not too large. If the complex roots are plotted in the complex plane we get a distribution something like that of Figure 2, which shows the points for $n \le 15$. It will be seen that the points appear to lie either on a line perpendicular to the real axis or on a C–shaped curve. Biggs, Damerell and Sands [2] give a similar plot, but at that time there was no proof that the roots did, in fact, approximate to the line or the curve, and no indication of what the C–shaped curve might be.

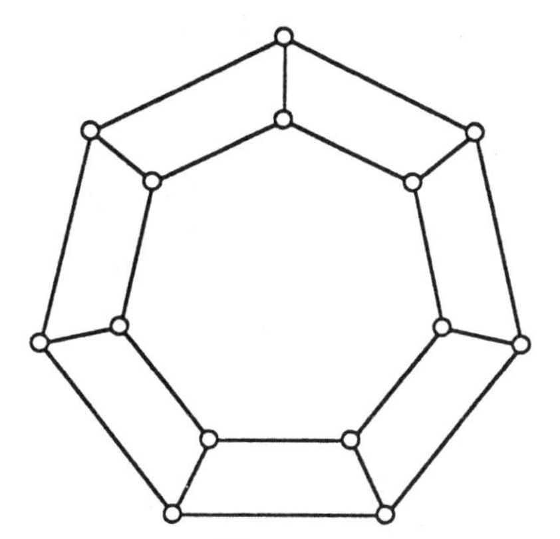

Figure 1

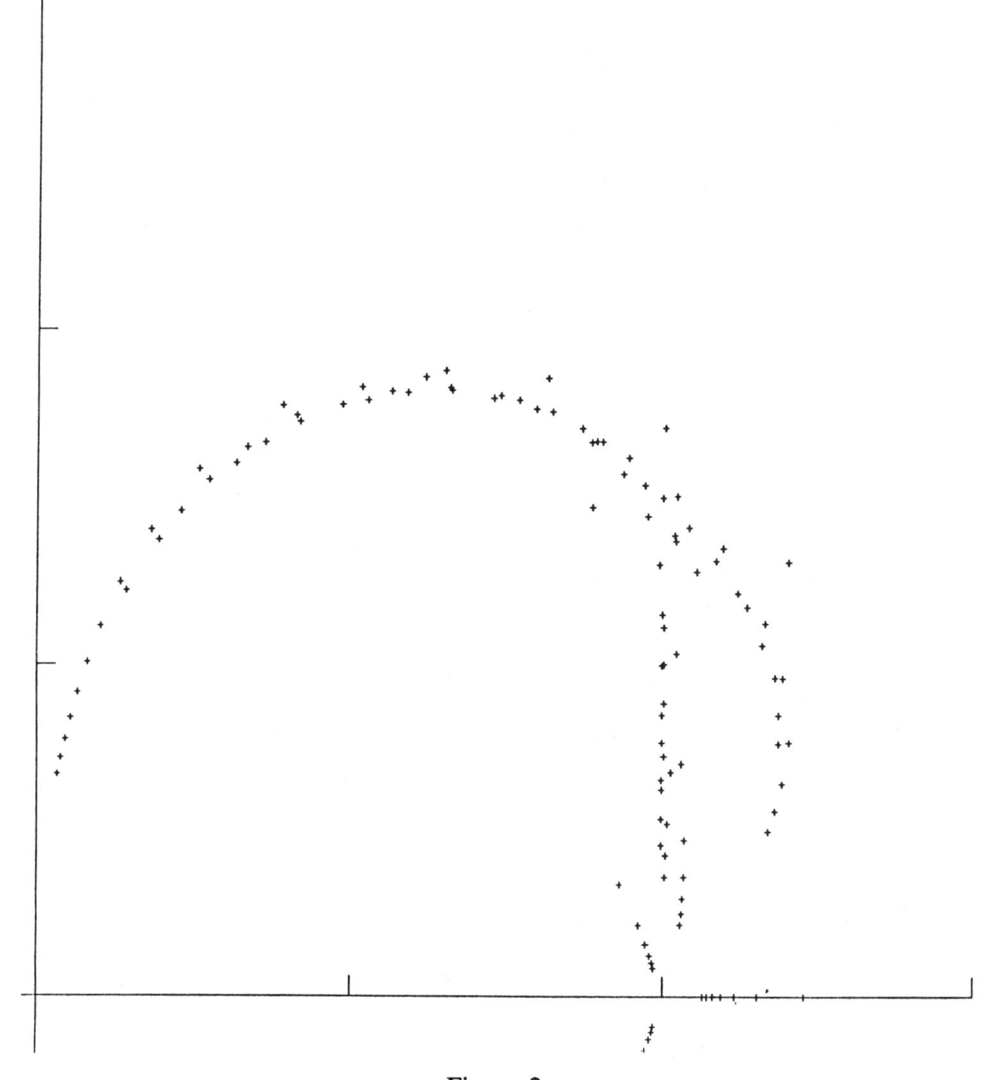

Figure 2

3. Limit points of zeros

Much of the behaviour of the chromatic roots for recursive families of graphs can be explained by a recent result of Beraha, Kahane and Weiss. Suppose we have a family of polynomials $\{P_n(z)\}$ given by

$$P_n(z) = \sum_{i=1}^{k} \alpha_i(z)\,[\lambda_i(z)]^n .$$

A complex number ζ is defined to be a limit point of zeros of this family if there exists a sequence $\{z_i\}$ tending to ζ, such that for each i, z_i is a root of $P_i(z)$. We then have the following theorem.

Theorem *(Beraha – Kahane – Weiss 1980). Under the non–degeneracy conditions that $\{P_n(z)\}$ does not satisfy a lower order recurrence, and $\lambda_i(z)/\lambda_j(z)$ is not identically a constant of unit modulus for any $i \neq j$, the complex number ζ is a limit point of zeros of $\{P_n(z)\}$ if and only if, at $z = \zeta$, one of the following two conditions holds:*

1. *One of the roots $\lambda_i(z)$ dominates all the other roots (that is, has larger modulus), and the corresponding $\alpha_i(z) = 0$.*

or

2. *Two or more of the roots $\lambda_i(z)$ are of equal modulus, and dominate the others.*

For the proof of this theorem, and more information, see [1].

Note that 1. will give isolated points, and that these points will depend on the functions $\alpha_i(z)$, while 2. will give curves of points, and that these curves will be independent of the coefficients $\alpha_i(z)$. Thus these curves will depend only on the recurrence for the polynomials $P_n(z)$ and not on the initial conditions; this means that they depend only on the coefficients in the auxiliary equation. In this paper we shall be interested only in the curves of limit points of zeros and therefore confine our attentions to condition 2.

4. Applications to ladders

For the case of ladders we have the following α_i and λ_i.

$$\alpha_1(z) = 1 \qquad \lambda_1(z) = z^2 - 3z + 3$$

$$\alpha_2(z) = z - 1 \qquad \lambda_2(z) = 3 - z$$

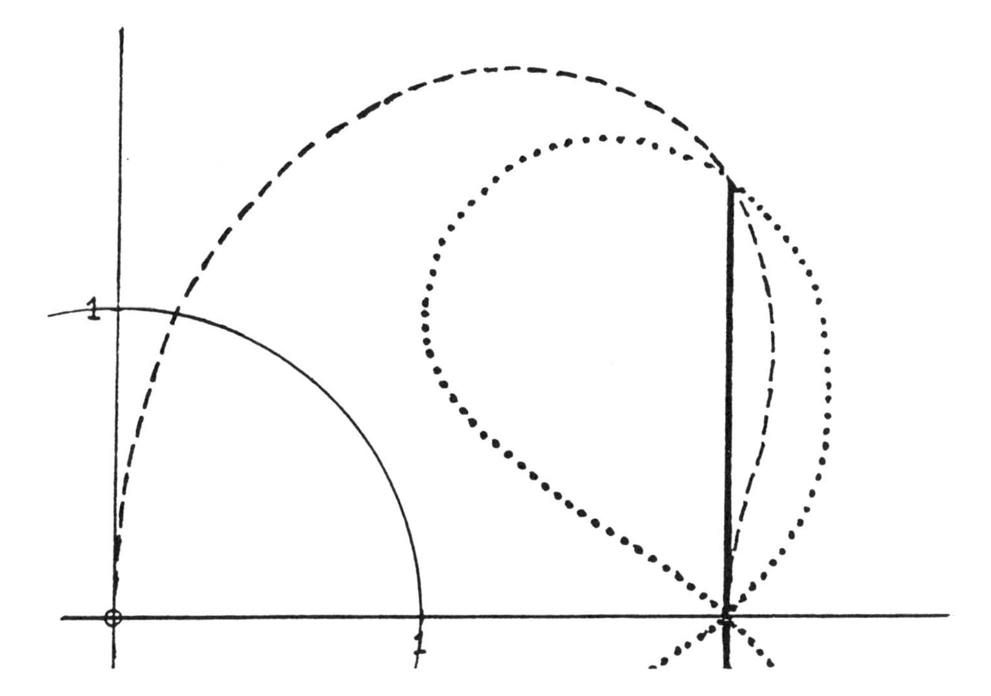

Figure 3

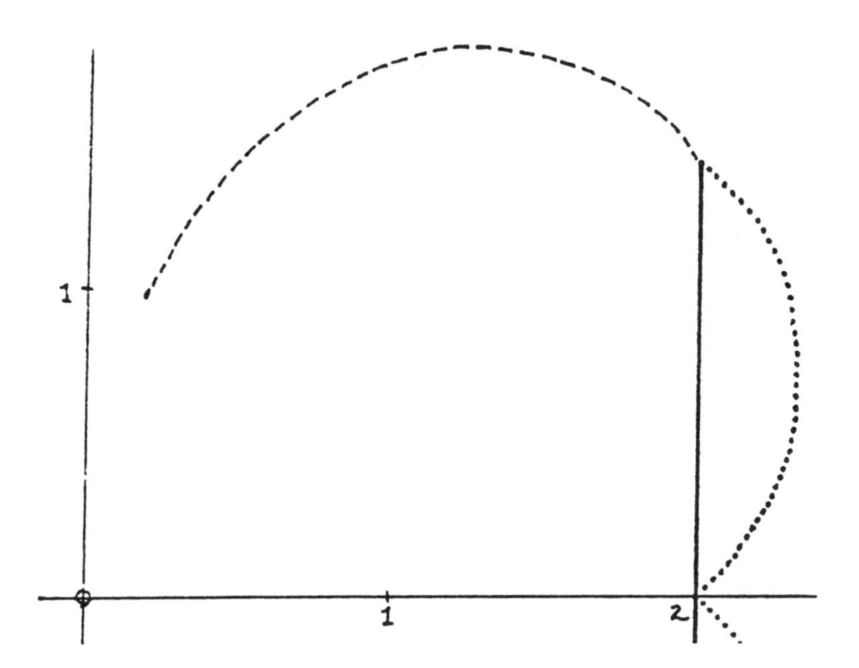

Figure 4

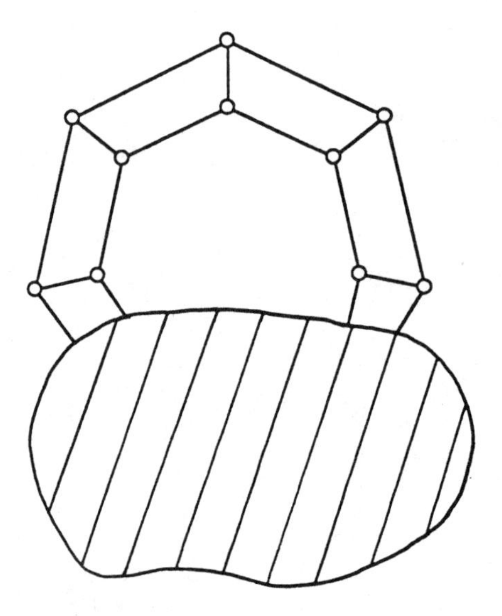

Figure 5

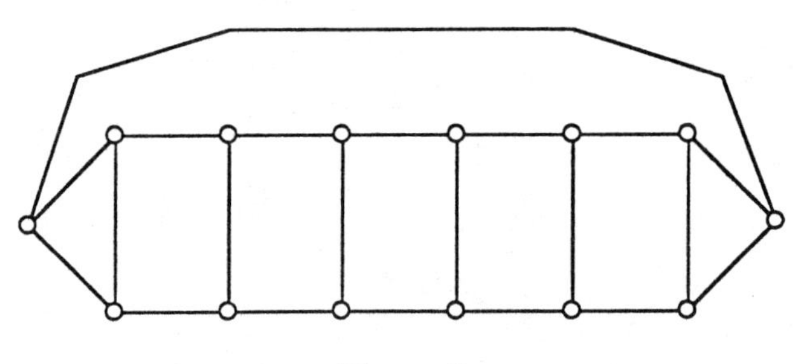

Figure 6

$$\alpha_3(z) = z - 1 \qquad \lambda_3(z) = 1 - z$$
$$\alpha_4(z) = z^2 - 3z + 1 \qquad \lambda_4(z) = 1.$$

It is easy to verify that, applying the Beraha–Kahane–Weiss theorem we get the following three portions of curves:

(i) $|\lambda_2| = |\lambda_3|$ gives the line $\Re(z) = 2$ between the points $2 - i\sqrt{2}$ and $2 + i\sqrt{2}$.

(ii) $|\lambda_1| = |\lambda_2|$, that is, $|z^2 - 3z + 3| = |3 - z|$ gives the quartic curve whose equation is

$$y^4 + 2(x^2 - 3x + 1)y^2 + (x^4 - 6x^3 + 14x^2 - 12x) = 0$$

where in order to satisfy the requirement for domination the points must be outside the unit circle and to the left of the line $\Re(z) = 2$

(iii) $|\lambda_1| = |\lambda_3|$ gives the quartic curve with equation

$$y^4 + 2(x^2 - 3x + 1)y^2 + (x^4 - 6x^3 + 14x^2 - 16x + 8) = 0$$

to the right of $\Re(z) = 2$.

These curves are shown in full in Figure 3, together with the portion of the straight line. Figure 4 shows only those parts of the curves which satisfy the requirements of the theorem, and it will be seen that there is excellent agreement between these portions of curves and the positions of the roots shown in Figure 2.

This problem of explaining the behaviour of the chromatic roots of ladders was briefly discussed by one of us as part of a talk given at the combinatorial conference in Barbados in January of this year [8]. We now report on more recent developments.

5. Some observations

We have already noted that the curves of limit points of zeros depend only on the recurrence satisfied by the family of graphs, and not on the coefficients $\alpha_i(z)$ in the solution. Now the recurrence for ladders can be obtained by means of operations on the graph – deletion of edges and identification of vertices – applied to a small portion of the graph involving just two or three "rungs". It follows from these remarks that if we consider a family of graphs such as is shown in Figure 5, where the shaded portion denotes an arbitrary graph and the successive members of the family are obtained by increasing the number of rungs in the ladder–like portion of the graph, then the chromatic polynomials of the graphs in this family will satisfy the same recurrence as do ladders.

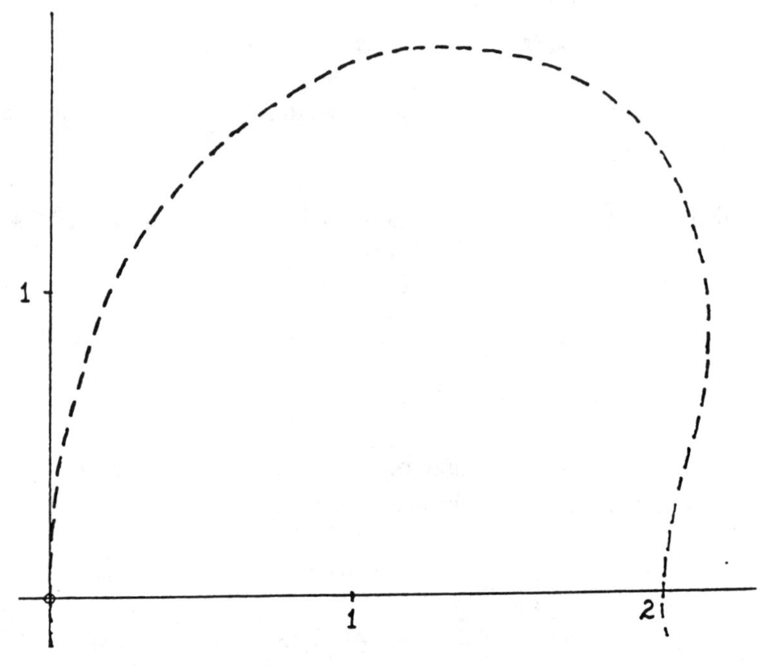

Figure 7

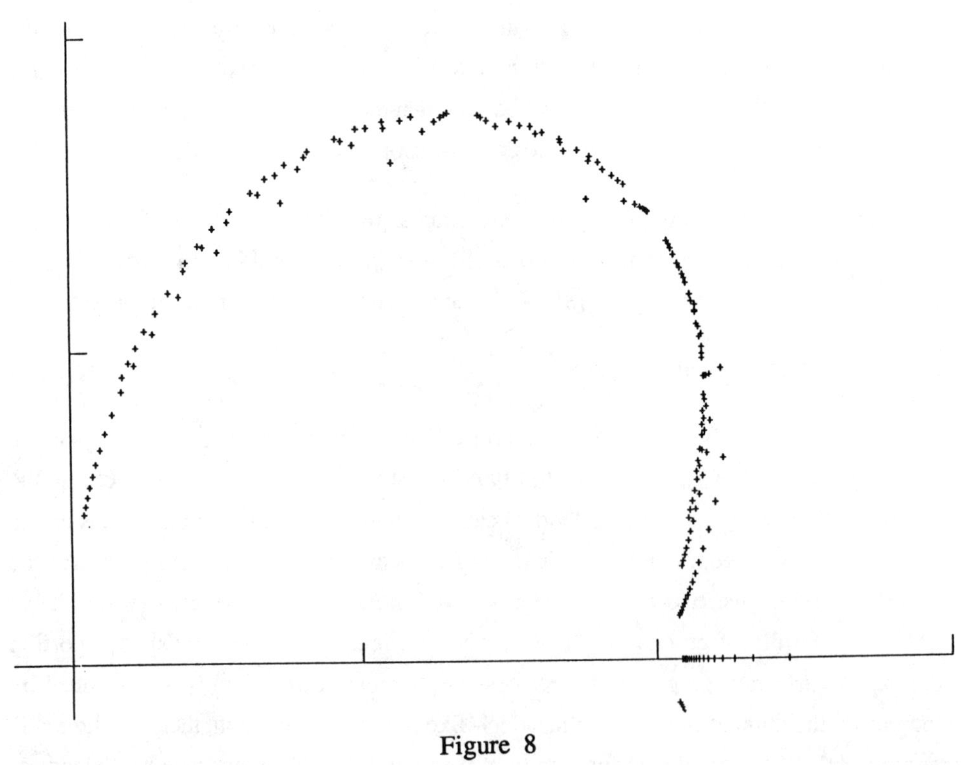

Figure 8

In particular, the Möbius ladders obtained from an ordinary ladder by crossing the two sides in between a pair of rungs, satisfy the same recurrence, and hence have the same curves of the limit points of zeros.

Note however that it is possible for such a family to satisfy a *simpler* recurrence. As an example of this consider the family of graphs of which a typical member is that shown in Figure 6, graphs which we call "rope ladders". It can be shown that these satisfy a second order recurrence and that the polynomial for the rope ladder on 2n vertices is given by

$$R_{2n}(z) = (z-1)^2(z-2)^2(z^2-3z+3)^{n-2} - 2(z-1)(z-2)(3-z)^{n-2}$$

In this case we have only two roots to the auxiliary equation. They are $\lambda_1(z)$ and $\lambda_2(z)$ in the notation used above. Thus there is only one curve, given by $|\lambda_1| = |\lambda_2|$, and since there are no other roots to dominate, every point of this curve is a limit point of zeros. This curve is shown in Figure 7, and a plot of the chromatic roots of the first few graphs of this family is shown in Figure 8.

6. Problems

It might seem that the Beraha–Kahane–Weiss theorem tells us the whole story of what happens to the chromatic roots of a family of graphs as the number of vertices increases. Unfortunately, this is far from being the case.

Consider the family of bipyramids – the graphs obtained by joining each of two non–adjacent vertices to all the vertices of a cycle, as shown in Figure 9. It is easy to show that the chromatic polynomial of the bipyramid on n + 2 vertices is given by

$$z(z-1)(z-3)^n + z(z-2)^n + (-1)^n z(z^2-3z+3)$$

thus we have $\lambda_1 = z-3$; $\lambda_2 = z-2$ and $\lambda_3 = -1$. Applying the Beraha–Kahane–Weiss theorem we see that

$$|\lambda_1| = |\lambda_2|$$

gives a portion of the line $\Re(z) = 2.5$.

$$|\lambda_1| = |\lambda_3|$$

gives an arc of the circle centre 2 radius 1.

$$|\lambda_2| = |\lambda_3|$$

gives an arc of the circle centre 3 radius 1.

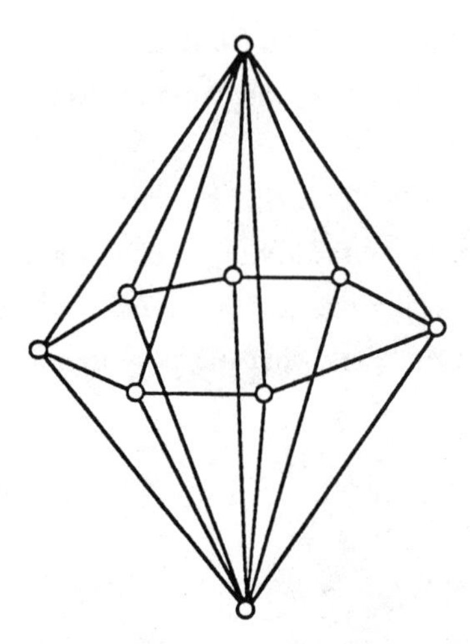

Figure 9

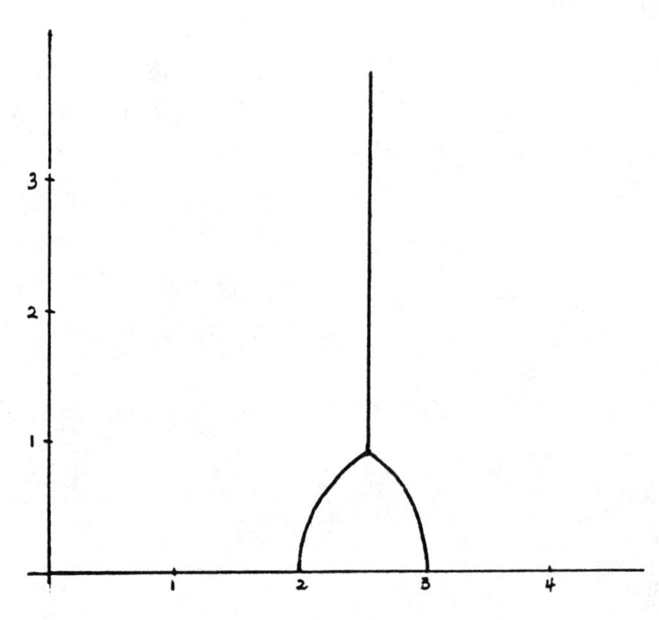

Figure 10

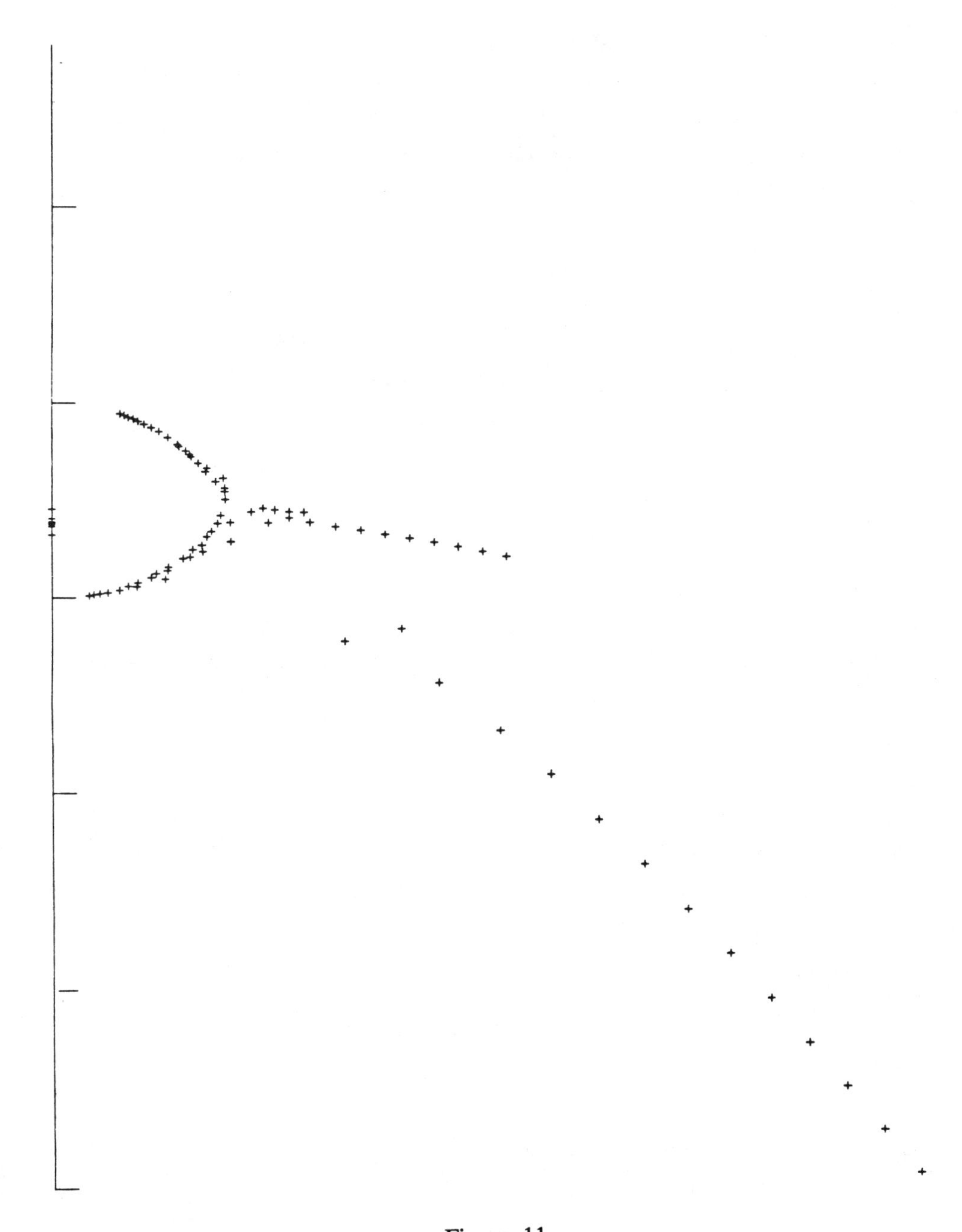

Figure 11

The relevant parts of this line and the circles are shown in Figure 10. Figure 11 shows the actual locations of the chromatic roots of the first few members of the family. It will be seen that there is excellent agreement with the two arcs of the circles, to which a large number of the roots approximate very closely. However, the remaining roots seem to arrange themselves in a number of "sprays", three of which can be seen in Figure 11. Presumably, if we could extend the calculation to very large graphs of this family, there would be further sprays getting closer and closer to the theoretical line of limit points of zero. Thus the theoretical result is not contradicted by the plot of the roots, but a new problem is introduced, namely that of explaining the behaviour of these sprays of roots. They appear to define smooth curves, and it would be of interest to know what these curves are; but up to now we have no answers to questions of this kind.

Thus even for recursive families of graphs there are still some open problems.

7. Non–recursive families

In this section we consider a somewhat different aspect of the problem of the location of chromatic roots. We concentrate on complete classes of graphs on a fixed number of vertices, rather than infinite families of directly related graphs. Unfortunately, when it comes to chromatic roots of non–recursive families of graphs there is, as yet, no theory at all to guide us. We therefore merely draw attention to some interesting cases that we have observed.

A natural class of graphs to examine is that of regular graphs. As the problem is trivial for graphs of valency 1 and 2, we shall start by considering cubic graphs. A complete catalogue of all cubic graphs up to 20 vertices has been prepared [4], and the following table gives their numbers.

Number of vertices	4	6	8	10	12	14	16	18	20
Number of vertices	1	2	5	19	85	509	4060	41301	510489

The algorithm described in [7] was used to compute the chromatic polynomials (in the tree basis) of all the cubic graphs on up to 16 vertices, and a standard library routine used to find the roots of these polynomials. The results obtained when these roots were plotted on the complex plane are shown in Figures 12, 13 and 14 for 12, 14 and 16 vertices respectively. Once again the roots clearly display a non–random pattern, tending to cluster about $(n-4)/2$ points in the complex plane arranged roughly equidistantly on an ellipse–like curve in the upper right quadrant (here we are ignoring the complex conjugates of the roots shown and the real roots). Moreover, the clusters for the 14 vertex cubic graphs start slightly to the left of those for the 12 vertex cubic graphs and interleave with them; similarly the 16 vertex clusters interleave with the 14 vertex clusters.

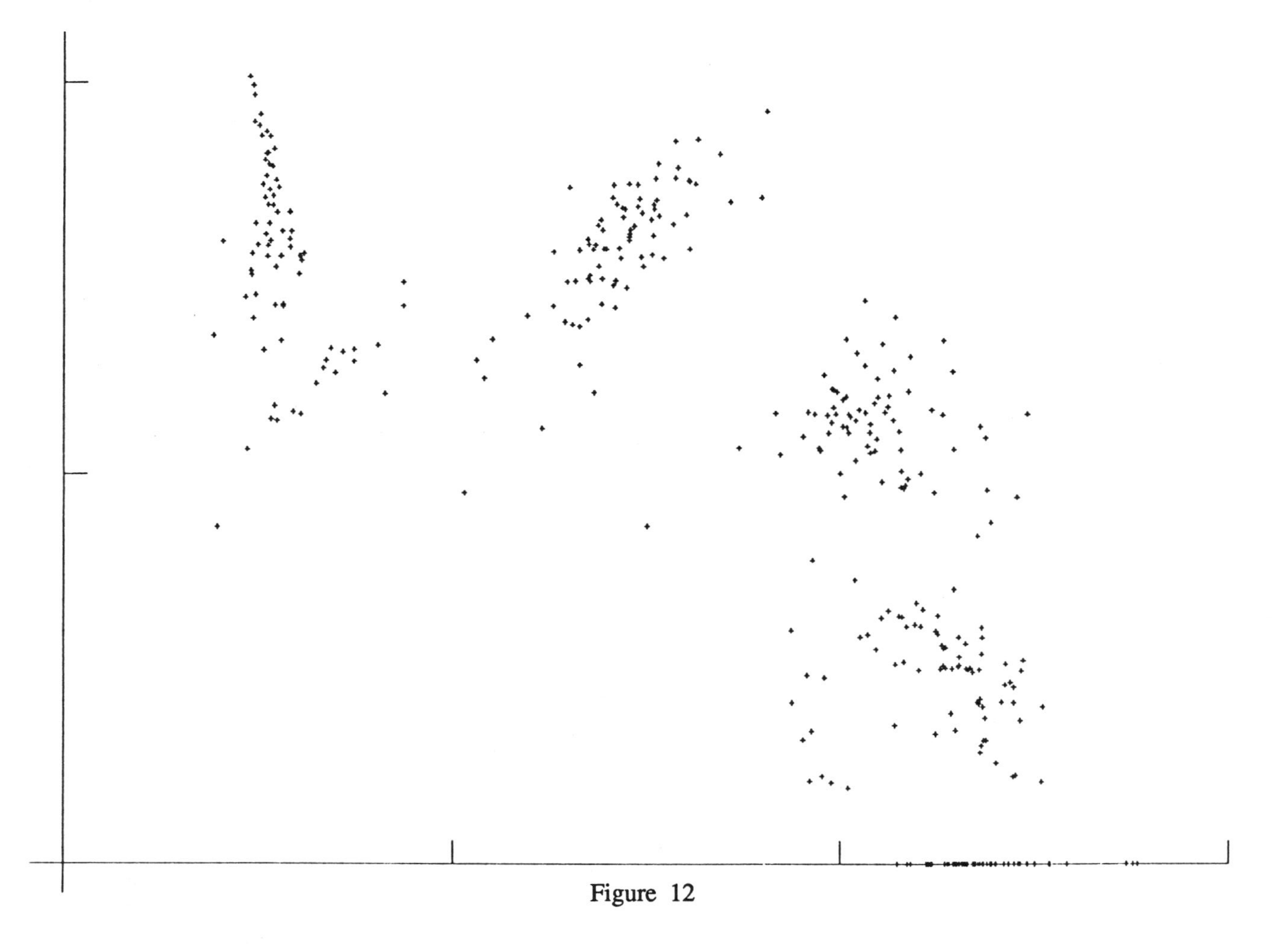

Figure 12

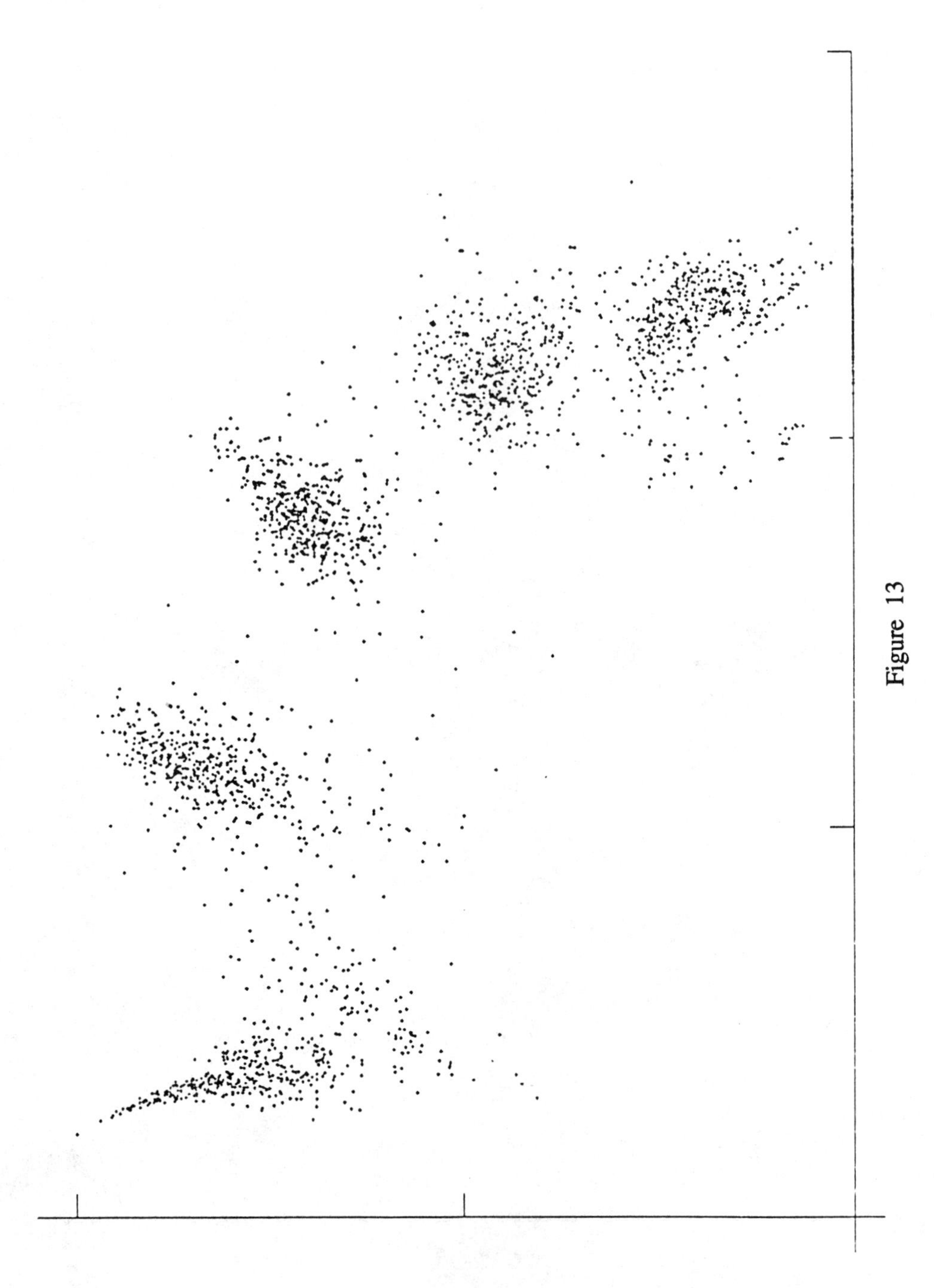

Figure 13

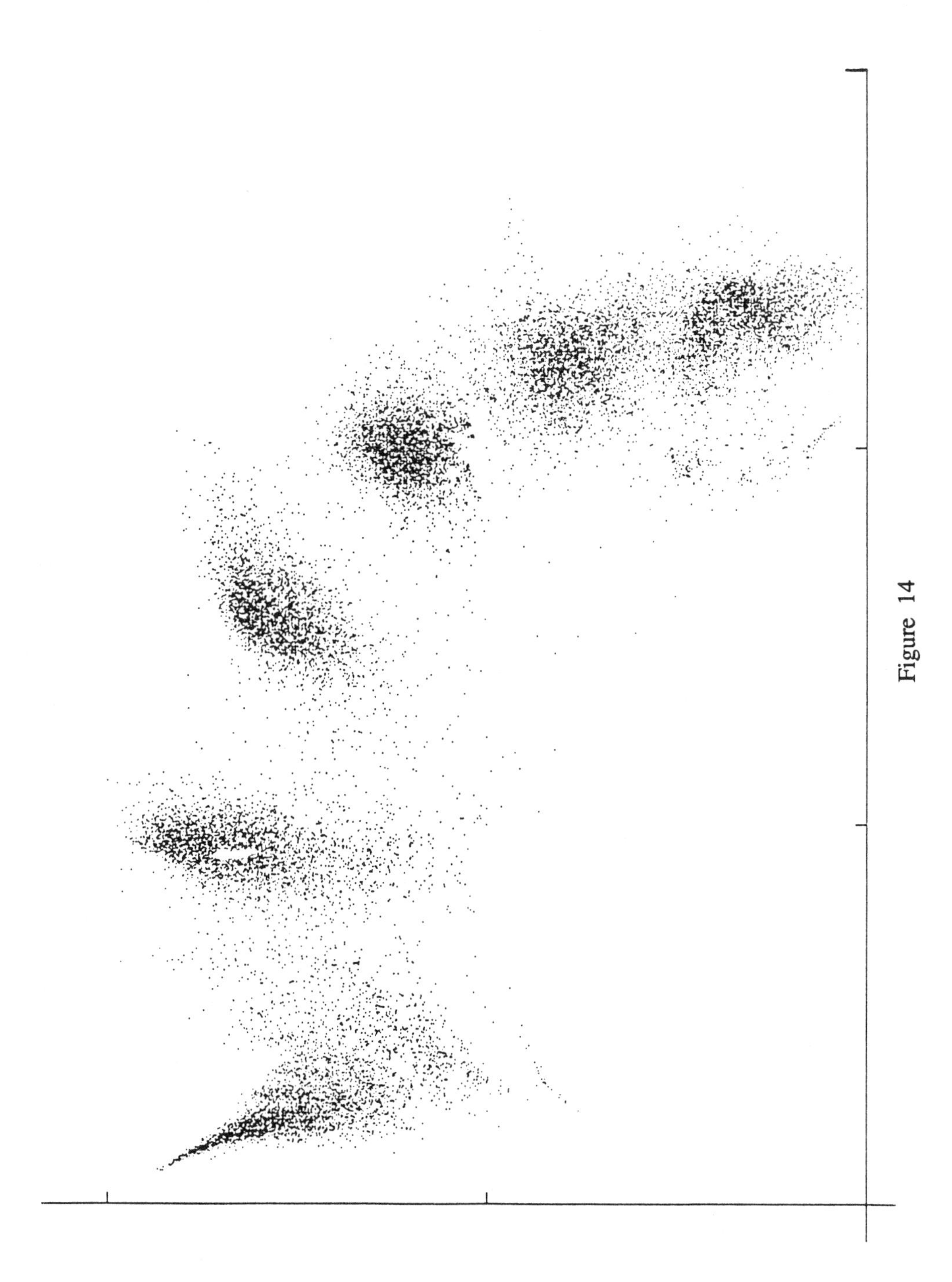

Figure 14

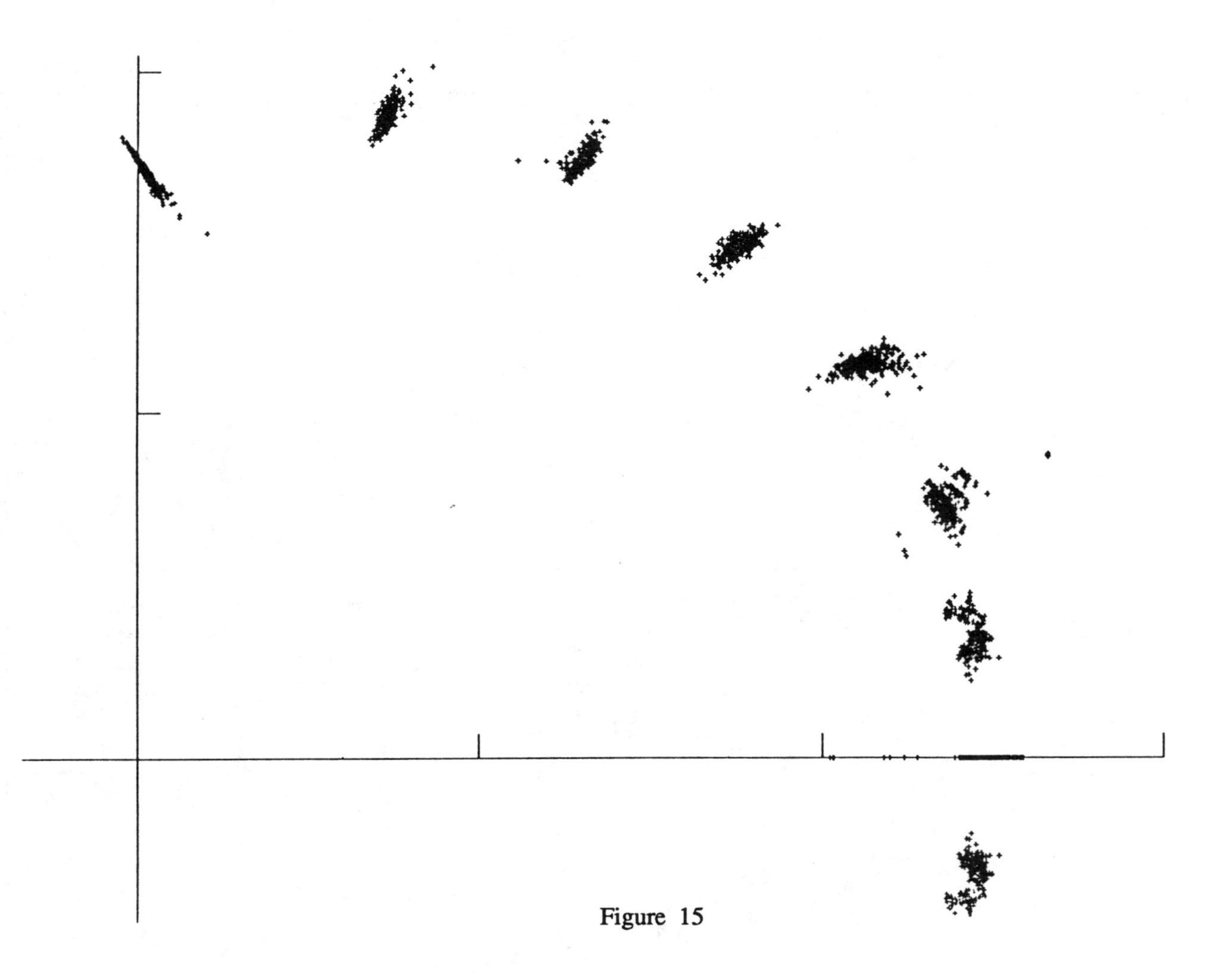

Figure 15

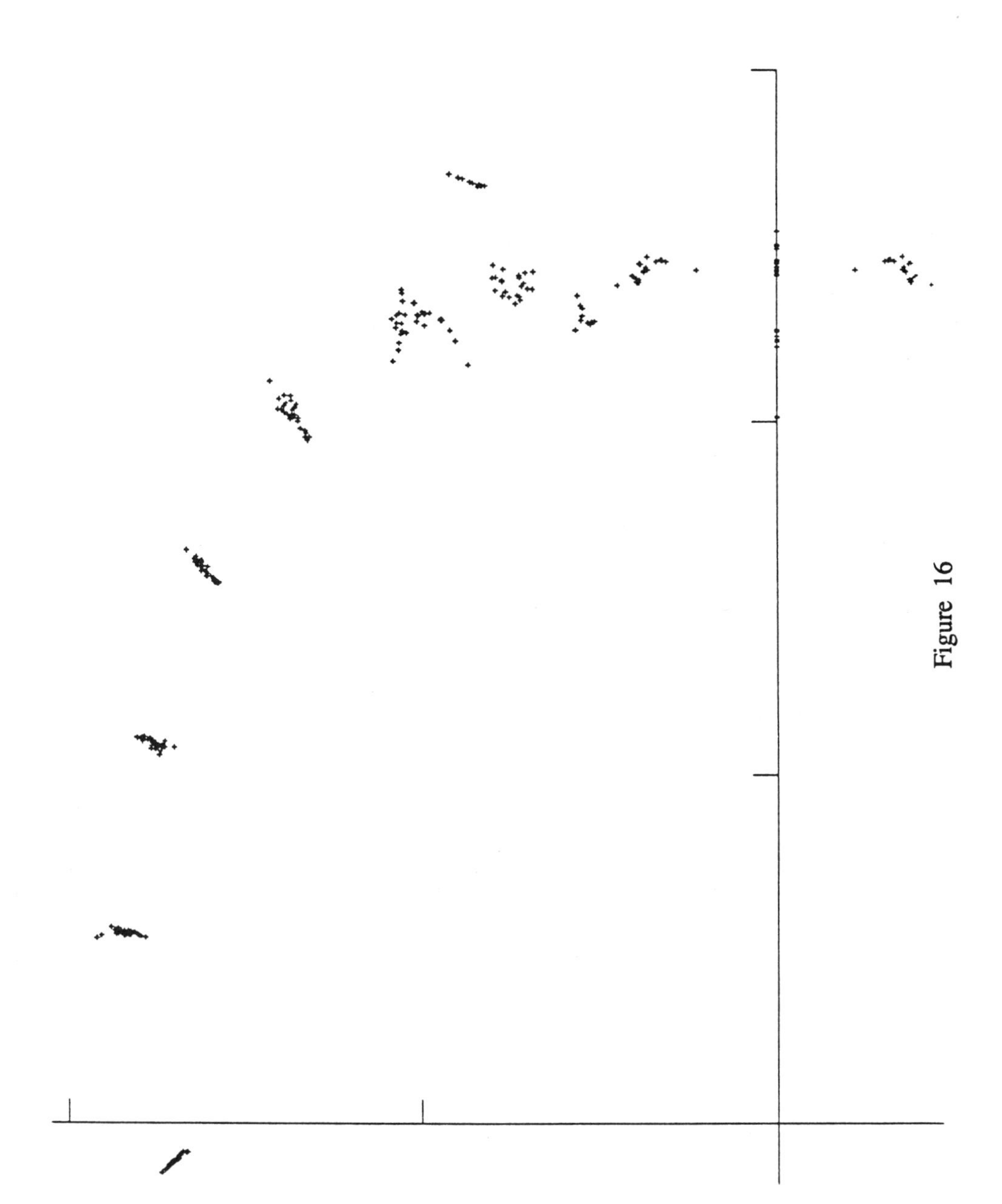

Figure 16

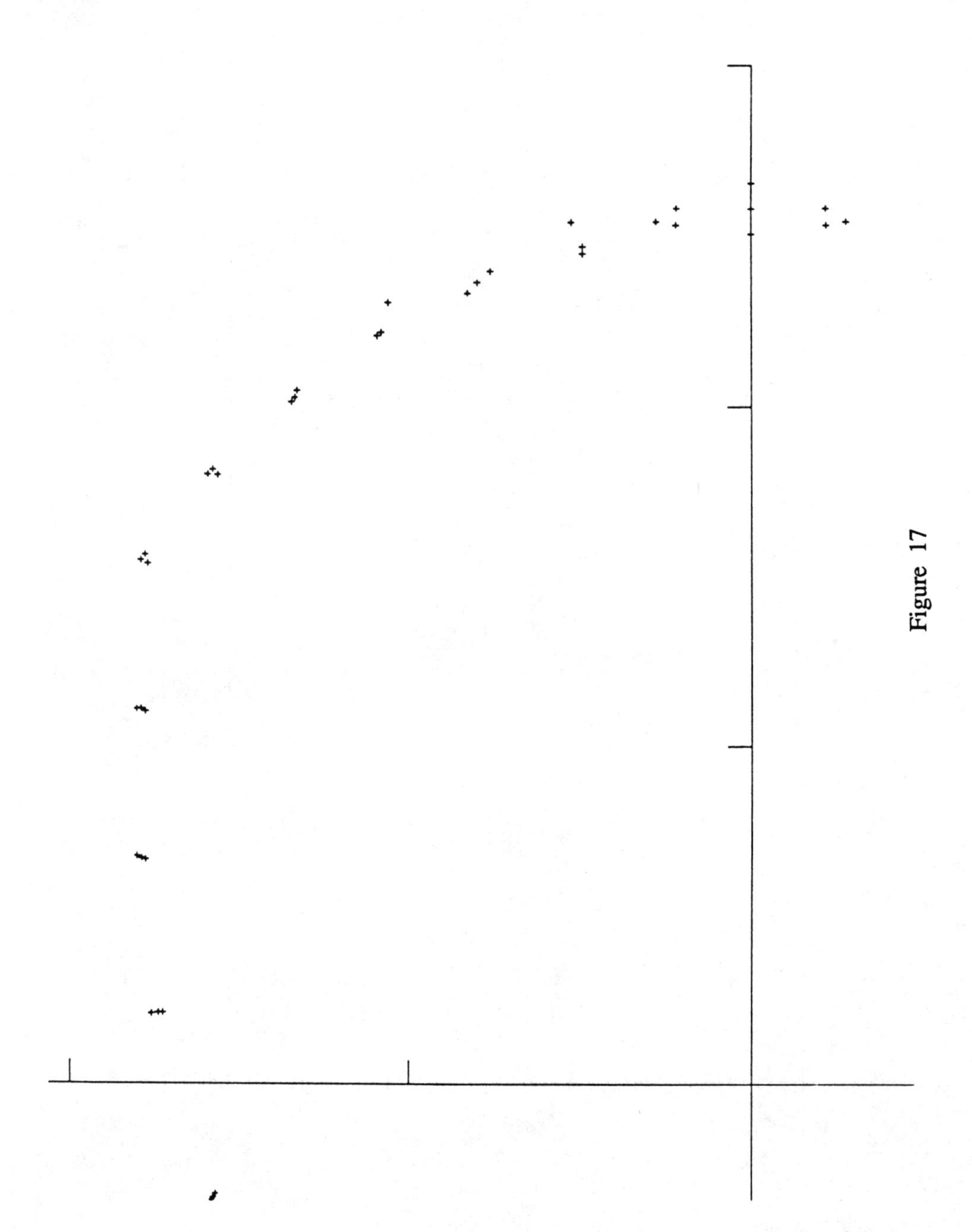

Figure 17

8. Two conjectures

We now describe two conjectures that have apparently been widely believed to hold; see [2], [3], [10].

Conjecture 1 For cubic graphs, the chromatic roots z all have modulus $|z| \leq 3$.

In fact this is a special case of the more general conjecture

Conjecture 1' There is a function $B\colon R \to R$, such that the chromatic roots z of k–regular graphs all have modulus $|z| \leq B(k)$.

Thus Conjecture 1 states that $B(3) = 3$.

Our results clearly support this conjecture, and as we have tested a large number of graphs, and (more importantly) have checked *every* graph of the classes in question, we regard this as strong evidence in favour of this conjecture. The second conjecture stems from the fact that all previously known chromatic roots have positive real parts.

Conjecture 2 For all chromatic roots z, $\Re(z) \geq 0$.

Below we demonstrate that this conjecture is false. In [10] the dodecahedron is presented as a counterexample to Conjecture 2. However the chromatic polynomial is given incorrectly there, and although the conjecture *is* false, the dodecahedron is not a counterexample.

As remarked above, the leftmost cluster of roots is closer to the imaginary axis for the 16 vertex cubic graphs than the 14 vertex cubic graphs. We wish to know the behaviour of this cluster as the number of vertices is increased still further, to determine whether it actually crosses the imaginary axis or, as Conjecture 2 suggests, whether it 'squeezes–up' against the imaginary axis.

However, it is not possible to test all the cubic graphs on 18 and 20 vertices because computing chromatic polynomials is time–consuming even for graphs of this size. Therefore we wish to determine which graphs are contributing the roots of smallest real part, and what property distinguishes them. Examination of the extreme graphs on 14 and 16 vertices leads us to the conclusion that the important factor is the girth of the graph; in particular graphs with high girth contribute the roots of smallest real part. It is not surprising that girth is a significant factor because the leading coefficients of the chromatic polynomial take their maximum absolute values when the girth is high. In fact if

$$P_G(z) = \sum_{i=0}^{n} (-1)^i a_{n-i} \lambda_{n-i}$$

and q is the number of edges of G, then $a_{n-s} \le \binom{q}{s}$. Furthermore, $a_{n-s} = \binom{q}{s}$ if the girth of the graph is greater than $s + 1$, and $a_{n-s} = \binom{q}{s} - p$, if the girth is precisely $s + 1$, where p is the number of circuits of length $s + 1$ [5]. Thus one would expect such graphs to be extreme in some sense.

This is useful because it is simple to extract the cubic graphs of high girth from the catalogues on 18 and 20 vertices, and in fact further catalogues of all the maximal girth cubic graphs on up to 30 vertices have been produced [9].

Figure 15 shows the chromatic roots of the high girth (≥5) cubic graphs on 18 vertices demonstrating the smallest known counterexample to Conjecture 2, Figure 16 shows the chromatic roots of the cubic graphs of girth ≥ 6 on 20 vertices and Figure 17 shows the chromatic roots of the cubic graphs of girth ≥ 7 on 26 vertices. Despite the small number of graphs the pattern is clear, and there are roots with quite large negative real parts.

Once again, further examination is halted by the prohibitive time required to directly compute the chromatic polynomials of even a small number of graphs with a large number of vertices. One immediate question raised by this is determining just how large a negative real part a chromatic root can have. To this end it would be particularly interesting to examine the behaviour of the *cages* – cubic graphs with a minimum number of vertices for a given girth – if their chromatic polynomials could be found.

REFERENCES

[1] S. Beraha, J. Kahane and N. J. Weiss, Limits of chromatic zeros of some families of maps. *J.Comb. Theory Ser. B.* **28** (1980) 52–65.

[2] N. L. Biggs, R. M. Damerell and D. A. Sands, Recursive families of graphs. *J.Comb. Theory Ser. B.* **12** (1972) 123-131.

[3] E. J. Farrell, Chromatic roots – some observations and conjectures. *Discrete Math.* **29** (1980) 161–167.

[4] B. D. McKay and G. F. Royle, Constructing the cubic graphs on up to 20 vertices. *Ars Combinatoria* **21–A** 129–140.

[5] G. H. J. Meredith, Coefficients of chromatic polynomials. *J. Comb. Theory Ser. B*, **13** (1972) 14-17.

[6] R. C. Read, Introduction to chromatic polynomials. *J. Comb. Theory* **4** (1968) 52-71.

[7] R. C. Read, An improved method for computing the chromatic polynomials of sparse graphs. *Research Report* CORR 87-20, C&O Dept., U.of Waterloo (1987).

[8] R. C. Read, Recent Advances in Chromatic Polynomial Theory. *Proceedings of the 5th Caribbean Conference on Combinatorics and Computing*. Barbados, 1988/ To appear.

[9] G. F. Royle, Constructive enumeration of graphs. PhD Thesis. University of Western Australia (1987).

[10] D. A. Sands, Dichromatic polynomials of linear graphs. PhD Thesis. Royal Holloway College, London (1972).

From Garbage to Rainbows: Generalizations of Graph Coloring and Their Applications

Fred S. Roberts
Rutgers University

ABSTRACT

We describe a variety of applications of ordinary graph coloring and some generalizations and variations of ordinary graph coloring which are motivated by these applications. The generalizations and variations discussed are called T-colorings, H–colorings, I–colorings, J–colorings, n–tuple colorings, consecutive colorings, set colorings in general, and list colorings. We then discuss combinations of some of these variants, describing set T–colorings, n–tuple T–colorings, list T–colorings, and set list T–colorings.

1. Introduction

The basic message of this paper is that graph coloring has many applications and that these applications have often led to very interesting generalizations of ordinary graph coloring. The paper starts with a grab–bag of applications of ordinary graph coloring. It then considers a variety of generalizations of ordinary graph coloring which are motivated by these applications. For each generalization, some recent results are mentioned and usually an open problem.

The generalizations to be discussed in this paper are called T–colorings, H-colorings, I–colorings, J–colorings, n–tuple colorings, consecutive colorings, set colorings in general, and list colorings. Some of these ideas are then combined with the first one, to obtain set T–colorings, n–tuple T–colorings, list T–colorings, and set list T–colorings.

Recall that if $G = (V, E)$ is a graph, a *coloring* of G is an assignment of a color $F(x)$ to each vertex x so that for all $x \neq y$ in V,

$$\{x, y\} \in E \rightarrow F(x) \neq F(y).$$

The *chromatic number* of G, $\chi(G)$, is the smallest k so that G has a coloring using k colors. We begin in the next section with a collection of applications of graph coloring.

2. Applications of Graph Coloring

Graph coloring has a long history, and from the beginning it has been closely tied to one well–known application, namely map coloring. Modern applications of graph coloring are made to many important applied problems. We mention some of these problems here.

Scheduling Meetings of Committees in a State Legislature

Graph coloring has been used in the New York State Assembly (see Bodin and Friedman [1971]). In this application, we wish to assign a meeting time to each legislative committee. If two committees have a member in common, they must get a different meeting time. We build a schedule graph whose vertices are the committees and which has an edge between x and y if and only if x and y have a member in common. Then the colors are the meeting times. The chromatic number gives the smallest number of meeting times required.

Channel Assignment Problem

In radio and television applications, we sometimes wish to assign a channel to each t.v. or radio transmitter. Some transmitters interfere. In the simplest version of this problem, they interfere if and only if they are within a certain distance of each other. However, interference can take place for reasons other than proximity. Transmitters which interfere must get different channels. We build an interference graph whose vertices are the transmitters and which has an edge between x and y if and only if x and y interfere. Then the colors are the channels. We often wish to use as little of the spectrum as possible. Thus, we are interested in the chromatic number of the interference graph. This problem, with various complications, has been studied at the Federal Communications Commission, AT&T Bell Laboratories, the National Telecommunications and Information Administration, the Department of Defense, and elsewhere. The idea of studying channel assignment graph–theoretically goes back to Metzger [1970] and Zoellner and Beall [1977], and the formulation of the channel assignment problem we shall study below is due to Hale [1980].

Garbage Collection

A *tour* of a garbage truck is a schedule of sites it visits on a given day. If two tours visit the same site, they must be run on different days. We wish to assign to each tour a day of the week on which it will be scheduled so that if two tours visit a given site, they get a different day. We build a tour graph with vertices the set of tours and with an edge between x and y if and only if tours x and y visit a common site. Then the colors are the days. If we wish to use no more than the 6 days of the week (excluding Sunday) to schedule garbage collection, then we are interested in determining whether or not the tour graph is 6–colorable. This problem arises as part of a larger garbage truck scheduling problem posed by the New York City Department of Sanitation and solved heuristically by Beltrami and Bodin [1973] and by Tucker [1973]. In the heuristic solution, it is necessary to solve the tour graph coloring problem over and over again.

Mobile Radio Frequency Assignment

We wish to assign frequencies to mobile radio telephones in vehicles. Each vehicle in a given region is to receive a frequency set over which it can transmit. Some regions interfere because of reasons of proximity, meteorological reasons, etc. Interfering regions must get different frequency sets. We build an interference graph by taking as vertices the regions and including an edge between x and y if and only if x and y interfere. Then the colors are the frequency sets. If we wish to minimize the number of frequency sets, we seek the chromatic number of the interference graph. This problem was introduced by Gilbert [1972] at Bell Labs and has been studied by a wide variety of others.

Fleet Maintenance

A maintenance facility awaits vehicles (trucks, planes, ships, etc.). Each vehicle has a time period during which it will be in the facility for regularly scheduled maintenance. We want to be sure that we can assign a space in the facility to each vehicle. If two vehicles are in the facility in overlapping time periods, they must get different spaces. We build an overlap graph whose vertices are the vehicles and which has an edge between x and y if and only if x and y are in the facility in overlapping time periods. Then the colors are the spaces. If we wish to minimize the total space in the facility, we want the chromatic number of the overlap graph. This problem has been studied for ship maintenance (with linear facilities or docks) by Alan Hoffman and Ellis Johnson at IBM (see Golumbic [1980] and Opsut and Roberts [1981]).

Traffic Phasing

There is a stream of requests for use of a facility (classroom, computer, traffic intersection). Each use is to be assigned a period of time, a green or "up" time, during which it may use the facility. Certain uses are incompatible with each other. Then we wish to assign green times so that incompatible uses receive different green times. We build an incompatibility graph whose vertices are the traffic streams and which has an edge between x and y if and only if x and y are incompatible. Then the colors are the green times. If we wish to minimize the total amount of time used, we seek the chromatic number of the incompatibility graph. This problem was introduced by Stoffers [1968] in the study of traffic lights and in the more general interpretation of traffic phasing by Opsut and Roberts [1981].

Task Assignment

A large task has been divided into subtasks. Some subtasks are incompatible because they require the same resources, tools, space, workers, etc. The problem is to schedule the subtasks so that incompatible subtasks get different time periods. We build an incompatibility graph whose vertices are the subtasks and which has an edge between x and y if and only if x and y are incompatible. Then the colors are the time periods. If we seek to minimize the total amount of time taken in all of the subtasks, we seek the chromatic number of the incompatibility graph. (This problem is a common one in the scheduling literature. See for instance Opsut and Roberts [1981].)

There are many other applications of graph coloring. For a discussion of some of these, and an elaboration of some of those above, see Golumbic [1980], Roberts [1976, 1978, 1984], or Bondy and Murty [1976].

Most of the problems posed above have various complications which imply that ordinary graph coloring is a simplification. These complications have given rise to interesting generalizations of graph coloring, which we describe in this paper. Before discussing these generalizations, we need two graph–theoretic preliminaries, which we introduce in the next section.

3. Intersection Graphs

We say that a graph G is an *intersection graph* of certain kinds of sets if there is an assignment of a set $S(x)$ of the given kind to each $x \in V(G)$ so that for all $x \neq y$,

$$\{x, y\} \in E(G) \leftrightarrow S(x) \cap S(y) \neq \phi.$$

The function S is called an *intersection assignment* for G. G is an *interval graph* if each set $S(x)$ is a real interval. For a general introduction to the theory of intersection graphs, and in particular of interval graphs, the reader is referred to Roberts [1976, 1978] or Golumbic [1980] or Fishburn [1985]. Intersection graphs in general, and interval graphs in particular, play an important role in what follows.

4. T-Colorings

The first generalization of graph coloring which we study arises from the channel assignment problem. Let us think of channels as positive integers. One complication in channel assignment is that certain separations between interfering transmitters are disallowed. Let T be a set of nonnegative integers representing disallowed separations between interfering transmitters. Assume $0 \in T$. We want to find a function $F: V(G) \to Z^+$ such that for all $x \neq y$ in $V(G)$,

$$\{x, y\} \in E \to |F(x) - F(y)| \notin T.$$

Such an F is called a *T-coloring* of G. For instance, suppose $T = \{0\}$. Then a T-coloring is an ordinary coloring. To give another example, suppose $T = \{0, 1\}$. Then, interference implies that channels are different and nonadjacent. The sets T can get rather complicated. For instance, in UHF television, sets such as $T = \{0, 7, 14, 15\}$ arise.

The formulation of the channel assignment problem as a T–coloring problem is due to Hale [1980]. Hale points out that there are several criteria for efficiency of a T-coloring. Let us define the *order* of a T–coloring F as the number of different colors F(x) and the *span* of F as the maximum of $|F(x) - F(y)|$ over all x, y. Then we might wish to minimize the order or the span. The *T–chromatic number* of G, denoted $\chi_T(G)$, is defined to be the minimum order of a T–coloring of G, while the *T–span* of G, denoted $sp_T(G)$, is defined to be the minimum span of a T–coloring of G. The difference between these two concepts can be illustrated by considering the complete graph $G = K_3$ and the set $T = \{0, 1, 4, 5\}$. Suppose we color one vertex of this graph with color (channel) 1. If we try to color a second vertex with the lowest possible channel, we are forced to use channel 3. But then the lowest possible channel available for the third vertex is channel 9. We have a T–coloring of G with order 3 and span 8. Using the colors 1, 4, and 7 gives us the same order, but a smaller span, 6. There are examples of graphs where it is not possible to optimize on both criteria at once. For instance, consider the 5–cycle C_5 with the set $T = \{0, 1, 4, 5\}$. In this graph, a minimum span assignment uses successively around the cycle the colors 1, 4, 2, 5, and 3. However, a minimum order assignment uses successively the colors 1, 4, 1, 4, 7.

Roberts [1986] is a recent survey paper on the theoretical work on T–colorings. Here, we summarize a number of the important results.

Note that if $T = \{0\}$, then $\chi_T = \chi$ and $sp_T = \chi - 1$. Therefore, the problems of computing χ_T and sp_T are NP–complete. In general, as Cozzens and Roberts [1982] observe, $\chi_T = \chi$, so sp_T is the interesting new number.

Theorem 1 *(Cozzens and Roberts 1982) If $\omega(G)$ is the size of the largest clique of G, then*

$$sp_T(K_{\omega(G)}) \leq sp_T(G) \leq sp_T(K_{\chi(G)}).$$

Corollary *If G is weakly γ–perfect, i.e., if $\omega(G) = \chi(G)$, then*

$$sp_T(G) = sp_T(K_{\chi(G)}).$$

Many of the known results about T–colorings are obtained either using special assumptions about T or special assumptions about G. We begin with some results under special assumptions about T. A set $T = \{0, 1, ..., r\} \cup S$ is called *r–initial* if S contains no multiple of $r + 1$. $T = \{0, 1, 2, 5\}$ is an example of a 2–initial set.

Theorem 2 *(Cozzens and Roberts 1982) If T is r–initial, then for all graphs G,*

$$sp_T(G) = (r + 1)\,[\chi(G) - 1] = sp_T(K_{\chi(G)}).$$

A set T is a *k–multiple of s set* if $T = \{0, s, 2s, ..., ks\} \cup S$, where $s \geq 1, k \geq 1$, and S is contained in $\{s + 1, s + 2, ..., ks - 1\}$.

Theorem 3 *(Raychaudhuri 1985, 1988) If T is a k–multiple of s set, then for all graphs G,*

$$sp_T(G) = sp_T(K_{\chi(G)})$$

$$= \begin{cases} st + skt - sk - 1 & \text{if } \chi(G) = st \\ st + skt + p - 1 & \text{if } \chi(G) = st + p \end{cases}.$$

The following *greedy algorithm* sometimes gives us $\chi_T(G)$ and $sp_T(G)$. Order V(G) as $x_1, ..., x_v$. Let $F(x_1) = 1$. Having assigned $F(x_1), ..., F(x_k)$, let $F(x_{k+1})$ be the smallest channel so that $F(x_1), ..., F(x_{k+1})$ do not violate the requirements. We used this algorithm earlier in obtaining the coloring of K_3 with colors 1, 3, 9. It is a basic open question to determine for what graphs and what sets T there is an ordering for which the greedy algorithm correctly computes χ_T, and similarly for sp_T.

We turn next to results under special assumptions about the graph G. Even the case where G is a complete graph is nontrivial. Note that $sp_T(K_q)$ is fundamental because for many graphs, $sp_T(G) = sp_T(K_{\chi(G)})$.

Wang [1985], Cozzens and Wang [1984] and Tesman [1988] obtain many results about this problem for different sets T and integers q. However, $sp_T(K_q)$ remains unknown for such simple sets T as {0, 1, 4, 5}, {0, 1, 4, 6}, and {0, 1, 4, 7}. It also remains open to characterize sets T and integers q such that the greedy algorithm gives $sp_T(K_q)$. The answer to this question would have wide applicability, as the next theorem shows:

Theorem 4 *(Roberts 1986) Suppose G is weakly γ-perfect. Then there is an ordering of G such that the greedy algorithm gives $sp_T(G)$ if and only if the greedy algorithm gives $sp_T(K_{\chi(G)})$.*

Probably the most important class of graphs for which to study T–colorings is the class of R–unit sphere graphs. An *R–unit sphere graph* is the intersection graph of closed spheres in R–space of unit diameter. The channel assignment problem is of particular interest for these graphs, because when transmitters are located in R–space (R = 1, 2, or 3), interference sometimes takes place if and only if two transmitters are within m miles. In this case, the interference graph is an R–unit sphere graph. (These graphs are studied by Maehara [1984] and by Havel [1982], Havel, Kuntz, and Crippen [1983], and Havel, et al. [1983] in connection with biochemistry.) Unfortunately, we do not know how to characterize R–unit sphere graphs except when R = 1. Moreover, even for 2–unit sphere graphs, computation of the ordinary chromatic number $\chi(G)$ is already an NP–complete problem (James Orlin, unpublished). In general, we do not even know, when $R > 1$, good heuristic or approximate methods for computing $\chi_T(G)$ or $sp_T(G)$ for G an R–unit sphere graph.

The case R = 1 is an interesting special case. The 1–unit sphere graphs are sometimes called unit interval graphs or indifference graphs. They have been characterized by Roberts [1969] and others. The interference graph is a 1–unit sphere graph if the transmitters are located in a linear corridor and interference corresponds to being within m miles.

Theorem 5 *(Cozzens and Roberts 1982) Suppose G is a 1–unit sphere graph. Then for a special vertex ordering called compatible (which always exists), the greedy algorithm computes $\chi_T(G)$. Moreover, if T is an r–initial set, the algorithm computes $sp_T(G)$ for such an ordering. Even if the ordering is not given, the algorithm has complexity $O(v^2 t)$ where v is the number of vertices of G and $t = |T|$. If T is r-initial, the complexity is $O(v^2)$.*

The T–coloring problem has also been studied for chordal graphs and for perfectly orderable graphs, which are defined in Golumbic [1980] and in Chvátal [1984], respectively. The reader can also refer to these papers for the definitions of the orderings in the following theorems.

Theorem 6 *(Raychaudhuri 1985, 1988) Suppose G is a chordal graph. The same results as in Theorem 5 hold for the reverse of a perfect elimination ordering and for either r-initial sets or k–multiple of s sets.*

Theorem 7 *(Raychaudhuri 1985, 1988) Suppose G is a perfectly orderable graph. The same results as in Theorem 5 hold for an admissible (perfect) ordering of G and for either r–initial sets or k–multiple of s sets, except that the complexity results only hold if the admissible ordering is given.*

5. H–Colorings

Recall that a *homomorphism* from a graph G to a graph H is a mapping F from V(G) to V(H) so that for all $x \neq y$ in V(G),

$$\{x, y\} \in E(G) \rightarrow \{F(x), F(y)\} \in E(H).$$

An *H–coloring* of G is defined to be a homomorphism from G into H. Note that an ordinary graph coloring with q colors is a K_q–coloring, i.e., a homomorphism from G into K_q.

Here we note that a T–coloring is just an H–coloring where H has a set of integers as its vertices and for all $x \neq y$ in V(H),

$$\{x, y\} \in E(H) \leftrightarrow |x - y| \notin T.$$

The H–coloring problem is the problem of deciding if a graph is H–colorable. If $H = K_q$, the H–coloring problem is of course NP–complete. If H is bipartite, it is easy to see that the problem is polynomial. However, Hell and Nesetril [1986] have proved that for all other graphs H, the H–coloring problem is NP–complete.

6. Set Colorings

In many of the problems we have talked about, it is more appropriate to think of assigning more than one color to a vertex. In channel assignment or mobile radio frequency assignment, a transmitter might be allowed to transmit over more than one frequency. In traffic phasing or task assignment, a use of a facility or a subtask of a task might be given more than one time period during which it can take place. In fleet maintenance, a vehicle in the maintenance facility might be given more than one space.

Suppose S is a function which assigns to each vertex x of graph G a set S(x) of colors. We call S a *set assignment* for G. We call S a *set coloring* for G if for all $x \neq y$ in V(G),

$$\{x, y\} \in E \rightarrow S(x) \cap S(y) = \phi.$$

A *(set) intersection assignment* is a set assignment S(x) so that for all $x \neq y$ in V(G),

$$\{x, y\} \in E \leftrightarrow S(x) \cap S(y) \neq \phi.$$

The terminology set assignment and set coloring is due to Roberts [1979].

There are different parameters we might try to optimize for a set assignment, in particular for a set coloring. The *order* of a set assignment S is the size of $\cup S(x)$ over all x in V. We might wish to minimize the order. The *score* of a set assignment S is the sum of the sizes of the sets $S(x)$ over all x. We might wish to maximize the score, subject to the order not being too large. For instance, in channel assignment, the order of a set coloring is the total number of channels used altogether, and the score is a count of the total number of channels assigned counting multiplicity. The latter is some measure of the efficiency of the assignment. In traffic phasing, the order of a set coloring is the total amount of time used to complete all the activities, while the score is the sum of the green times, a measure of how well the facility is utilized. To give an example, consider the graph C_4. If we label the vertices in order around the cycle as x_1, x_2, x_3, x_4, then one set coloring takes $S(x_1) = \{a, b, c\}$, $S(x_2) = \{d, e\}$, $S(x_3) = \{a, b, c\}$, and $S(x_4) = \{f\}$. The order of this set coloring is $|\cup S(x)| = |\{a, b, c, d, e, f\}| = 6$. The score of this set coloring is $\Sigma\, |S(x)| = 3 + 2 + 3 + 1 = 9$.

With no restrictions on the sizes of the sets $S(x)$ other than that they be nonempty, the minimum order of a set coloring of C_4 is 2: Simply let $S(x_1) = S(x_3) = \{a\}$, $S(x_2) = S(x_4) = \{b\}$. Moreover, the maximum score of a set coloring if the order is at most 5 is now 10. To see why, note that $|S(x_1) \cup S(x_2)| \leq 5$ and $|S(x_3) \cup S(x_4)| \leq 5$, which implies that the score of any set coloring is at most 10. However, the score can be as high as 10: Take $S(x_1) = S(x_3) = \{a, b, c\}$, $S(x_2) = S(x_4) = \{d, e\}$.

A basic observation relates the minimum order or maximum score of a set coloring of a graph to the minimum order or maximum score of a set intersection assignment of spanning subgraphs of the complement. Let $\chi_\lambda(G)$ be the minimum order of a set coloring of G where all the sets $S(x)$ satisfy some property λ. Let $i_\lambda(G)$ be defined analogously for set intersection assignments. These numbers are undefined if no assignments of the specified types exist. The following result relates these numbers.

Theorem 8 *(Roberts 1979) For all conditions λ of interest in this paper, $\chi_\lambda(G)$ is the minimum $i_\lambda(H)$ for all spanning subgraphs H of G^c for which there is a set intersection assignment satisfying condition λ.*

Let $\chi^{\lambda,N}(G)$ be the maximum score over all set colorings of G which use sets satisfying condition λ and which have order at most N. Let $i^{\lambda,N}(G)$ be defined analogously for set intersection assignments. Again, these numbers are undefined if no assignment of the specified type exists. Also as above, these two numbers are related:

Theorem 9 *(Roberts 1979)* $\chi^{\lambda,N}(G)$ *is the maximum of* $i^{\lambda,N}(H)$ *for all spanning subgraphs* H *of* G^c *for which there is a set intersection assignment satisfying condition* λ *and having order at most* N.

6.1. n–tuple Colorings

The simplest example of a set coloring arises when condition λ says that each set $S(x)$ is a set of n colors. We call a set coloring satisfying this condition λ an *n–tuple coloring*. Such colorings were introduced by Gilbert [1972] in connection with the mobile radio frequency assignment problem. If we use the notation for the vertices introduced above, a 2–tuple coloring of C_4 is given by taking $S(x_1) = S(x_3) = \{a, b\}$ and $S(x_2) = S(x_4) = \{c, d\}$.

In the mobile radio frequency assignment problem, we are often interested in minimizing the total number of frequencies used, i.e., minimizing the order, i.e., minimizing $|\cup S(x)|$. This minimum is $\chi_\lambda(G)$, where λ is the condition that each set $S(x)$ is a set of n elements. Here, $\chi_\lambda(G)$ is called the *n–tuple chromatic number of* G and is denoted $\chi_n(G)$.

Obviously, $\chi_2(C_4) = 4$. Thus, $\chi_n(G) = n\chi(G)$. If G is bipartite (and has an edge), $\chi_n(G) = 2n$, i.e., $\chi_n(G) = n\chi(G)$. Is $\chi_n(G)$ always equal to $n\chi(G)$? The answer is: lots of the time, but not always.

Theorem 10 *(Roberts 1978) If* G *is weakly* γ*-perfect, then* $\chi_n(G) = n\chi(G)$.

However, the theorem is false in general. It is easy to show that $\chi_2(C_5) = 5$.

A useful early result about n–tuple colorings, which is used to prove Theorem 10, is the following:

Theorem 11 *(Stahl 1976) For every graph* G,

$$\chi_n(G) = \chi(G[K_n]),$$

where $G[K_n]$ *is the lexicographic product of* G *with* K_n.

There was a flurry of activity about the computation of χ_n in the 1970's. Very little has been done since. Two sample results follow.

Theorem 12 *(Irving 1983) It is NP–complete to determine if* $\chi_n(G) \leq 2n + 1$.

Theorem 13 *(Raychaudhuri 1985):* $\chi_n(G)$ *can be calculated as the solution to an integer programming problem in which the variables correspond to the maximal cliques of* G^c.

The maximum score of an n–tuple coloring is trivial to compute since $\Sigma\, |S(x)| = vn$ for all n–tuple colorings.

6.2. Consecutive Colorings

Many variations of n–tuple colorings have been studied. One variation is to allow the sets $S(x)$ to have different sizes. Here, $i_\lambda(G)$, the minimum order of a set intersection assignment, is called the *intersection number* of the graph. This concept was introduced by Erdös, Goodman, and Pósa [1966] and has been widely studied. The minimum order and maximum score of a set coloring in this situation have been studied by Roberts [1979], Opsut and Roberts [1981], and Opsut [1984].

Another variation of n–tuple colorings is to let the sets have different sizes, but for each to have a certain minimum size.

If the sets $S(x)$ are finite, there is no loss of generality in thinking of them as being sets of integers. Another variation on n–tuple colorings is to let $S(x)$ be a set of consecutive integers. This is a natural variation in connection with the various applications, especially the scheduling applications. This variation has been studied by Roberts [1979], Opsut and Roberts [1981], and Opsut [1984].

If we ask for sets of consecutive integers, it is also natural to ask that each set $S(x)$ have a certain minimum size, r_x. We call a set coloring which satisfies these conditions a *consecutive coloring*. Let $\chi_{c,r_x}(G)$ denote the minimum order of a consecutive coloring. An upper bound on this number is given by the following theorem:

Theorem 14 *(de Werra and Hertz 1988)*

$$\chi_{c,r_x} \le \max_x\, [r_x + \{\Sigma\, r_y : x, y \text{ adjacent}\}].$$

As Gilbert [1972] points out in another context, it is easy to see that a lower bound on the minimum order of a consecutive coloring is given by

$$\chi_{c,r_x} \ge \omega(r_x) = \max\{\sum_{x \in K} r_x : K \text{ is a clique}\}.$$

Graphs for which this inequality is an equality for all sets of nonnegative integers r_x are called *superperfect*.[1] Alan Hoffman has shown that all comparability graphs are

[1]Technically, superperfection was originally defined using the notion of I–coloring defined in the next section. Then G is defined to be superperfect if for all sets of nonnegative real numbers r_x, $\chi'_{I \ge r_x} = \omega(r_x)$, where $\chi'_{I \ge r_x}(G)$ is like $\chi_{I \ge r_x}(G)$ as defined in the next section except that all intervals are of length exactly r_x. It is easy to show that $\chi'_{I \ge r_x} = \chi_{I \ge r_x}$ and that if all r_x are integers, these two numbers equal χ_{c,r_x}. Moreover, Golumbic [1980] observes that $\chi'_{I \ge r_x} = \omega(r_x)$ for all sets r_x of nonnegative reals if and only if this is true for all sets r_x of nonnegative integers. Hence, the two notions of superperfection are equivalent.

superperfect. Characterization of the superperfect graphs is still an open problem. For more on superperfection, see Golumbic [1980].

To illustrate some of these ideas, consider first the graph C_4 and take $r_x = 2$ for all x. Then if the vertices around the cycle are in order x_1, x_2, x_3, x_4, we find that the following is a consecutive coloring: $S(x_1) = S(x_3) = \{1, 2\}$, $S(x_2) = S(x_4) = \{3, 4\}$. It follows that for C_4, $\chi_{c,r_x} = \omega(r_x) = 4$. However, for C_5 with $r_x = 2$ for all x, $\omega(r_x)$ still equals 4, while it is easy to see that $\chi_{c,r_x} > 4$.

6.3. I–Colorings

For some problems, it makes sense to take the set S(x) to be a real interval. For instance, in traffic phasing or task assignment, S(x) is an interval of time, in channel assignment and mobile radio frequency assignment, S(x) is a frequency band, and in fleet maintenance, S(x) is a space along a linear dock (as in the shipbuilding problem). A set coloring in which each S(x) is a real interval is called an *I–coloring*. A set intersection assignment in which each S(x) is a real interval is called an *I–intersection assignment*. Figure 1 shows a graph and an I–coloring.

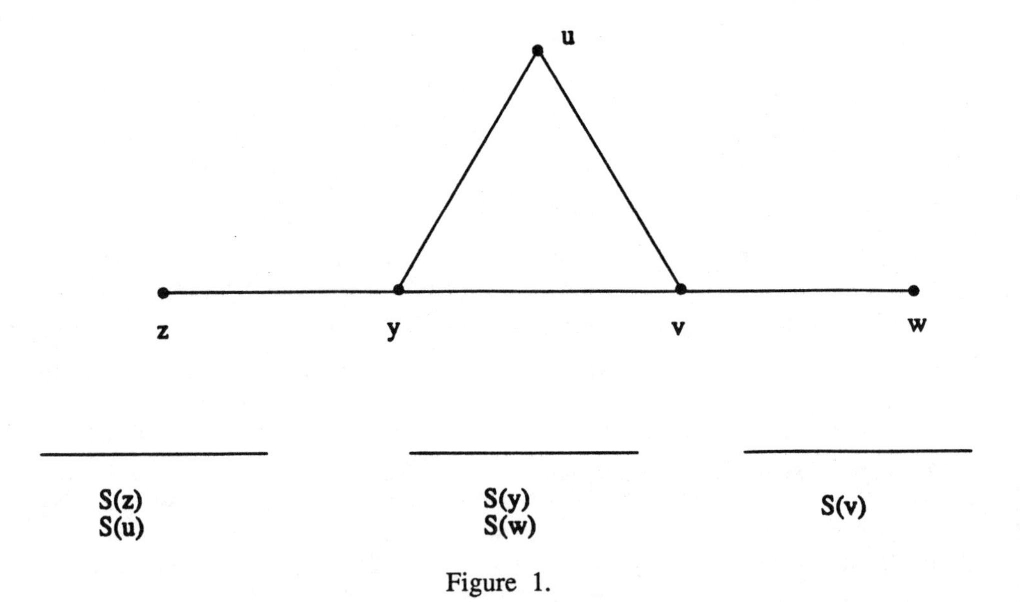

Figure 1.

In dealing with set assignments using real intervals, it makes sense to interpret the size of a set as its measure or length rather than its cardinality and to compute the infimum order rather than the minimum order. For set colorings, this infimum is always 0: we take all the intervals to be very small. The problem is more interesting, and more realistic, if we require (as in Section 6.2) that the interval assigned to x have a length which is at least as large as a specified minimum r_x. The requirement of specified minimum length makes sense in applications such as traffic phasing or fleet maintenance.

We define $\chi_{I \geq r_x}(G)$ to be $\chi_\lambda(G)$ if λ is the property that each set $S(x)$ is a real interval of length at least r_x. We define $i_{I \geq r_x}(G)$ analogously in terms of $i_\lambda(G)$. Thus, $\chi_{I \geq r_x}(G)$ is the infimum of the measures of $\cup S(x)$ over all set colorings S such that each $S(x)$ is a real interval of length $\geq r_x$.

To give an example, consider the graph of Figure 1 and suppose $r_u = 1, r_z = 1, r_y = 2, r_v = 1, r_w = 1$. By using the I–coloring shown in the figure and taking the lengths of the intervals assigned to z, u, and v as 1 and the lengths of the intervals assigned to y and w as 2, we see that $\chi_{I \geq r_x}(G) \leq 4$. It is clear that $\chi_{I \geq r_x}(G) = 4$.

In general, computation of $\chi_{I \geq r_x}(G)$ is an NP–complete problem. By a result of Larry Stockmeyer (see Golumbic [1980]), it is NP–complete even if all r_x are 1 or 2 and G is an interval graph.

However, by the following theorem, $i_{I \geq r_x}(G)$ can be computed in polynomial time for those graphs for which it is defined, namely the interval graphs.

Theorem 15 *(Opsut and Roberts 1983b) The number $i_{I \geq r_x}(G)$ can be computed by solving a linear program whose variables depend on the maximal cliques of G. Hence, $i_{I \geq r_x}(G)$ can be computed in polynomial time for all interval graphs.*

The second part of the theorem follows because for an interval graph, the maximal cliques can be found in polynomial time (see Golumbic [1980]). Since linear programs can be solved in polynomial time, the result follows.

For the reader who is interested, the linear program mentioned in Theorem 15 is defined as follows. If $K_1, \ldots, K_p$ are the maximal cliques of G, $i_{I \geq r_x}(G)$ is the solution to the following linear programming problem:

$$\text{Minimize} \quad \sum_{i=1}^{p} d_i$$

$$\text{Subject to} \quad \sum_{i: x \in K_i} d_i \geq r_x \qquad \text{(for all } x \in V(G))$$

$$d_i \geq 0.$$

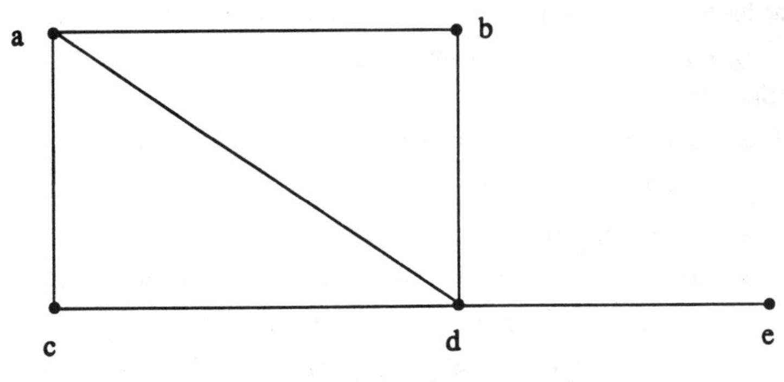

Figure 2.

Consider for example the graph of Figure 2 where $r_a = 2$, $r_b = 1$, $r_c = 3$, $r_d = 1$, and $r_e = 2$. The maximal cliques are $K_1 = \{a, b, d\}$, $K_2 = \{a, c, d\}$, $K_3 = \{d, e\}$. Then we have the linear programming problem

$$\begin{array}{lll} \text{Minimize} & d_1 + d_2 + d_3 & \\ \text{Subject to} & d_1 + d_2 \geq 2 & (x = a) \\ & d_1 \geq 1 & (x = b) \\ & d_2 \geq 3 & (x = c) \\ & d_1 + d_2 + d_3 \geq 1 & (x = d) \\ & d_3 \geq 2 & (x = e). \end{array}$$

The solution to this problem is: $d_1 = 1$, $d_2 = 3$, $d_3 = 2$, $i_{I \geq r_x}(G) = 6$.

Recall the observation of Theorem 8 that $\chi_{I \geq r_x}(G)$ can be computed as the minimum $i_{I \geq r_x}(H)$ over spanning subgraphs H of G^c which have a set intersection of the kind described. Thus, $\chi_{I \geq r_x}(G)$ can be computed by solving these linear programs for each such H. Unfortunately, there might be lots of these linear programs. However, this observation leads to some useful conclusions, such as:

Theorem 16 *(Opsut and Roberts 1983a)* *If G^c is an interval graph, then $\chi_{I \geq r_x}(G)$ is $i_{I \geq r_x}(G^c)$ and hence can be computed in polynomial time.*

There are analogous results for scores. For scores with assignments of intervals, it again makes sense to interpret size as measure or length and it now makes sense to take the supremum rather than the maximum. Thus, let $\chi^{I \geq r_x, N}(G)$ be the supremum score of any set coloring where each set S(x) is a real interval of length at least r_x and where the total order is at most N. Let $i^{I \geq r_x, N}(G)$ be defined analogously for set intersection assignments.

Theorem 17 *(Opsut and Roberts 1983b) The numbers $i^{I \geq r_x, N}(G)$ can be computed by solving a linear program whose variables depend on the maximal cliques of G. Hence, $i^{I \geq r_x, N}(G)$ can be computed in polynomial time for all interval graphs.*

It follows from the observation of Theorem 9 that $\chi^{I \geq r_x, N}(G)$ can be computed by solving these linear programs for each spanning subgraph H of G^c which has a set intersection assignment of the kind described. One corollary of this observation is that $\chi^{I \geq r_x, N}(G)$ can be computed in polynomial time (indeed $O(v^2)$ time) for G the complement of an interval graph. (See Opsut and Roberts [1983a]).

It still remains an open problem to determine if $\chi_{I \geq r_x}(G)$ and $\chi^{I \geq r_x, N}(G)$ are computable in polynomial time for G the complement of a chordal graph.

6.4. J–Colorings

Sometimes it makes more sense to think of each set S(x) as a union of intervals. This occurs, for example, in traffic phasing, task assignment, channel assignment, or mobile radio frequency assignment. We call a set coloring in which each S(x) is a union of real intervals a *J–coloring*. A *J–intersection assignment* is defined analogously.

It also makes sense to study J–colorings and J–intersection assignments in which $\cup S(x)$ has total length at least a specified number r_x. We denote this condition by $J \geq r_x$. Then the numbers $\chi_{J \geq r_x}$, $i_{J \geq r_x}$, $\chi^{J \geq r_x, N}$, and $i^{J \geq r_x, N}$ are defined in the usual manner as χ_λ, i_λ, $\chi^{\lambda, N}$, and $i^{\lambda, N}$, respectively. Note that the second number is always defined because every graph is the intersection graph of sets each of which is a union of real intervals (Griggs and West [1980], Trotter and Harary [1979]).

Theorem 18 *(Raychaudhuri 1985) For any graph G,*

$$\chi_{J \geq r_x}(G) = i_{J \geq r_x}(G^c)$$

and both numbers can be calculated by a linear program whose variables are defined in terms of the maximal cliques of G^c.

Note that this theorem does not say that $\chi_{J \geq r_x}(G)$ is computable in a polynomial number of steps for every G. It is necessary first to show that the maximal cliques of G^c can be indentified in a polynomial number of steps. The latter is true for G^c a chordal graph and hence for G^c an interval graph (see Golumbic [1980]).

The next theorem relates I–colorings to J–colorings.

Theorem 19 *(Raychaudhuri 1985) If G is the complement of an interval graph, then*

$$\chi_{J \geq r_x}(G) = \chi_{I \geq r_x}(G).$$

Thus, we have the surprising result that for complements of interval graphs, going to unions of intervals does not allow us to improve the total order of a coloring.

Raychaudhuri obtains analogous results for scores.

We close this subsection by mentioning a number of open problems. An important one is to find ways to compute the minimum order and maximum score for *D–colorings*, colorings in which each set S(x) is a union of at most two real intervals (the union being of at least a certain minimum size r_x). One difficulty here is that the characterization of the intersection graphs of families of sets each of which is a union of at most two intervals, the so–called *double interval graphs*, remains an open question. (See Trotter and Harary [1979] for partial results.)

Another open problem is to investigate the minimum order or maximum score when the sets S(x) are rectangles in the plane (of certain minimum measure). This question arises in the fleet maintenance problem if each vehicle is assigned a rectangular space in the floor of a maintenance facility.

7. List Colorings

In many practical coloring problems a choice of color to assign to x is restricted. A set or *list* R(x) of possible colors to be assigned to x is specified, and we seek a graph coloring F(x) so that F(x) always belongs to R(x). We say that G can be *list–colored* if there is such a coloring for G. List colorings arise for instance in the channel assignment problem (when we specify possible acceptable channels) or the traffic phasing problem (when we specify possible green times say for use of a classroom). Erdös, Rubin, and Taylor [1979] introduced the idea of considering when a graph G can be list–colored for every assignment of lists R(x) of k colors each to the vertices. If G can always be list–colored for every such assignment, we say that G is *k-choosable*. We define the *choice number* of G, ch(G), to be the smallest k so that G is k–choosable. (Bollobas and Harris [1985] consider similar concepts for edge colorings.) To illustrate this concept, we note that C_5 is not 2–choosable: Let R(x) = {1, 2} for every x. However, C_4 is 2–choosable.

Among the known results about graph list colorings is a characterization of 2-choosability. To present the characterization, let us say that a graph is called a *θ-graph* if it consists of two distinguished vertices u and v and three chains joining u to v which are otherwise vertex–disjoint. The graph $\theta_{i,j,k}$ is the θ–graph in which the chains between u and v have lengths i, j, and k. The *core* of a graph is obtained by successively removing vertices of degree 1 until it is impossible to do so.

Theorem 20 *(Erdös, Rubin, and Taylor 1979) A connected graph is 2–choosable if and only if its core is either a single vertex, an even cycle, or the graph* $\theta_{2,2,2m}$ *for some* $m \geq 1$.

A more recent result about list colorings is the following:

Theorem 21 *(Tesman 1988, 1989) If G is chordal, the choice number of G is* $\chi(G)$.

By way of open problems, we mention two conjectures which still seem to be unsettled:

Conjecture *(Erdös, Rubin, and Taylor 1979) Every planar graph is 5–choosable.*

Conjecture *(Erdös, Rubin, and Taylor 1979) There is a planar graph which is not 4–choosable.*

8. Set T–Colorings

We have described so far a variety of generalizations or variants of ordinary graph coloring. It seems reasonable to consider as well different combinations of these. In what follows, we describe some results about combinations of set coloring and T-coloring and about combinations of list coloring and T–coloring, and we suggest the possibility of combining all three ideas. A *set T–coloring* of G is an assignment of a set S(x) of positive integers to each vertex x of G so that for all $x \neq y$ in V(G),

$$\{x, y\} \in E \rightarrow |F_x - F_y| \notin T$$
$$\text{for all } F_x \in S(x), F_y \in S(y).$$

We shall concentrate on *n–tuple T–colorings*, set T–colorings in which each set is a set of n positive integers.

The *order* of an n–tuple T–coloring is the number of distinct integers used in all of the sets S(x). The *span* of such a coloring is the difference between the largest and smallest integers used in any of the sets S(x). The minimum order of an n–tuple T-coloring of G is denoted $\chi_T^n(G)$. The minimum span over all n–tuple T-colorings of G is denoted $sp_T^n(G)$. Of course, if $T = \{0\}$, $\chi_T^n = \chi_n$.

Many of the early Cozzens–Roberts [1982] results about ordinary T–colorings go over for n–tuple T–colorings.

Theorem 22 *(Tesman 1988)* $\chi_T^n(G) = \chi_n(G)$.

Theorem 23 *(Tesman 1988)* $sp_T^n(K_{\omega(G)}) \le sp_T^n(G) \le sp_T^n(K_{\chi(G)})$.

Theorem 24 *(Tesman 1988) The greedy algorithm on a compatible ordering and the reverse of a perfect elimination ordering computes* χ_T^n and sp_T^n *for* 1*-unit sphere graphs and chordal graphs, respectively, if* T *is* $\{0, 1, ..., r\}$.

However, this greedy algorithm fails for general r–initial sets, even for complete graphs. It also fails for simple k–multiple of s sets {0, s, 2s, ..., ks}, even for complete graphs.

One case where the Cozzens–Roberts results do not go over easily is captured in the following theorem.

Theorem 25 *(Tesman 1988, Füredi, Griggs, and Kleitman 1988) Suppose* $T = \{0, 1, ..., r\}$. *Then if* G *has an edge,*

$$(r+1)[\chi(G)-1] + 2(n-1) \le sp_T^n(G) \le (n+r)\chi(G) - (r+1).$$

The following conjecture presents an open problem in this area:

Conjecture *(Füredi, Griggs, and Kleitman 1988) All values in the interval in the above theorem are attained by suitable graphs* G.

Theorem 26 *(Füredi, Griggs, and Kleitman 1988) The conjecture is true when* $n = 2$.

9. List T–Colorings

In this section, we combine the idea of a list coloring with the idea of a T–coloring. Given lists R(x) for G and a set T of nonnegative integers, a *list T–coloring* of G is a T–coloring F in which each F(x) belongs to the set R(x). This idea is mentioned in Hale [1980] and Roberts [1986]. We say that G is *T–k–choosable* if there is a T–coloring for every assignment of lists R(x) in which each list has k elements. The *T–choice number* of G, T–ch(G), is the smallest k so that G is T–k–choosable.

For instance, suppose T = {0, 1}. Then K_3 is not T–4–choosable. For, suppose the vertices of K_3 are x, y, and z and we take R(x) = R(y) = R(z) = {1, 2, 3, 4}. Then it is easy to show that we cannot find F(x), F(y), F(z) from R(x), R(y), R(z), respectively, with

$$|F(x) - F(y)| \ne 0, 1$$

$$|F(x) - F(z)| \neq 0, 1$$
$$|F(y) - F(z)| \neq 0, 1.$$

On the other hand, suppose again that $T = \{0, 1\}$. Then K_3 is T–7–choosable. To see why, pick $F(x) = a$. Thus, $F(y) \neq a, a-1, a+1$. Since $|R(y)| = 7$, we can pick $F(y) = b \neq a, a-1, a+1$. Now we need $F(z) \neq a, a-1, a+1, b, b-1, b+1$. Hence, since $|R(z)| = 7$, we can pick $F(z) = c$ for some $c \in R(z)$.

The following is a recent result about list T–colorings.

Theorem 27 *(Tesman 1988, 1989) For G chordal and* $T = \{0, 1, ..., r\}$,
$$(r+1)[\chi(G)-1]+1 \leq T\text{-}ch(G) \leq (2r+1)[\chi(G)-1]+1.$$

It is easy to show that both bounds in Theorem 27 are tight, the lower bound for arbitrary χ and the upper bound at least for $\chi = 2$; moreover, the lower bound is tight for 1–unit sphere graphs. However, it remains an open question to determine if the upper bound is tight for 1–unit sphere graphs.

10. Set List T–Colorings

If we combine three ideas, we can think of set list T–colorings, at least n–tuple list T–colorings. There are two possible variants of this idea, one where the lists R(x) consist of acceptable single elements and the other where the lists R(x) consist of acceptable sets (n–element sets). To the best of the author's knowledge, no work has been done on either of these variants, and they appear to be attractive variants for consideration.

Acknowledgement: The author thanks Pey–chun Chen, Garth Isaak, Suh–ryung Kim, and Barry Tesman for their helpful comments and the Air Force Office of Scientific Research for its support under grant number AFOSR–85–0271 to Rutgers University.

REFERENCES

Beltrami, E., and Bodin, L., "Networks and Vehicle Routing for Municipal Waste Collection," *Networks*, **4** (1973), 65 – 94.

Bodin, L.D., and Friedman, A.J., "Scheduling of Committees for the New York State Assembly," Tech. Report USE No. 71–9, Urban Science and Engineering, State University of New York, Stony Brook, 1971.

Bollobas, B., and Harris, A.J., "List Colourings of Graphs," *Graphs and Combinatorics*, **1** (1985), 115 – 127.

Bondy, J.A., and Murty, U.S.R., *Graph Theory with Applications*, Elsevier, New York/MacMillan, London, 1976.

Chvátal, V., "Perfectly Ordered Graphs," in C. Berge and V. Chvátal (eds.), *Topics on Perfect Graphs,* North–Holland, Amsterdam, 1984, 63 – 65.

Cozzens, M.B., and Roberts, F.S., "T–Colorings of Graphs and the Channel Assignment Problem," *Congressus Numerantium,* **35** (1982), 191 – 208.

Cozzens, M.B., and Wang, D–I., "The General Channel Assignment Problem," *Congressus Numerantium,* **41** (1984), 115 – 129.

de Werra, D., and Hertz, A., "Consecutive Colorings of Graphs," *Zeitschrift für Operations Research,* **32** (1988), 1 – 8.

Erdös, P., Goodman, A., and Pósa, L., "The Representation of a Graph by Set Intersections," *Can. J. Math.,* **18** (1966), 106 – 112.

Erdös, P., Rubin, A.L., and Taylor, H., "Choosability in Graphs," *Congressus Numerantium,* **26** (1979), 125 – 157.

Fishburn, P.C., *Interval Orders and Interval Graphs,* Wiley, New York, 1985.

Füredi, Z., Griggs, J.R., and Kleitman, D.J., "Pair Labellings with Given Distance," mimeographed, Institute for Mathematics and its Applications, University of Minnesota, Minneapolis, 1988. To appear in *SIAM J. on Discr. Math.*

Gilbert, E.N., Unpublished Technical Memorandum, Bell Telephone Laboratories, Murray Hill, New Jersey, 1972.

Golumbic, M.C., *Algorithmic Graph Theory and Perfect Graphs,* Academic Press, New York, 1980.

Griggs, J.R., and West, D.B., "Extremal Values of the Interval Number of a Graph," *SIAM J. Alg. Disc. Math.,* **1** (1980), 1 – 7.

Hale, W.K., "Frequency Assignment: Theory and Applications," *Proc. IEEE,* **68** (1980), 1497 – 1514.

Havel, T., "The Combinatorial Distance Geometry Approach to the Calculation of Molecular Conformation," *Congressus Numerantium,* **35** (1982), 361 – 371.

Havel, T., Kuntz, I.D., and Crippen, G.M., "The Combinatorial Distance Geometry Approach to the Calculation of Molecular Conformation I: A New Approach to an Old Problem," *J. Theor. Biol.,* **104** (1983), 359 – 381.

Havel, T., Kuntz, I.D., Crippen, G.M., and Blaney, J.M., "The Combinatorial Distance Geometry Approach to the Calculation of Molecular Conformation II: Sample Problems and Computational Statistics," *J. Theor. Biol.,* **104** (1983), 383 – 400.

Hell, P., and Nesetril, J., "On the Complexity of H–Coloring," Tech. Rep. TR–86–4, Department of Computing Science, Simon Fraser University, Burnaby, British Columbia, 1986.

Irving, R.W., "NP–Completeness of a Family of Graph–Colouring Problems," *Discr. Appl. Math.,* **5** (1983), 111 – 117.

Maehara, H., "Space Graphs and Sphericity," *Discr. Appl. Math.,* **7** (1984), 55 – 64.

Metzger, B.H., "Spectrum Management Technique," paper presented at 38th National ORSA Meeting, Detroit, MI, 1970.

Opsut, R.J., *Optimization of Set Assignments for Graphs*, Ph.D. Thesis, Department of Mathematics, Rutgers University, New Brunswick, N.J., May 1984.

Opsut, R.J., and Roberts, F.S., "On the Fleet Maintenance, Mobile Radio Frequency, Task Assignment, and Traffic Phasing Problems," in G. Chartrand, Y. Alavi, D.L. Goldsmith, L. Lesniak–Foster, and D.R. Lick (eds.), *The Theory and Applications of Graphs,* Wiley, New York, 1981, 479 – 492.

Opsut, R.J., and Roberts, F.S., "I–colorings, I–phasings, and I–intersection Assignments for Graphs, and their Applications," *Networks,* **13** (1983), 327 – 345. (a)

Opsut, R.J., and Roberts, F.S., "Optimal I–intersection Assignments for Graphs: A Linear Programming Approach," *Networks,* **13** (1983), 317 – 326. (b)

Raychaudhuri, A., *Intersection Assignments, T Coloring, and Powers of Graphs,* Ph.D. thesis, Department of Mathematics, Rutgers University, New Brunswick, N.J., 1985.

Raychaudhuri, A., "Further Results on T–Colorings and Frequency Assignment Problems," mimeographed, Department of Mathematics, College of Staten Island (CUNY), Staten Island, New York, 1988.

Roberts, F.S., "Indifference Graphs," in F. Harary (ed.), *Proof Techniques in Graph Theory,* Academic Press, New York, 1969, 139 – 146.

Roberts, F.S., *Discrete Mathematical Models, with Applications to Social, Biological, and Environmental Problems,* Prentice–Hall, Englewood Cliffs, N.J., 1976.

Roberts, F.S., *Graph Theory and its Applications to Problems of Society,* CBMS–NSF Monograph No. 29, SIAM, Philadelphia, 1978.

Roberts, F.S., "On the Mobile Radio Frequency Assignment Problem and the Traffic Light Phasing Problem," *Annals New York Acad. Sci.,* **319** (1979), 466 – 483.

Roberts, F.S., *Applied Combinatorics*, Prentice–Hall, Englewood Cliffs, N.J., 1984.

Roberts, F.S., "T–Colorings of Graphs: Recent Results and Open Problems," Research Rep. RRR 7–86, Rutgers Center for Operations Research, Rutgers University, New Brunswick, N.J., May 1986. To appear in *Annals of Discrete Mathematics.*

Stoffers, K.E., "Scheduling of Traffic Lights – A New Approach," *Transportation Res.,* **2** (1968), 199 – 234.

Stahl, S., "n–tuple Colorings and Associated Graphs," *J. Comb. Theory,* **20** (1976), 185 – 203.

Tesman, B., Ph.D. Thesis, Department of Mathematics, Rutgers University, New Brunswick, New Jersey, in preparation, 1988.

Tesman, B., "Vertex List T–Colorings of Graphs," mimeographed, Department of Mathematics, Rutgers University, New Brunswick, NJ 1989.

Trotter, W.T., and Harary, F., "On Double and Multiple Interval Graphs," *J. Graph Theory,* **3** (1979), 205 – 211.

Tucker, A.C., "Perfect Graphs and an Application to Optimizing Municipal Services," *SIAM Rev.,* **15** (1973), 585 – 590.

Wang, D–I., "The Channel Assignment Problem and Closed Neighborhood Containment Graphs," Ph.D. Thesis, Northeastern University, Boston, MA, 1985.

Zoellner, J.A., and Beall, C.L., "A Breakthrough in Spectrum Conserving Frequency Assignment Technology," *IEEE Trans. on Electromag. Comput.,* **EMC–19** (1977), 313 – 319.

EDGE DOMINATING NUMBERS OF COMPLEMENTARY GRAPHS

Seymour Schuster

ABSTRACT

An edge dominating set in a graph G is a set S of edges of G such that every edge of G belongs to S or is adjacent to an edge of S. The edge dominating number (also called the edge-edge-covering number) $\alpha_{11}(G)$ of a graph G is the cardinality of a smallest dominating set of edges in G. Inequalities of the Nordhaus-Gaddum type are established, providing best possible upper and lower bounds for $\alpha_{11}(G) + \alpha_{11}(\overline{G})$ and $\alpha_{11}(G) \cdot \alpha_{11}(\overline{G})$, where $\overline{G}$ is the complement of G.

1. Introduction

In 1956, Nordhaus and Gaddum [9] established the following inequalities involving the chromatic number $\chi(G)$ of a graph G of order p and the chromatic number $\chi(\overline{G})$ of its complement $(\overline{G})$:

$$2\sqrt{p} \leq \chi(G) + \chi(\overline{G}) \leq p + 1$$

$$p \leq \chi(G) \cdot \chi(\overline{G} \leq \lfloor \frac{p+1}{2} \rfloor^2.$$

Inequalities of this "Nordhaus-Gaddum type" have also been found for a host of other graphical parameters: achromatic number by Gupta [7]; edge chromatic number by Alavi and Behzad [2] and by Vizing [10]; domination number by Jaeger and Payan [8]; connectivity and edge connectivity by Alavi and Mitchem [3]; diameter, girth, circumference, covering number and edge covering number by Xu [11]; independence and edge independence number by Chartrand and Schuster [5] and by Erdös and Schuster [6].

In the current paper, we study the edge domination number (which has also been called the edge-edge covering number (with the aim of obtaining Nordhaus-Gaddum inequalities for this graphical parameter. We begin by presenting some necessary definitions and accompanying notation. A set of vertices is called *independent* if no two vertices in the set are adjacent. The *vertex independence number* $\beta_0(G)$ of a graph G is the cardinality of a largest independent set of vertices of G. A *vertex dominating set* S for a graph G is the set of vertices such that every vertex of G belongs to S or is adjacent to a vertex of S. The cardinality of a smallest vertex dominating set is called the *vertex dominating number* $\alpha_{00}(G)$ *of* G. For euphonic consideration, the word "dominating" is occasionally replaced by "domination". Also, the vertex dominating number has been called the *vertex-vertex covering number* since the dominating set S covers the vertex set $V(G)$ of G. The *edge independence number* $\beta_1(G)$, and *edge dominating set* and the *edge dominating number* $\alpha_{11}(G)$ are defined analogously.

2. Relations Between Edge Domination and Edge Independence

Since the principal focus of our study is on the edge dominating number α_{11}, we begin with some immediate observations concerning edge domination.

Observation 1 If S is a smallest edge dominating set in G and $uv \in E(G)$, then at least one of u and v belong to an edge of S.

This is so, for otherwise uv would not be covered by S.

Observation 2 A smallest edge dominating set cannot possess three consecutive edges. This is true, for if v_1v_2, v_2v_3, v_3v_4 are in a dominating set, then v_1v_2 and v_3v_4 already cover the edges covered by v_2v_3. So edge v_2v_3 is superfluous in the dominating set.

Observation 3 For any graph G of order $p, \alpha_{11}(G) \leq \beta_1(G) \leq \lfloor \frac{p}{2} \rfloor$.

Proof Let T be a maximal independent set of edges of G. Then T covers $E(G)$; otherwise, there would exist $e \in (E(G) - T)$ that would not be adjacent to any edge of T, and this would contradict the maximality of T. Hence, $\alpha_{11}(G) \leq |T| \leq \beta_1(G) \leq \lfloor \frac{p}{2} \rfloor$. See [1].

Observation 4 If K_p is the complete graph of order p, then $\alpha_{11}(K_p) = \lfloor \frac{p}{2} \rfloor$.

Observation 5 For the bipartite graph $K_{m,n}$ with $m \leq n, \alpha_{11})(K_{m,n}) = m$.

Theorem 1 For any graph G, there exists a smallest edge dominating set in G that is also a maximal independent set of edges.

Proof Let S be a smallest edge dominating set that is not an independent set. Then there are two adjacent edges $v_1v_2, v_2v_3 \in S$. The vertex v_3 is not an endvertex (i.e., deg $v_3 > 1$); otherwise v_2v_3 would be superfluous in S. Also, we assert that none of the other edges of G that are incident with v_3 can belong to S, for otherwise Observation 2 would be contradicted. We observe, still further, that if all of these other edges incident with v_3 were adjacent to an edge of S other than v_2v_3, then again v_2v_3 would be superfluous in S. Hence, there is an edge v_3v_4 that is not adjacent to any edge of S other than v_2v_3. We therefore form a new edge-cover

$$S' = (S - \{v_2v_3\}) \cup \{v_3v_4\}$$

Clearly, S' is also a smallest dominating set, and S has one fewer pair of adjacent edges than S. Repeating this process of constructing smallest dominating sets while reducing the number of adjacencies among the edges, we obtain a smallest dominating set that is simultaneously an independent set of edges. Moreover, this cover is a maximal independent set of edges.

In consequence of this theorem, we obtain the following result of Chartrand and Lesniak [4, p. 249]:

Corollary 1 If $\beta_1^*(G)$ is the minimum cardinality among the maximal independent sets of edges of any graph G, then $\alpha_{11}(G) = \beta_1^*(G)$.

Another result - the analogue of a theorem which Allan and Laskar proved for vertex domination in [1] - also follows immediately.

Corollary 2 For any graph G, the independence edge dominating number $\alpha'_{11}(G)$, which is the cardinality of a smallest independent dominating set of edges, equals the edge dominating number of G. That is, $\alpha'_{11}(G) = \alpha_{11}(G)$.

3. The Nordhaus-Gaddum Inequalities

We first establish inequalities of the Nordhaus-Gaddum type that hold for graphs of any order $p \geq 2$; later, we shall see that these may be sharpened in accordance with the number-theoretic character of p.

Theorem 2 For any graph G of order $p \geq 3$,

$$\lfloor \frac{p}{2} \rfloor \leq \alpha_{11}(G) + \alpha_{11}(\overline{G}) \leq 2\lfloor \frac{p}{2} \rfloor. \tag{1}$$

Proof Theorem 1 guarantees the existence of a set S of independent dominating set of edges. If $|S| = k$, then the edges of S are incident with exactly $2k$ vertices. If $u, v \in V(G)$ are not among these $2k$ vertices, then $uv \notin E(G)$ for uv is not adjacent to any edge of S. Hence, the remaining $p - 2k$ vertices induce a subgraph H of G which must be empty. Therefore, $\overline{G} \supset K_{p-2k}$. Since $\alpha_{11}(Kp - 2k) = \lfloor \frac{p-2k}{2} \rfloor$ (by Observation 4), we have

$$\alpha_{11}(G) + \alpha_{11}(\overline{G}) \geq k + \lfloor \frac{p-2k}{2} = \lfloor \frac{p}{2} \rfloor,$$

which establishes the lower bound.

Applying Observation 3 to G and $\overline{G}$, we get

$$\alpha_{11}(G) + \alpha_{11}(\overline{G}) \leq \lfloor \frac{p}{2} \rfloor + \lfloor \frac{p}{2} \rfloor = 2\lfloor \frac{p}{2} \rfloor,$$

which produces the upper bound.

Theorem 3 For every graph G of order $p \geq 2$

$$0 \leq \alpha_{11}(G) \cdot \alpha_{11}(\overline{G}) \leq \lfloor p2 \rfloor^2.$$

Proof Applying the arithmetic-geometric mean inequality to $\alpha_{11}(G)$ and $\alpha_{11}(\overline{G})$ together with Theorem 2 gives

$$\sqrt{\alpha_{11}(G) \cdot \alpha_{11}(\overline{G})} \leq \lfloor p2 \rfloor. \tag{2}$$

The question of whether Theorems 2 and 3 are "best possible" will be discussed in our next section.

4. Realizability

It is easy to find graphs for which the upper and lower bounds in Theorems 2 and 3 are attained provided $p \equiv 0, 1,$ or 3 (mod 4): the lower bounds of the theorems are attained if $G = K_p$; if $p \equiv 0 \pmod 4$, the upper bounds in (1) and (2) are attained when $G = K_{\frac{p}{2},\frac{p}{2}}$. If $p \equiv 1$ or 3, the upper bounds in (1) and (2) are attained when $G = K_{\lceil \frac{p}{2} \rceil, \lfloor \frac{p}{2} \rfloor}$.

Attaining the upper bound if $p \equiv 2 \pmod 4$ is a different matter, for the likely candidate $G = K_{\frac{p}{2},\frac{p}{2}}$ fails to yield the upper bound of (1)-hence, of (2) as well. This is so, for $\overline{G} = 2K_{\frac{p}{2}}$ with $\frac{p}{2}$ odd; hence $\alpha_{11}(\overline{G}) = 2\lfloor \frac{p}{4} \rfloor = 2\lfloor \frac{p}{2} \rfloor - 1 < \frac{p}{2}$. Therefore, if $p \equiv 2 \pmod 4$ then $\alpha_{11}(K_{\frac{p}{2},\frac{p}{2}}) + a_{11}(\overline{K}_{\frac{p}{2},\frac{p}{2}}) = \frac{p}{2} + (\frac{p}{2} - 1) = p - 1$, which leads us to our next result.

Theorem 4 For every graph G of order p with $p \equiv 2 \pmod 4$,

$$\frac{p}{2} \leq \alpha_{11}(G) + \alpha_{11}(\overline{G}) \leq p - 1 \tag{3}$$

$$0 \leq \alpha_{11}(G) \cdot \alpha_{11}(\overline{G}) \leq \lfloor (\frac{p-1}{2}) \rfloor^2. \tag{4}$$

Proof Let us write $p = 4k + 2$, where k is a non-negative integer. Then $|E(G)| + |E(\overline{G})| = \frac{p(p-1)}{2} = \frac{(4k+2)(4k+1)}{2} = (2k + 1)(4k + 1)$, which is odd. Thus, $|E(G)|$ and $|E(\overline{G})|$ are of different parity.

Since $\alpha_{11}(G) \leq \frac{p}{2}$, we establish (3) by proving that $\alpha_{11}(G)$ and $\alpha_{11}(\overline{G})$ cannot both equal $\frac{p}{2}$ simultaneously. We shall do this by showing that $\alpha_{11}(G) = \frac{p}{2}$ only if $|E(G)|$ is odd (in which case $\alpha_{11}(\overline{G}) < \frac{p}{2}$ since $|E(\overline{G})|$ must be even).

We suppose that $\alpha_{11}(G) = \frac{p}{2}$ and that a smallest edge dominating set S consists of the $\frac{p}{2}$ independent edges $v_1v'_1, v_2v'_2, ..., v_{\frac{p}{2}}v'_{\frac{p}{2}}$. With $\equiv 2$ (mod 4), we know that $\frac{p}{2}$ is odd.

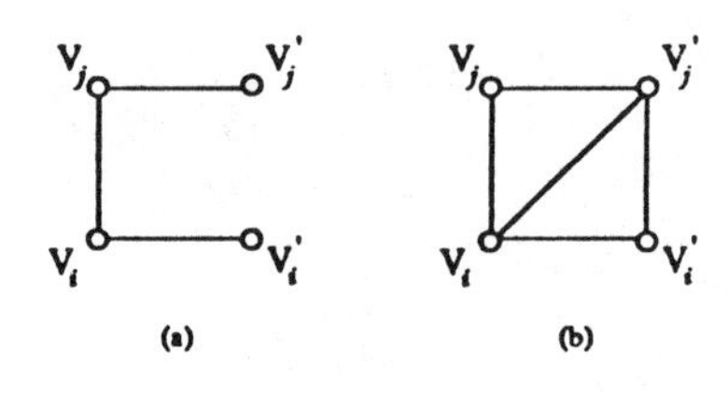

Figure 1

In order to determine the parity of $|E(G)|$, we examine the graphs induced by pairs of edges of S, namely the graph $< v_i, v'_i, v_j, v'_j >$ where $1 \leq i < j \leq \frac{p}{2}$. If any of these graphs have an odd number of edges, then it must be isomorphic to one of the graph in Figure 1. In both of these cases, we would have a contradiction of the fact that S is a smallest dominating set, for the two edges $v_iv'_i$ and $v_jv'_j$ could be replaced by a single edge (v_iv_j in (a) or $v_iv'_j$ in (b)) to obtain a smaller dominating set. Thus the induced graphs have 0, 2 or 4 edges in addition to the two in S, namely $v_iv'_i$ and $v_jv'_j$. Since the number of edges in S is odd and the number not in S is even, we have that $|E(G)|$ is odd. Of course, this implies that $|E(\overline{G})|$ is even so $\alpha_{11}(\overline{G}) < \frac{p}{2}$. This establishes the upper bound of (3).

The remaining bounds follow immediately from Theorems 2 and 3.

REFERENCES

[1] R. B. Allen and R. Laskar, "On domination and independent domination numbers of a graph," *Discrete Math.* 23 (1978), 73-76.

[2] Y. Alavi and M. Behzad, "Complementary graphs and edge chromatic numbers," *SIAM J. Appl. Math.* 20 (1971) 161-163.

[3] Y. Alavi and J. Mitchem, "Connectivity and line-connectivity of complementary graphs," in *Recent Trends in Graph Theory*, M. Capobianco, J. B. Frechen and M. Krolik, eds., Springer, New York, 1971.

[4] G. Chartrand and L. Lesniak, *Graphs and Digraphs*, 2nd ed., Wadsworth, Inc., Belmont, CA, 1986.

[5] G. Chartrand and S. Schuster, "On the independence number of complementary graphs," Trans. New York. Acad. Sci. Ser. II, 36 (1974) 247-251.

[6] P. Erdös and S. Schuster, "Existence of complementary graphs with specified independence numbers," on *Theory and Applications of Graphs*, G. Chartrand, et al., eds., John Wiley and Sons, Inc., 1981, 343-349.

[7] R. P. Gupta, "Bounds on the chromatic and achromatic numbers of complementary graphs," in *Recent Progress in Combinatorics*, W. T. Tutte, ed., Academic Press, New York, 1969, 229.

[8] F. Jaeger and C. Payan, "Relations du type Nordhaus-Gaddum pour le Nombre d'absorption d'un graphe simple," C. R. Acad. Sci. Paris, Ser. A, t. 274 (1972) 728-730.

[9] E. Nordhaus and J. W. Gaddum, "On complementary graphs," *Amer. Math. Monthly* 63 (1956) 175-177.

[10] V. G. Vizing, "The chromatic class of a multigraph," *Cybernetics* 1 (3) (1965) 32-41.

[11] S.-j. Xu, "Some parameters of graph and its complement," *Discrete Math.* 65 (1987) 197-207.

MINIMUM DOMINATING, OPTIMALLY INDEPENDENT VERTEX SETS IN GRAPHS

W. J. Selig

NASA/Marshall Space Flight Center
Huntsville, AL

P. J. Slater

The University of Alabama in Huntsville

ABSTRACT

As one example of a new class of problems (namely, evaluating a "single set, prioritized multiproperty" parameter) we introduce the following problem. Find the minimum possible number of edges in the subgraph induced by a minimum dominating set for a graph G. We show this to be an NP-hard problem in general, and we present a linear time algorithm to compute this parameter for a tree.

1. Introduction

Given the graph $G = (V, E)$, vertex set $S \subseteq V$ is independent if no two vertices in S are adjacent, and S is a dominating set if each vertex in $V - S$ is adjacent to at least one vertex in S. Let $\beta_0(G)$ denote the maximum cardinality of an independent vertex set in G, and let $\gamma(G)$ be the minimum cardinality of a dominating set. Determining if $\beta_0(G) \geq K$ is an NP-complete problem (Garey,

*Research supported in part by the U.S. Office of Naval Research Grant N00014-86-K-0745.

Johnson and Stockmeyer [3]), as is deciding if $\gamma(G) < k$ (Garey and Johnson [2]). Finding a maximum independent set (aβ_0-set) and finding a minimum dominating set (a γ-set) are examples of finding a single set that is extremal relative to a single property.

Independence and domination have been widely studied, and extensive bibliographies on the subjects have been compiled by P. L. Hammer (RUTCOR) and S. T. Hedetniemi and R. Laskar (Clemson University), respectively.

On the other hand we can form a partition of the vertex set $V = S_1 \bigcup S_2 \bigcup \cdots \bigcup S_k$ where each S_i has some property P. The chromatic number $\chi(G)$ is the minimum number of vertex sets in a partition of V into independent sets, and the domatic number $do(G)$ is the maximum number of vertex sets in a partition of V into dominating sets. Deciding if $\chi(G) < K$ is NP-complete (Karp [11]), as is determining if $do(G) \geq K$ (Garey, Johnson and Tarjan [4]).

An intermediate type of problem is to optimize a parameter involving a fixed number (such as two) of sets. In Grinstead and Slater [6] we introduced this type of problem, specifically asking for two minimum dominating sets with minimum possible intersection. It was that simply determining if there exist two disjoint minimum dominating sets is an NP-hard problem for arbitrary bipartite graphs. Further results on this particular problem appear in these proceedings [7], and further treatment of the "multiset single parameter" problem type will appear in [5,8].

Here we introduce another problem type, the "single set, prioritized multiproperty" problem type. As an example, we consider the two properties of domination and independence with priority on the domination parameter. Thus we seek a minimum dominating set (a γ-set) which will be "as independent as possible."

We first note that the fairly well studied parameter $i(G)$, the independent domination number of G, is not really a multiproperty parameter. We defined $\gamma(G)$ as the minimum number of vertices in a (minimal) dominating set, and $\Gamma(G)$ has been defined as the maximum number of vertices in a minimal dominating set. Similarly, $\beta_0(G)$ is the maximum number of vertices in a (maximal) independent set, and $i(G)$ is the minimum number of vertices in a maximal independent set (see Beyer, et al [1]).

Because each independent set is maximal if and only if it is a dominating set, $i(G)$ equals the minimum number of vertices in a dominating set that is independent. But the single property of independence suffices to define $i(G)$.

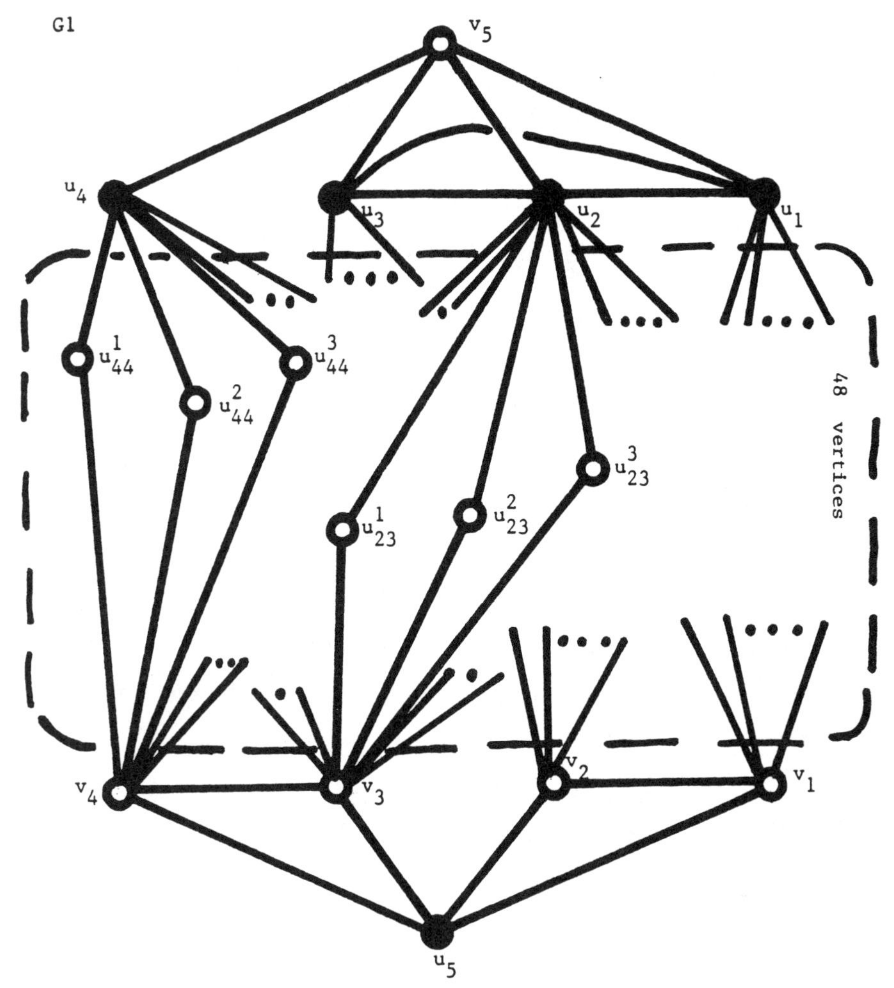

Figure 1. Graph $G1$ with $\gamma(G1) = 5$ and $\gamma(G1)$ sets $\{u_1, u_2, u_3, u_4, u_5\}$ and $\{v_1, v_2, v_3, v_4, v_5\}$

Several measures of what "as independent as possible" means could be used. We illustrate two of these in Figure 1. The graph $G1$ has 58 vertices with $V(G1) = \{u_1, u_2, u_3, u_4, u_5, v_1, v_2, v_3, v_4, v_5\} \bigcup \{u_{ij}^k \ : 1 \le i, j \le 4, 1 \le k \le 3\}$. The edge set contains $\{u_1u_2, u_2u_3, u_1u_3, u_1v_5, u_2v_5, u_3v_5, u_4v_5, v_1v_2, v_3v_4, v_1u_5, v_2u_5, v_3u_5, v_4u_5\}$. Further, each u_{ij}^k is a vertex of degree two adjacent to u_i and v_j. That is, each of the sixteen pairs $\{u_i, v_j\}$ share three common neighbors, each of degree two. It is fairly obvious that $\gamma(G1) = 5$, and the only γ-sets are $D1 = \{u_1, u_2, u_3, u_4, u_5\}$ and $D2 = \{v_1, v_2, v_3, v_4, v_5\}$. Vertex set $D1$ can be considered to be more independent than $D2$ because $< D1 >$, the subgraph generated by $D1$, has two isolated vertices while $< D2 >$ only has one. On the other hand, $D2$ can be considered to be more independent than $D1$ because $< D2 >$ has only two edges while $< D1 >$ has three. It is this latter measure we will use here.

2. The NP-Hard Dipset Problem

Given a graph $G = (V, E)$ we are interested in finding a minimum Dominating set D which is as Independent as Possible under the criterion that $< D >$ has as few edges as possible. We define $\text{DIP}(G)$ to be the minimum number of edges in a subgraph induced by a minimum dominating set of G. In particular, $\text{DIP}(G) = 0$ if and only if $\gamma(G) = i(G)$. For example consider the tree $T2$ in Figure 2. We have $i(G) = 9$ and {2, 8, 10, 13, 15, 17, 20, 24, 27} is an $i(G)$-set. Also, $\gamma(G) = 8$ and $\{2, 8, 10, 14, 17, 20, 24, 27\} = D1$ is a $\gamma(G)$-set with $< D1 >$ containing four edges. $D1$ is the minimum dominating set obtained by the "anti-greedy" algorithm applied to $T2$ with vertex 17 as the root, and $D1$ is "locally optimal" in the sense that any $\gamma(T2)$-set $D2$ satisfying $D2 = D1 - v + u$ where $v \epsilon D1$ has at least four edges in $< D2 >$. Note however that $D = D1 - 17 - 14 + 16 + 13$ is a $\gamma(T2)$-set with only two edges in $< D >$. In fact, DIP(T2) = 2. We call a $\gamma(G)$-set, D, a DIP-set if $< D >$ has DIP(G) edges.

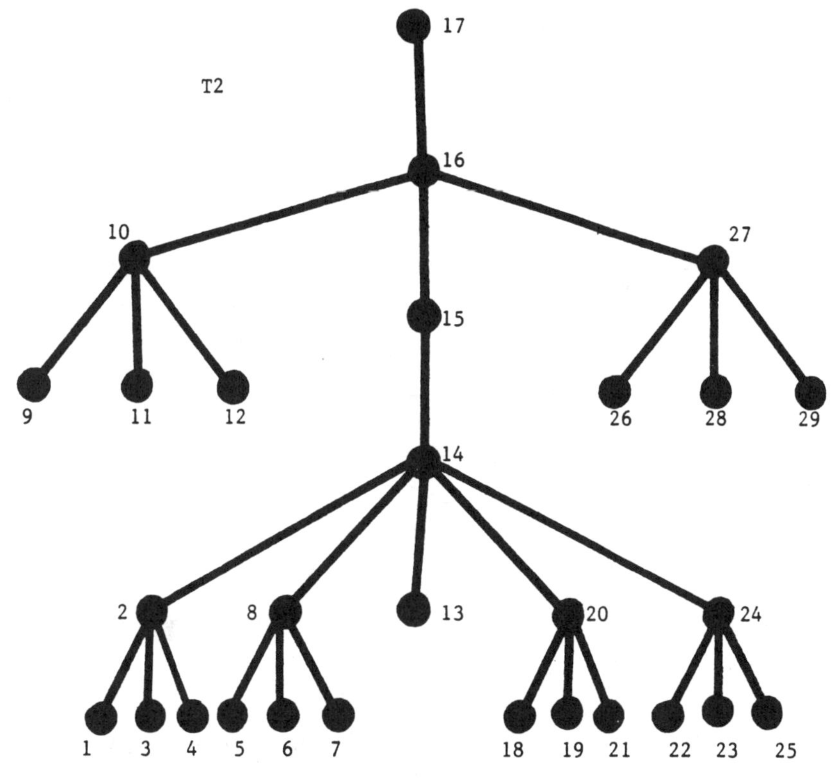

Figure 2. Tree T2 with $\gamma(\text{T2}) = 8$ and $\text{DIP(T2)} = 2$

We next show that determining whether or not $\text{DIP(G)} < \text{K}$ for a non-negative inter K is NP-hard. We describe a polynomial time reduction from 3-Satisfiability, one of the six basic NP-complete problems in Garey and Johnson [2], implying that our problem is NP-hard. We note that DIPSET is in NP^{NP}, but it appears not to be in NP because verification of a proposed solution set D includes verifying that D includes verifying that D is a *minimum* dominating set.

3-Satisfiability

Instance: Collection $C = \{c_1, c_2, \cdots, c_m\}$ of clauses on a finite set $U = \{u_1, u_2, \cdots, u_n\}$ of variables such that $|c_i| = 3$ for $1 \leq i \leq m$.

Question: Is there a truth assignment for U that satisfies all the clauses in C?

Dipset

Instance: Graph $G = (V, E)$, positive integer K.

Question: Is DIP(G) $\leq$ K? Given an instance of 3-satisfiability we construct a graph G as follows. First, for each variable $u_i \epsilon U$ we construct the 7-vertex, 9-edge graph H_i illustrated in Figure 3.

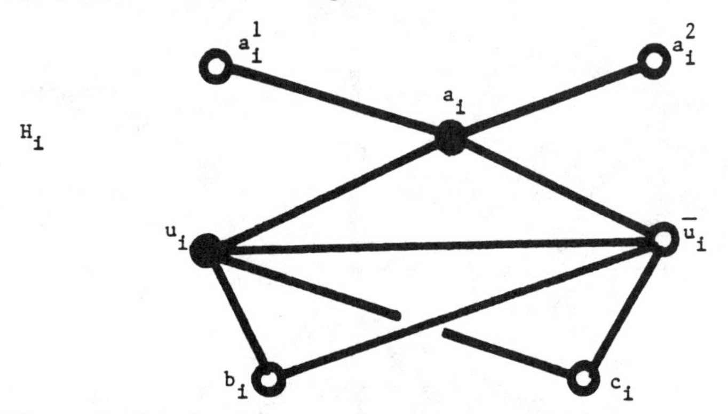

Figure 3. Graph H_i corresponding to variable v_i.

Second, for each clause $c_j \epsilon C$ construct the 21-vertex, 24-edge graph M_j illustrated in Figure 4.

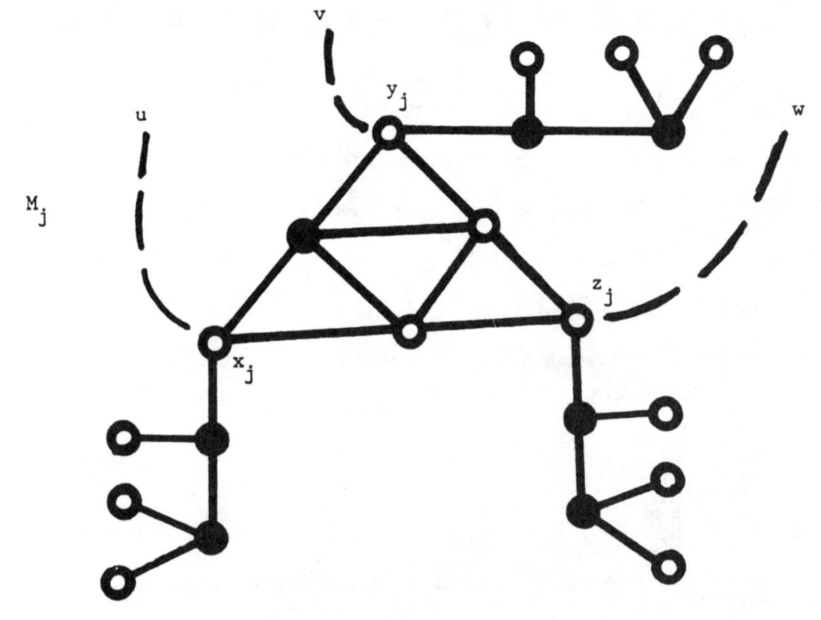

Figure 4. Graph M_j corresponding to clause c_j.

Finally, if $c_j = (u \vee v \vee w)$ where each of u, v, w is a literal u_k or $\overline{u}_k$ then add edges $x_j u, y_j v$, and $z_j w$ as indicated in Figure 4. The graph G so constructed has $7n + 21m$ vertices and can be constructed from C in polynomial time. We have $\gamma(G) = 2n + 7m$. (One can use each a_i and u_i and each set of seven darkened vertices from each M_j to form a $\gamma(G)$-set). Because each $\gamma(G)$-set must contain each a_i and either u_i or $\overline{u}_i$, we always have DIP(G) $\geq$ n. It remains only to see, as can be easily verified, that there exists a truth assignment for U if and only if DIP(G) $\leq$ n.

Indeed, a referee has improved our result and shown the stronger result that *zero-dipset* is NP-hard.

Zero-Dipset

Instance: Graph $G = (V, E)$.

Question: Is DIP(G) = 0? (That is, does $\gamma(G) = i(G)$?)

3. A Linear Algorithm for Determining DIP(T) and a Corresponding DIP-set for a Tree T

In this section we will describe a linear algorithm for determining DIP(T), the minimum cardinality of the edge set of the induced subgraph of $d \subseteq V(T)$, where D is a minimum dominating set of T. We note that this algorithm can be extended to allow us to find such a set D, also in linear time. Throughout this section we will use $\delta(T)$ to denote DIP(T).

Without loss of generality, it will be assumed that all trees are rooted at some vertex that may be chosen arbitrarily. This will enable us to use recursive representations of trees. Given a rooted tree T, we will represent T by the number of nodes in T, say n, an endnode list $EL = (v_1, v_2, \cdots, v_n)$, and an associated parent list $PA = (u_1, u_2, \cdots, u_{n-1})$. The endnode list is any enumeration of the nodes of T in which each node precedes its parent. In the associated parent list, each u_i is the parent of v_i in T. Note that PA has length $n - 1$ and not n, since the root v_n has no parent. For example, the tree of Figure 2 may be represented by n = 29, EL = (1, 5, 18, 22, 23, 19, 6, 3, 4, 7, 21, 25, 13, 2, 9, 8, 26, 20, 11, 24, 28, 14, 12, 15, 29,

10, 27, 16, 17) and PA = (2, 8, 20, 24, 24, 20, 8, 2, 2, 8, 20, 24, 14, 14, 10, 14, 27, 14, 10, 14, 27, 15, 10, 16, 27, 16, 16, 17).

These lists can be constructed for a tree of n nodes in time O(n) (see [12] and [14]), so requiring them does not increase the order of execution time for our algorithm.

We will also make use of the following notation. As the vertex v is reached in left-to-right processing of the endnote list, let Tv be the subtree induced by v and all of its descendants. (Note that all of the descendants of v have already been processed since they precede v in EL.) Let u be the parent of v (which is determined using PA), and let Tu' be the subtree induced by u, the children of u that precede v in EL, and the descendants of all such children. Finally, let tu be composed of Tu', Tv and the edge (u, v). Note that Tu does not necessarily contain all of the descendants of u since there may be children of u which appear after v in EL. See Figure 5.

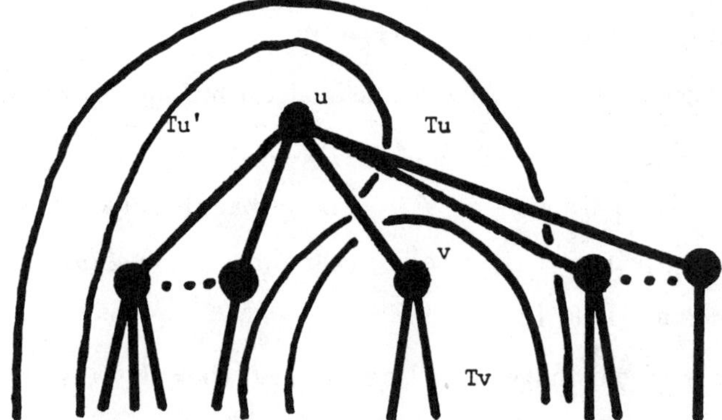

Figure 5. Illustration of Tv, Tu and Tu' notation

The first three parameters we introduce are used to insure that we have a minimum dominating set. Note that $\gamma(T)$ and the corresponding minimum dominating set D can be found without the use of these parameters by using an "anti-greedy" algorithm, but we employ them here because they will be used later in this section in determining $\delta(T)$. For a vertex $u \epsilon V(T)$, define

$$\gamma_y(Tu) = \text{MIN } \{|D| : u \in D, D \text{ dominates } Tu\}$$

$$\gamma_y(Tu) = \text{MIN } \{|D| : u \notin D, D \text{ dominates } Tu\}$$

$$\gamma_{\overline{n}}(Tu) = \text{MIN } \{|D| : u \notin D, D \text{ dominates } Tu - u, \; D \text{ does not dominate } u\}$$

where $D \subseteq V(tu)$. (Note that we need to require $D \subseteq V(Tu)$ only for the case $\gamma_n(Tu)$ where u is an endnode. In this case we will define $\gamma_n(Tu) = \infty$.) That is, let $\gamma_y(Tu)$ be the minimum cardinality of a set containing u that dominates Tu; let $\gamma_n(Tu)$ be the minimum cardinality of a set not containing u that dominates Tu; and let $\gamma_{\overline{n}}(Tu)$ be the minimum cardinality of a set not containing u that dominates $Tu - u$ but specifically does not dominate u itself. This third parameter will be useful when u is to be dominated by its parent or by an as yet unprocessed child.

Similar parameters have been used in other work dealing with minimum dominating sets. See Grinstead and Slater [6]. Note that our parameters differ slightly from theirs in that they used $\gamma\overline{n}(Tu)$ to mean it does not matter whether u is dominated or not. We, however, use it to mean that u specifically *does not* get dominated.

If D is a minimum dominating set of Tu and if $u \epsilon D$, then we may write $D = U \cup V$ where U is a minimum dominating set of Tu', and V is a minimum set of vertices in $V(Tv)$ which dominated $Tv - v$ (since v is already dominated by u). The set V may contain v, not contain v but dominate it, or not contain v and not dominate it.

$$\gamma_y(Tu) = \gamma_y(Tu') + \text{ MIN } \{\gamma_y(Tv), \gamma_n(Tv), \gamma_{\overline{n}}(Tv)\} \tag{1}$$

Similarly, it is straightforward to derive the following:

$$\gamma_n(Tu) = \text{ MIN } \{\gamma_n(Tu') + \gamma_n(Tv), \gamma_n(Tu') + \gamma_y(Tv), \gamma_{\overline{n}}(Tu') + \gamma_y(Tv)\} \tag{2}$$

$$\gamma_{\overline{n}}(Tu) = \gamma_{\overline{n}}(Tu') + \gamma_n(Tv) \tag{3}$$

To see how these parameters should be initialized, consider a subtree consisting of a single vertex v. The minimum number of vertices needed to dominate the subtree using v is one; it is not possible to dominate the subtree without sign v; and zero vertices are required to dominate the subtree minus v. Thus for any endnote v_i of a tree T we may initialize $\gamma_y(Tv_i) = 1, \gamma_n(Tv_i) = \infty$, and $\gamma_{\overline{n}}(Tv_i) = 0$. For any internal node u_i of T we initialize its parameters similarly. After this initialization, we may proceed left-to-right through the endnode list evaluating equations (1) through (3) for u_i, where u_i is the parent of the current endnode list entry v_i. Once v_n is reached, we can determine $\gamma(T) = \text{MIN } \{\gamma_y(Tv_n), \gamma_n(Tv_n)\}$.

As previously stated, similar parameters have been used before. We now present an additional three parameters which are used to determine $\delta(T)$, the minimum cardinality of the edge set of the induced subgraph of $D \subseteq V(T)$, where D is a minimum dominating set (MDS) of T. While the domination parameters are used to maintain information about possible minimum dominating sets, these new parameters are used to maintain information about the "non-independence" of those sets. For a vertex $u \epsilon V(T)$, define

$$\delta_y(Tu) = \text{MIN } \{|E(< D >)| : u \in D, D \text{ is an MDS of } Tu\}$$

$$\delta_n(Tu) = \text{MIN } \{|E(< D >)| : u \notin D, D \text{ is an MDS of } Tu\}$$

$$\delta_{\overline{n}}(Tu) = \text{MIN } \{|E(< D >)| : u \notin D, D \text{ is an MDS of } Tu - u \quad D \text{ does not dominate } u\}$$

That is, let $\delta_y(Tu)$ be the minimum cardinality of the edge set of the subgraph induced by a minimum dominating set of Tu that contains vertex u; let $\delta_n(Tu)$ be the minimum cardinality of the edge set of the subgraph induced by a minimum dominating set of Tu that does not contain vertex u; and let $\delta_{\overline{n}}(Tu)$ be the minimum cardinality of the edge set of the subgraph induced by a minimum dominating set of $Tu - u$ that does not contain vertex u and specifically does not dominate u.

These new parameters are used to keep track of the order of the edge set induced by a dominating set, but in this case we only consider minimum dominating sets as determined by the γ-parameters. Thus, the calculation of the δ-parameters closely parallels the calculation of the δ- parameters but only involves those terms corresponding to the terms in the γ-parameters that lead to a minimum for that γ-parameter, as follows.

For a parent vertex $u \epsilon V(T)$, let $M1 = \text{MIN } \{\gamma_y(Tv), \gamma_n(Tv), \gamma_{\overline{n}}(Tv)\}$ from equation (1). We calculate $\delta_y(Tu)$ as follows.

$$\begin{aligned}
&\text{IF } \gamma_y(Tv) = M1 \text{ THEN } A = \delta_y(Tu') + 1 + \delta_y(Tv) \\
&\qquad\qquad\qquad\quad \text{ELSE } A = \infty \\
&\text{IF } \gamma_n(Tv) = M1 \text{ THEN } B = \delta_y(Tu') + \delta_n(Tv) \\
&\qquad\qquad\qquad\quad \text{ELSE } B = \infty \\
&\text{IF } \gamma_{\overline{n}}(Tv) = M1 \text{ THEN } C = \delta_y(Tu') + \delta_{\overline{n}}(Tv) \\
&\qquad\qquad\qquad\quad \text{ELSE } C = \infty \\
&\delta_y(Tu) = \text{ MIN}\{A, B, C\}
\end{aligned}$$

Similarly, let $M2 = \gamma_n(Tu)$ from equation (2) and calculate $\delta_n(Tu)$ as follows.

$$\begin{aligned}
&\text{IF } \gamma_n(Tu') + \gamma_n(Tv) = M2 \text{ THEN } A = \delta_n(Tu') + \delta_n(Tv) \\
&\qquad\qquad\qquad\qquad\qquad\quad \text{ELSE } A = \infty \\
&\text{IF } \gamma_n(Tu') + \gamma_y(Tv) = M2 \text{ THEN } B = \delta_n(Tu') + \delta_y(Tv) \\
&\qquad\qquad\qquad\qquad\qquad\quad \text{ELSE } B = \infty \qquad (5) \\
&\text{IF } \gamma_{\overline{n}}(Tu') + \gamma_y(Tv) = M2 \text{ THEN } C = \delta_{\overline{n}}(Tu') + \delta_y(Tv) \\
&\qquad\qquad\qquad\qquad\qquad\quad \text{ELSE } C = \infty \\
&\delta_n(Tu) = \text{ MIN}\{A, B, C\}
\end{aligned}$$

Finally, to calculate $\delta_{\overline{n}}(Tu)$

$$\delta_{\overline{n}}(Tu') + \delta_n(Tv) \qquad (6)$$

In a manner similar to that for the domination parameters, the initializations for these new parameters for a vertex v_i of T are $\delta_y(Tv_i) = 0, \delta_n(Tv_i) = \infty$ and $\delta_{\overline{n}}(Tv_i) = 0$. After initializing all six parameters, we may proceed through the endnode list as before evaluating equations (1) through (6) for u_i, where u_i is the parent of the current endnode list entry v_i. Once v_n is reached, we can determine $\gamma(T)$ as before and $\delta(T)$ as follows.

$$M = \gamma(T)$$

$$\begin{aligned} \text{IF } \gamma_y(Tv_n) = M \text{ THEN } & A = \delta_y(Tv_n) \\ \text{ELSE } & A = \infty \\ \text{IF } \gamma_n(Tv_n) = M \text{ THEN } & B = \delta_n(Tv_n) \\ \text{ELSE } & B = \infty \end{aligned}$$

$$\delta(T) = \text{ MIN } \{A, B\}$$

These six parameters let us determine the parameters $\gamma(T)$ and $\delta(T)$. We note than one can find a minimum dominating set, D, corresponding to $\gamma(T)$ and $\delta(T)$ in one additional right-to-left scan of the endnode list if one calculates some additional information during the initial left-to-right scan. Details are available in [13].

REFERENCES

[1] T. Beyer, A. Proskurowski, S. Hedetniemi, and S. Mitchell, "Independent domination in trees," Proc. 8th S. E. Conference on Combinatorics, *Graph Theory and Computing*, Utilitas Mathematica, Winnipeg, 1977, 321-328.

[2] M. R. Garey and D. S. Johnson, *Computers and Intractability*, W. H. Freeman and Company, San Francisco, 1979.

[3] M. R. Garey, D. S. Johnson, and L. Stockmeyer, "Some simplified NP-complete graph problems," *Theor. Comp. Sci.*, 1, 1976, 237-267.

[4] M. R. Garey, D. S. Johnson, and R. E. Tarjan, 1976, unpublished.

[5] D. L. Grinstead, Ph.D. thesis, in preparation.

[6] D. L. Grinstead and P. J. Slater, "On minimum dominating sets with minimum intersection," *Annals of Discrete Math*, to appear.

[7] D. L. Grinstead and P. J. Slater, "On minimum dominating sets with minimum intersection in series-parallel graphs," these proceedings.

[8] D. L. Grinstead and P. J. Slater, "A multiset, single property problem: maximum independent sets with minimum intersection," in preparation.

[9] P. L. Hammer, "A bibliography on domination in graphs," personal communication.

[10] S. T. Hedetniemi and R. Laskar, "A bibliography on domination in graphs," personal communication.

[11] R. M. Karp, "Reducibility among combinatorial problems," in R. E. Miller and J. W. Thatcher (eds.), *Complexity of Computer Computations,* Plenum Press, New York, 1972, 85-103.

[12] D. E. Knuth, "The Art of Computer Programming, Vol. 1: Fundamental Algorithms," Addison-Wesley, Reading (1968) 334-338.

[13] W. J. Selig and P. J. Slater, "Minimum dominating, optimally independent vertex sets in graphs," The University of Alabama in Huntsville Mathematical Sciences Technical Report, 1988-81.

[14] R. E. Tarjan, "Depth-first search and linear graph algorithms," *SIAM J. Computing* 1 (1972) 146-160.

Duality Theorems for the Maximum Concurrent Flow Problem

Farhad Shahrokhi
Department of Computer Science
New Mexico Institute of Mining and Technology

ABSTRACT

The maximum concurrent flow problem (MCFP) is a multicommodity flow problem. The objective in solving the MCFP is to maximize the portion of the demands which could be realized in the network, subject to the capacity constraints. In this paper, we develop duality theorems similar to the max–flow min cut–theorem for the MCFP.

1. Introduction

The single commodity flow problem [FF62] is a well known problem with many applications to the other combinatorial problems [FF62, PS82]. The elegant maximum–flow minimum–cut theorem has two important features: (i) it relates the maximum flow problem to an important dual combinatorial problem, the minimum cut problem, and (ii) it provides for the efficient algorithms which simultaneously solve both the maximum flow problem and the minimum cut problem [PS82]. Unfortunately, not much of the elegant structure of the single commodity flow problem has been discovered for the multicommodity flow problems. That is, the max–flow min–cut theorem, in general, does not hold for the multicommodity flow problems. (For very restricted multicommodity flow problems the max–flow min–cut theorem holds [Hu69, OS81, Se81].)

The maximum concurrent flow problem (MCFP) [Ma85] is the optimization version of the feasibility problem in multicommondity flows [Hu69], [On70], [Ir71]. The objective is to maximize the ratio of the flow supplied for every commodity to the demand for that commodity, subject to the capacity constraints. The ratio of the utilized flow to the demand for a commodity is termed the throughput and is the same for all the commodities. In [SM88] we developed an efficient approximation algorithm and constructed combinatorial duals for the MCFP. One of these duality theorems combined the distance dual [Ir71], [On70] to the concept of a *k–partite* cut and provided for a more natural combinatorial path–cut dual. The duality results in [MS88] have been verified only for the MCFP with uniform capacity, that is, when all the edge capacities are the same. The main contribution of this paper is to extend the duality results for the MCFP with uniform capacity in [SM88] to the general MCFP. Our first duality theorem (which extends the results in [On70], [Ir71] for feasibility problem to the optimization problem) relates the concurrent flow problem to the problem of associating costs (distances) to the edges so as to maximize the minimum cost of routing the concurrent flows. Our effort to obtain a natural combinatorial "cut dual" for the MCFP is, then, embodied in a second duality theorem which is our main result. Our duailty proofs are short and employ the dulaity theory of the linear programming. Our work emphasizes on utilizing the duality theory of linear programming as a tool to explore and analyze the combinatorial structure of the problem. The work in this paper is non–algorithmic and has focussed on exploring the combinatorial structure of the MCFP.

It is worth mentioning that although the linear programming had been initially devised for solving the continuous optimization problems, it has been proven to be fundamental to the study of many strictly combinatorial problems. The idea of applying the LP duality to a network problem and explore the structure of the problem has been useful in the desing of specialized algorithms for many combinatorial optimization problems such as, the single source shortest paths problem, the single commodity flow problem, the minimum cost flow problem, and the weighted matching proble [PS82].

Our paper is organized as follows: Section 2 formally defines the MCFP and the related concepts. In Section 3 tow duality theorems are presented. Finally, in Section 4 we give our concluding remarks regarding the computational aspects of our duality results.

2. The Maximum Concurrent Flow Problem (MCFP)

We assume the reader is familiar with basic graph terminology as in [CL86]. Throughout this paper we assume that graph $G = \langle V, E \rangle$ has the vertex set $V = \{1, 2, ..., n\}$. Let $G = \langle V, E \rangle$ be a graph. We denote by P_e the set of all paths containing the edge e, for each edge $e \in E$, and by P_{ij} the set of all paths between end vertices i, j (termed ij paths), for each distinct pair of vertices $i, j \in V$. Also let P denote the set of all paths in G. A capacity function on G is a function $C : E \rightarrow R^+$. A demand fucntion D on G is a function $D : V \times V \rightarrow R^+$ such that $D(i, j) = D(j, i)$ and $D(i, i) = 0$ for all $i, j \in V$. To eliminate the trivial cases we assume, $D(i, j) > 0$ for some $i, j \in V$.

A *concurrent flow* of throughput Z in G is a function $f: P \rightarrow R^+$ such that

i) $\Sigma_{p \in P_{ij}} f(p) = Z\, D(i, j)$ for all distinct pairs of vertices $i, j \in V$ and

ii) $\Sigma_{p \in P_e} f(p) \leq C(e)$ for every edge $e \in E$,

where D and C are the given demand and capacity function on G. We extend the domain of a concurrent flow f to $P \cup E$ using

$$f(e) = \sum_{p \in P_e} f(p) \quad \text{for all } e \in E.$$

$P' = \{p \mid f(p) > 0\}$ denotes the set of active paths for the concurrent flow f. The *maximum concurrent flow problem* (MCFP) is to find the largest throughput $\hat{Z}$ achieved by a concurrent flow function f which is explicitly defined on the set of active paths P'. The MCFP can be expressed as the linear program,

$$\begin{cases} \text{Maximize } Z, \text{ subject to} & \\ \Sigma_{p \in P_{i,j}} f(p) - Z\, D(i, j) = 0 & \text{for all } i, j \in V \text{ with } i < j, \\ \Sigma_{p \in P_{i,j}} f(p) \leq C(e) & \text{for all } e \in E, \\ f(p) \geq 0 & \text{for all } p \in P, \end{cases} \tag{2.1}$$

where D, C are the demand and capacity function on $G = \langle V, E \rangle$. The set of *saturated edges* for f is $\{e \mid f(e) = C(e)\}$.

A distance (toll) function on $G = \langle V, E \rangle$ is a function $d : E \rightarrow R^+$ such that $d(e) > 0$ for at least one edge e. Let d be a distance function; we set,

$d(p) = \Sigma_{e \in p}\, d(e)$ for all $p \in P$.

$d(i, j) = \min_{p \in P_{ij}} \{d(p)\}$ for all $i, j \in V$.

A shortest distance ij path $p \in P_{ij}$ has $d(p) = d(i, j)$ and the expression

$$\left\{ \frac{\Sigma_{e \in E}\, d(e) C(e)}{\Sigma_{i<j}\, d(i, j) D(i, j)} \right\}$$

which is denoted by Z_d is termed the *distance ratio* of d. Let $(A, \bar{A})$ be a cut in G. Let $C(A, \bar{A}) = \Sigma_{e \in (A, \bar{A})} C(e)$ and $D(A, \bar{A}) = \Sigma_{i \in A,\, j \in \bar{A}} D(i, j)$. The expression $\frac{C(A, \bar{A})}{D(A, \bar{A})}$ is termed density of $(A, \bar{A})$, and is denoted by *den* $(A, \bar{A})$. A cut of minimum density is termed a *sparsest cut*. The following lemma is easy to verify.

Lemma 2.1 *[Ma85], [SM87] Let G, D, and C be an instance of the MCFP. Then any maximum concurrent flow $\hat{f}$ of throughput $\hat{Z}$ satisfies*

*(i) **[Cut Upper Bound]** $\hat{Z} \leq den(A, \bar{A})$, where $(A, \bar{A})$ is any cut in G with $D(A, \bar{A}) > 0$, and*

*(ii) **[Distance Upper Bound]** $\hat{Z} \leq Z_d$, where Z_d is the distance reatio correpsonding to any distance function d on G with $\Sigma_{i<j}\, d(i, j) D(i, j) > 0$.*

3. Duality for the MCFP

It is interesting to observe that the cut upper bound in Lemma 2.1 will not provide a dual to the MCFP, that is, the largest throughput does not necessarily equal to the density of a sparest cut. To see this, consider the MCFP for the graph $K_{2,3}$ with unit capacities on the edges and unit demands between all vertex pairs. This instance of the MCFP has an optimal throughput of 3/7 with a sparest cut of density 1/2 [Se81], [Ma85]. Therefore, in general, the cut upper bound is subject to an inherent gap. Our first duality result establishes the more gneral distance upper bound of Lemma 2.1 as the true "dual", meaning that for some distance fucntion the distance upper bound inequality becomes equality.

It is sometimes more convenient to treat maximzing the throughput of the concurrent flow as minimizing the utilized capacity to realize the demands. For a given G, D and C, the minimum capacity ultilization problem (MCUP) is

$$\begin{cases} \text{Minimize } u, \text{ subject to} & \\ \Sigma_{p\in P_{i,j}} f(p) = D(i,j) & \text{for all } i,j \in V \text{ with } i<j, \\ uC(e) - \Sigma_{p\in P_e} f(p) \geq 0 & \text{for all } e \in E, \\ f(p) \geq 0 & \text{for all } p \in P, \end{cases} \tag{3.1}$$

Note that in the MCUP we are seeking smallest percentage of (uniform) increase on the edge capacities such that all demands are satisfied. Observe that if the optimal values of objective functions for (3.1) and (2.1) are denoted by u^*, $\hat{Z}$ respectively, then $\hat{Z} = 1/u^*$. In this paper we refer to the maximum concurrent flow problem to include both MCFP and MCUP, with the application or algorithm dictating the preferred formulation.

Let G, D, C be an instance of the MCFP. The *Min Distance Ratio Problem* (MDRP) is to determine a distance function $\hat{d}$ such that

$$Z_{\hat{d}} = \min \{Z_d \mid d \text{ is a distance function on } G\} \tag{3.2}$$

Our first duality theorem is the following.

Theorem 3.1 *[Max–Throughput Min–distance Ratio Theorem.] Let G, D, C be an instance of the MCFP with the optimal throughput $\hat{Z}$. Then*

$$\hat{Z} = Z_{\hat{d}}.$$

Proof We will show this, using the dulaity theory of linear programming. Consider the dual of (3.1) which is the LP,

$$
\begin{cases}
\text{maximize } \Sigma_{i<j\in V} Q(i,j)D(i,j) \text{ subject to,} & \\
\Sigma_{e\in p} Q(e)C(e) \le 1, & \\
Q(i,j) - \Sigma_{e\in p} Q(e) \le 0, & \text{for all } ij \text{ paths } p,\ i, j \in V,\ i<j \\
Q(e) \ge 0 & \text{for all } e \in E \\
Q(i,j) <> 0 & \text{for all } i, j \in V,\ i<j
\end{cases}
\tag{3.3}
$$

($Q(i,j) <> 0$ indicates that $Q(i,j)$ is unconstrained in sign.) Let $\hat{Q}$ be an optimal solution to (3.3) and assume $\bar{d}$ is the distance function identified by $\bar{d}(e) = \hat{Q}(e)$ for all $e \in E$. Let p be any ij path with $\bar{d}(p) = \bar{d}(i,j)$ and $D(i,j) > 0$. We have, $\hat{Q}(i,j) \le \Sigma_{e\in p} \bar{d}(e) = \bar{d}(i,j)$, implying that $\bar{d}$ is also an optimal solution to (3.3). Thus

$$\frac{1}{\hat{Z}} = u^* = \sum_{i<j} D(i,j)\bar{d}(i,j) \text{ or } \hat{Z} = \frac{1}{\Sigma_{i<j}\bar{d}(i,j)D(i,j)} = Z_{\bar{d}}.$$

(Note that $\Sigma_{e\in E} \bar{d}(e)c(e) = 1$, since $\bar{d}$ is optimal.) However, by Lemma 2.1, $\hat{Z} \le Z_{\bar{d}}$. Thus, $Z_{\bar{d}}$ is the smallest distance ratio which implies, $\hat{Z} = Z_{\bar{d}} = Z_{\hat{d}}$. ❑

The following simple observation which derives from Theorem 3.1 is important; let f, d be a concurrent flow and a distance function feasible in (3.1) and (3.3) respectively, then by the complementary slackness of the linear programming f, d are jointly optimal if and only if

$$f(p)(d(p) - d(i,j)) = 0 \text{ for all } p \in p_{ij},$$

[*every active ij path is a shortest ij path*] (3.4)

and

$$d(e)(uC(e) - f(e)) = 0 \text{ for all } e \in E,$$

[*non–saturated edges are assigned zero distances*] (3.5)

In order to present our second duality theorem, additional terms must be defined.

Let $G = <V, E>$ be a graph; denote by $(A_1, A_2, ..., A_k)$, for some fixed $k \ge 2$, the set of all edges in G whose end vertices are in distinct elements of the partition $\{A_1, A_2, ..., A_k\}$ of V. We term $(A_1, A_2, ..., A_k)$ a *k–partitie cut* in G. Let $(A_1, A_2, ..., A_k)$ denote any k–partite cut of G. For $1 \le i, j \le k$, let $C(A_i, A_j)$ denote the total capacity of all edges between A_i and A_j, and let $D(A_i, A_j)$ denote the total demand between vertices of A_i and A_j. Also, define a weight function to be any function

$$w : \{1, 2, ..., k\} \times \{1, 2, ..., k\} \to R^+$$

satisfying $w(i,j) + w(j,l) \ge w(i,l)$ and $w(i,j) = w(j,i)$ for any $1 \le i, j, l \le k$, with $\Sigma_{1\le i<j\le k}\, w(i,j)D(A_i, A_j) \ne 0$. Let $den(A_1, A_2, ..., A_k)$ denote the *density* of the k-partite cut $(A_1, A_2, ..., A_k)$ defined by

$$\text{den}(A_1, A_2, \ldots, A_k) = \min_{w} \frac{\sum_{1\le i<j\le k} w(i,j)C(A_i,A_j)}{\sum_{1\le i<j\le k} w(i,j)D(A_i,A_j)} \tag{3.6}$$

where the minimum is over all weight functions. Note that for any 2–partite cut $(A, \overline{A})$ of G, the density is simply

$$\text{den}(A,\overline{A}) = \frac{C(A,\overline{A})}{D(A,\overline{A})} = \frac{\Sigma_{e\in(A,\overline{A})} C(e)}{\Sigma_{i\in A,\, j\in\overline{A}} D(i,j)},$$

which is the density of $(A, \overline{A})$.

The set of *critical edges* in G consists precisely of all those edges that are saturated under any maximum concurrent flow function in G. The following theorem explores the structure of the critical edges and will be used in our main result.

Theorem 3.2 *[Ma85], [SM88] The set of critical edges is a k–partite cut.*

We now present our main result which exhibits the "cut dual" of the MCFP.

Theorem 3.3 *[Max concurrent flow Min k–partite cur Theorem] For any G, D and C, the maximum throughput of any concurrent flow equals the minimum density of any k–partite cut, and in particular, equals the density of the k-partite cut of critical edges.*

Proof Assume that, $(A_1, A_2, \ldots, A_j)$, $2 \le j \le n$ is any j–partite cut. Let w' be a weight function minimizing (3.6). We can construct a distance function d', such that $d'(e) = w'(i,j)$ whenever the end vertices of e are in different parts A_i, A_j, and $d'(e) = 0$ otherwise. The shortest distance function extension of d' in G to $V \times V$ must have $d'(u,v) \ge w'(i,j)$ for $u \in A_i$ and $v \in A_j$, $i \ne j$, since w' satifies the triangle inequality. Furthermore, $d'(u,v) = w'(i,j)$ if some edge $e \in E$ joins a vertex in A_i to a vertex in A_j. Note that, if there are no edges between A_i and A_j, then $C(A_i, A_j) = 0$, hence $\Sigma_{e\in e} d'(e)C(e) = \Sigma_{1\le i<j\le k} w'(i,j)C(A_i, A_j)$. But $d'(u,v) \ge w'(i,j)$ for $u \in A_i$, $v \in A_j$, thus we have $w'(i,j) = d'(u,v)$ since w' minimizes the right hand side of (3.6). Thus $\text{den}(A_1, A_2, \ldots, A_j) = Z_{d'} \ge \hat{Z}$.

Conversely, observe that by (3.5) any optimal distance function d must assign non–zero distances to the critical edges, which compose the k–partite cut $E_c = (A_1, A_2, \ldots, A_k)$. Furthermore, for any $e_1 = u_1v_1$ and $e_2 = u_2v_2$, with $u_t \in A_i$ and $v_t \in A_j$, $t = 1, 2$, it is not difficult to verify that $d(e_1) = d(e_2) = d(u_1,v_1) = d(u_2,v_2)$. Now let $u \in A_i$ and $v \in A_j$ and set $w(i,j) = d(u,v)$, for all $1 \le i, j \le k$, $i \ne j$, and obtain $\text{den}(A_1, A_2, \ldots, A_k) = \hat{Z}$, which implies the result. ❑

4. Computational Aspects of the MCFP

The general instances of MCFP can be solved in polynomially bounded time [Ta86], [SM88], however, the corresponding algorithms are not efficient. In [SM88, Sh89] we have developed an efficient fully polynomial time approximation algorithm [PS82] for the MFCP. It is interesting to note that the correctness of the algorithm in [SM88] can be verified using the duality results presented in this paper (this fact was not obvserved in [SM88]). In particular, the algorithm in [SM88] computes a concurrent f and a distance d such that (3.4) and (3.5) are almost satisfied (here, almost means with small error). To (almost) satisfy (3.5), the algorithm assigns the saturated edges much larger distances thant the distances assigned to those edges which are not saturated. To (almost) satisfy (3.4), the algorithm reroutes the flow from the long ij paths to the short ij paths and recomputes the distances on the edges. It can be verified, then, that after polynomially many iteration, both (3.4) and (3.5) will be (almost) satisfied. This, however, implies that the near optimal solutions to the MCFP and its dual are obtained [SM88].

Finally, the feasibility version of the MCFP problem [Hu69, On70] in planar networks has been solved using efficient combinatorial algorithms [MNS85, MNS86]. These algorithms are based on the elegant duality theorems in [Se81] and [Os81]. The duality results in [Se81, Os81] indicate that for the MCFP in certain planar networks, the density of a sparsest cut equals to the largest throughput.

REFERENCES

[CL86] Chartrand, G., and Lesniak, L., Graphs and Digraphs: Wadsworth and Books/Cole Mathematics series 1986.

[FF62] Ford, L.R., and Fulerson, D.R., Flows in Networks, Princeton: Princeton University Press, 1962.

[Hu69] Hu, T.C., Integer Programming and Network Flows. Reading, Mass.: Addison–Wesley.

[Ir71] Iri, M., On an Extension of the Maximum–Flow Minimum–Cut Theorem to the Multicommodity Flows, J. of Operation Research Society of Japan, vol. 5, no. 4, Dec. 1967, pp. 691 – 703.

[Ma85] Mastula, D.W., Concurrent Flow and Concurrent Connectivity in Graphs, in Graph Theory and its Applications to Algorithms and Computer Science, ed. Alavi, Y., et al., Wiley, New York, 1985, pp. 543 – 559.

[MNS85] Matsumoto, K., Nishizeki, T., and Saito, N., An Efficient Algorithm for Finding Multicommodity Flows in Planar Networks, SIAM J. on Compt., 14, 1985, pp. 289 – 301.

[MNS86] Matsumoto, K., Nishizeki, T. and Saito, N., Planar Multicommodity Flows, Maximum Matchings and Negative Cycles, SIAM J. on Compt., 15, 1986, pp. 495 – 510.

[OS81] Okamura, H., Seymour, P.D., Multicommodity Flows in Planar graphs, J. Combin. Theory B, 31 (1981), pp. 75 – 81.

[On70] Onaga, K., A Multicommodity Flow Theorem, Trans. on IECE, Japan, vol. 53–A, No. 7, 1970, pp. 350 – 356.

[PS82] Papadimitriou, H., and Stieglitz, K., Combinatorial Optimization: Algorithms and Complexity, Englewood Cliffs: Prentice Hall, 1982.

[Se81] Seymour, P.D., On Odd Cuts and Planar Multicommodity Flows, Proc. London, Math. Soc., (3) 42 (1981), pp. 178 – 192.

[SM87] Shahrokhi, F., and Matula, D.W., The Maximum Concurrent Flow Problem, to appear in JACM.

[Sh89] Shahrokhi, F., Approximation Algorithms for the Maximum Concurrent Flow Problem, to appear in ORSA J. on Computing.

[Ta86] Tardos, E., A Strongly Polynomial Algorithm to Solve Combinatorial Linear Programs, J. Operations Research, Vol. 34, No. 2, 1986.

Who is the Best Doubles Tennis Player?

An Introduction to k–tournaments

Paul K. Stockmeyer
College of William and Mary

ABSTRACT

The graph–theoretic concept of a (round–robin) tournament, *which models a competition in which each of* n *participants plays a match with each of the others, can be generalized to team competitions. Thus a* k–tournament of order n *can be defined in the obvious way to model a competition in which each of the* $\binom{n}{k}$ *k–subsets of a set of* n *participants plays a match with each of the* $\binom{n-k}{k}$ *k-subsets disjoint from it.*

In this paper we begin to study the structure of k–tournaments, examining questions such as the one implied by the title: under what circumstances is it possible to rank a set of individual participants based upon team performances?

1. Introduction

Suppose that four people with the rather unimaginative names of A, B, C, and D wish to determine their relative talents in doubles tennis. They first play a match with A and B competing against C and D, which we may assume without loss of generality is won by A and B. Next A and C team up against B and D, and by symmetry we may as well suppose that A and C win. Finally, A and D compete against B and C. Here the situation is not symmetric. Prior to this third match, A and D have win–loss records of 2–0 and 0–2, respectively, while B and C are both 1–1.

If the team of A and D wins this third match, then A has a final record of 3–0, while B, C, and D all end up with records of 1–2. On the other hand, if B and C defeat A and D, then A, B, and C tie for first place with records of 2–1 while D comes in last at 0–3. Moreover, except for possible renaming, these two outcomes are the only ones possible. Either one player will dominate, winning all three of his matches, with the other three players tie for last place, or else one player will lose all three of his matches while the others tie for first place. It is not possible to linearly order the players based on team competitions; depending on the relative strengths of the players, it is possible to determine either the best player or the worst player, but not both. These sets of matches illustrate the concept of a k–tournament, which we now define.

Definition A *k–tournament of order* n is a triple $\langle P, T, M \rangle$ consisting of

1). a set P of n players,

2). the set T of all k–element subsets (teams) of P, and

3). a set M of ordered pairs (matches) of disjoint teams such that exactly one of the pairs $\langle t_i, t_j \rangle$ and $\langle t_j, t_i \rangle$ is in M for each disjoint pair of teams t_i and t_j. If $\langle t_i, t_j \rangle \in$ M, we say that t_i *dominates* t_j; this will occasionally be denoted $t_i > t_j$.

There are two comments that should be made here. First, a traditional graph–theoretic (round–robin) tournament is the same as a 1–tournament if we identify players with 1–person teams. Thus k–tournaments are true generalizations of standard tournaments. Second, we allow the possibility that $n < 2k$. When $n < k$, a k–tournament of order n consists of players only, with no teams. For $k \le n \le 2k$, a k–tournament has teams but no matches. Such k–tournaments are called *trivial*; a k–tournament with $n \ge 2k$ is called *non–trivial.*

Although the concept of a k–tournament seems quite natural, it does not appear to have been studied previously. In this paper we begin an exploration of these structures, presenting some elementary results and suggesting areas for further investigation.

2. Enumeration

There are several elementary counting results for k–tournaments that we list here for future reference. The proofs are immediate.

Theorem 1 *In any k–tournament of order n,*

a). there are exactly $\binom{n}{k}$ *teams;*

b). each team participates in $\binom{n-k}{k}$ *matches;*

c). there are $\dfrac{\binom{n}{k}\binom{n-k}{k}}{2}$ *total matches;*

d). each player participates in $\binom{n-1}{k-1}\binom{n-k}{k}$ *matches. Moreover,*

e). there are exactly $2^{\binom{n}{k}\binom{n-k}{k}/2}$ total (labeled) k–tournaments of order n.

In the case of 2–tournaments of order 4, we have already seen that there are 6 teams; each team and each player participates in all 3 matches. There are $2^3 = 8$ possible outcomes for this set of 3 matches, and hence 8 (labeled) 2–tournaments. They partition into 2 equivalence classes under the relation of isomorphism, which we now formally define.

Definition Two k–tournaments $\mathbf{T}_1 = \langle P_1, T_1, M_1 \rangle$ and $\mathbf{T}_2 = \langle P_2, T_2, M_2 \rangle$ are *isomorphic* if there exists a bijection σ from P_1 to P_2 such that

$$\langle \{p_{i_1}, p_{i_2}, ..., p_{i_k}\}, \{p_{j_1}, p_{j_2}, ..., p_{j_k}\} \rangle \in M_1$$

if and only if

$$\langle \{\sigma(p_{i_1}), \sigma(p_{i_2}), ..., \sigma(p_{i_k})\}, \{\sigma(p_{j_1}), \sigma(p_{j_2}), ..., \sigma(p_{j_k})\} \rangle \in M_2.$$

Such a mapping σ is an *isomorphism* from $\mathbf{T}_1$ to $\mathbf{T}_2$.

The number of unlabeled k–tournaments, or more precisely, the number of isomorphism classes of k–tournaments, can be determined by making fairly standard use of Pólya–type techniques, as described, for example in [1]. Specifically, the number of nonisomorphic k–tournaments of order n is computed by modifying the cycle index of the symmetric group S_n of degree n. Each term in the cycle index is replaced by the number of (labeled) k–tournaments that are left fixed by the permutation of the players indicated by that term; evaluating the resulting numerical expression yields the desired results. Relatively simple formulas exist for carrying this out in the case of the standard 1–tournaments (see, for example, [2, p. 85 – 89], but the details get increasingly more complex as k gets larger. Table 1 gives the number of non–isomorphic k–tournaments of all orders $n \leq 6$.

k \ n	1	2	3	4	5	6
1	1	1	2	4	12	56
2				2	324	48, 869, 983, 488
3						13

Table 1. The number of k–tournaments of order n.

As with most graphical enumeration problems, the first term dominates all of the others. This leads to the following asymptotic result.

Theorem 2 *For each positive integer* k*, the number* $f_k(n)$ *of non–isomorphic* k*-tournaments of order* n *satisfies*

$$f_k(n) \sim \frac{1}{n!} \, 2^{\binom{n}{k}\binom{n-k}{k}/2}$$

as n *tends to infinity.*

3. Automorphisms

As with any graphical structure, a k–tournament posses automorphisms. In this case, they form a permutation group acting on the set P of players.

Definition An *automorphism* of a k–tournament **T** is an isomorphism from **T** to itself. The *automorphism group* of **T** is the group of all automorphisms of **T** under the operation of composition.

Each of the two non–ismorphic 2–tournaments of order 4 has the group $E_1 + S_3$ of order 6 as its automorphism group. Amoug the non–isomorphic 1–tournaments of order 4, two of them have group $E_1 + C_3$, while the other two have group E_4.

The number of k–tournaments with each possible automorphism group can in theory be determined by the methods in [3], although the required knowledge of the lattice of subgroups of S_n makes it pratical only for fairly small orders. In Table 2 we give the number of non–isomorphic k–tournaments of order 5 with each possible auto–morphism group, for $k = 1$ and 2.

	k	
Group	1	2
E_5	7	232
$E_3 + S_2$		76
$E_2 + C_3$	4	2
C_5	1	2
$E_2 + S_3$		8
$C_2 + C_3$		4
Total	12	324

Table 2. The number of k–tournaments of order 5 with each possible automorphism group.

An intriguing problem is to determine which groups can serve as the automorphism group of a k–tournament, either isomorphically as an abstract group or identically as a concrete permutation group. The first interpretation of the problem has been solved for the case of 1–tournaments in the following theorem of Moon [2, p. 74 – 78].

Theorem 3 *A finite group G is abstractly isomorphic to the automorphism group of a 1–tournament if and only if the order of G is odd.*

As with most graphical structures, the group that most often occurs as the automorphism group of a k–tournament is the identity group, E_n. Let $g_k(n)$ be the number of non–isomorphic k–tournaments of order n with group E_n. The next theorem is proved in essentially the same way as Theorem 2.

Theorem 4 *Almost every k–tournament has the identity group as its automorphism group, in the sense that for each positive integer* k, $f_k(n) \sim g_k(n)$ *as* n *tends to infinity.*

4. Self–Converse k–tournaments

The concept of the converse of a standard tournament generalizes in an obvious fashion.

Definition The *converse* of the k–tournament $\langle P, T, M\rangle$ is the k–tournament $\langle P, T, M'\rangle$, where

$$\langle t_1, t_2\rangle \in M' \text{ if and only if } \langle t_2, t_1\rangle \in M.$$

A k–tournament is *self–converse* if it is isomorphic to its converse.

Intuitively, the converse of a k–tournament is obtained by reversing the outcome of all of the matches. Each of the two non–isomorphic 2–tournaments of order 4 is the converse of the other, so there are no self–converse tournaments with these parameters. Self–converse k–tournaments can be enumerated by using techniques similar to those used for self–converse digraphs, as illustrated in [1]. The trick in each case is to determine the number of structures that are mapped onto their converse by each permutation in S_n. The next theorem summarizes what is known about the number of self–converse k–tournaments.

Theorem 5

a). There exist self–converse 1–tournaments of all orders. The number of such tournaments of order n *for* $n = 1, 2, ..., 8$ *is 1, 1, 2, 2, 8, 12, 88, and 176, respectively.*

b). There exist no non–trivial self–converse 2–tournaments of any order.

c). There exist exactly 7 self–converse 3–tournaments of order 6.

Proof Parts a) and c) are obtained through fairly simple application of standard counting results. For part b), suppose there exists some permuation σ that maps a 2-tournament **T** onto its converse. It follows that if $\langle t_1, t_2\rangle \in M$, then $\langle \sigma^j(t_1), \sigma^j(t_2)\rangle \in M$ when j is even, and $\langle \sigma^j(t_2), \sigma^j(t_1)\rangle \in M$ when j is odd. We consider 4 cases.

Case 1. The disjoint cycle decomposition of σ contains a cycle π of the form $\pi = (p_1, p_2, ..., p_{2j+1})$ for some $j \geq 2$. Let team $t_1 = \{p_1, p_2\}$, and $t_2 = \{p_3, p_4\}$. Then $\sigma^{2j+1}(t_i) = t_i$ for $i = 1$ and 2, which is impossible.

Case 2. The disjoint cycle decomposition of σ contains a cycle π of the form $\pi = (p_1, p_2, ..., p_{4j})$ for some $j \geq 1$. Let team $t_i = \{p_1, p_{j+1}\}$, and $t_2 = \{p_{2j+1}, p_{3j+1}\}$. Then $\sigma^{2j}(t_1) = t_2$, and $\sigma^{2j}(t_2) = t_1$, which is impossible.

Case 3. The disjoint cycle decompostion of σ contains cycle π of the form $\pi = (p_1, p_2, ..., p_{4j+2})$ for some $j \geq 1$. Let team $t_1 = \{p_1, p_{2j+2}\}$, and $t_2 = \{p_2, p_{2j+3}\}$. Then $\sigma^{2j+1}(t_i) = t_i$ for $i = 1$ and 2, which is impossible.

Case 4. All of the cycles in the disjoint cycle decomposition of σ have length ≤ 3. The various subcases can all easily be shown to be impossible.

Thus there is no possible permutation σ that can map a 2–tournament onto its converse. ❑

Similar investigations have confirmed that there do not exist any self–converse 4-tournaments of order 8, suggesting that the following conjecture may be true.

Conjecture *There exist no non–trivial self–converse k–tournaments when k is even.*

5. Score Sequences

The score sequence is another concept from standard tournament theory that we want to extend to k–tournaments. Although we could assign scores to either the teams or the players, the second alternative seems more natural.

Definition The *score* of player p_i in the k–tournament $\mathbf{T}$ of order n is the number s_i of matches in which p_i is a member of the dominant team. The *score sequence* of $\mathbf{T}$ is the n–tuple $\langle s_1, s_2, ..., s_n \rangle$, with the players numbered so that $s_1 \leq s_2 \leq ... \leq s_n$.

The two 2–tournaments of order 4 have score sequences $\langle 0, 2, 2, 2 \rangle$ and $\langle 1, 1, 1, 3 \rangle$. Two obvious questions arise: which n–tuples are in fact the score sequences of k–tournaments, and how many distinct score sequences exist for a given n and k? The second question is quite difficult to answer even in the case $k = 1$, and we will not pursue it here. The first question, though, has a very nice answer for $k = 1$, as presented for example in [2, p. 61 – 62].

Theorem 6 *The sequence* $\langle s_1, s_2, ..., s_n \rangle$, *with* $s_1 \le s_2 \le ... \le s_n$ *is the score sequence for some 1–tournament* **T** *if and only if*

$$\sum_{i=1}^{j} s_i \ge \binom{j}{2}$$

for $j = 1, 2, ..., n$, *with equality when* $j = n$.

The necessity of these conditions follows from the fact that the right hand side of the inequality expresses the number of matches that take place totally within a set of j players and hence the minimum total score of any such set. Similar necessary conditions can be derived for other values of k by selecting a set S of j players and imagining a k–tournament in which each match is won by the team with fewer players from S. The sum of the scores of the players in S is then a sharp lower bound for the sum of the scores of the j worst players. For $k = 2$ and $n = 4$, for instance, we have $s_1 \ge 0$, $s_1 + s_2 \ge 2$, $s_1 + s_2 + s_3 \ge 3$, and $s_1 + s_2 + s_3 + s_4 = 6$.

Unfortunately, these necessary conditions are not sufficient to insure that a sequence is in fact the score sequence of some 2–tournament **T**. The sequence $\langle 1, 1, 2, 2 \rangle$ satisfies the conditions but is not the score sequence of either 2–tournament. Furthermore, these inequalities can not be sharpened. We are forced to conclude that no set of conditions based on lower bounds for the sum of the j smallest scores can be both necessary and sufficient in this case. Similar complications arise for larger values of n and k. Thus the problem of characterizing legitimate score sequences of k–tournaments appears to be quite difficult for $k \ge 2$.

A *regular* k–tournament is one in which the scores of all players are equal. In this case each player must win exactly half of his matches, so an obvious necessary condition for the existence of a regular k–tournament is that each player competes in an even number of matches. For $k = 1$, this means that n must be odd, and it is well–known that there exist regular 1–tournaments of all odd orders. The following generalization is an easy exercise.

Theorem 7 *The number*

$$\binom{n-1}{k-1}\binom{n-k}{k} = \frac{(n-1)!}{k!(k-1)!(n-2k)!}$$

of matches in which each player in a k*–tournament of order* n *participates is even except when* k *is a power of* 2 *and* n *is a multiple of* $2k$.

It appears likely that this necessary condition is also sufficient.

Conjecture *There exist regular* k*–tournaments of order* n *in all cases except those excluded by Theorem 7.*

6. Other Topics

We conclude with a short list of other question about k–tournaments that we consider worthy of study. Almost nothing is known about any of them.

A 1–tournament is called *transitive* if $t_1 > t_2$ and $t_2 > t_3$ imply $t_1 > t_3$ for any three (one person) teams in the tournament. There is exactly one transitive 1–tournament of each order, up to isomorphism. How should this concept be extended to more general k–tournaments? Using exactly the same definition does not work, as t_1 and t_3 will not always be disjoint and hence $t_1 > t_3$ will not be possible. There are many equivalent characterizations of transitivity in 1–tournaments, (see [2, p. 15], for example), but none of them seem to properly capture the concept for higher values of k. Being free of directed cycles among the teams, for example, does not seem to possess the property we desire to capture. On the other hand, demanding that the scores be all distinct seems overly restricitve. What is the most appropriate way to extend the concept of transitivity?

At the other extreme are the strong touranments. Here again, there are many equivalent characterizations of this concept in the case of 1–tournaments, and their extensions are not all equivalent for larger values of k. What is the best way to extend the concept of strength? Can the set of players in a k–tournament be partitioned into strong components that can be contracted to form the set of players for a new, transitive k–tournament?

What do random k–tournament look like? Suppose the outcome of each match is determined by an independent flip of a fair cion. We know from Theorem 4 that such a k–tournament will almost certainly have the identity group E_n as its automorphism group. But how transitive or strong is it likely to be? What is the expected value of the largest score?

Are k–tournaments of order n ever reconstructable (uniquely determined up to isomorphism) from their sub–k–tournaments of order $n-1$? It is known (see [4], for example), that the 1–tournaments of order 7 are so reconstructable, but that there are non–reconstructable 1–tournaments of all other orders from 2 to 10. Moreover, there are non–reconstructable 1–tournaments for infinite number of orders. What can be said for other values of k? Since there are only two distinct 2–tournaments of order 4, there are only 6 possible decks that could serve as the collection of subtournaments of a 2-tournament of order 5. It is easy to verify that each of these 6 possible decks is in fact the deck derived from at least two of the 324 2–tournaments of order 5, so that none of these is reconstructable.

REFERENCES

[1] F. Harary and E.M. Palmer, *Graphical Enumeration,* Academic Press, New York, 1973.

[2] J.W. Moon, *Topics on Tournaments*, Holt, Rinehart and Winston, New York, 1968.

[3] P.K. Stockmeyer, *Enumeration of Graphs with Prescribed Automorphism Group*, Ph.D. Dissertation, University of Michigan, 1971.

[4] P.K. Stockmeyer, "A census of non–reconstructable digraphs, I: Six related families," *J. Combinatorial Theory, Series B* **31** (1981), 232 – 239.

Recent Results on Graph Embeddings

Carsten Thomassen

ABSTRACT

We survey some recent results on embeddings of graphs on surfaces. We indicate short graph theoretic proofs of the Jordan–Schönflies theorem and of the classification of the surfaces. We describe a polynomially bounded algorithm which determines the genus of a large class of graphs. We also explain why the graph genus problem for all graphs is NP–complete. Details can be found elsewhere.

1. Introduction

A *curve* in a topological space X is the image of a continuous 1–1 map f: [0,1] $\rightarrow$ X. We say that a graph G can be *embedded* into X if G can be represented in X such that the vertices of G are distinct elements (points) in X, and each edge xy in G is a curve in X joining the points corresponding to x and y. Moreover, two edges in X do not intersect except possibly at an end. There is an immense literature on graph embeddings. The main source of motivation was the following problem: Given a surface S, determine the smallest natural number h(S) such that every map on S can be colored in h(S) colors in such a way that no two neighboring countries receive the same color. When S is the sphere this is the *4–color–problem* and for other surfaces it is the *Heawood conjecture*. More recently, graph embeddings are also studied because of the applications to real world problems. In such problems there are often complicated additional constraints imposed which can make the embedding problems complicated even for "simple" topological spaces X. We shall here treat embeddings with no additional constraints.

If X is the Euclidean space $\mathbf{R}^3$, then the embedding problem is uninteresting because every graph can be embedded into X: Simply put every vertex on the moment curve $\{(t, t^2, t^3) \mid t \in [0,1]\}$ and join every pair of adjacent vertices by a straight line segment. If X is $\mathbf{R}^2$ (or the sphere in $\mathbf{R}^3$), then the embedding problem is solved by

Kuratowski's theorem: A graph G can be embedded into $\mathbf{R}^2$ if and only if G contains no subdivision of any of the Kuratowski graphs K_5 or $K_{3,3}$. In view of this it seems natural to focus on topological spaces which are "smaller" than $\mathbf{R}^3$ but "larger" than $\mathbf{R}^2$. One such example is the *k-book* which is obtained from k disjoint squares of side length 1 (the pages) by choosing one side (the spine) on each page and identifying these. Clearly, a graph can be embedded into a k–book ($k \leq 2$) if and only if the graph is planar. A surprising result of G. Atneosen (see [2]) says that every (other) graph can be embedded in the 3–book. This was a corollary of a more general topological result. A trivial proof is given in [2]. So, also the k–books are uninteresting from the general embedding point of view.

The most interesting spaces from a graph embedding point of view are probably the surfaces. (A *surface* is a compact connected space which is locally homeomorphic to $\mathbf{R}^2$). When dealing with graph embeddings one makes often use of the *classification theorem*: Every surface is homeomorphic to the surface S_g obtained from the sphere S_0 by adding g handles, or the the surface N_k obtained from the sphere by adding k crosscaps. Another fundamental result needed for embeddings is the *Jordan curve theorem* or the stronger *Jordan–Schönflies theorem*: If f is a homeomorphism of a closed curve C_1 in $\mathbf{R}^2$ onto a closed curve C_2 in $\mathbf{R}^2$, then f can be extended to a homeomorphism of the whole plane $\mathbf{R}^2$. We shall here indicate the ideas behind simple graph theoretic proofs of the Jordan–Schönflies theorem and the classification theorem. Detailed proofs are given in [5].

2. The Jordan–Schönflies Curve Theorem

We recall that a *curve* in the plane $\mathbf{R}^2$ is the image of a continuous 1–1 map f: $[0,1] \to \mathbf{R}^2$. We say that the curve *joins* $f(0)$ and $f(1)$. A *closed curve* is defined analogously except that $f(0) = f(1)$. A (closed) polygonal curve is a (closed) curve which is the union of a finite number of straight line segments. If Ω is an open set in $\mathbf{R}^2$, then a *region* Ω' of Ω is a maximal subset of Ω such that any two points of Ω' are joined by a curve (in Ω). We say that Ω is *connected* if it has only one region. A *planar graph* is a graph that can be embedded in $\mathbf{R}^2$. The following statements are easy exercises (see [1, 5]).

(1) If $\Omega \subseteq \mathbf{R}^2$ is open and connected, then any two points of Ω can be connected by a polygonal curve in Ω.

(2) If G is a planar graph, then G can be drawn (embedded) in the plane such that all edges are polygonal curves.

(3) If C is a closed polygonal curve in $\mathbf{R}^2$, then $\mathbf{R}^2 \setminus C$ has two regions each of which has C as boundary.

Using (1), (2), (3) one easily proves

(4) K_5 and $K_{3,3}$ are nonplanar.

Also (5) below has a short graph theoretic proof [5].

(5) If P is a curve in $\mathbf{R}^2$, then $\mathbf{R}^2 \setminus P$ is connected.

We now indicate how (1) – (5) imply

The Jordan Curve Theorem *If C is a closed curve in the plane $\mathbf{R}^2$, then $\mathbf{R}^2 \setminus C$ has precisely two regions each of which has C as boundary.*

Proof Let L_1 and L_2 be the vertical straight lines both intersecting C such that C is in the right (respectively left) closed half plane of L_1 (respectively L_2). Let P_1, P_2 be two curves on C joining q_1 and q_2 where q_i is the top point of $C \cap L_i$ for i = 1, 2. Consider a vertical straight line between L_1 and L_2. That line contains a segment P_3 joining P_1 and P_2 and having only its ends in common with C. Let P_4 be a horizontal straight line segment joining L_1 and L_2 above C. Now P_3 and P_4 must be in distinct regions of $\mathbf{R}^2 \setminus C$. Otherwise, a curve in $\mathbf{R}^2 \setminus C$ from P_3 to P_4 together with $C \cup P_3 \cup P_4$ and segments of L_1, L_2 would form an embedding of $K_{3,3}$ in $\mathbf{R}^2$ contradicting (4). Hence $\mathbf{R}^2 \setminus C$ has at least two regions.

To see that $\mathbf{R}^2 \setminus C$ has at most two regions, let us assume that q_1, q_2, q_3 are points in distinct regions of $\mathbf{R}^2 \setminus C$. Let p_1, p_2, p_3 be any three points on C. Let D_1, D_2, D_3 be pairwise disjoint discs around p_1, p_2, p_3, respectively, such that none of them contains any of q_1, q_2, q_3. Let P_i be a segment of C inside D_i (i = 1, ,2 ,3). By (5), $R_i = \mathbf{R}^2 \setminus (C \setminus P_i)$ is connected for i = 1, 2, 3. Any curve from q_1 to q_2 in R_i interesects P_i. Hence $\mathbf{R}^2 \setminus C$ contains a polygonal curve from q_i to D_j for i = 1, 2, 3 and j = 1, 2, 3. We can assume that these nine curves do not intersect except at q_1, q_2, q_3. Now we obtain a plane representation of $K_{3,3}$ from these nine curves by adding straight line segments in $D_1 \cup D_2 \cup D_3$. This contradicts (5) and shows that $\mathbf{R}^2 \setminus C$ has precisely two regions. ❑

The unbounded region of $\mathbf{R}^2 \setminus C$ is called the *exterior* of C and is denoted ext(C). The other region of $\mathbf{R}^2 \setminus C$ is the *interior* int(C). A point p on C is *accessible* from int(C) if, for some (and hence each) point q in int(C), there is a polygonal curve from q to p having only p in common with C. The proof of the Jordan Curve Theorem shows that the points on C which are accessible from int(C) form a dense set on C.

We shall now sketch a simple graph theoretic proof of the Jordan–Schönflies Theorem which extends the Jordan Curve Theorem and which is normally regarded as a difficult result. A detailed proof can be found in [5].

The Jordan–Schönflies Theorem *If f is a homeomorphism of a closed curved C_1 onto a closed curve C_2 (both in the plane), then f can be extended to a homeomorphism of the whole plane.*

Sketch of Proof Without loss of generality we can assume that C_2 is a closed polygonal curve. We shall extend f to $\text{int}(C_1)$. (The extension to $\text{ext}(C_1)$ is done in a similar way). Let A be a countable set of accessible points on C_1 (from $\text{int}(C_1)$) such that A is dense on C_1. We shall consider a sequence of plane 2–connected graphs Γ_0, $\Gamma_1, \Gamma_2, \ldots$ such that $C_1 \subseteq \Gamma_0 \subseteq \Gamma_1 \subseteq \ldots$ and such that each Γ_k consists of C_1 (which is a cycle in Γ_k) together with polygonal curves in $C_1 \cup \text{int}(C_1)$. Furthermore, $\Gamma_k \setminus C_1$ is connected for each k. Finally we shall construct the sequence $\Gamma_0, \Gamma_1, \ldots$ such that $V(\Gamma_0) \cup V(\Gamma_1) \cup \ldots$ contains A and is dense in $\text{int}(C_1)$. At the same time we construct a sequence of graphs $\Gamma'_0, \Gamma'_1, \ldots$ such that Γ'_k consists of C_2 and polygonal curves in $\text{int}(C_2) \cup C_2$ and such that there exists a graph isomorphism $g_k : \Gamma_k \to \Gamma'_k$. Moreover, $\mathbf{R}^2 \setminus \Gamma_k$ and $\mathbf{R}^2 \setminus \Gamma'_k$ have a finite number of regions each of which has a cycle of Γ_k (or Γ'_k) as boundary. Finally, a cycle S in Γ_k is such a boundary if and only if also $g_k(S)$ is the boundary of a region in $\mathbf{R}^2 \setminus \Gamma'_k$.

We construct Γ_0 as follows: We pick two points p, q in A and join them by a polygonal curve P such that $P \setminus \{p,q\} \subseteq \text{int}(C_1)$. Then we put $\Gamma_0 = C \cup P$ and consider Γ_0 as a graph with vertices p, q joined by three edges. An easy extension of the Jordan Curve Theorem shows that $\mathbf{R}^2 \setminus \Gamma_0$ has three regions whose boundaries are the three cycles of Γ_0. We let Γ_0' be obtained from C_2 by adding a polygonal curve joining $f(p)$ and $f(q)$. Suppose we have already defined $\Gamma_0, \Gamma_1, \ldots, \Gamma_k$ and Γ'_0, $\Gamma'_1, \ldots, \Gamma'_k$ and the isomorphisms $g_0, g_1, \ldots, g_k$. Then we select a region of $\mathbf{R}^2 \setminus \Gamma_k$ bounded by the cycle S, say. In int(S) we add a polygonal curve joining two points on S on edges a and b, say. In $\text{int}(g_k(S))$ we add a corresponding curve joining two points on $g_k(a)$ and $g_k(b)$ This results in Γ_{k+1} and Γ'_{k+1}, respectively. The isomorphism g_{k+1} is defined in the obvious way. We extend successively f such that it is defined on C_1 and the vertex set of Γ_k and agrees with g_k.

Since A is countable we can construct $\Gamma_0, \Gamma_1, \ldots$ such that each point of A is a vertex of some Γ_k. We can also assume that each point of A is a vertex of some Γ_k. We can also assume that each point in $\text{int}(C_1)$ which has rational corrdinates is in some Γ_k. Hence f will be defined on a dense set in C_1 and a dense set in $\text{int}(C_1)$. Furthermore, with a little additional care we can make sure that the following holds: For every point p in $\text{int}(C_1)$ (respectively $\text{int}(C_2)$) there is a cycle S in some Γ_k (respectively Γ_k') such that $p \in \text{int}(S)$ and S has arbitrarily small diameter. This will guarantee that f when considered as a function defined on $C_1 \cup V(\Gamma_0) \cup V(\Gamma_1) \cup \ldots$ can be extended to a function (which is also denoted f) on $\text{int}(C_1)$ which is 1–1 and maps $\mathrm{int}(C_1)$ onto $\mathrm{int}(C_2)$ and which is continuous on $\mathrm{int}(C_1)$ and has a continuous inverse on $\mathrm{int}(C_2)$. (Note that the construction of f implies that f maps $\mathrm{int}(S)$ into $\mathrm{int}(g_k(S))$ and $\mathrm{ext}(S)$ into $\mathrm{ext}(g_k(S))$ and hence S into $g_k(S)$ when S is the above cycle in Γ_k. This also holds if S intersects C_1). It only remains to show that f is continuous on C_1. For this we consider a point p on C_1 and a sequence $q_1, q_2, \ldots$ of points in $C_1 \cup \mathrm{int}(C_1)$ converging towards p. We shall prove that $f(q_n) \to f(p)$ as $n \to \infty$. Suppose therefore that this is not the case. Then we can assume (by considering a subsquence if necessary) that $f(q_n) \to f(q) \neq f(p)$ as $n \to \infty$. Since f is continuous in $\mathrm{int}(C_1)$ and also continuous on C_1 (when considered as a function defined on C_1) we can assume that $q_n \in \mathrm{int}(C_1)$ and $q \in C_1$. In some Γ_k there is a polygonal curve P joining two points p_1, p_2 on C_1 such that one of the segments of C_1 from p_1 to p_2 contains p and the other contains q. Let S be the cycle in $C_1 \cup P$ containing p. Then $q_n \in \mathrm{int}(S)$ for n sufficiently large. Then $f(q_n) \in \mathrm{int}(f(S))$, by the above parenthetical remark. But this contradicts the assumption that $f(q_n) \to f(q)$ as $n \to \infty$. ❑

3. The Classification of Surfaces

In Section 2 it is shown how graph theory can be an efficient tool in topology. Conversely, the classification theorem is a topological result of great importance to graph embeddings. If we wish to draw (embed) a given graph G on a given surface S, then instead of starting with S and trying to draw G on it we start out with G and use it as the "skeleton" of a surface obtained by pasting discs together on G. It may not be immediately obvious whether or not the resulting surface we get is homeomorphic to S. But that can be decided using the classification theorem (combined with Euler's formula). We shall here indicate a simple graph theoretic proof of the classification theorem. The advantage of the present proof is that it depends only on local considerations and it is very easy to make it rigorous. Here only the ideas are described. Again, the details are in [5].

A surface is defined as a connected compact topological space which is locally homeomorphic to a disc. The sphere S_0 is a surface. Adding a crosscap to the sphere means that we cut out a disc and identify diametrically opposite points on its boundary. Adding a *handle* to a disc may be described as cutting out two disjoint discs and identifying the boundary of one with the boundary in the other such that the orientations of these boundaries are the same. If the orientations do not agree, we say that we add a *twisted handle*. Let S_g and N_k be obtained from the sphere S_0 by adding g handles, respectively k crosscaps. It is an easy exercise to show that adding a twisted handle amounts to the same as adding two crosscaps and that adding a handle after a crosscap has been added amounts to the same as adding a twisted handle (i.e. two crosscaps). So, if we add handles, twisted handles and crosscaps to the sphere, we obtain one of the surfaces S_g, N_k.

There is another natural way of forming a surface. Let us consider a collection of disjoint triangles of side length 1 in the plane. We identify each side of a triangle with precisely one side in another triangle. This results in a compact topological space S and a graph G whose vertices and edges are the corners and sides, respectively, of the triangles. If G (and hence also S) is connected and S is locally homeomorphic to a disc (i.e., G is locally isomorphic to a wheel), then S is a surface. If G has no vertices of degree 2, then we will call G a *triangulation of S*. Thus the smallest triangulation is obtained by pasting four triangles together such that G is K_4 and S is the tetrahedron (or, equivalently, the sphere). It is well–known that every surface S can be triangulated. Intuitively, S can be triangulated as follows: For every point p on S we consider a small disc containing p. In that disc we consider a closed curve having p in the interior. Since S is compact there is a finite collection of such curves such that the union of their interior is S. In [5] it is shown, using the Jordan–Schöflies theorem combined with elementary graph theory, that the above curves can be chosen such that only finitely many points on S are on two or more of the curves. Then we triangulate the interior of each of them. Using this we may proceed to:

The Classification Theorem *Every surface S is homeomorphic to one of the surfaces S_g $(g \geq 0)$ or N_k $(k \geq 1)$.*

Sketch of proof Let G be a triangulation of S with n vertices, e edges, and f regions which are all bounded by triangles of G. We shall prove that S is homeomorphic to S_g or N_k. We shall do this by contradiction assuming that g or k is minimal. Euler's formula (which we haven't proved yet) says that

$$n - e + f = 2 - 2g \text{ or } 2 - k.$$

Therefore, we assume that the counter example S, G is chosen such that

(i) $2 - n + e - f$ is minimum. (An easy exercise shows that $2 - n + e - f$ is nonnegative, see [5]).

Among those counter examples we assume that

(ii) n is minimal

and that

(iii) the smallest vertex degree of G is minimal.

If G has a vertex v of degree 3, then also $G - v$ triangulates S and we have a contradiction to (ii).

If G has a vertex v of minimal degree ≥ 4 and with neighbors $v_1, v_2, ..., v_n$ in that cyclic order, then we can assume that G contains the edges $v_1 v_3, v_2 v_4, ...$ For if $v_1 v_3$ is missing then $G - v v_2 + v_1 v_3$ is a triangulation of S contradicting (iii).

Now we consider the triangle T: $v v_1 v_3$ in G. Note that T is not the boundary of a region of $S \setminus G$ and that T does not separate G or S since $v_2 v_4$ is an edge in G. Then we form a new surface S_1 as follows: We cut along T such that each vertex and edge of T is cut into two vertices (or edges). This results in a "near–surface" S' and a graph G' which has $n' = n + 3$ vertices and $e' = e + 3$ edges. The triangle T in G corresponds in G' either to two disjoint triangles T_1, T_2 or a cycle C of length 6. Note that S' is not locally homeomorphic to a disc on $T_1 \cup T_2$ or C. We extend S' into a surface S_1 by pasting discs on each of T_1, T_2 or on C. Then S_1 is a surface with $f' = f + 2$ or $f + 1$ regions. In case G' contains C, we extend G' into a triangulation G" of S_1 by adding a vertex inside C and joining it to the six vertices of C. Now (S_1, G') or $(S_1, G")$ cannot be a counterexample to the theorem because that would contradict (i). So, S_1 is one of the surfaces S_g, N_k. Since S is obtained from S_1 by adding a handle, or a twisted handle or a crosscap, the proof is complete. ❑

Note that the above proof also implies Euler's formula for triangulations. With a little additional reasoning, it can be extended to include Euler's formula in its full generality. It also implies the invariance of the genus (i.e. the fact that all the surfaces $S_0, S_1, ..., N_1, N_2, ...$ are pairwise nonhomeomorphic). For details, see [5].

4. Finding the Genus of a Graph

The *genus* g(G) of a graph G is the smallest number g such that G can be embedded into S_g. The *crosscap number* of G is defined anaolgously. We shall here concentrate on the genus and we shall only consider connected graphs G. Let $V(G) = \{v_1, v_2, ..., v_n\}$ and let π_i be a cyclic permutation (which we shall refer to as a *clockwise orientation*) of the edges inicident with v_i, $i = 1, 2, ..., n$. Suppose we walk towards vertex v_i along the edge a. If we continue walking away from v_i along $\pi_i(a)$ we say that we turned *sharp left* at v_i. If we continue walking sharp left at every vertex we encounter, we obtain a *facial walk*. Let $e = |E(G)|$ and f the number of facial walks. For each of the f facial walks we select a disc and identify the boundary of the disc with the facial walk. This results in a surface S. One can show, for example using methods in the previous section that $S = S_g$ where g satisfies Euler's formula

$$n - e + f = 2 - 2g.$$

Furthermore, one can show that every embedding of G on a surface of minimum genus is of this form. So, determining g(G) amounts to specifying clockwise orientations (called a *rotation system*) such that f is maximized. This is a purely combinatorial problem and it makes sense to speak of the computational complexity of the problem.

The genus has been determined for some graphs of a high degree of symmetry, such as complete graphs, complete bipartite graphs, and also of graphs obtained by using special graph operations. For g fixed, a polynomially bounded algorithm for deciding if a graph can be embedded into S_g was found by Filotti, Miller and Reif and is also a special case of a more general algorithm in the Robertson–Seymour theory on graph minors. These results are surveyed in various books or papers, for example [2].

We shall here confine ourselves to the description of an algorithm which determines the genus of a large class of graphs without any symmetry properties and with unbounded genus. This algorithm and the theory behind it is described purely combinatorially in [3]. Here we shall describe it partly in topological terms in order to emphasize the ideas behind the algorithm. If G is a graph embedded in S_g then a cycle C in G is *contractible* if $S_g \setminus G$ has two connected components, one of which is homeomorphic to a disc. (That disc is called the *interior* of C). Otherwise, C is *noncontractible*. It is very easy to decide if a given cycle is noncontractible (see [3]). Moreover, it is shown in [3] that the shortest noncontractible cycle can be found as follows: Pick a vertex v in G and grow a breadth–first tree (distance tree) T from v. Look at each fundamental cycle with respect to T and let C_v be a shortest noncontractible one. (It is possible that C_v does not exist). Repeat this for every vertex. The shortest cycle found in this way is a shortest noncontractible cycle.

Now define the *edge–width* $ew(G)$ of the embedded graph G as the length of a shortest noncontractible cycle. We say that the embedding has *large edge–width* if all facial walks have length $\leq ew(G)$ and we say that G is *LEW–embedded*. This concept, which was introduced by J. Hutchinson, is very useful for studying embeddings since LEW–embeddings share many properties with planar embedding as demonstrated in [3]. Let G be a connected graph and C a cycle in G. Following W.T. Tutte we define a *C–bridge* as a chord of C (together with its ends) or a connected component of $G - V(C)$ together with all edges joining it to C and the ends of these edges. If H is a C–bridge, then $H \cap C$ are the *vertices of attachment* of C. Two bridges H_1, H_2 *avoid one another* if C has two edge–disjoint paths P_1, P_2 such that H_i has all its vertices of attachment on P_i for $i = 1, 2$. Otherwise, H_1 and H_2 *overlap*. If G is drawn in the plane and two C–bridges overlap, then one of them is in ext(C) and the other is in int(C). The *overlap graph* $O(G,C)$ has the C–bridges as vertices such that two vertices are adjacent iff the corresponding bridges overlap. So if G is planar, then $O(G,C)$ is bipartite. We say that C is *induced* and *nonseparating* if G has only one C-bridge. Tutte proved that the facial cycles in a 3–connected planar graph are precisely the induced nonseparating cycles. This extends Whitney's result that a 3–connected planar graph has only one embedding in the plane. These results were generalized in [3].

Theorem 1 *If G is a 3–connected LEW–embedded graph, then the embedding is the unique minimum genus embedding of G. The facial walks are precisely the induced nonseparating cycles of length < ew(G).*

Theorem 2 *If G is a cycle of length < ew(G) in a 3–connected LEW–embedded graph, then O(G,C) is connected and bipartite. Moreover, if G is nonplanar, then G has precisely one C–bridge which together with C forms a nonplanar graph.*

Theorems 1 and 2 are structural results which imply the following algorithmic result.

Theorem 3 *There exists a polynomially bounded algorithm for describing an LEW–embedding of an arbitrary 3–connected graph G or deciding that G has no such embedding.*

By Theorem 1, the algorithm determines the genus of a large class of graphs, namely those that are 3–connected and have LEW–embeddings. In order to describe the embedding it is sufficient to describe the (unique) rotation system or the induced nonseparating cycles of length < ew(G). The problem is that we do not know ew(G) in advance. We can overcome this problem by using Theorem 2. We consider any edge e = vu and, by Theorem 1, all we have to do is to find the shortest and second shortest induced nonseparating cycle through e. Also, this is an unsolved problem in general, (see Problem 1 below). So, let C be any shortest cycle through e. If G has an LEW–embedding, then O(G,C) is bipartite and G has only one C-bridge H such that $H \cup C$ is nonplanar. We consider all the C–bridges in the partite class of O(G,C) not containing H. These can be drawn (in polynomial time) as a planar graph inside C. In this way we find one of the two shortest induced nonseparating cycles containing e (if G has an LEW–embedding). In order to find the other one we select any edge e' incident with v (and going out from the above drawing of C) and we let C' be a shortest cycle through e and e'. We repeat the above procedure with C' instead of C. We have already found the successor or predecessor (say the latter) of e around v. When e' is the successor of e the following will happen in the above procedure: The interior of C' contains either all or none of the edges incident with v and distinct from e, e'. When this happens we have found the successor and predecessor of e around v. If something goes wrong (for example if O(G,C) is not bipartite or has more than one nonplanar bridge or some edge has no successor), then G has no LEW–embedding and the algorithm stops.

If a graph G is embedded on S_g where $g = g(G)$, then it is easy to subdivide G and add vertices and edges such that the resulting graph is 3–connected and LEW–embedded. This suggests that Theorem 3 might be extended to a larger class of graphs although it is not clear how to add the new vertices and edges. The following result of [4], which answers one of the basic questions on NP–completeness raised by Garey and Johnson, indicates that Theorem 3 can probably not be extended to all graphs.

Theorem 4 *The following problem is NP–complete: Given a connected graph G and a natural number m, is $g(G) \leq m$?*

We shall here indicate how the well–known NP–complete problem of finding the independence number $\alpha(G)$ of a graph G can be reduced, in polynomial time, to that in Theorem 4. If $|V(G)| = n$ and $|E(G)| = e$, then it is easy to see that

$$g(G) \leq e - n + 1.$$

First draw a spanning tree of G on the sphere. Then add successively the remianing $e - n + 1$ edges by adding a new handle at each step. The above upper bound is in general a poor one because G does not really use the handles fully. Let G_k by the graph obtained from G by replacing each edge xy by a cycle of length k, i.e., we delete the edge xy and add instead a cycle of length k joined completely to x and y. The above argument shows that also G_k has genus at most $e - n + 1$ (let the new cycles of length k go around the handles). In [4] it is shown that, if $k > 3n^2$, then G_k has genus precisely $e - n + 1$. We now add to G_k a new vertex v and join v to precisely one vertex in each of the new cycles of length k. It is shown in [1] that the resulting graph G'_k has genus $e - \alpha(G)$. Now the problem: Is $\alpha(G) \geq m$? (which is NP–complete) reduces to the problem: Is $g(G'_k) \leq e - m$?

5. Unsolved Problems

Theorem 3 would follow quickly from Theorem 1 (without the use of Theorem 2) if we could find, in polynomial time, the shortest and second shortest induced nonseparating cycle through a given edge in a 3-connected graph. W.T. Tutte has described an algorithm for finding two distinct induced nonseparating cycles through a given edge but they need not be short. So, we suggest the following.

Problem 1 Does there exist a polynomially bounded algorithm for finding a shortest induced nonseparating cycle through a given edge in a 3–connected graph?

In [3] is described a general algorithm for finding shortest cycles of many different types, for example a shortest noncontractible cycle in an embedded graph. J. Hutchinson raised the following question which was not covered in [3].

Problem 2 Does there exist a polynomially bounded algorithm for finding a shortest contractible cycle in an embedded graph?

We shall here point out that the answer is affirmative in the important special case where all vertices have degree at least 3. We first find all facial walks. Among those which are cycles we select a shortest one. Then we consider all cycles of length 3, 4, 5. If one of them is contractible we select a shortest contractible one. This algorithm is clearly polynomially bounded. We claim that it produces a shortest contractible cycle if there is one. For suppose that C is a shortest contractible cycle and that C is not a facial walk and that C has length ≥ 6. Then C and its interior is a planar graph and all vertices in int(C) have degree ≥ 3. It is now an easy exercies to use Euler's formula and find a facial cycle in int(C) which is shorter than C.

REFERENCES

[1] C. Thomassen, *Kuratowski's theorem*, J. Graph Theory 5 (1981), 225 – 241.

[2] C. Thomassen, *Embeddings and minors*, in "Handbook of Combinatorics (eds. R.L. Graham, M. Grötschel and L. Lovász)," North–Holland (to appear).

[3] C. Thomassen, *Embeddings with no short noncontractible cycles*, J. Combinatorial Theory, Ser. B (to appear).

[4] C. Thomassen, *The graph genus problem is NP–complete*, J. Algorithms (to appear).

[5] C. Thomassen, *The Jordan–Schönflies theorem and the classification of surfaces* (to appear).

Symmetric Embeddings of Cayley Graphs in Nonorientable Surfaces

Thomas W. Tucker*
Colgate University

ABSTRACT

An embedding of a Cayley graph for a group A in an orientable or nonorientable surface S is symmetric if the natural action of the group A on the Cayley graph extends to the surface S. It is shown that every irredundant Cayley graph for the group A has a symmetric nonorientable embedding if and only if A is not a 2-group. Parameters analogous to the genus and symmetric genus of a group are introduced for the nonorientable case and are computed for a few examples. The projective plane and klein bottle groups are classified. Some inequalities relating the orientable and nonorientable parameters are summarized.

1. Introduction

Given a group A and a generating set X for A, the *Cayley graph* C(A, X) has the elements of A as vertices, and edges from a to ax for each $a \in A$ and $x \in X$. If x has order two, the pair of edges from a to ax and from ax to axx = a is sometimes identified to a single edge, sometimes not; we allow both possibilities, even in the same Cayley graph, with one restriction -- all pairs of edges corresponding to the same generator must be treated the same. Left multiplication by an element of A induces an automorphism of any Cayley graph for A. This *natural action* of a group on any of its Cayley graphs is transitive and fixed–point free on the vertex set. In fact, if a group A acts transitively and fixed–point freely on the vertex set of a graph, the graph must be a Cayley graph for A (see [S] or [GT]). If the edge from a to ax is directed and labeled x for each a and x, then the natural action preserves edge directions and labels.

An embedding of a Cayley graph C(A, X) in a closed surface S is *symmetric* if the natural action of A on the Cayley graph C(A, X) extends to an action by a group of homeomorphisms of the surface S (all embeddings in this paper are 2–cell: the interior

* Supported by NSF Contract DMS–8601760.

of each face is an open disk). If, in addition, the surface S is orientable and the extended action preserves the orientation of S, then the embedding is *strongly symmetric*. For orientable surfaces, the symmetry of an embedding simply means that the embedding looks locally the same at every vertex: that is, the cyclic ordering of edge labels and directions encountered in a small "clockwise" trip around any vertex is the same (strongly symmetric) or sometimes the same and sometimes the reverse (symmetric but not strongly symmetric). For nonorientable surfaces, the symmetry of an embedding is more technical and less intuitive. It is explained in the next section.

The *genus* of a graph is the minimum genus of an orientable surface containing an embedding of the graph. If the graph is a Cayley graph and embeddings are restricted to being symmetric (respectively, strongly symmetric), we have the *symmetric* (respectively, *strong symmetric*) *genus* of a Cayley graph. For nonorientable surfaces, one can likewise define a *crosscap number* for any graph and a *symmetric crosscap number* for a Cayley graph. White [WI] has defined the *genus of a group* A, denoted $\gamma(A)$, to be the minimum genus of any of its Cayley graphs. We have defined similarly [T2] the *symmetric* (respectively, *strong symmetric*) *genus* of a group A, denoted $\sigma(A)$ (respectively, $\sigma^{\circ}(A)$) to be the minimum symmetric (respectively, strong symmetric) genus of any Cayley graph for A. For nonorientable surfaces, there are corresponding definitions of the *crosscap number of a group*, denoted $\overline{\gamma}(A)$, and *symmetric crosscap number of a group,* denoted $\overline{\sigma}(A)$. The last has not appeared explicitly in the literature.

Generally speaking, the most convenient parameter for surfaces is not the genus g or number of crosscaps c, but rather its Euler characteristic χ (for orientable surfaces $\chi = 2 - 2g$ and for nonorientable surfaces $\chi = 2 - c$). For that reason, let us define some Euler characteristic parameters for a group. Let $\chi(A)$ (respectively, $\Sigma(A)$, $\Sigma^{\circ}(A)$) denote the maximal Euler characteristic of any orientable surface containing an embedding (respectively, symmetric, strong symmetric) of a Cayley graph for A. Let $\overline{\chi}(A)$ (respectively, $\overline{\Sigma}(A)$) be the maximal Euler characteristic of any nonorientable surface containing an embedding(respectively, symmetric) of a Cayley graph for A. Thus $\chi = 2 - 2\gamma$, $\Sigma = 2 - 2\sigma$, and $\Sigma^{\circ} = 2 - 2\,\sigma^{\circ}$. Similarly, $\overline{\chi} = 2 - \overline{\gamma}$ and $\overline{\Sigma} = 2 - \overline{\sigma}$, where we agree that $\overline{\sigma} \geq 1$ (for graphs one usually allows the crosscap number to be 0, even though a surface with no crosscaps is orientable).

In [T2], it is proved that a group A acts on a surface S if and only if there is a Cayley graph for A symmetrically embedded in S. Thus symmetric genus etc. can be defined without any reference to Cayley graphs or generating sets. In this guise, the strong symmetric genus has a long history, which goes back to Burnside [B] and which has strong connections to groups of conformal automorphisms of Riemann surfaces (see for example [M] or [LM]). For more information about the genus and symmetric genus of a group, see [W2], [T2], or [GT].

The groups in this paper are all finite. The genus of an infinite group [L] is either 0 or ∞, so in that sense the genus of an infinite group is less interesting. For a partial classification of infinite groups of genus 0, see [MPTW].

The remainder of this paper is organized as follows. Section 2 considers nonorientable symmetric embeddings in general. Section 3 relates such embedding to the orientable double covering of a nonorientable surface and classifies groups with $\bar{\Sigma} = 1$. Section 4 recalls the Riemann–Hurwitz equation and classifies groups with $\bar{\Sigma} = 0$. Section 5 computes $\bar{\Sigma}$ for some examples. Section 6 summarizes the known inequalities relating the various χ and Σ parameters. Section 7 presents some problems for further study.

2. Nonorientable Symmetric Embeddings

To understand nonorientable symmetry, it is best to take the "band decomposition" viewpoint of a graph embedding, as given in [GT]. An embedding of a graph G in the surface S is pictured as a thickened neighborhood of the graph in the surface S: a small disk around each vertex, a long narrow band for each edge, and a hole in the surface for each face (which can be capped off with a disk to give the closed surface S). For an orientable surface, one can choose a local orientation for each vertex–disk and edge–band so that the orientation of each edge–band is consistent with the orientation of the vertex–disks at its endpoints. Thus to describe an orientable embedding it suffices to give the cyclic ordering or *rotation* of the edges at each vertex given by a single fixed orientation of the surface. The rotation at each vertex tells how to attach the incident edge–bands to the vertex–disk, automatically creating the thickened neighborhood of the graph in the embedding surface.

For nonorientable surfaces, each vertex–disk can be assigned some local orientation, but these orientations may not be consistent with that of the edge–bands. Given a choice of orientation for the vertex–disks, call an edge–band *type–0* if the directions induced on the boundary of the edge–band by the orientation of its endpoint vertex disks agree; call the edge–band *type–1* otherwise. Figure 1 illustrates a type–0 and a type–1 edge–band. Notice that if the vertex–disks at the end of an edge–band are both placed flat in the plane so that both vertex–disk local orientations are, say, clockwise, then a type–1 edge–band has a half–twist, while a type–0 edge–band lies flat in the plane. Thus, one often thinks of a type–1 band as "twisted", but please note that this twisting is only an artifact of how one chooses to view the vertex–disks. If the two vertex–disks are laid in the plane so that one orientation is clockwise and the other is counterclockwise, then a type–0 edge–band is twisted while a type–1 band is not! The type of an edge–band is thus not an intrinsic property of the embedding, but rather depends on the arbitrary choice of local orientations of the vertex–disks.

Figure 1. A type–0 edge–band and a type–1 edge–band.

The presence of a type–1 edge does not even guarantee that the embedding surface is nonorientable. Although vertex–disk orientations of an orientable embedding can be chosen so that all edge–bands are type–0, that does not mean that vertex–disk orientations must be chosen that way. For example, if one chooses the vertex–disk orientations for a 4–cycle embedded in the sphere so that they alternate clockwise, counterclockwise around the 4–cycle, then every edge–band is type–1 even though the embedding surface, the sphere, is orientable. The criterion for deciding when an embedding surface, given in terms of rotations and edge types, is orientable is not hard: the surface is nonorientable if and only if there is a cycle in the graph containing an odd number of type–1 edges.

Suppose that a Cayley graph C(A, X) is symmetrically embedded in a possibly nonorientable surface S. Since the natural action of A preserves edge labels and directions, in order to map one vertex–disk onto another the rotation at each vertex (in terms of the X–label and direction of each edge) must be the same or reversed. Choose any vertex and fix one of the two orientations for that vertex–disk and now choose the orientation of every other vertex–disk to be the same. For this choice of vertex–disk orientations, some edge–bands may be type–0 and some type–1, but since the action of A preserves edge labels, in order for the action of A to extend to the embedding surface, all edges corresponding to the same generator $x \in X$ must have the same type: a surface homeomorphism cannot preserve vertex–disk orientations and take a type–0 edge–band to a type–1 edge–band or vice versa.

Conversely, if vertex–disk orientations of an embedding of the Cayley graph C(A, X) in the surface S can be chosen so that every rotation is the same and all edges corresponding to the same generator have the same type, then the natural action of A on the Cayley graph C(A, X) extends to the surface S. Just define the extended action first for the vertex–disks, then for edge–bands, then the face–holes. The consistency of rotations and edge types guarantees that no obstructions arise. To summarize this discussion we state the following:

Proposition 2.1 *An embedding of a Cayley graph $C(A, X)$ in the surface S is symmetric if and only if vertex–disk orientations can be chosen so that every rotation is the same and every edge corresponding to the same generator $x \in X$ has the same type. Given such a choice of vertex–disk orientations, the given surface S is nonorientable only if there is a relator in the generating set X in which some generator appears an odd number of times.*

Proof The first statement follows from the previous discussion. The surface S is nonorientable if and only if there is a cycle in the Cayley graph C(A, X) containing an odd number of type–1 edges. Every cycle in C(A, X) corresponds to a relator in the generating set X, that is, a finite sequence of elements of X or inverses of elements of X, whose product is the identity in the group A. Since every edge corresponding to a given generator (or its inverse) has the same edge type, in order for a cycle to contain an odd number of type–1 edges, the corresponding relator must have an odd number of occurrences of some generator (the inverse of a generator counts as an occurrence). ❑

Proposition 2.1 has the surprising consequence that some Cayley graphs have no symmetric nonorientable embeddings at all. For example, let $A = Z_4$, the cyclic group of order 4, and let x be a generator for A. Then C(A, {x}) has no symmetric nonorientable embedding since the only relator is x^4. Is there a group A having no symmetric nonorientable embeddings for any of its Cayley graphs? No, because one can always include the identity element in the generating set X and declare all edge–bands for that element to have type–1. If one doesn't like multiple edges or loops, one can always introduce a redundant generator $z = x^2$ or $z = xy$, where $x, y \in X$, and declare all edge types for z to be type–1 while all other edge types are type–0 (the resulting graph has no multiple edges unless A has order 2 or 3). For example, the Cayley graph $C(Z_4, \{x, x^2\})$ has a symmetric nonorientable embedding obtained by making all the x^2 edges type–1; the resulting surface is the projective plane, if multiple edges corresponding to the generator x^2 are identified, and the nonorientable surface of Euler characteristic –2, otherwise.

But what if we do not allow redundant generators? Call a generating set X for the group A *irredundant* if no proper subset of X generates A. Call the Cayley graph C(A, X) irredundant if X is irredundant. Then no irredundant Cayley graph for Z_4 has a symmetric nonorientable embedding. In general, we have the following somewhat mysterious characterization of 2–groups:

Theorem 2.2 *The group* A *has a symmetric nonorientable embedding for at least one of its irredundant Cayley graphs if and only if the order of* A *is not a power of* 2.

Proof Suppose that no irredundant Cayley graph for A has a symmetric nonorientable embedding. Then no irredundant generating set for A has an element of odd order by Proposition 2.1, since a symmetric nonorientable embedding can be obtained by declaring all edges for an odd order generator to be type–1. For each prime p dividing the order of A, let B_p be a p–Sylow subgroup of A. Let X be the union of the subgroups B_p. Then the subgroup B of A generated by X has order divisible by the highest power of p dividing the order of A for each prime p, since B contains B_p. Thus X generates A. The generating set X contains an irredundant generating set X'. Since no irredundant generating set for A contains any element of odd order, X' must be contained in B_2. But then $A = B_2$ and hence the order of A is a power of 2.

Conversely, suppose the order of A is a power of 2. Let F be the Frattini subgroup of A, namely the intersection of all maximal subgroups of A. The following facts about F are well–known (see [G] for example):

(1) F is normal in A with quotient group A/F a direct product of Z_2's.

(2) if $X \cup F$ generates A, then so does X.

Let X be an irredundant generating set for A and suppose $C(A, X)$ has a symmetric nonorientable embedding. Then some generator x must appear in some relator an odd number of times. The image of x in the quotient group $A/F = Z_2 \times ... \times Z_2$ therefore appears exactly once in a relator in A/F involving the images of the other elements of X. Thus x can be expressed in terms of the other elements of X and the elements of F. By fact 2, this means $X - \{x\}$ generates A, contradicting the irredundancy of X. We conclude that every symmetric embedding of $C(A, X)$ is orientable. ❑

The second half of the proof of this theorem is essentially contained in [PTW]. Part of the proof also applies to groups other than 2–groups. If A has a homomorphism onto the direct product of n Z_2's and X is a generating set for A of n elements, then $C(A, X)$ has no symmetric nonorientable embedding.

3. Orientable Double Coverings

Let $\tilde{S}$ be an orientable surface. Call a homeomorphism t of $\tilde{S}$ to be an *antipodal map* if the order of t is two, t has no fixed points, and t reverses the orientation of $\tilde{S}$. If $\tilde{S}$ is embedded in R^3 symmetrically with respect to the origin, then the map of R^3 taking (x, y, z) to $(-x, -y, -z)$ is an antipodal map of $\tilde{S}$ when restricted to $\tilde{S}$. In fact, all antipodal maps of $\tilde{S}$ are conjugate to this one in the group of homeomorphism of $\tilde{S}$. Given an antipodal map t of the orientable surface $\tilde{S}$, one can form the quotient space $\tilde{S}/t = S$ by identifying t and t(u) for each point $u \in \tilde{S}$. It follows from standard topology [T2] that the natural projection of $\tilde{S}$ onto S is a 2–sheeted, unbranched covering and that S is a closed, nonorientable surface of half the Euler characteristic of $\tilde{S}$. Every nonorientable surface can be obtained in this way. The surface $\tilde{S}$ is called the *orientable double covering of* S *with antipodal map* t. The following is the fundamental theorem about orientable double coverings and group actions:

Theorem 3.1 *The group A acts on the nonorientable surface S if and only if A acts on the orientable double covering $\tilde{S}$ so that the action preserves orientation and commutes with the antipodal map of $\tilde{S}$ (that is, $Z_2 \times A$ acts on $\tilde{S}$ with the A factor preserving orientation and the Z_2 factor generated by an antipodal map).*

Proof The "only if" direction is Theorem 15 of [T2]. Conversely, suppose such an action of A on $\tilde{S}$ exists. Thus for each $a \in A$ there is an orientation preserving homeomorphism $\tilde{h}_a : \tilde{S} \to \tilde{S}$ such that $\tilde{h}_a t = t\tilde{h}_a$, where t is the antipodal map of $\tilde{S}$. Define a map $h_a : S \to S$ as follows. Let $p : \tilde{S} \to S$ be the natural projection. For each $u \in S$, let $\tilde{u}$ be one of the two points of $p^{-1}(u)$. Then $h_a(u) = p\tilde{h}_a(u)$. We must check that h_a is well defined: $p\tilde{h}_a(t(\tilde{u})) = pt\tilde{h}_a(\tilde{u}) = p\tilde{h}_a(\tilde{u})$. It follows that h_a is a homeomorphism because $\tilde{h}_a$ is, and that $h_{ab} = h_a h_b$ because $\tilde{h}_{ab} = \tilde{h}_a\tilde{h}_b$. Thus we have an action of A on the surface S. ❑

Determining when an orientation preserving action commutes with an antipodal map is not easy. This is related to symmetries of Riemann surfaces [Sn] and reflexible maps [CM]. One case at least is well understood: finite group actions on the sphere. These groups are just the automorphism groups (or subgroups) of the regular prisms and platonic solids. They are generated by standard rotations of the unit sphere in R^3. Thus the following classification of groups with $\tilde{\Sigma} = 1$ is probably a folk theorem to many topologists. We denote the cyclic group of order m by Z_m and the dihedral group of order 2m by D_m.

Theorem 3.2 *(Classification of projective plane groups). $\tilde{\Sigma}(A) = 1$ if and only if A is Z_m, D_m, the symmetric group on 4 symbols, or the alternating groups on 4 or 5 symbols.*

Proof By Theorem 3.1 and the discussion of Section 2, the group A has a Cayley graph symmetrically embedded in the projective plane if and only if A acts on the sphere (the orientable double covering of the projective plane) preserving orientation and commuting with an antipodal map. The groups given in the statement of this theorem are precisely the groups which act on the sphere preserving orientation (see Theorem 6.3.1 of [GT]). Each of these actions can be given as a group of orthogonal 3×3 matrices acting on the unit sphere in R^3. The antipodal map of the unit sphere is represented by the negative of the identity matrix and hence commutes with all these actions. ❑

4. The Classification of Klein Bottle Groups

We wish to classify as well groups A such that $\bar{\Sigma}(A) = 0$. To do so we need first the Riemann–Hurwitz equation. An action of a group A on the closed surface S is *pseudo–free* if the number of points left fixed by some element of A is finite. If A acts pseudo–freely on the closed surface S then the quotient surface S/A is a closed surface and the projection $p: S \to S/A$ is a regular branched covering. Each branch point u in S/A is the image of some fixed point of some element of A acting on S. The elements of A leaving a point in $p^{-1}(u)$ fixed form a subgroup of A. Subgroups leaving different points in $p^{-1}(u)$ fixed are conjugate, and by standard facts about surfaces the subgroup is cyclic. Its order is called the *order* of the branch point u and is denoted here r(u). By simply counting vertices, edges, and faces for a graph embedded in S/A whose vertices include all branch points, one obtains the following *Riemann–Hurwitz equation* relating the Euler characteristics $\chi(S)$ and $\chi(S/A)$:

$$\chi(S) = |A| \, (\chi(S/A) - \sum_u (1 - 1/r(u))).$$

The sum in this equation is over all branch points.

Not all actions on surfaces are pseudo–free. If h is a homeomorphism of finite order of the surface S, there are basically two possibilities. Either h has a finite number of fixed points or h has order two and leaves some embedded simple closed curve fixed. In the latter case for orientable surfaces, h also reverses orientation (see Proposition 1 of [T2]), and the curves left invariant by h separate S into "halves" which are switched by h. In [T2], such an h is called a *reflection* and we retain that terminology for nonorientable surfaces, but if should be noted that for nonorientable surfaces reflections do not have to interchange halves of a surface. For more about reflections of surfaces see [B, Sn]. The important observation here is that if h has odd order it cannot be a reflection, and if A has odd order, then no action of A on any surface can contain any reflections.

With this brief discussion of pseudo–free actions, we are ready to classify groups with $\bar{\Sigma} = 0$. The klein bottle has a simple closed curve dividing the surface into two projective planes with a disk removed. Since the cyclic group Z_n and dihedral group

D_m act on the projective plane it is not hard to construct actions $Z_2 \times Z_n$ and $Z_2 \times D_m$ on the klein bottle. Surprisingly enough, these are the only possibilities.

Theorem 4.1 *(Classification of klein bottle groups)* $\overline{\Sigma}(A) = 0$ *if and only if* A *is* $Z_2 \times Z_{2m}$ *or* $Z_2 \times D_{2m}$.

Proof By Theorem 3.1, if $\overline{\Sigma}(A) = 0$ then $Z_2 \times A$ acts on the double cover of the klein bottle, namely the torus, with the Z_2 factor generated by an antipodal map t and A preserving orientation. The groups which act on the torus preserving orientation are the finite quotients of the five Euclidean space groups that are orientation preserving. The corresponding partial presentations from [GT] are:

a) $\langle x, y : x y x^{-1} y^{-1} = 1, \ldots\rangle$
b) $\langle x, y, z : y^2 = z^2 = 1, (xy)^2 = (xz)^2 = 1, \ldots\rangle$
c) $\langle x, y : x^3 = y^3 = (xy)^3 = 1, \ldots\rangle$
d) $\langle x, y : x^4 = y^2 = (xy)^4 = 1, \ldots\rangle$
e) $\langle x, y : x^3 = y^2 = (xy)^6 = 1, \ldots\rangle$

Each of the group actions on the torus given by presentations a) – e) can be obtained by taking the corresponding Euclidean space group C and projecting it onto the torus by "rolling up" the Euclidean plane in two independent directions. The space group C in this case preserves orientation and is of the type p1, p2, p3, p4, p6 (see [L] or [CM]). It is generated by translations h_1 and h_2 by independent vectors V_1 and V_2 together, in cases b) – e), with a rotation of order 2, 3, 4, 6, respectively. The set T of all translations, which is generated by h_1 and h_2, is an abelian normal subgroup of C and its quotient is the cyclic group generated by the added–on rotation (or in fact by any rotation in C of maximal order). The finite group A is then the quotient A/N where N is an normal subgroup of C. The intersection of N with the translation subgroup T must have finite index in T and hence is generated by translations in independent directions. Since C is generated by rotations in cases b) – e), if N contains a rotation, C/N is trivival, or cyclic of low order, or dihedral, none of which are of interest here. Thus N contained in T.

The antipodal map t, which commutes with the action of A on the torus, also is the projection of some isometry g of the Euclidean plane. The only orientation reversing Euclidean isometries are reflection in a line or a glide along a line (a translation in the direction of the line composed with a reflection in the line). Since a reflection leaves a line fixed and t has no fixed points, g must be a glide. Since t has order 2, g^2 is in N. Let f be the reflection part of the glide g. Then if h is a translation in C, $g h g^{-1} = f h f^{-1}$ because the translation part of g commutes with h. Elementary geometry shows that if h is translation by the vector U, then $f h f^{-1}$ is translation by the reflected vector $f(U)$.

We can immediately eliminate presentations c), d), and e). It follows from elementary geometry that if r is a rotation by θ about the point p, then $g r g^{-1}$ is rotation by the angle $-\theta$ about the point $g(p)$. Since the projection of g on the torus, namely the antipodal map t, commutes with the action of A, it follows that only rotations by $180°$ are allowed in the group C. Since presentation c) has $120°$ rotations, d) has $90°$ rotations, and e) has $60°$ rotations, none of these are possibilities for the group C.

For the remaining cases a) and b), we first claim that the quotient group T/N is $Z_2 \times Z_{2m}$ or Z_n. Again assume that T is generated by translations h_1 and h_2 by vectors V_1 and V_2, respectively. Suppose the glide g is in the direction of $aV_1 + bV_2$ where a and b are relatively prime integers. If c and d satisfy $ad - bc = 1$, then translations by $aV_1 + bV_2$ and $cV_1 + dV_2$ also generate T. Thus we can and will assume that the glide g is in the direction of V_1. As before, let f denote the reflection part of g. Then $h_2 g h_2^{-1} g^{-1}$ is a translation by the vector $U = V_2 - f(V_2)$ and is in N since the projections of h_2 and g in the torus commute. By elementary geometry, $V_2 + f(V_2) = cV_1$ for some integer c. Suppose that c is even. Since $U = 2V_2 - cV_1$, the translation h'_2 by $(1/2)U$ is in T. Since $V_2 = (1/2)U + (c/2)V_1$, it follows that h_1 and h'_2 generate T. Moreover, translation by U is in N and hence $(h'_2)^2$ is in N. Therefore T/N is $Z_2 \times Z_k$ for some integer k. If k is odd, then T/N is cyclic. This establishes our claim when c is even. Suppose instead that c is odd. Then the translation h''_2 by the vector $(1/2)(U + V_1)$ is in T. Again since $V_2 = (1/2)(U + V_1) + (c - 1)/2\ V$, if follows that h_1 and h''_2 generate T. Moreover $h_1(h''_2)^{-2}$ is translation by $-U$ and hence is in N. Therefore in T/N the images of h_1 and $(h''_2)^2$ are equal, so h''_2 generates T/N and hence T/N is cyclic, as claimed.

The cases a) and b) are now easily analyzed. In case a), there are no rotations and $C = T$. Thus $A = C/N = T/N$ has the desired form $Z_2 \times Z_{2m}$ or is cyclic, in which case $\bar{\Sigma}(A) = 1$. In case b) the space group C is obtained by adding to T a rotation by $180°$. Conjugation by such a rotation takes any translation to its inverse. Thus if $T/N = Z_2 \times Z_{2m}$, the group $A = C/N$ has a presentation

$$\langle x, y, z : x^{2m} = y^2 = z^2 = 1, y x y = x, z x z = x^{-1}, z y z = y \rangle.$$

The subgroup generated by x and z is just the dihedral group D_{2m}, and since y commutes with x and z, the group A is just $Z_2 \times D_{2m}$. If $T/N = Z_n$ instead, then A has the presentation $\langle x, z : x^n = 1, z x z = x^{-1} \rangle$, which is just the dihedral group D_n, in which case $\bar{\Sigma}(A) = 1$. ❑

5. Some Examples

In order to understand $\bar{\Sigma}$ and its relationship to the orientable parameters Σ, $\Sigma°$, and χ, we present the following examples. We restrict our attention mostly to groups of odd order so that group actions are pseudo–free and the Riemann–Hurwitz equation can be applied. The idea is as follows. If A has odd order, then no generator of A has order 2 and any action of A on a surface is pseudo–free. Thus if a Cayley graph for A is symmetrically embedded in a surface S, that Cayley graph projects to a bouquet of circles embedded in the quotient surface S/A. Each face of that quotient embedding contains a branch point of order equal to the order of the group element of A given by taking the product of the generators encountered in tracing the boundary of the face. By considering possible quotient embeddings, we can use the Riemann–Hurwitz equation to compute the possible symmetric embeddings of C(A, X). The voltage graph construction [GT] actually constructs the given symmetric embedding.

Example 1 Let $A = Z_n \times Z_n$, n odd and $n > 3$. Clearly $\Sigma° = \Sigma = \chi = 0$. Thus $\bar{\chi} = 0$ or -1; it is not known which. Consider $\bar{\Sigma}$. Any Cayley graph for A involves at least two generators, so we need to consider bouquets of at least two circles in various quotient surfaces. A bouquet of two circles in the projective plane has two faces. If x and y are the generators, those face boundaries correspond to group elements $x^{\pm 1}$, $y^{\pm 1}$, $x^{\pm 1}y^{\pm 2}$ or $x^{\pm 2}y^{\pm 1}$ all of which have order n. By the Riemann–Hurwitz equation, the embedding surface has Euler characteristic

$$\chi(S) = |A| (1 - 2(1 - 1/n)) = |A| (-1 + 2/n).$$

A similar computation when S/A is the klein bottle, gives $\chi(S) = |A| (-1 + 1/n)$. If $\chi(S/A) \leq -1$, then $\chi(S) \leq |A| (-1)$. Therefore, the best we can do is when S/A is the projective plane, and hence $\bar{\Sigma}(A) = |A| (-1 + 2/n)$. ❑

Observe that in this example, $\bar{\Sigma}$ can be made to differ by an arbitrary amount from $\Sigma°$, Σ, χ, and $\bar{\chi}$. Also, because a one face embedding for a degree 4 Cayley graph for A has Euler characteristic $|A| - 2|A| + 1$, the given value for $\bar{\Sigma}(A)$ is not far from being the "worst" it can be. As n approaches infinity, the ratio $\bar{\Sigma}(A)/|A|$ approaches -1, which means that the ratio of the number of faces to the order of A for the best nonorientable symmetric embedding approaches 0.

When n is even in Example 1, the situation becomes more complicated. If the generating set χ has only two elements, then no relator uses either generator an odd number of times so by Proposition 2.1 all symmetric embeddings are orientable. Thus Cayley graphs of degree 5 or greater must be used, driving $\bar{\Sigma}$ down to be even more negative. On the other hand, group actions may now involve reflections, which generally drives $\bar{\Sigma}$ back up again.

It is probably feasible to compute $\bar{\Sigma}$ as well for odd order abelian groups of rank greater than two. The parameters χ and $\bar{\chi}$ have been computer for most abelian groups [JW, PW], and Σ° has been computed for all abelian groups [M] (and for odd groups $\Sigma^{\circ} = \Sigma$).

Example 2 Let A be the metacyclic group of order 27 with the presentation $\langle x, y: y^3 = x^9 = 1, y^{-1} x y = x^4 \rangle$. Then $\bar{\chi}(A) = -3$ and $\chi(A) = -6$ by [BRS]. It can be shown that A is not generated by elements of order 3. By analyzing bouquets of two or more circles, it follows that if A acts on the surface S, then the quotient surface S/A must have 3 or more branch points at least two of which have order 9 when S/A is the sphere, 2 or more branch points at least one of which has order 9 when S/A is the projective plane, and at least one branch point when S/A is the torus or klein bottle. We therefore conclude that

$$\Sigma(A) = |A|\,(2 - (1 - 1/3) - 2(1 - 1/9)) = -12$$

and $$\bar{\Sigma}(A) = |A|\,(1 - (1 - 1/3) - (1 - 1/9)) = -15.\ \square$$

This group is the first known one for which χ and $\bar{\chi}$ differ by more than 1. We suspect also that the ratio $\Sigma / \bar{\chi}$ is the largest of any group with $\bar{\chi} < 0$. Cayley graphs for which χ and $\bar{\chi}$ differ by an arbitrary amount are given in [PTW].

The other, more famous group of order 27, namely $Z_3 \times Z_3 \times Z_3$, can also be analyzed. By [MPSW] and [BS], we know that $\chi = -12$ and $\bar{\chi} = -11$ or -12. The Riemann–Hurwitz equation can be used once again to show that $\chi = -18$ and $\bar{\Sigma} = -27$.

Example 3 Let A be the alternating group on n symbols, n > 167. Conder [C] has shown that A has a presentation of the form

$$\langle x, y, z : x^2 = y^2 = z^2 = 1, (xy)^2 = (yz)^3 = (xz)^7 = 1, \ldots \rangle.$$

If we choose the same rotation at every vertex of C(A, {x, y, z}) and every edge is type–1, we obtain an embedding in which every face corresponds to one of the relators $(xy)^2$, $(yz)^3$, or $(xz)^7$. Counting faces, we find the embedding surface has Euler characteristic $-|A| / 84$. Hurwitz's theorem [GT] states the $\Sigma^{\circ}(B) \le -|B| / 42$ for any group B. Since the alternating group A is simple, it has no index 2 subgroup and hence any action of A on an orientable surface preserves orientation. Thus the given embedding surface of Euler characteristic $-|A|/84$ must be nonorientable. In addition, the action of A on the induced double covering achieves the Hurwitz bound. Thus $\bar{\Sigma}(A) = -|A| / 84$ and $\Sigma^{\circ}(A) = \Sigma(A) = -|A| / 42$. The Hurwitz theorem for embedded Cayley graphs [T3] implies that $\bar{\chi}(A) = -|A| / 84$ as well. On the other hand, $\chi(A)$ is still unknown. $\square$

By the double covering Theorem 3.1, we know that $\bar{\Sigma} \leq \Sigma^{\circ}/2$, but in most of the examples we have given $\bar{\Sigma}$ is less than Σ°, sometimes a lot less. In addition, $\bar{\Sigma}$ has also been much less than Σ, χ, and $\bar{\chi}$. This example shows that symmetric nonorientable embeddings can be large, in fact as large as the Hurwitz bound allows. In particular, it is possible to have $\bar{\Sigma} > \chi$, $\bar{\Sigma} = \Sigma^{\circ}/2$, and $\bar{\Sigma} = \bar{\chi}$.

6. Inequalities

There are a number of inequalities relating the various parameters $\chi, \Sigma, \Sigma^{\circ}, \bar{\chi}, \bar{\Sigma}$ to each other and to the order of the given group. Some of these are trivial. Some are classical and deep. Some are new, difficult and not well understood. We begin with the easy ones.

Easy Inequalities $\Sigma^{\circ} \leq \Sigma \leq \chi, \bar{\Sigma} \leq \bar{\chi}, \bar{\chi} \geq \chi - 1, \bar{\Sigma} \leq \Sigma^{\circ}/2.$ ❑

The next to last inequality comes from the observation that one can always twist a single edge in any orientable embedding of any graph (other than a tree) to obtain a nonorientable embedding at the expense of decreasing the Euler characteristic by at most 1. The last inequality follows of course from Theorem 3.1.

The next collection of inequalities are the Hurwitz inequalities relating our various parameters to the order of the group. Whenever the order of a group is sufficiently large compared to the Euler characteristic of the embedding surface, the structure of the group becomes restricted by the presence of the short relators necessary to construct the many small faces of the embedding. This leads to some complicated refinements of the basic Hurwitz inequality. The following notation and terminology is helpful. Call A a *(p, q, r)° group* if A has a presentation of the form

$$\langle x, y, z : x^2 = y^2 = z^2 = 1, (xy)^p = (yz)^q = (xz)^r = 1, \ldots \rangle.$$

Call a (p, q, r) group A *proper* if the subgroup generated by xy and yz is a proper subgroup (it is easily shown to have index at most 2); call it *improper* otherwise. Call A a *(p, q, r)° group* if it has a presentation of the form

$$\langle u, v : u^p = v^q = (uv)^r = 1, \ldots \rangle.$$

Hurwitz Inequalities Let A be a group with $\chi(A) < 0$. Then $\chi, \Sigma, \bar{\chi}$, and $\bar{\Sigma}$ are all bounded above by $-|A|/84$. This bound is achieved if and only if A is a proper (2, 3, 7) group (for χ and Σ) or an improper (2, 3, 7) group (for $\bar{\chi}$ and $\bar{\Sigma}$). If A is not a (2, 3, 7) group, then $\chi, \Sigma, \bar{\chi}, \bar{\Sigma}$ are all bounded above by $-|A|/48$ and this bound is achieved only for proper and improper (2, 3, 8) groups. Similarly, Σ° is bounded above by $-|A|/42$, with equality only when A is a $(2, 3, 7)^{\circ}$ group. ❑

Hurwitz's original inequality [H] is the one for Σ° and appears in the context of the conformal automorphisms of Riemann surfaces. The version for embedded Cayley graphs is first given in [T1]. There are many further and important refinements [T4], which we have not presented here.

The inequalities we give in this last collection are new. Proofs will appear elsewhere [T5]. These inequalities also will have a number of refinements, including applications to the genus of a quotient group. We state here only the simplest versions.

New Inequalities Let A be any group with $\chi(A) < 0$. Then the following inequalities hold:

a) $\chi > 6\bar{\chi}$, $\Sigma > 6\bar{\chi}$, $\Sigma > 6\chi$

b) $\chi \geq 4\bar{\chi}$, $\Sigma \geq 4\bar{\chi}$, $\Sigma \geq 4\chi$ if A has no Cayley graph of degree 5 or less

c) $\Sigma^\circ > 6\bar{\chi}$, $\Sigma^\circ > 6\chi$, $\Sigma^\circ > 6\Sigma$ unless A is a (2, 3, 7) or (2, 3, 8) group. ❑

It is expected that the numbers 6 and 4 can be lowered in many cases in some of the inequalities. Notice however for the b) inequalities that $\Sigma = 4\bar{\chi}$ in Example 2, and for the c) inequalities, it can be shown that $\Sigma^\circ = 8\Sigma$ for the group of genus two [T5].

7. Problems

We conclude with some problems for further study. Some are purely computational, others are wider ranging.

Problem 1 Compute $\bar{\Sigma}(Z_n \times Z_n)$ for n even.

Problem 2 Compute $\bar{\Sigma}$ for various 2–groups.

Problem 3 Determine all groups A with $\bar{\Sigma} = -1$. By work similar but easier to that in [T4], it can be shown that if $\bar{\chi} \geq -1$, then $\chi \geq 0$. Also since $\Sigma(Z_2 \times A) = -2$, the computations in [T7] can be applied to $Z_2 \times A$, and it seems that $|A| \leq 24$. The Riemann–Hurwitz equation can be used to show that there are no pseudo–free group actions of order greater than 6 on the surface of Euler characteristic −1. We conjecture that there are no groups with $\bar{\Sigma} = -1$.

Problem 4 Find a lower bound for $\bar{\Sigma}$ in terms of χ, $\bar{\chi}$, Σ, or Σ°. The New Inequalities say nothing about $\bar{\Sigma}$. None of the methods used in [T5] apply directly to $\bar{\Sigma}$. Notice in Example 2 that $\bar{\Sigma} = 5\bar{\chi}$ and for the group of genus two we believe that $\bar{\Sigma} = 9\Sigma = 9\chi$.

Problem 5 Determine all the surfaces containing a symmetric Cayley graph embedding for a given group. Graph embeddings satisfy an interpolation theorem: if a graph has a 2–cell embedding in orientable surfaces of genus g_1 and g_2, then it has embeddings in all orientable surfaces of genus between g_1 and g_2. Symmetric embeddings do not satisfy this result, mostly because there are so few symmetric embeddings of a Cayley graph. For example, a Cayley graph of degree three only has 4 or 8 possible symmetric embeddings, depending on whether the generating set has 2 or 3 elements (there are only two possible rotations and one is the reverse of the other, so the only choice is edge types).

Problem 6 Classify irredundant Cayley graphs embeddable in the klein bottle. Theorem 4.1 only classifies symmetric embeddings. The full classification of groups with $\chi = 0$ given in [T4] implies that any irredundant Cayley graph embeddable in the klein bottle embeds symmetrically in the torus, but the original klein bottle embedding of that Cayley graph could be nonsymmetric. We conjecture that $\overline{\chi} = 0$ if and only if $\overline{\Sigma} = 0$. A first step would be to show that $\overline{\chi}(Z_m \times Z_n) = -1$, where m divides n and $m > 2$. This problem is solved for the projective plane: the only group with $\overline{\chi} = 1$ but $\overline{\Sigma} < 1$ is $Z_3 \times Z_3$ (see Exercises in [GT]).

REFERENCES

[BRS] M.G. Brin, D.E. Rauschenberg, and C.C. Squier, On the genus of the semidirect product of Z_9 by Z_3, J. Graph Theory, to appear.

[BS] M.G. Brin and C.C. Squier, On the genus of $Z_3 \times Z_3$, Europ. J. Combin., to appear.

[BSn] E. Bujalance and D. Singerman, the symmetry type of a Riemann surface, Proc. London Math Soc. 51 (1985), 501 – 509.

[B] W. Burnside, *Theory of Groups of Finite Order*, Cambridge Univ. Press, Cambridge, 1897.

[CM] H.S.M. Coxeter and W.O.J. Moser, *Generators and Relations for Discrete Groups*, 4th Ed., Springer–Verlag, New York, 1980.

[G] D. Gorenstein, *Finite Groups,* Chelsea, New York, 1980.

[GT] J.L. Gross and T.W. Tucker, *Topological Graph Theory*, Wiley–Interscience, New York, 1987.

[H] A. Hurwitz, Uber algebraische Gebilde mit eindeutigen Transformationen in sich, Math. Ann. 41 (1892), 403 – 442.

[JW] M. Jungerman and A.T. White, On the genus of finite abelian groups, Europ. J. Combin. 1 (1978), 243 – 251.

[L] H.M. Levinson, On the genera of graphs of group presentations, Ann. New York Acad. Sci. 175 (1970), 277 – 284.

[LM] H.M. Levinson and B. Maskit, Special embeddings of Cayley diagrams, J. Combin. Theory Ser. B 18 (1975), 12 – 17.

[Ly] R.C. Lyndon, *Groups and Geometry*, Lon. Math. Soc. Notes 101, Cambridge Univ. Press, Cambridge, 1985.

[M] C. Maclachlin, Abelian groups of automorphisms of compact Riemann surfaces, Proc. Lond. Math Soc. 15 (1965), 699 – 712.

[MPSW] B. Mohar, T. Pisanski, M. Skoviera, and A.T. White, The cartesian product of three triangles can be embedded into a surface of genus 7, Discrete Math. 56 (1985), 87 – 89.

[MPTW] B. Mohar, T. Pisanski, T.W. Tucker, and M.E. Watkins, On the classification of infinite planar groups, in preparation.

[PTW] T. Pisanski, T.W. Tucker, D. Witte, The nonorientable genus of some metacyclic groups, preprint.

[PW] T. Pisanski and A.T. White, Nonorientable embeddings of groups, Europ. J. Combin. 9 (1988), 445 – 461.

[S] G. Sabidussi, On a class of fixed–point free graphs, Proc. Amer. Math. Soc. 9 (1958), 800 – 804.

[Sn] D. Singerman, Symmetries of Riemann surfaces with large automorphism group, Math Ann. 210 (1974), 17 – 32.

[T1] T.W. Tucker, The number of groups of a given genus, Trans. Amer. Math Soc. 258 (1980), 167 – 179.

[T2] T.W. Tucker, Finite groups acting on surfaces and the genus of a group, J. Combin. Theory Ser. B 34 (1983), 82 – 98.

[T3] T.W. Tucker, A refined Hurwitz theorem for imbeddings of irredundant Cayley graphs, J. Combin. Theory Ser. B (1984), 244 – 268.

[T4] T.W. Tucker, There is one group of genus two, J. Combin. Theory Ser. B (1984), 269 – 275.

[T5] T.W. Tucker, Bounds on the genus parameters of a group and its quotient groups, in preparation.

[W1] A.T. White, *Graphs, Groups, and Surfaces*, North–Holland, Amsterdam, 1973 (rev. ed. 1984).

[W2] A.T. White, The genus parameter for groups, Scientia, to appear.

The Number of Isomorphism Classes of Spanning Unicyclic Subgraphs of a Graph

Preben Dahl Vestergaard
Aalborg University
Denmark

ABSTRACT

In [3], B. Zelinka proved that the spanning trees of a graph containing n disjoint circuits can be partitioned into at least n + 1 isomorphism classes.

We generalize this result in several directions: if a graph contains n disjoint circuits, then it contains at least n isomorphism classes of spanning unicyclic graphs (Corolllary 1), and we give a lower bound depending on the girth of G (Theorem 1), also we determine the extremal graphs (Theorem 3).

Definitions

A *tree* is a connected graph with no circuit. A *rooted graph* (G, v) is a graph G together with a distinguished vertex v, called the root, from V(G). If it is clear which root $v \in V(G)$ is intended, we may just write the rooted graph G. A *unicyclic graph* G is a connected graph with precisely one circuit C, a component of G – E(C) is called a *pendant tree,* and it is said to be *attached* to C at the unique vertex which it has in common with C. In a graph a *block* is defined to be a maximal subgraph spanned by a set of edges, any two of which belong to a common circuit. An edge which belongs to no circuit is a block. A *cactus* is a connected graph in which each block is either a circuit or an edge. An *n–cactus* is a cactus with precisely n circuits. A *spanning subgraph* of G is G itself, or a subgraph of G obtained by deleting edges from G. The *distance* between two subsets of V(G) is the length of a shortest path joining a vertex from one subset to a vertex from the other subset. Two graphs are said to be *isomorphic,* if there exists a bijection from one graph onto the other which preserves incidence. Two rooted graphs are said to be *root–isomorphic,* if there exists an isomorphism between the graphs which maps one root onto the other root. $(v_1, v_2, \ldots, v_r)$ denotes a circuit with the vertices in that order. $\lceil t \rceil$ denotes the least integer not less than t, and $\lfloor t \rfloor$ denotes the largest integer not greater than t.

A circuit C in a cactus G is called an *internal* circuit if $G - E(C)$ has at least two connected components each containing a circuit, if $G - E(C)$ has at most one connected component containing a circuit, then C is said to be *external.* For a unicyclic graph H with circuit C, we define the weight of H to be

$$w(H) = \sum_{v \in V(C)} \sum_{x \in V(G)} d(x,v)$$

Theorem 1 *If the graph G contains a spanning n–cactus, $n \geq 2$, with an external circuit of length r, $r \geq 3$, then G contains at least $n - 2 + \frac{r}{2}$ pairwise non–isomorphic spanning unicyclic subgraphs.*

Corollary 1 *If a connected graph G contains n pairwise disjoint circuits, $n \geq 1$, then G contains at least n pairwise non–isomophic spanning unicyclic subgraphs.*

Proof of Corollary 1 Any connected graph G which contains a circuit C also contains a spanning unicyclic subgraph: delete if necessary edges from $G - E(C)$ until all other circuits are destroyed and such that the resulting graph remains connected.

This proves the Corollary for $n = 1$. For $n \geq 2$ we see that $r \geq 3$ implies that an integer $\geq n - 2 + \frac{r}{2}$ must be at least n. Since a connected graph with n disjoint circuits contains a spanning n–cactus, we can apply Theorem 1 to obtain the desired reult that G contains at least n pairwise non-isomorphic spanning unicyclic subgraphs. This proves Corollary 1. ❑

Proof of Theorem 1 Theorem 1 is only claimed to hold for $n \geq 2$, and in fact the statement of Theorem 1 does not hold for $n = 1$, $r \geq 5$, because then $\lceil n - 2 + \frac{r}{2} \rceil \geq 2$, but if G is unicyclic, then by definition G contains only one spanning unicyclic subgraph, namely G itself.

It is enough to prove Theorem 1 for an n–cactus, so we shall suppose that G is an n–cactus and we shall do the proof by induction on n.

Let $n = 2$, and let G be a 2–cactus with the two circuits C_1 and C_2. Let the vertices of C_1 be $v_1, v_2, v_3, \ldots, v_r$ in that order, and let v_1 be the uniquely determined vertex on C_1 having minimum distance in G to C_2. Suppose r is odd. We shall now list $\frac{r+1}{2}$ spanning unicyclic subgraphs of G, no two of which are isomorphic to each other, namely:

$$G - (v_1, v_2), G - (v_2, v_3), \ldots, G - (v_{\frac{r-1}{2}}, v_{\frac{r+1}{2}}), G - (v_{\frac{r+1}{2}}, v_{\frac{r+3}{2}})$$

Each of the $\frac{r+1}{2}$ graphs above contains all vertices of G, contains C_2 as its only circuit and is connected, so each one is a unicyclic spanning subgraph of G. No two of the graphs are isomorphic to each other because no two of them have the same weight.

It is easy to see that the weights of the graphs above form a strictly decreasing sequence, because as i increases from 1 to $\frac{r+1}{2}$ access from C_2 through v_1 to vertices round $C_1 - (v_i, v_{i+1})$ happens through paths some of whose lengths are shorter and some of whose lengths are unchanged. E. g., the distance from v_1, are hence from vertices on C_2, to v_2 is $r - 1$ in $G - (v_1, v_2)$ but only 1 in $G - (v_2, v_3)$.

If r is even we can analogously enumerate $\frac{r}{2}$ pairwise non–isomorphic spanning unicyclic subgraphs for G. This proves Theorem 1 for $n = 2$ and for all $r \geq 3$.

Let $n \geq 3$. Suppose that Theorem 1 holds for $n - 1$ and all $r \geq 3$, we shall then prove that it also holds for n and all $r \geq 3$.

Let G be an n–cactus and let $C_1 = (v_1, v_2, \ldots, v_r)$ be an external circuit in G. Since C_1 is external, it contains a uniquely determined vertex, say v_1, through which any path joining the other vertices of C_1 to another ciruict of G must pass. Suppose r is even, the case r odd can be treated analogously. The graph $G - (v_1, v_2)$ by induction hypothesis contains at least $n - 1$ pairwise non–isomorphic spanning unicyclic subgraphs. Among them all let H be one with minimum weight. Denote the graph $H + (v_1, v_2)$ by H'. We shall now list $\frac{r}{2} - 1$ spanning unicyclic subgraphs of G, all of which have weights which are pairwise distinct and also less than the weight of H. This will give us at least $(n-1) + (\frac{r}{2}-1)$ pairwise non–isomorphic spanning unicyclic subgraphs for G as desired. The $\frac{r}{2} - 1$ graphs are:

$$H' - (v_2, v_3), H' - (v_3, v_4), \ldots, H' - (v_{\frac{r}{2}}, v_{\frac{r}{2}+1}).$$

Their weights form a stricly decreasing sequence of numbers less than $w(H)$ because for each i, $i = 2, 3, \ldots, \frac{r}{2}$, the distance from v_1 to v_i is shorter in $H' - (v_i, v_{i+1})$ than in $H' - (v_{i-1}, v_i)$. This proves Theorem 1. ❑

The following example shows that Theorem 1 is best possible:

Example Let G be the n–cactus consisting of n triangles v, v_{2i-1}, v_{2i}, $i = 1, 2, \ldots, n$, with the common vertex v. Each triangle is an external circuit of length $r = 3$ and G must by Theorem 1 contain $\geq n - 2 + 1\frac{1}{2}$, i. e., at least n, pairwise non–isomorphic spanning unicyclic subgraphs, and we can verify that G in fact contains exacly n

pairwise non–isomorphic spanning unicyclic subgraphs: a spanning unicyclic subgraph for G consists of one triangle with a vertex v from which emanates 2i paths of length 1 and n – 1 – i paths of length 2, where i is one of the numbers 0, 1, 2, ..., n – 1.

We shall show in Theorem 3 that the graphs which are extremal for Theorem 1 are all quite similar to G in the Example above.

For the proof of Theorem 3 we shall need the following Theorem 2 and Lemma 1.

Theorem 2 *If a graph contains a spanning 2–cactus and has exactly two isomorphism classes of spanning unicyclic subgraphs then*

either (1) G can be constructed thus:

1a. Let v_1, v_2, v_3 *and* w_1, w_2, w_3 *be two 3–circuits which are disjoint except that possibly* $v_1 = w_1$. *If* v_1 *and* w_1 *are distinct, then they are joined by a* v_1w_1*–path which has nothing but* v_1 *and* w_1 *in common with the two 3–circuits.*

1b. Attach to each of v_2, v_3, w_2, w_3 *a copy of a rooted tree A.*

1c. If $v_1 = w_1$, *then attach to* v_1 *a rooted tree B. Otherwise, let the* v_1w_1*–path be* $t_1, t_2, ..., t_n$ *with* $t_1 = v_1$ *and* $t_n = w_1$. *For each* $i = 1, 2, ..., \lfloor\frac{n+1}{2}\rfloor$ *attach to each of* t_i, t_{n-i+1} *a copy of a rooted tree* B_i.

or (2) G can be constructed thus:

2a. Let v_1, v_2, v_3, v_4 *and* w_1, w_2, w_3, w_4 *be two 4–circuits which are disjoint except that possibly* $v_1 = w_1$. *If* v_1 *and* w_1 *are distinct, then they are joined by a* v_1w_1*–path which has nothing but* v_1 *and* w_1 *in common with the two 4–circuits.*

2b. Attach to each of v_2, v_4, w_2, w_4 *a copy of a rooted tree A. Attach to each of* w_3, w_3 *a copy of a rooted tree B.*

1c. If $v_1 = w_1$, *then attach to* v_1 *a rooted tree C. Otherwise, let the* v_1w_1*–path be* $t_1, t_2, ..., t_n$ *with* $t_1 = v_1$ *and* $t_n = w_1$. *For each* $i = 1, 2, ..., \lfloor\frac{n+1}{2}\rfloor$ *attach to each of* t_i, t_{n-i+1} *a copy of a rooted tree* C_i.

Theorem 2 can be derived from results in [2] and it states that if a graph G with a spanning 2–cactus has exactly two isomorphism classes of spanning unicyclic subgraphs then G is a symmetric 2–cactus with circuit length 3 or 4. ❑

Lemma 1 *If G contains a spanning n–cactus, $n \geq 3$, and has exactly n isomorphism classes of spanning unicyclic subgraphs and if e is an edge of an external circuit of the cactus then $G - e$ has exactly $n - 1$ isomorphism classes of spanning unicyclic subgraphs.*

Proof of Lemma 1 $G - e$ contains a spanning $(n - 1)$–cactus and by Theorem 1 has at least $n - 1$ isomorphism classes of spanning unicyclic subgraphs. Suppose $G - e$ has n or more isomorphism classes of spanning unicyclic subgraphs then as in the proof of Theorem 1 we take H, a spanning unicyclic subgraph of $G - e$ with maximum/minimum weight and on the external circuit containing e we can find another edge f such that $H + e - f$ is another spanning unicyclic subgraph of G with larger/smaller weight than H and thus G will have at least $n + 1$ isomorphism classes of spanning unicyclic subgraphs. This contradicts the hypothesis about G and Lemma 1 is proven. ❑

We are now ready to state Theorem 3:

Theorem 3 *If the graph G contains a spanning n–cactus, $n \geq 3$, and if G contains exactly n isomorphism classes of spanning unicyclic subgraphs, then G can be obtained by the following construction:*

1. *Let (M, v) denote a unicyclic graph M with root v and with the circuit having length 3 or 4. In both cases, the circuit has exactly two vertices x an y which are equidistant from v. Further, let the pendant trees in M at x and y, respectively, be root–isomorphic to each other.*

2. *Form the disjoint union of n copies of (M, v) with v identified, and*

3. *Attach to v a rooted tree (T, v), possibly $T = v$.*

Theorem 3 obviously does not hold for $n = 1$ because Theorem 3 prescribes a circuit length of 3 or 4, but also a unicyclic graph circuit with length 5 or more has exactly one spanning unicyclic subgraph. Neither is Theorem 3 quite true for $n = 2$, but that case is completely described by Theorem 2.

Examples of G obtained by the construction of Theorem 3:

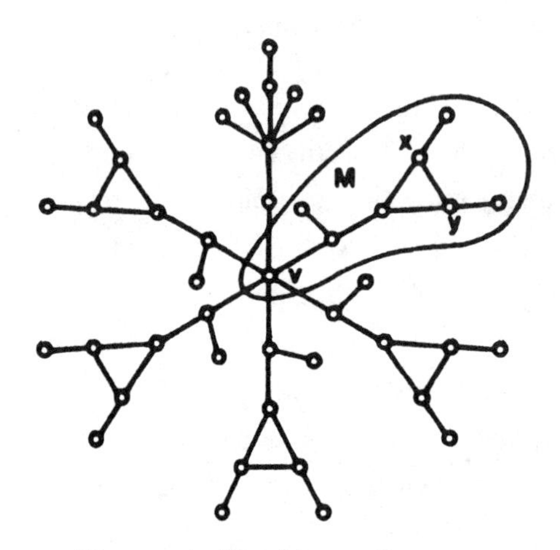

The graph G with $n = 5$.
M has a circuit of length 3.

The graph G with $n = 6$.
M has a circuit of length 4.

Proof of Theorem 3 First we shall prove Theorem 3 under the assumption that G is an n–cactus. We shall use induction on n. Let $n = 3$. Any 3–cactus has at least two external circuits, so G has at most one internal circuit and one of the following two cases occur:

Case 1. G has one internal circuit.

Case 2. G has no internal circuit.

Suppose Case 1 occurs, then G contains one internal circuit, C_2, and two external circuits, C_1, C_3. Let $v_1 \in V(C_1)$ have minimum distance from C_2, and let $C_1 = (v_1, v_2, ..., v_r)$. For r odd, let $e = (v_{\frac{r+1}{2}}, v_{\frac{r+3}{2}})$, and for r even, let $e = (v_{\frac{r}{2}}, v_{\frac{r}{2}+1})$. $G - e$ is a 2–cactus, and, by Lemma 1, $G - e$ has exactly two isomorhpism classes of spanning unicyclic subgraphs, and, by Theorem 2, $G - e$ is a symmetric 2–cactus, i. e., the pendant trees to C_3 are the same as those to C_2.

Let $f = (v_1, v_2)$, and apply now Lemma 1 and Theorem 2 to the graph $G - f$, then we also obtain that the pendant trees to C_3 are the same as those to C_2. but for $G - e$ and $G - f$ the pendant trees of C_3 remain unchanged while those of C_2 do change. This is a contradiction which proves that Case 1 cannot occur.

Thus Case 2 must occur, i. e. all 3 circuits C_1, C_2, C_3 in G are external. By Lemma 1, $G - e$ has for any $e \in E(C_1)$ exactly two isomorphism classes of spanning unicyclic subgraphs. Hence $G - e$, by Theorem 2, is a symmetric 2–cactus.

Since this holds for all edges of C_1, we can conclude that the tree containing $C_1 - e$ necessarily is attached to the central vertex of the unique path joining C_2 to C_3. considering $G - f$ for $f \in E(C_i)$, $i = 2, 3$, we find that G has the desired structure and Theorem 3 is proven for $n = 3$.

Let $n \geq 4$. Suppose Theorem 3 holds for $n - 1$, we shall then prove that it also holds for n.

Let C_1 be an external circuit for G. For any $e \in E(C_1)$, $G - e$ is an $(n - 1)$–cactus which by Lemma 1 has exactly $n - 1$ isomorphism classes of spanning unicyclic subgraphs. By induction hypothesis $G - e$ is a symmetric $(n - 1)$–cactus. By considering $G - f$, where f belongs to a circuit of $G - E(C_1)$ we find that G has the desired structure and Theorem 3 is proven under the assumption that G is a cactus.

To finish the proof we only have to demonstrate that if a graph G satisfies the hypothesis of Theorem 3 and hence contains a spanning cactus G' then $G = G'$.

Suppose $E(G') \subset E(G)$ and let $e \in E(G) \setminus E(G')$, then it is easy to see that $G' + e$ and hence G contains at least $n + 1$ isomorphism classes of spanning unicyclic subgraphs. This completes the proof of Theorem 3. ❑

Problem 1 *How does the number of spanning unicyclic subgraphs depend on circuit length?*

It is probably not true that the number of spanning unicyclic subgraphs in some sense increases monotonically with the girth or with, say, the sum of lengths of all circuits in the graph, but Theorem 1 shows that at least for special graphs as the cacti circuit length plays a role.

Problem 2 *What can be said about the number of spanning 2–cacti, the number of spanning subgraphs with cyclomatic number* $r = 0, 1, 2, 3, \ldots$?

REFERENCES

[1] B. L. Hartnell, Some problems on spanning trees, *Proc. Fifth Manitoba Conf. Numer. Math.* (1975) 375–384.

[2] P. D. Vestergaard, On graphs with two isomorphism classes of spanning unicyclic subgraphs, in preparation.

[3] Bohdan Zelinka, The number of isomorphism classes of spanning trees of a graph, *Math. Slovaca.* **28** (1978) no. 4, 385–388.

A Survey of Snarks

John J. Watkins
The Colorado College

Robin J. Wilson
The Open University, England

ABSTRACT

A snark *is a connected bridgeless cubic graph with edge–chromatic number 4 — that is, it cannot be 3–edge–colored. In this paper we survey the work that has been done on snarks, including their history, various constructions, and possible generalizations. This paper updates the survey of snarks given in [10].*

1. Introduction

The story of snarks began in 1880. P.G. Tait, a mathematical physicist at the University of Edinburgh, was attempting to find a short proof of the four–color theorem. An accepted 'proof' had been given by A.B. Kempe in 1879, but P.J. Heawood found a flaw in it in 1890. Tait translated the problem of coloring maps to one of coloring the edges of cubic graphs [29]. In particular, he showed that *the four–color theorem is equivalent to the statement that every bridgeless cubic planar graph is 3–edge–colorable* (see [14], pp. 26–27).

Tait believed that all cubic graphs are 3–edge–colorable, and was thus convinced that he had proved the four–color theorem. However, he overlooked some cubic graphs that cannot be 3–edge–colored: those with bridges, such as the graph in Figure 1, and certain non–planar graphs, such as the Petersen graph (Figure 2).

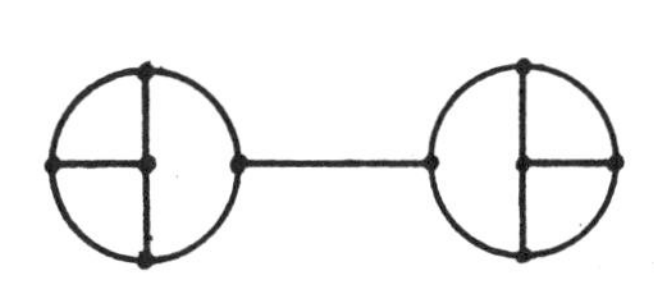

Figure 1

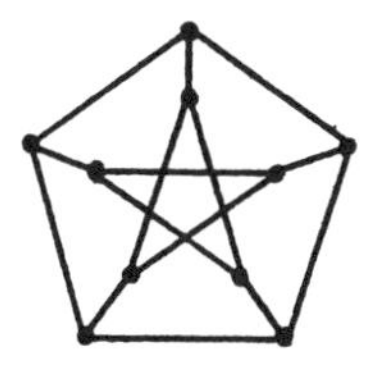

Figure 2

Non–trivial cubic graphs that cannot be 3–edge–colored, such as the Petersen graph, captured the fancy of Martin Gardner and led him to introduce the term *snark* in a Scientific American column in 1976 [17].

2. Examples of Snarks

The earliest snark to appear was the Petersen graph, discussed by Petersen [25] in 1898; Figure 3 shows an alternative drawing, due to Kempe [22] some years earlier.

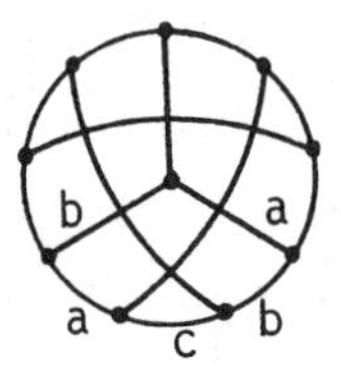

Figure 3

To see that this graph cannot be 3–edge–colored, we assume the contrary, and begin (without loss of generality) by 3–coloring any 5–cycle a, b, a, b, c as shown. This immediately determines the colors of five additional edges, and leads to an impasse since we are then forced to color two adjacent edges with the same color.

The next snark appeared in 1946, in a chemical paper of D. Blanuša [2]. We now know that this type of snark comes in two varieties (Figure 4), each of which is composed from two 'copies' of the Petersen graph, in a sense to be explained later.

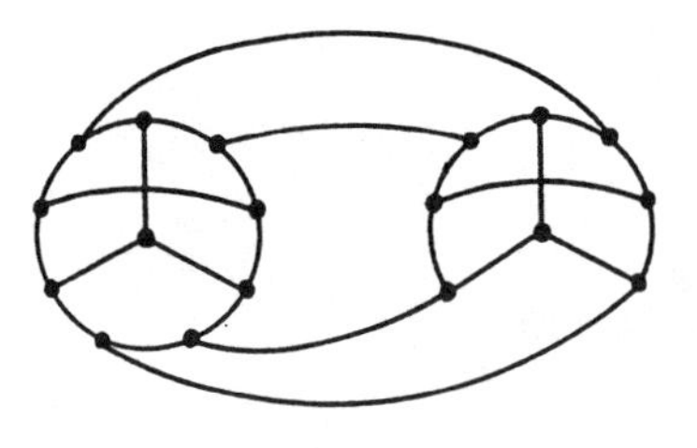

Type 1

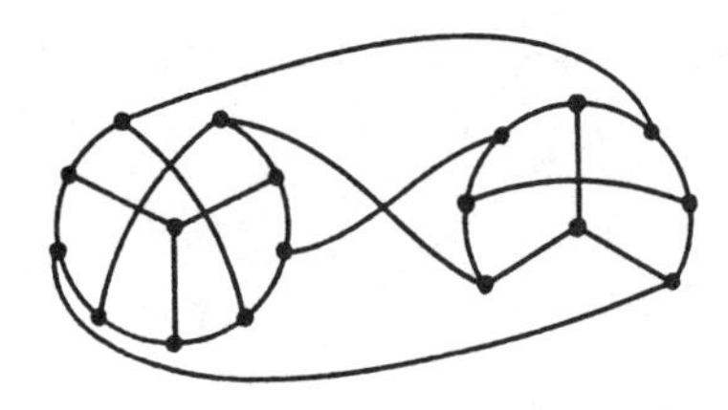

Type 2

Figure 4

Another snark was found by Blanche Descartes [12] in 1948; it has 210 vertices, and was constructed by adding a 'star configuration' to each edge of the Petersen graph (see [14]). A further snark was found in 1973 by G. Szekeres [28]. As shown in Figure 5, it is composed of five 'copies' of the Petersen graph.

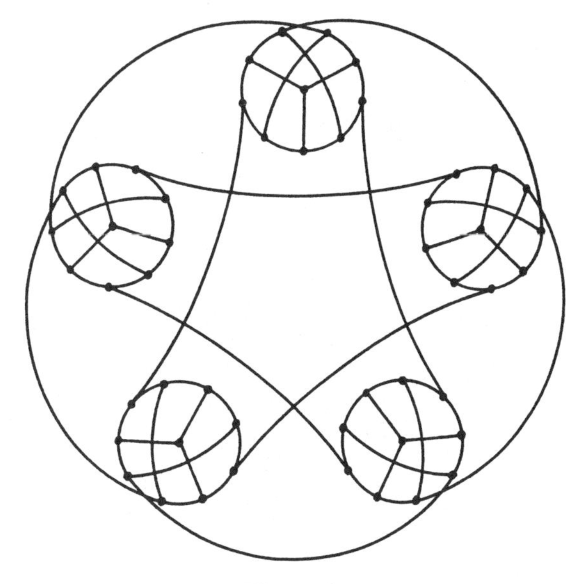

Figure 5

Such painstakingly slow progress in the discovery of snarks underwent a dramatic change when an infinite family of snarks was discovered, by E. Grinberg in 1972, and independently by R. Isaacs in 1974 [19]; they are known as the *flower snarks* $J_3, J_5, J_7, \ldots$ of orders 12, 20, 28, … . The first flower snark J_3 is really just the Petersen graph, with the central vertex replaced by a triangle. We show the first two flower snarks in Figure 6.

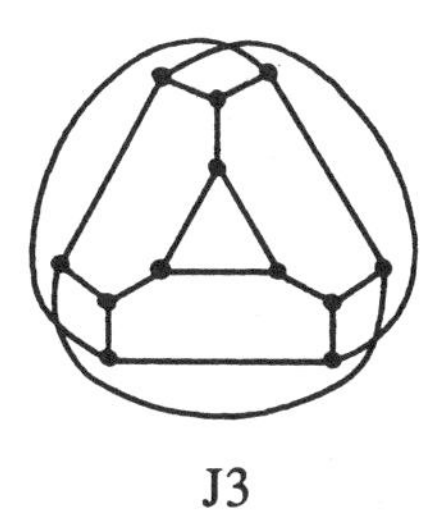

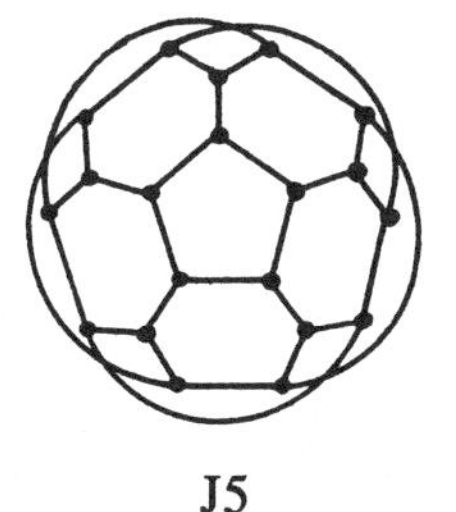

J3 J5

Figure 6

Isaacs also presented a method for joining two snarks to get a new snark; for example, two copies of the Petersen graph combine in this way to give the two types of

Blanuša snark. In this process, two snarks G_1 and G_2 are joined to form a new snark $G_1 \cdot G_2$, called a *dot product*, as follows:

(1) remove any two non–adjacent edges ab and cd from G_1;

(2) remove any two adjacent vertices x and y from G_2;

(3) join the vertices, as shown in Figure 7.

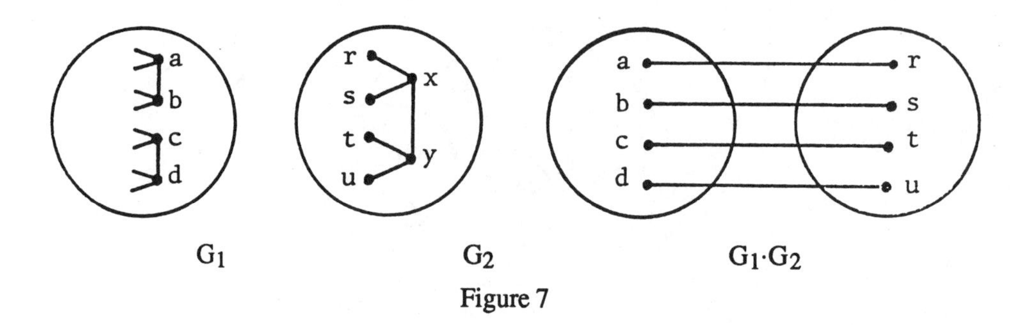

Figure 7

A dot product depends both on G_1 and G_2, and on the choices made in (1) and (2). For example, the two types of Blanuša snark arise from the Petersen graph because there are essentially two different choices in (1).

With his remarkable paper, Isaacs transformed snark hunting. Where before there had been essentially four snarks, there were now infinitely many snarks: the flower snarks and those formed by dot products. Furthermore, these two families are distinct and contain all of the previously known snarks.

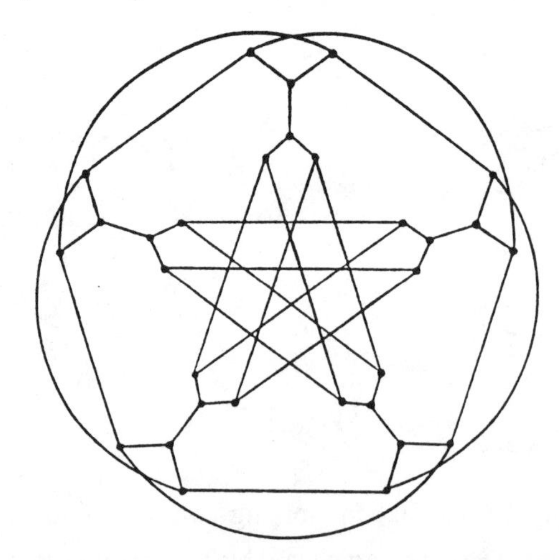

Figure 8

Isaacs [19] also discovered another snark which is not a member of either of these infinite families. This is the *double–star snark* D(5,2), shown in Figure 8; it has five pieces in each of the inner and outer 'rims', and the connections from the outer rim are made to every second piece of the inner rim. The double–star graph D(n,k) is defined similarly. A.G. Chetwynd [8] has shown that D(n,k) is a snark if and only if it is D(5,2) or D(n,k), where $\frac{1}{3}n$ is an integer and $\gcd(n,k) = \frac{1}{3}n$. (However, the latter snarks are often considered trivial, a notion we discuss in the next section.) Surprisingly, a similar generalization of the Petersen graph yields no additional snarks [5].

3. Trivial Snarks

Before describing some further constructions of snarks, we must deal with questions of triviality. Some snarks are trivial modifications of others, and we should eliminate such examples. This topic has attracted considerable discussion in the literature (see, for example, [10], [17], [19], [27], and [32]). But exactly which modifications of snarks are we to consider trivial?

If we take any snark and replace any vertex by a triangle, then the resulting cubic graph still cannot be 3–edge–colored. Similarly, a digon (a pair of multiple edges) can be inserted into any edge, as in Figure 9. Conversely, contracting a triangle to a vertex does not affect 3–edge–colorability, nor does contracting a digon to an edge. Thus we may assume that snarks have girth 4 or more.

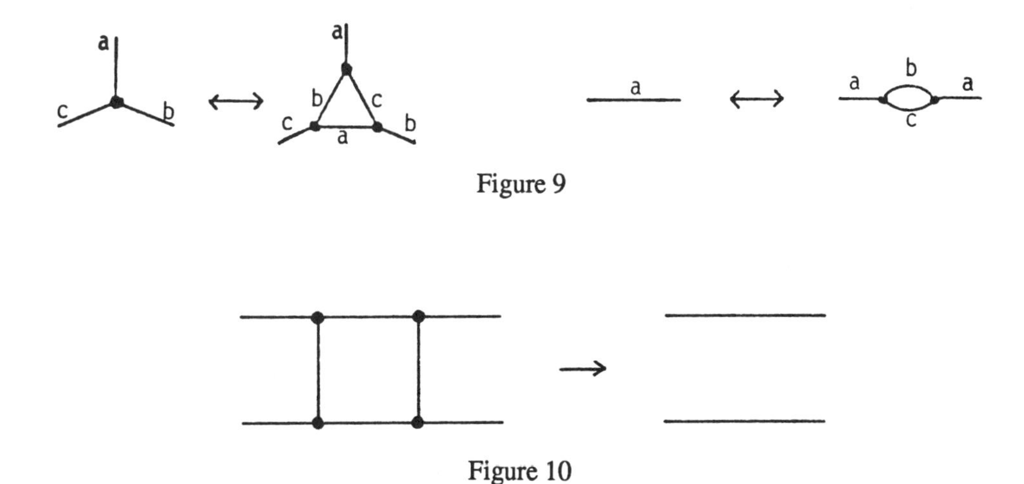

Figure 9

Figure 10

We can go further, since a quadrilateral in a snark may be replaced by two edges, as shown in Figure 10. This produces a smaller snark, since any 3–coloring of the smaller graph would immediately yield a 3–coloring of the original snark, which is impossible. On the other hand, inserting a quadrilateral along two edges of a snark *can* affect the colorability. For example, a quadrilateral inserted along any two non–adjacent edges of the Petersen graph yields a 3–edge–colorable graph. The status of quadrilaterals is thus somewhat different from that of digons and triangles. Nonetheless, it has been customary to exclude quadrilaterals and require that snarks have girth 5 or more.

It has also been customary to require a snark to be cyclically 4–edge–connected; that is, deleting fewer than four edges does not disconnect the graph into two components each containing a cycle. For if a cubic graph has a bridge, then it is not 3–edge–colorable; if it has a cutset with two edges, then in any 3–edge–coloring of the graph these two edges must be colored the same, and thus we can take a snark and a cubic graph and construct a new snark by cutting any edge in each graph and joining the resulting semi–edges; similarly, if it has a cyclically–disconnecting cutset with three edges, then in any 3–edge–coloring of the graph these three edges are colored differently, and thus we can take a snark and a cubic graph and construct a new snark by deleting a vertex from each graph and joining the three pairs of semi–edges.

It has recently been shown that similar results hold for snarks with cutsets of size 4 and 5 (see [4], [18] and [27]). Thus these snarks can also be considered to be trivial, because they can easily be constructed from smaller snarks. However, it is *not* usual to make this restriction.

It seems, therefore, that the elusive notion of triviality has escaped us. We could impose the restriction that snarks be cyclically 6–edge–connected, but there may also be a decomposition result for cutsets of size 6. It therefore seems premature to exclude such snarks at present.

An alternative approach is to focus on the structure of snarks, leaving the definition of a snark as broad as possible — simply, a connected bridgeless cubic graph with chromatic index 4. The goal is then to specify a collection of 'prime' snarks, and a list of basic constructions with which every snark can be built up from prime snarks. We cannot yet offer a definition of a prime snark. However, a prime snark should contain no multiple edges, triangles or quadrilaterals and (apart from the Petersen graph) should be cyclically 6–edge–connected.

4. Some Specific Constructions

The dot product is a very general and powerful method for generating new snarks from known snarks. In this section we narrow our focus to some *specific* constructions for snarks; when these constructions yield cyclically 4–edge–connected snarks, there is automatically an alternative construction using the dot product.

Loupekine Snarks

If we remove a path of length 3 from a snark, we obtain a graph with five semi–edges. In any coloring of this graph, one pair of semi–edges has the same color (a match) and another pair has different colors (a mismatch). The color of the fifth edge does not concern us.

This idea was used by F. Loupekine to construct an infinite family of snarks (see [10], [20] and [32]). For example, a snark results if we join an *odd* number of these graphs, as in Figure 11; here we have removed a path of length 3 from the Petersen graph.

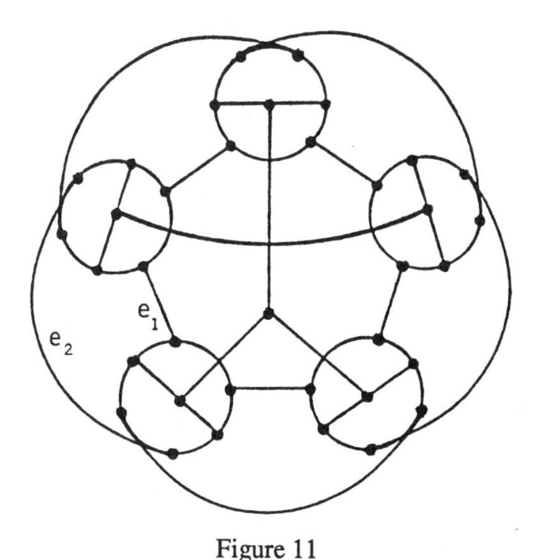

Figure 11

In this construction, any snark may be used, we may cross some edges (such as e_1 and e_2 in Figure 11), or we may insert arbitrary graphs into a pair of connecting edges. We may also treat the central edges in any fashion we like, since they are not involved in considerations of colorability.

Loupekine also obtained snarks by joining an *even* number of these graphs together. Here, the matched and mismatched pairs work out, but he created snarks by attaching the central semi–edges to appropriate graphs (see [32]). It turns out that successive pairs of central semi–edges match and can easily be joined to something to make a snark. For example, if we join three successive central edges to a single vertex, and join the remaining edges in any appropriate way, then we obtain a snark. Figure 12 shows such an example. Note that this snark is cyclically 4–edge–connected and could also be derived using the dot product.

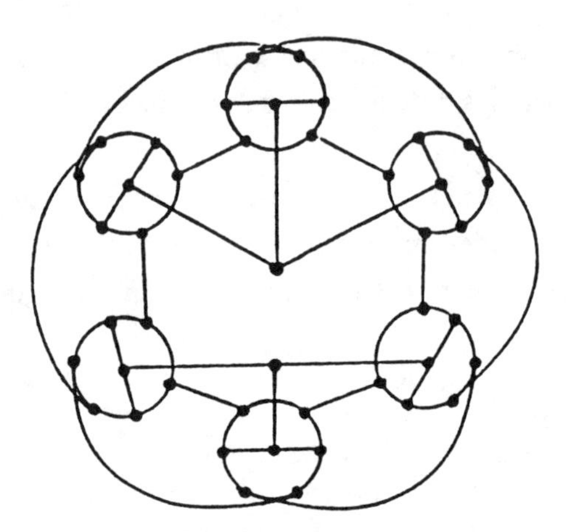

Figure 12

Another construction of Loupekine uses the notion of a 'critical' cycle. Let C be a cycle in a snark G, and construct $G - C$, keeping all the semi–edges. There are two possible cases:

(1) if $G - C$ is not 3–edge–colorable, then we can construct a variety of snarks by adding whatever we like to $G - C$;

(2) if $G - C$ is 3–edge–colorable, then we call the cycle C *critical*. An example of a critical cycle is any 5–cycle of the Petersen graph.

Loupekine [32] has found a general method for constructing snarks, which he calls *cartwheel snarks*, using an odd number of copies of $G - C$, when C is critical.

Goldberg Snarks

M. Goldberg [18] has constructed an infinite family of snarks, which can be used to give infinitely many counter–examples to the critical graph conjecture [11]. We illustrate two of these snarks in Figure 13. The Goldberg snarks are all cartwheel snarks.

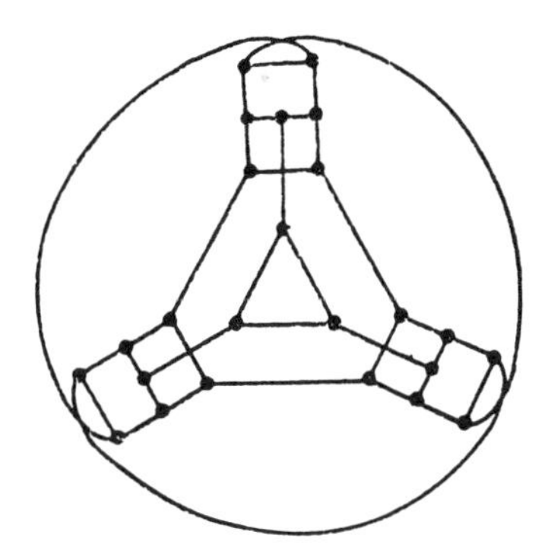
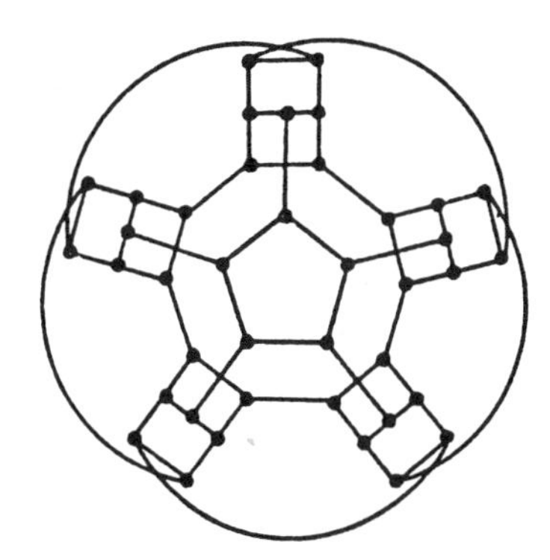

Figure 13

Celmins and Swart Snarks

Several additional snarks have been constructed by Celmins and Swart [7] from planar configurations which are not D–reducible, in the sense of the four–color theorem. Two of these snarks are shown in Figure 14.

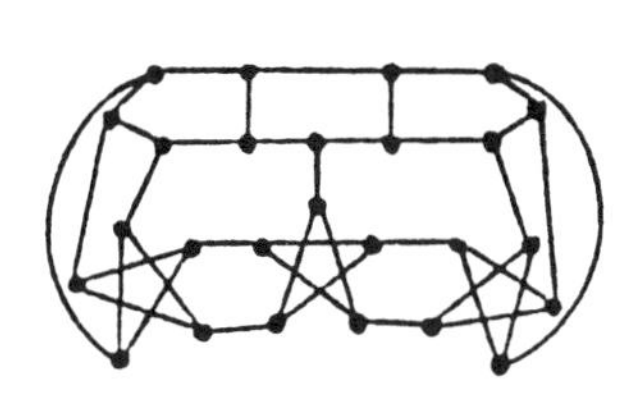
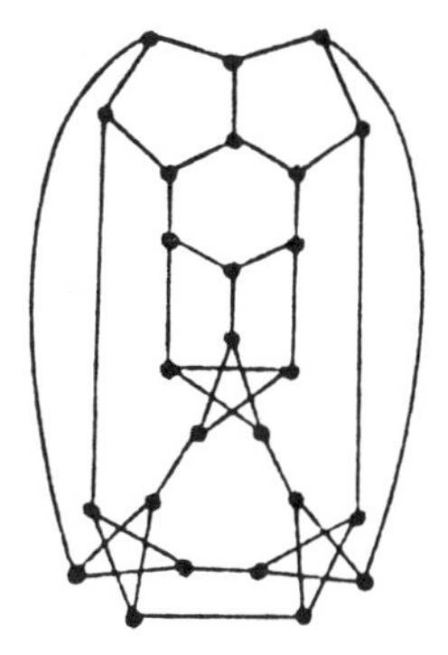

Figure 14

Watkins Snarks

In 1983, J. Watkins [31] constructed two specific families of snarks of orders 10, 18, 26, ..., containing the two Blanuša snarks. The idea is to join graphs with mutually incompatible coloring properties. We illustrate the member of order 42 from each family in Figure 15. These snarks can also be derived using the dot product.

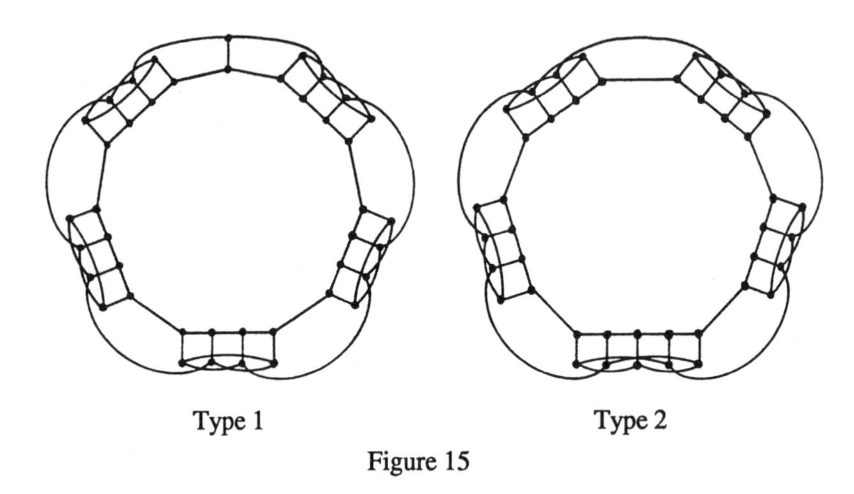

Figure 15

Two specific families of snarks of orders 50, 90, 130, ... were also constructed by Watkins. One family includes the Szekeres snark; the other gives a different snark of order 50, shown in Figure 16. Moreover, these families can be mixed to yield additional snarks; for example, there are six snarks of order 50 that can be formed in this way. Again, the dot product could also be used.

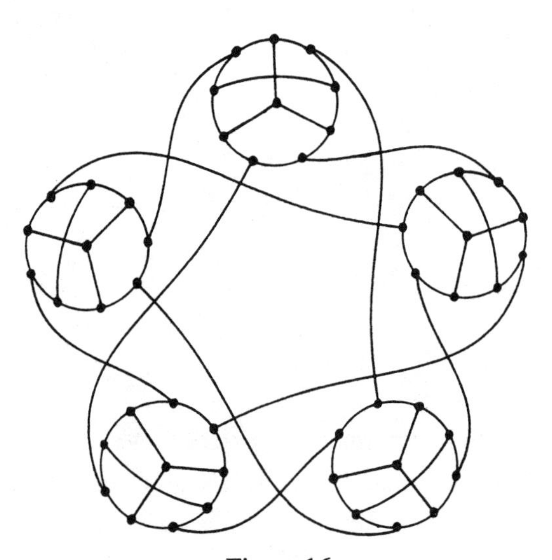

Figure 16

There are two ways of removing two non–adjacent edges from the Petersen graph. Watkins observed that replacing an edge by either of these two graphs does not affect 3–edge–colorability. This process is a special case of the dot product, but provides an easy and specific way to construct snarks. For example, if each 'spoke' of the Petersen graph (Figure 2) is replaced by one of these graphs, then the result is the Szekeres snark. If each 'spoke' is replaced by the other graph, then the result is the snark of Figure 16.

5. Small Snarks

In their 1981 paper [10], Chetwynd and Wilson listed ten snarks of order not exceeding 30. Many more have now been constructed (see, for example, Fouquet et al. [16]). The Petersen graph is the only snark of order 10, and there are no snarks of orders 12, 14 or 16 (see, for example, [6], [13], [14] and [15]). It has been shown independently by a number of people that there are just two snarks of order 18 (see, for example, M. Preissmann [26]); they were incorrectly depicted in [10]. For other small orders the situation has changed dramatically. For example, Chetwynd and Wilson presented only one snark of order 20 (the flower snark J_5) and two snarks of order 22 (Loupekine snarks). There are precisely six snarks of order 20 (see Figure 17 for the five additional ones), and precisely twenty snarks of order 22 (see [1] and [31]). A recent paper of Leizhen Cai [3] presents a snark of order 24; also included in this paper is a proof of a conjecture of Isaacs that snarks exist for all even orders greater than or equal to 18.

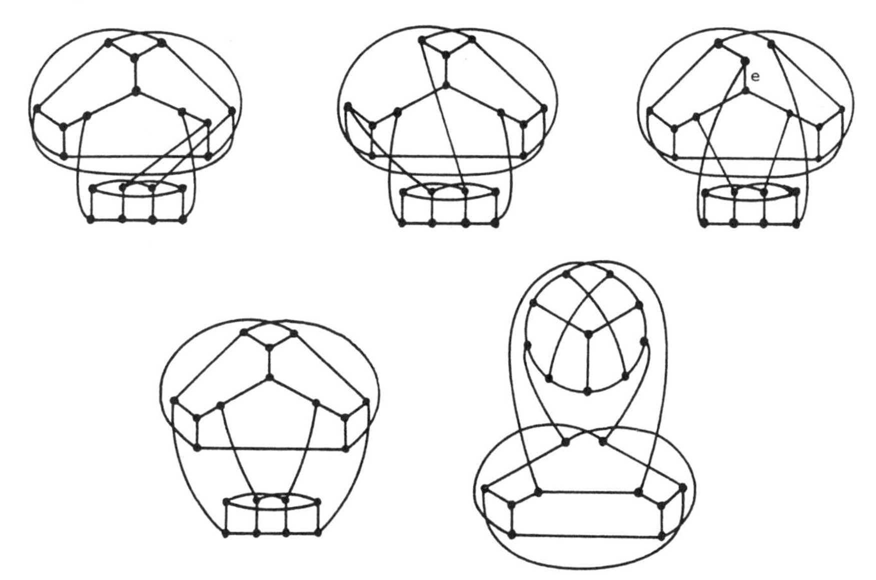

Figure 17

Computational work on small snarks has recently been carried out by F. Holroyd (unpublished). He defined an *antipodal snark* as a snark formed by identifying antipodally opposite ends of a plane cubic tree. A computer search showed that, up to order 20, the Petersen graph is the only antipodal snark. He also defined a *snarkmaker* to be an even cycle with a semi–edge incident with each vertex, and a pairing defined on the semi–edges such that, for every 3–edge–coloring of the graph, at least one of the defined pairs is colored alike; it is so–called, because if each semi–edge is joined to a new vertex, and if the new vertices are joined in *any* way to produce a cubic graph with girth not exceeding 5, then a snark is produced. A computer search showed that, up to order 14, the only snarkmaker is as shown in Figure 18 with antipodal pairing, giving rise to the Petersen graph. It will be interesting to see which other snarks are produced by this approach.

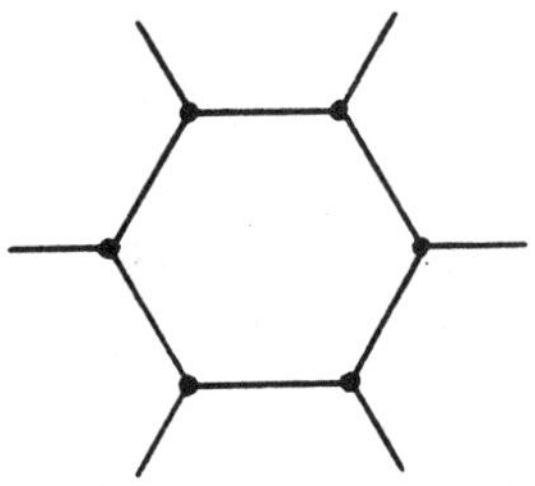

Figure 18

Holroyd has also defined a *bicycle snark* of order n; this is a snark consisting of two n–cycles and n 'spokes', each joining a vertex of one n–cycle to a vertex of the other. A computer search correctly yielded the Petersen and Blanuša snarks, confirmed the non–existence of bicycle snarks of order 14, showed the non–existence of bicycle snarks of order 22, and produced over 50 bicycle snarks of order 26.

6. k–Snarks

We have seen that a snark is essentially a 3–regular graph with edge–chromatic number 4. It is natural to generalize this idea, and define a *k–snark* (for $k \geq 3$) to be a k–regular graph with edge–chromatic number $k + 1$. Thus a 3–snark is simply a snark.

As with snarks, there are a number of 'trivial' graphs which we exclude from our consideration. In addition, since every k–regular graph of odd order automatically has edge–chromatic number $k + 1$, we exclude them and concentrate only on regular graphs of even order.

As was shown by Kotzig (unpublished), an important class of 4–snarks is obtained by taking 3–snarks and forming their line graphs. For example, the line graphs of the flower snarks of orders 12, 20, 28, ... are 4–snarks of orders 18, 30, 42, ... As shown in [10] and [13], these 4–snarks can be used to obtain further counter–examples of the critical graph conjecture. A 4–snark of order 18 is shown in Figure 19.

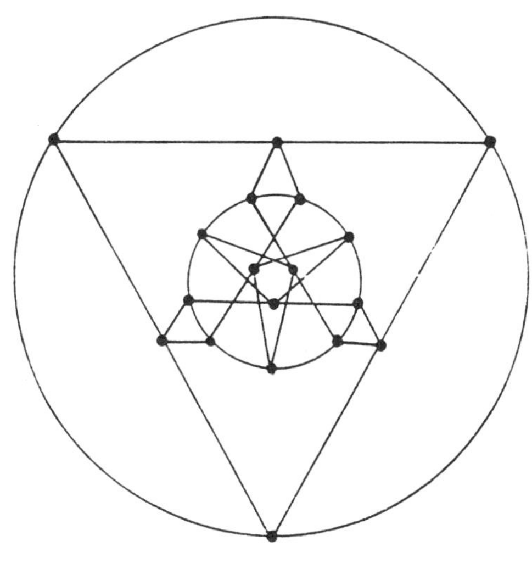

Figure 19

Meredith [24] has given a construction which yields a k–snark for each $k \geq 3$. To construct such a k–snark, we take ten copies of the complete bipartite graph $K_{k,k-1}$, and join them as in Figure 20.

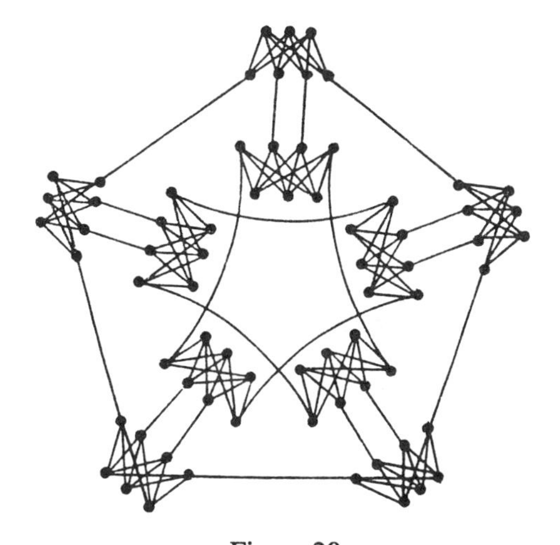

Figure 20

Some new constructions for k–snarks and multisnarks have recently been given by Chetwynd and Hilton [9]. Further information about k–snarks in general can be found in [10].

7. Open Problems

We conclude this survey by mentioning some unsolved conjectures. These conjectures have been around for some time, and appeared in [10].

The first conjecture, due to Tutte, relates to the apparent omnipresence of the Petersen graph in the study of snarks:

Tutte's conjecture. Every snark contains

(a) a subgraph homeomorphic to the Petersen graph;

(b) a subgraph contractible to the Petersen graph.

Another celebrated conjecture relates to the fact that all snarks so far constructed are of girth 5 or 6. For example, the Petersen graph has girth 5, whereas the flower snarks have girth 6:

The girth conjecture. Every snark has a 5–cycle or a 6–cycle.

Much of the interest and importance of snarks arises from the fact that the following general conjectures can (without loss of generality) be restricted to snarks — more precisely, any minimal counter–example must necessarily be a snark. Celmins [6] has shown that all three conjectures are true for the flower snarks and the double–star snark D(5,2):

The 5–flow conjecture. Every bridgeless graph has a 5–flow.

The double cover conjecture. Every bridgeless graph can be twice–covered with cycles.

Fulkerson's conjecture. Every cubic bridgeless graph can be twice–covered with 1–factors.

The relationships between these conjectures and their connections with snarks are discussed in the recent survey of Jaeger [21].

REFERENCES

[1] R. E. L. Aldred, B. Sheng, D. A. Holton and G. F. Royle, All the small snarks, to appear.

[2] D. Blanuša, Problem ceteriju boja (The problem of four colors), *Hrvatsko Prirodoslovno Društvo Glasnik Mat.–Fiz. Astr. Ser. II*, **1** (1946), 31–42.

[3] L. Cai, A snark of order 24, to appear.

[4] P. J. Cameron, A. G. Chetwynd and J. J. Watkins, Decomposition of snarks, *J. Graph Theory* **11** (1987), 13–19.

[5] F. Castagna and G. Prins, Every generalized Petersen graph has a Tait coloring, *Pacific J. Math.* **40** (1972), 53–58.

[6] U. A. Celmins, A study of three conjectures on an infinite family of snarks, *Department of Combinatorics and Optimization, University of Waterloo, research report* CORR 79–19.

[7] U. A. Celmins, and E. R. Swart, The construction of snarks, *Department of Combinatorics and Optimization, University of Waterloo, research report* CORR 79–18.

[8] A. G. Chetwynd, Ph.D. thesis, The Open University, England.

[9] A. G. Chetwynd and A. J. W. Hilton, Snarks and k–snarks, *Ars Combinatoria* **25C** (1988), 39–54.

[10] A. G. Chetwynd and R. J. Wilson, Snarks and supersnarks, *The Theory and Applications of Graphs,* John Wiley & Sons, New York, (1981), 215–241.

[11] A. G. Chetwynd and R. J. Wilson, The rise and fall of the critical graph conjecture, *J. Graph Theory*, **7** (1983), 154–157.

[12] B. Descartes, Network colourings, *Math Gazette*, **32** (1948), 67–69.

[13] M. A. Fiol, Contribución a la teoria de grafos regulares, Universidad Politécnica de Barcelona, 1979.

[14] S. Fiorini and R. J. Wilson, Edge–Colourings of Graphs, *Research Notes in Math.* **16**, Pitman, London, (1977).

[15] J.–L. Fouquet, Note sur la non existence d'un snark d'ordre 16, *Discrete Math.* **38** (1982), 163–171

[16] J.–L. Fouquet, J.–L. Jolivet and M. Rivière, Graphes cubiques d'indice trios, graphes cubiques isochromatiques, graphes cubiques d'indice quatre, *J. Combinatorial Theory (B),* **31** (1981), 262–281.

[17] M. Gardner, Mathematical Games: Snarks, Boojums and other conjectures related to the four–color–map theorem, *Scientific American*, **234(:4)** (April 1976), 126–130.

[18] M. K. Goldberg, Construction of class 2 graphs with maximum vertex degree 3, *J. Combinatorial Theory (B)*, **31** (1981), 282–291.

[19] R. Isaacs, Infinite families of non–trivial trivalent graphs which are not Tait colorable, *Amer. Math. Monthly*, **82** (1975), 221–239.

[20] R. Isaacs, Loupekine's snarks: A bifamily of non–Tait–colorable graphs, unpublished.

[21] F. Jaeger, Nowhere–zero flow problems, *Selected Topics in Graph Theory*, **3** (ed. L. W. Beineke and R. J. Wilson), Academic Press, London, (1988).

[22] A. B. Kempe, A memoir on the theory of mathematical form, *Phil. Trans. Roy. Soc. London*, **177** (1886), 1–70.

[23] F. Loupekine and J. J. Watkins, Cubic graphs and the four–color theorem, *Graph Theory and its Applications to Algorithms and Computer Science,* John Wiley & Sons, New York, (1985), 519–530.

[24] G. H. J. Meredith, Regular n–valent, n–connected, non–Hamiltonian, non–n–edge–colorable graphs, *J. Combinatorial Theory (B)*, **14** (1973), 55–60.

[25] J. Petersen, Sur le théoreme de Tait, *Intermed. Math.*, **15** (1898), 225–227.

[26] M. Preissmann, Snarks of order 18, *Discrete Math.*, **42** (1982), 125–126.

[27] M. Preissmann, C–minimal snarks, *Annals of Discrete Math.,* **17** (1983), 559–565.

[28] G. Szekeres, Polyhedral decompositions of cubic graphs, *Bull. Austral. Math. Soc.*, **8** (1973), 367–387.

[29] P. G. Tait, Remarks on the colouring of maps, *Proc. Roy. Soc. Edinburgh*, **10** (1880), 729.

[30] F. C. Tinsley and J. J. Watkins, A study of snark embeddings, *Graphs and Applications,* John Wiley & Sons, New York (1985), 317–332.

[31] J. J. Watkins, On the construction of snarks, *Ars Combinatoria*, **16B** (1983), 111–123.

[32] J. J. Watkins, Snarks, *Proc. First China–U.S.A. Graph Theory Conf.*, to appear.

Graphical Combinatorial Families and Unique Representations of Integers

Herbert S. Wilf*

University of Pennsylvania

In 1978 I published ([1], [2]) two papers on a unified approach to several different algorithmic problems in combinatorics, which grew out of joint work [3] (particularly chapter 13) with A. Nijenhuis. Aside from these algorithmic uses, however, the method has purely mathematical consequences, and these have not been sufficiently brought out. So I would like to take this opportunity to review the unified approach here, this time discussing the graph theoretical side of the story, and its implications, more fully.

Let G be a strongly connected acyclic digraph. By a *terminal vertex* of G we mean a vertex of outdegree 0. We assume that G has exactly one terminal vertex, τ. We assume further that the edges of G are numbered, computer–science style. That is to say, if v is a fixed vertex, with ρ outgoing edges, then these edges are numbered 0, 1, ..., $\rho - 1$.

Such a digraph will be called a graphical combinatorial family.

Let $v \in V(G)$. By a *combinatorial object of order* v we mean a walk from v to τ. Associated with each combinatorial object of order v is its edge code word, which is the sequence of edge numbers that occur in the walk. Evidently the order v of the walk, and the code word determine the walk uniquely.

Many garden–variety combinatorial families are special cases of these structures. Take $V(G)$ to the set of lattice points (n, k) $(n \geq k \geq 0)$ in the plane. If $v = (n, k)$ with $n > k \geq 1$ then out of v there go two edges, one, numbered 0, goes to $(n - 1, k)$, and the other, numbered 1, goes to $(n - 1, k - 1)$. If $n = k \geq 1$ there is just one edge, numbered 0, and it goes to $(n - 1, k - 1)$. If $n > 0$ and $k = 0$ there is also just one edge, numbered 0, and it goes to $(n - 1, k)$. The unique terminal vertex is $\tau = (0, 0)$.

Now, for a fixed $v = (n, k)$, there is a 1–1 correspondence between the combinatorial objects of order v and the subsets of k elements chosen from n: given a

* Research supported by the United States Office of Naval Research

walk from v to (0, 0), insert into the set the x–coordinate of the initial vertex of every edge numbered 1 that occurs in the walk.

That was one example. There is an extensive list of combinatorial objects that can be brought into the fold, including set partitions, number partitions, permutations by cycles, permutations by runs, Young tableaux, etc. etc. Hence it is a worthwhile matter to study the general properties of these systems.

The original reason for studying them was to carry out four algorithms: listing, ranking, unranking, and random selection. It turns out that we can list the walks from v to τ in lex order in any such digraph. Hence we can list the objects in any such family. For example, we can make a list of all of the permutations of 19 letters that have 8 cycles, or of all of the partitions of [1..15] that have 8 classes, etc., by the same algorithm.

The ranking problem is this: given an object in a family of objects. Determine where it is on the list. More precisely, determine its position in the lex ordered list of all such objects. For instance, in the list of all 225 permutations of 6 letters that have 3 cycles, where is (14) (523) (6)?

Ranking problems can also be done by a general method, in the graph–theoretic setting of combinatorial families. The problem is the, given a walk from v to τ. Where is it in the lex ordered list of all walks from v to τ?

In order to present the new applications of the method, it will be helpful to review the ranking procedure. Suppose we are given a path from v to τ in a graphical combinatorial family, as shown in Fig. 1.

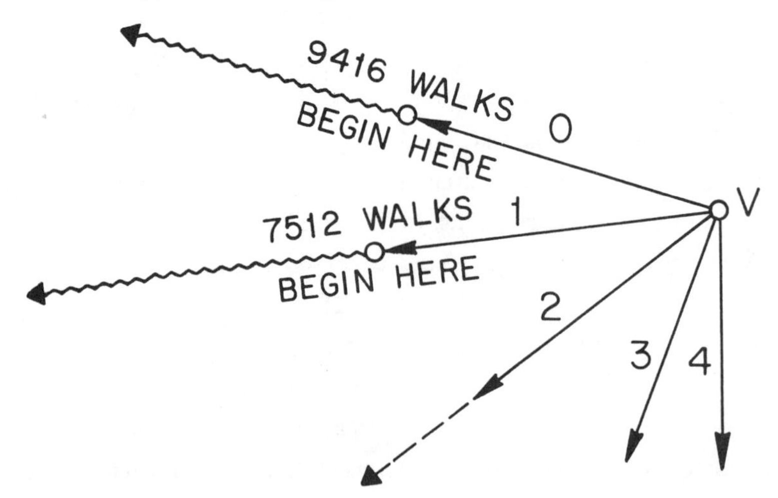

Fig. 1: This step costs $16,928.

Suppose the walk has arrived at v, and that it next uses edge number 2 outbound from v. It will then ignore all of the walks that would have used edges 0 or 1 next, instead. Furthermore, all of those walks *precede* the walk that uses edge 2 since the letters 0, 1 precede 2. Therefore, if a total of 9416 walks begin at the terminal vertex of edge 0, and if 7512 walks begin at the terminal vertex of edge 1, we will have jumped to a place in the alphabet that is below all 16, 928 of those walks. Hence, if edge 2 is in the walk that we are considering, its rank is augmented by 16, 928 on that single step.

Furthermore, the rank of a complete walk from v to τ is the sum of all of the numbers, like 16,928, that occur on every edge of the walk. We call 16,928 the *cost* of the edge. The costs of all the edges can be precomputed, without reference to some particular walk. The cost of an edge e, whose initial vertex is v and whose terminal vertex is w, is the sum of the numbers of walks from v' to τ, over all vertices v' such that $(v, v') \in E(G)$, and edge–number of edge (v, v') is $< e$.

If these costs have been precomputed, the *the rank of a walk is the sum of the costs of the edges in the walk.*

Corollary *In a graphical combinatorial family G, let v be a vertex, and suppose that there are exactly $b(v)$ walks from v to τ. Then every integer r, $0 \le r \le b(v) - 1$ is uniquely representable as the sum of the costs on the edges of one of these walks.*

Proof Every such integer is the rank of exactly one such walk. ❑

This observation contains as special cases a number of well known, and some new, unique representation theorems for integers.

Example 1. Fibonacci Numbers

Consider the graph G whose vertices are the nonnegative integers. There is an edge from m to $m - 1$, numbered 0, and from m to $m - 2$, numbered 1, for each $m \ge 2$, and edge (1, 0), numbered 0. The number of walks from m to 0 is F_m, and consequently *every integer r, $0 \le r \le F_m - 1$ is uniquely representable as* $F_{m_1} + F_{m_2} + \ldots$, *where all $m_{i+1} \le m_i - 2$.* ❑

Example 2. Partitions of Integers

Define the graph G as follows. Its vertices are the plane lattice points (n, k) with $n \ge k \ge 1$ and the terminal vertex $\tau = (0, 0)$. There is an edge from (n, k) to $(n - k, k)$, for $n \ge k \ge 1$, numbered 0, and from (n, k) to $(n - 1, k - 1)$, numbered 1. The number of walks from (n, k) to $(0, 0)$ is then $p(n, k)$, the number of partitions of n into parts $\le k$, because both $p(n, k)$ and the number of walks satisfy the same recurrence relation with the same initial conditions, viz.

$$p(n, k) = p(n-1, k-1) + p(n-k, k).$$

It follows that *every integer* r *such that* $0 \le r \le p(n, k) - 1$ *is uniquely expressible as*

$$r = p(n_0, k) + p(n_1, k-1) + \ldots$$

where, for all $j \ge 0$, $n_{j+1} = n_j - t_j(k-j) - 1$, $n_0 = n - (t+1)k$, *and* $t, t_0, t_1, \ldots$ *are nonnegative integers.* ❑

Example 3. Binomial Coefficients

Here the vertices of G are the lattice points (n, k) with $n \ge k \ge 0$. There is an edge from each (n, k), $n > k \ge 1$, to $(n-1, k)$, numbered 0, and an edge from (n, k) to $(n-1, k-1)$, numbered 1. If $n > 0$ and $k = 0$ there is one edge, numbered 0, from (n, k) to $(n-1, k)$. The terminal vertex is $(0, 0)$. The number of walks from (n, k) to τ is $\binom{n}{k}$.

We obtain at once the fact that *every integer* r, $0 \le r \le \binom{n}{k} - 1$, *is uniquely of the form*

$$r = \binom{n_1}{k} + \binom{n_2}{k-1} + \ldots + \binom{n_k}{1}$$

where $n > n_1 > \ldots \ge 0$. ❑

Example 4. Stirling Numbers

Let $S(n, k)$ be the number of partitions of $[n]$ into k classes (the Stirling numbers of the second kind.) The same method yields the result that *every integer* r, $0 \le r \le S(n, k) - 1$ *is uniquely expressible as*

$$r = j_1 S(n-1, k_1) + j_2 S(n-2, k_2) + \ldots$$

where for all i, $0 \le j_i \le k_i$, $2 \le k_i \le n - i$, *and*

$$k_{i+1} = \begin{cases} k_i & \text{if } j_i < k_i - 1 \\ k_i & \text{if } j_i = k_i, \end{cases}$$

and $k_1 = k$. ❑

Similar resluts hold for Stirling numbers of the first kind, q–binomial coefficients, hook–formula numbers, etc. All of these formulas simply express the fact that in a suitable lex ordered list, every element has a unique rank.

One other area of application of the method of combinatorial families deserves mention, because of interesting results that have been obtained by Bruce Sagan [4].

Let a sequence $\{c_j\}$ count certain sets $S_1, S_2, \ldots$ Suppose we want to prove that the sequence is log concave, i.e., that $c_{j+1}c_{j-1} < c_j^2$, for all j. One way to do that would be to exhibit an injection $S_{j+1} \times S_{j-1} \to S_j \times S_j$, for all j.

If the sets in question are representable as walks in a graphical combinatorial family, then we want an injection from a pair of walks with nearby initial vertices to a pair of walks with the same initial vertex. The geometric picture of walks on the plane lattice, which occurs in several important examples, opens the possibility of cutting and pasting pieces of the given pair of walks to obtain the final pair of walks, and yields injections that are natural to discuss in that context. Refer to [4] for the details, but here is one of the results that he obtained. It relates the concavity of the solution of a recurrence to the concavity of the coefficients of the recurrence.

Theorem *(Sagan [4]) Let* $\{t_{n,k}\}$ *satisfy a recurrence of the form*

$$t_{n,k} = c_{n,k}t_{n-1,k-1} + d_{n,k}t_{n-1,k} \quad (t, c, d, \geq 0)$$

where the $c_{n,k}$ *and the* $d_{n,k}$ *are themselves log concave in* k*, and further satisfy*

$$c_{n,k-1}d_{n,k+1} + c_{n,k+1}d_{n,k-1} \leq 2c_{n,k}d_{n,k}.$$

Then $\{t_{n,k}\}$ *is log concave in* k.

As a problem for future research, it would seem that the method has the capability of producing unimodality results, where log–concavity doesn't exist. This would require injections from some S_j to an S_{j+1}, that might easily be facilitated by cutting and pasting operations suggested by the lattice pictures.

REFERENCES

[1] Herbert S. Wilf, A unified setting for sequencing, ranking and random selection of combinatorial objects, Adv. Math. **24** (1977), 281 – 291.

[2] Herbert S. Wilf, —, II, Ann. Discr. Math. **2** (1978), 281 – 291.

[3] A. Nijenhuis and H.S. Wilf, Combinatorial Algorithms (2nd ed.), Academic Press, New York, 1978.

[4] Bruce Sagan, Inductive and injective proofs of log concavity results, Discr. Math. **68** (1988), 281 – 292.

Number Theory for Graphs

Robin J. Wilson
The Open University, England
and
The Colorado College

ABSTRACT

In this paper we show how the definitions of several standard number-theoretic functions, such as the Möbius function μ and the Euler phi-function ϕ, can be extended to connected graphs, using the operation of Cartesian product.

1. The Cartesian product of graphs

Let G_1 and G_2 be finite graphs with vertex–sets $V(G_1)$ and $V(G_2)$, respectively. The *Cartesian product* $G_1 \times G_2$ is the graph with vertex–set $V(G_1) \times V(G_2)$ in which the vertices (v_1, w_1) and (v_2, w_2) are joined by an edge if and only if

either v_1 is adjacent to v_2 in G_1, and $w_1 = w_2$

or $v_1 = v_2$, and w_1 is adjacent to w_2 in G_2.

For example:

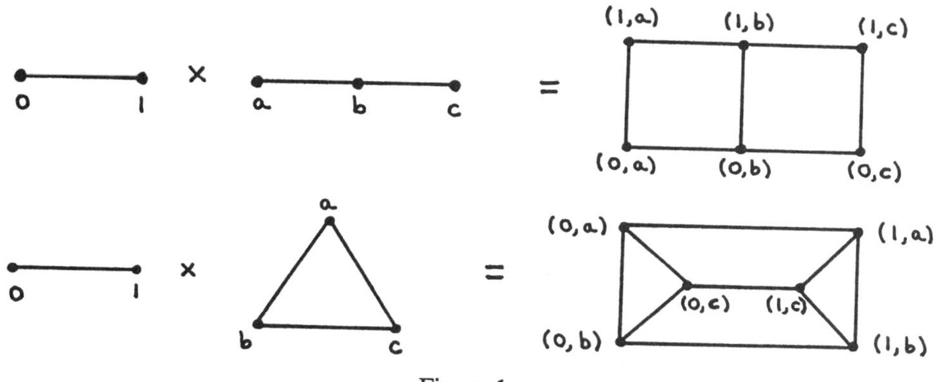

Figure 1

It is easy to check that the Cartesian product $\times$ is associative and commutative, and that the one–vertex graph K_1 is an identity element for $\times$.

2. Prime graphs

A connected graph is *prime* if it cannot be decomposed into non–trivial factors with respect to $\times$. (A factor is non–trivial if it is connected and has at least two vertices.) For example, the following graphs are all prime:

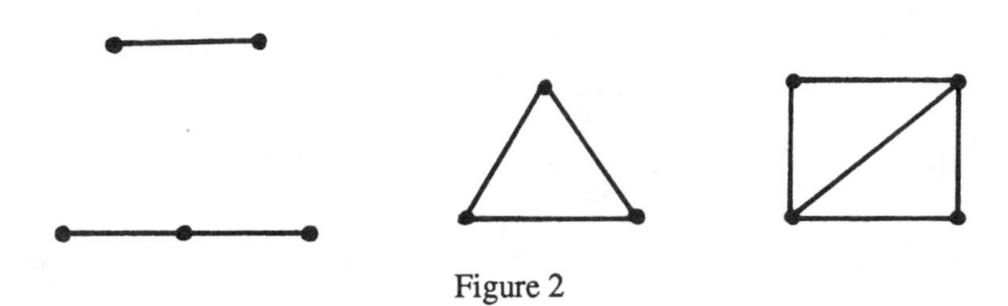

Figure 2

Note that, if G has a prime number of vertices, then G must be a prime graph, but the converse result is false. It is known (see, for example, Feigenbaum [2] or Winkler [10]) that almost all connected graphs are prime; in particular, Feigenbaum has proved that the proportion of graphs of order n that are non–trivially factorizable vanishes as 2^{-cn^2} as $n \to \infty$, where c is a positive constant.

The following table gives the numbers of connected graphs and prime graphs of order n, for $2 \leq n \leq 8$:

n	2	3	4	5	6	7	8
connected graphs	1	2	6	21	112	853	11117
prime graphs	1	2	5	21	110	853	11111

The main result on prime graphs is the following fundamental theorem, proved independently, and in various versions, by Sabidussi [8], Vizing [9], Miller [7], and Imrich [5], [6]:

Unique Factorization Theorem *Any connected graph G can be expressed uniquely as the Cartesian product of prime graphs:*

$$G = P_1^{k_1} \times P_2^{k_2} \times \ldots \times P_r^{k_r}.$$

Feigenbaum, Hershberger and Schäffer [3] and Winkler [10] have shown that there exists a polynomial algorithm for deciding whether a given connected graph G is a non–trivial Cartesian product, and for finding its prime decomposition if so. The corresponding statement for disconnected graphs is false; indeed, Feigenbaum [2] and Winkler [10] have shown that the factorizing problem for disconnected graphs is equivalent to the graph isomorphism testing problem.

3. Divisibility

If G and H are connected graphs, then H *divides* G (written $H \mid G$) if $G = H \times K$ for some graph K. The following results follow immediately from this definition; K_1 denotes, as usual, the one–vertex graph:

(i) for all connected graphs G, $G \mid G$ and $K_1 \mid G$;

(ii) if $H_1 \mid H_2$ and $H_2 \mid G$, then $H_1 \mid G$;

(iii) if $H_1 \mid G_1$ and $H_2 \mid G_2$, then $H_1 \times H_2 \mid G_1 \times G_2$.

If G and H are connected graphs, we define their greatest common divisor and least common multiple in the obvious way – namely:

$\gcd(G, H) = K$ if (i) $K \mid G$, $K \mid H$, and (ii) $L \mid G$, $L \mid H \Rightarrow L \mid K$;

$\text{lcm}(G, H) = L$ if (i) $G \mid L$, $H \mid L$, and (ii) $G \mid K$, $H \mid K \Rightarrow L \mid K$.

We can then derive such results as

$$\gcd(G, H) \times \text{lcm}(G, H) = G \times H.$$

If $\gcd(G, H) = K_1$, then G and H are *relatively prime*.

Just as for the set of positive integers, the set of connected graphs can be represented as a distributive lattice under divisibility, as shown below. It is the direct product of chains of the form $K_1 — P — P^2 — P^3 — \ldots$, where P is a prime graph.

Note that, if P is a prime graph, and if $P \mid G \times H$, then $P \mid G$ or $P \mid H$. Also, if $G = P_1^{k_1} \times \ldots \times P_r^{k_r}$, and if $P \mid G$, then $P = P_i$ for some $i = 1, \ldots, r$.

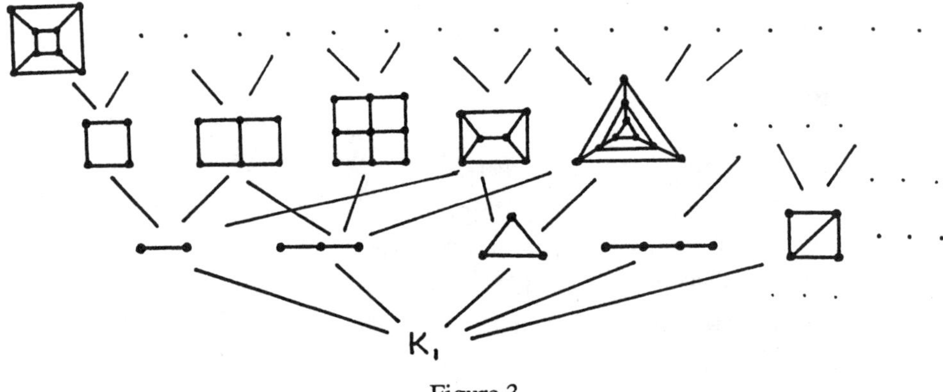

Figure 3

4. Multiplicative functions

An *arithmetical function* is a function f that associates with each connected graph G a real number f(G). Such a function f is *multiplicative* if $f(G \times H) = f(G)\, f(H)$, whenever $\gcd(G, H) = K_1$.

Examples of multiplicative functions are the following:

(1) *The Möbius function*

We define the *Möbius function* μ by

$$\mu(G) = \begin{cases} 1, \text{ if } G = K_1 \\ 0, \text{ if } P^2 \mid G, \text{ for some prime graph } P \\ (-1)^r, \text{ if } G = P_1 \times \ldots \times P_r \text{ (distinct primes).} \end{cases}$$

It is then a simple matter to derive the following basic properties of the Möbius function; we prove only (ii):

(i) μ *is multiplicative;*

(ii) $$\sum_{H \mid G} \mu(H) = \begin{cases} 1 \text{ if } G = K_1 \\ 0 \text{ if } G \neq K_1 . \end{cases}$$

Proof Since μ is multiplicative, it is sufficient to prove the result when G is a prime power, $G = P^k$. In this case,

$$\sum_{H|G} \mu(H) = \mu(1) + \mu(P) + \ldots + \mu(P^k)$$
$$= 1 - 1 + 0 + \ldots + 0$$
$$= 0, \text{ as required.}$$

(iii) *(The Möbius inversion formula) If f is an arithmetical function, and if*

$$F(G) = \sum_{H|G} f(H), \textit{ then } f(G) = \sum_{H|G} \mu(G/H)\, F(H),$$

where G/H is the graph K defined by $G = H \times K$.

(2) *The divisor function*

We define the *divisor function* $\tau(G)$ to be the number of graphs which divide G; thus, $\tau(G) = \sum_{H|G} 1$. It follows that τ is multiplicative, and that

$$\text{if } G = P_1^{k_1} \times \ldots \times P_r^{k_r}, \text{ then } \tau(G) = (k_1 + 1) \ldots (k_r + 1).$$

We can also use the multiplicativity to prove results such as the following:

(i) $\sum_{H|G} \mu(H)\, \tau(H) = (-1)^r$, and $\sum_{H|G} \mu(G/H)\, \tau(H) = 1$;

(ii) $\sum_{H|G} \tau(H)^3 = (\sum_{H|G} \tau(H))^2$.

Proof of (ii) Since τ is multiplicative, it is sufficient to prove the result when G is a prime power, $G = P^k$. In this case,

$$\sum_{H|G} \tau(H)^3 = \tau(1)^3 + \tau(P)^3 + \ldots + \tau(P^k)^3$$
$$= 1^3 + 2^3 + \ldots + (k+1)^3$$
$$= (1 + 2 + \ldots + (k+1))^2$$
$$= (\tau(1) + \tau(P) + \ldots + \tau(P^k))^2$$
$$= (\sum_{H|G} \tau(H))^2.$$

(3) *The Liouville function*

If $G = P_1^{k_1} \times \ldots \times P_r^{k_r}$, we define the *Liouville function* λ by $\lambda(G) = (-1)^{k_1 + \ldots + k_r}$. It follows that λ is multiplicative, and that

(i) $\sum_{H|G} \lambda(H) = \begin{cases} 1 & \text{if } G = K^2 \text{ for some graph } K \\ 0 & \text{if not.} \end{cases}$

(ii) $\sum_{H|G} \mu(H)\, \lambda(H) = 2^r$.

There are, however, no obvious analogues of the Euler ϕ–function and the sum–of–divisors function σ. In order to define these, we introduce the idea of a norm function.

5. Norm functions

A *norm function* is a function N that associates with each connected graph G a positive real number N(G), satisfying

$$N(K_1) = 1, \text{ and } N(G_1 \times G_2) = N(G_1)\, N(G_2) \text{ for all graphs } G_1, G_2.$$

It follows that $N(P_1^{k1} \times ... \times P_r^{kr}) = N(P_1)^{k1} ... N(P_r)^{kr}$, and hence that a norm function is specified completely by its effect on prime graphs. We normally assume that $N(G) > 1$ if $G \neq K_1$.

Examples of possible norm functions (some of which were suggested by M. Petkovsek and P. Winkler) are:

(i) N(G) = the number of vertices of G;

(ii) $N(G) = 2^d$, where d is the average vertex–degree in G;

(iii) $N(G) = 2^\Delta$, where Δ is the maximum vertex–degree in G;

(iv) list all the prime graphs in some order, and assign 2 to the first prime graph P_1, 3 to the second prime graph P_2, and so on; then define $N(G) = 2^{k1}\, 3^{k2} ...$ for $G = P_1^{k1} \times P_2^{k2} \times ...$.

Once we have found a norm function for graphs, we can easily define the Euler phi–function and the sum–of–divisors function as follows:

(1) *The Euler phi–function* ϕ

Rather than defining $\phi(G)$ to be the number of graphs H such that $N(H) < N(G)$ and $\gcd(G, H) = K_1$, generalizing the usual definition of the Euler phi–function $\phi(n)$, we choose to generalize the results

$$\sum_{d|n} \phi(d) = n \text{ and } \phi(n) = \sum_{d|n} \mu(n/d)\, d \text{ (obtained by Möbius inversion).}$$

We therefore define

$$\phi(G) = \sum_{H|G} \mu(G/H)\, N(H).$$

It follows that ϕ is multiplicative, and that

(i) $\sum_{H|G} \phi(H) = N(G)$;

(ii) $\phi(P^k) = N(P^k) - N(P^{k-1})$;

(iii) $\phi(P_1^{k1} \times \ldots \times P_r^{kr}) = N(G) \prod_i \{1 - (1/N(P_i))\}$.

(2) *The sum–of–divisors function* σ

We define the *sum–of–divisors function* $\sigma(G)$ by

$$\sigma(G) = \sum_{H|G} N(H).$$

It follows that σ is multiplicative, and that

$$\sigma(P_1^{k1} \times \ldots \times P_r^{kr}) = \prod_i \{(N(P_i)^{ki+1} - 1)/(N(P_i) - 1)\}.$$

6. Dirichlet series

If f is an arithmetical function, we define its formal *Dirichlet series* F by

$$F(s) = \sum_G f(G)\, N(G)^{-s}.$$

(We are not concerned here with questions of convergence.) For example, taking $f(G) = 1$ for all G gives the *Riemann zeta function*

$$\zeta(s) = \sum_G N(G)^{-s}.$$

When f is completely multiplicative, we also obtain the *Euler product formula*

$$F(s) = \prod_P (1 - f(P)\, N(P)^{-s})^{-1},$$

where the product extends over all prime graphs P; for example,

$$\zeta(s) = \prod_P (1 - N(P)^{-s})^{-1}.$$

The *product* of two Dirichlet series

$$F(s) = \sum_G f(G)\, N(G)^{-s} \text{ and } G(s) = \sum_G g(G)\, N(G)^{-s}$$

is the Dirichlet series H defined by

$$H(s) = \sum_G h(G)\, N(G)^{-s}, \text{ where } h(G) = \sum_{H|G} f(H)\, g(G/H).$$

We then have the following results:

(i) $1/\zeta(s) = \sum_G \mu(G)\, N(G)^{-s}$;

(ii) $\zeta(s-1)/\zeta(s) = \sum_G \phi(G)\, N(G)^{-s}$;

(iii) $\zeta^2(s) = \sum_G \tau(G)\, N(G)^{-s}$;

(iv) $\zeta(2s)/\zeta(s) = \sum_G \lambda(G)\, N(G)^{-s}$;

(v) $\zeta(s)\, \zeta(s-1) = \sum_G \sigma(G)\, N(G)^{-s}$.

The proofs of these results are elementary, and depend on the results of Sections 4 and 5; as an example, we prove (ii):

Proof of (ii) $\zeta(s) \sum_G \varphi(G)\, N(G)^{-s} = \sum_G \varphi(G)\, N(G)^{-s} \cdot \sum_G N(G)^{-s}$

$= \sum h(G)\, N(G)^{-s}$, where $h(G) = \sum_{H|G} \phi(H)$

$= \sum N(G)\, N(G)^{-s}$, by Section 5(1)(i)

$= \sum N(G)^{1-s} = \zeta(1-s)$.

7. Conclusion

In this paper we have shown how many of the standard definitions and results in number theory can be extended to connected graphs with a norm function defined on them. Using these ideas, we can obtain graph–theoretic analogues of many results in number theory, such as the prime number theorem. We hope to develop these ideas in future papers.

REFERENCES

All of the number–theoretic results in this paper can be found in [1] or [4].

[1] D.M. Burton, *Elementary Number Theory*, Allyn and Bacon, Boston, 1980.

[2] J. Feigenbaum, Product Graphs: Some Algorithmic and Combinatorial Results, Stanford University Technical Report STAN–CS–86–1121, Ph.D. Thesis, 1986.

[3] J. Feigenbaum, J. Hershberger and A. Schäffer, A polynomial time algorithm for finding the prime factors of Cartesian–product graphs, *Discrete Appl. Math.* 12 (1985), 123–130.

[4] G.H. Hardy and E.M. Wright, *An Introduction to the Theory of Numbers* (fifth edition), Oxford University Press, Oxford, 1979.

[5] W. Imrich, Über das schwache Kartesiche Produkt von Graphen, *J. Combinatorial Theory* 11 (1971), 1–16.

[6] W. Imrich, Embedding graphs into Cartesian products, to appear.

[7] D.J. Miller, Weak Cartesian product of graphs, *Colloq. Math.* 21 (1970), 55–74.

[8] G. Sabidussi, Graph multiplication, *Math Z.* 72 (1960), 446–457.

[9] V.G. Vizing, The Cartesian product of graphs (Russian), *Vyčisl. Sistemy* 9 (1963), 30–43; English translation in *Comp. Elec. Syst.* 2 (1966), 352–365.

[10] P. Winkler, Factoring a graph in polynomial time, *Europ. J. Combinatorics* 8 (1987), 209–212.

THE DESTRUCTIBILITY NUMBER OF A VERTEX

P. A. Winter

University of Natal

Durban, South Africa

ABSTRACT

A connected graph G *of order* $p = |V|$ *and size* $q = |E|$ *is said to be* $a_i, b_i)$*-destructible (with respect to* E_i *and* V_i *say) if* a_i, b_i *are integral factors of* p *and an* a_i*-set of edges* E_i *exists whose removal from* G *results in exactly* b_i *components isomorphic to* K_1*, i.e. whose removal from* G *isolates the vertices in a* b_i*-set* V_i*. Let* G *posses exactly* m *distinct* (a_i, b_i)*-destructions, each with respect to* E_i *and* V_i *and let* $\overline{V} = V \setminus \bigcup_{i=1}^{m} V_i$*. If* $v \in V$*, then the number of such distinct* (a_i, b_i)*-destructions with* $v \in V_i$ *is called the destructibility number of* v *and denoted by* $dest_G(v)$*. We define* $dest_G(v) = 0$ *if and only if* $v \in \overline{V}$*. The graph* G *is said to be destruction-regular of degree* k *if* $dest_G(v) = k$ *for each* $v \in V$*. In this paper we consider destruction-regular graphs of degree 0 and 1. We show that the only* **regular** *graphs which are destruction-regular of degree 1 are cycles of prime order and* K_2 *which is equivalent to a characterization of regular graphs reconstructible from their collection of uniquely destruction-labeled vertices.*

1. Introduction

The terminology and notation of [2] will be used throughout this paper. We will let G denote a simple, connected graph of order $p = |V|$ and size $q = |E|$. A cycle of length n will be denoted by C_n and, for $n \geq 2, P_n$ will denote a path of length n. The length of the longest path in G will be denoted by $\ell(G)$. Finally $\Delta(G) = max\{d_G(v)|v \in V\}, \delta(G) = min\{d_G(v)|v \in V\}$.

The end, or absence of a proof, will be denoted by $\square$.

2. Graphs which are *(a,b)*-destructible

The definition and subsequent investigation of the concept of (a, b)-destructibility of simple graphs was partially motivated by consideration of the reconstruction problem.

The Reconstruction Conjecture, first formulated by S.M. Ulam in collaboration with P. J. Kelly in 1941, may be stated as follows:

Every simple (finite) graph G on at least three vertices is determined, up to isomorphism, by its collection of vertex-deleted subgraphs, i.e. by the collection of all subgraphs $G_i = G - v_i, v_i \in V(G)$.

If G is uniquely determined by its collection of vertex-deleted subgraphs we say that G is *reconstructible* from this collection of subgraphs.

Advances made towards the solution of the problem and variations thereof have been extensively surveyed (cf. [1], [2], [4]). One variation involves the reconstruction of G from its n-vertex-deleted subgraphs, $n \geq 2$. Another variation involves the reconstruction of G from its m-edge- deleted subgraphs).

If we remove edges *and* vertices from K_2 we obtain $\emptyset$ (which is not a graph) and since K_2 is the only (simple) graph on one edge and two vertices, K_2 is uniquely reconstructible from $\emptyset$ together with the labels 1(edge) and 2(vertices) (i.e. $\emptyset$ labeled 1,2). Similarly K_3 is uniquely determined by $\emptyset$ labeled 3,3. The graph $K_n + \{u\} + \{uv\}(v \in V(K_n))$, is uniquely reconstructible from K_n labeled 1,1. The graph $K_4 - e(e \in E(K_4))$ is uniquely reconstructible from K_2 labeled 4,2 and K_3 labeled 2.1. The labels 1,2; 3,3; 1,1; 4,2; 2,1 involve factors of the order of each graph involved so that in [6] a further variation of the problem was introduced by considering the reconstruction of a graph G of order $p \geq 2$ from the subgraphs obtained by deleting a edges and b vertices where we imposed upon a, b the condition that they be factors of p - thus also providing a necessary bound on the number of edges and vertices removed which enabled us to make advances towards the solution of Ulam's problem. This concept generalizes the deletion of an end vertex and gave rise to a structural characterization of a particular class of prime-order graphs (cf. [7]), (m,n)- matroids (cf. [5]) and a new vertex chromatic number (cf. [8]). In this paper we consider a reconstruction problem involving labeled vertices.

Formally, if a, b are integral factors of p, G is defined to be (a, b)-*destructible* (with respect to E' and V') if an a-set E' of edges exists such that $G - E'$ contains exactly b isolated vertices (vertices incident with no edge of G) in a b-set V'. We denote $G - E' - V'$ (to $\oslash$) by $D_{a,b}(G)$ if $b \neq p$ (or $b = p$, respectively). The operation $D_{a,b}$ is called an *(a,b)-destruction* of G. The graph G is *stable* if it is no *(a,b)*- destructible for any a, b which divide p (cf. [6], [7]).

We say that the graph G is *immediately annihilable* if it is (q, p)-destructible. Clearly the only immediately annihilable graphs are those simple, connected graphs obtained by the insertion of a single edge into a tree of order $p \geq 3$ (graphs of *nullity one*), and K_2.

3. A Reconstruction Problem

Since each vertex v of a graph G can either be isolated by some (a, b)-destruction (and then removed) or not isolated by a destruction at all, we can associate each vertex with either the label a,b or the number 0, respectively. By assigning either this label a,b or 0 or v we say that v has been *non-zero destruction -labeled* or *zero destruction -labeled*, respectively. A vertex v is non-zero destruction -labeled for each distinct destruction which isolates v. Note that a vertex can have many (not necessarily different) non-zero destruction-labels associated with it and not vertex and have both a zero and a non-zero destruction-label. We now have the following reconstruction problem:

> Determine those graphs which are reconstructible from their collection of destruction-labeled vertices.

In this paper we shall determine those regular graphs which are reconstructible from their collection of uniquely non-zero destruction-labeled vertices. We formally discuss the above concept of destruction- labeled vertices in the next section.

4. The Destructibility Number of a Vertex

Let G possess exactly m distinct (a_i, b_i)-destructions each with respect to E_i and V_i and let $\overline{V} = V \backslash \cup_{i=1}^{m} V_i$. For a vertex v of G, the number of such distinct (a_i, b_i)-destructions with $v \in V_i$ is called the ***destructibility number*** of v and is denoted by $dest_G(v)$ (this number is the number of non-zero destruction-labels associated with v). We define $dest_G(v) = 0$ if and only if $v \in \overline{V}$. The graph G is said to be ***destruction-regular*** of degree k if $dest(v) = k$ for each $v \in V$. We say that G is destruction-regular if G is destruction-regular of degree k for some k i.e., for $k \neq 0$ each vertex of G has the same number of non- zero destruction-labels associated with it and for $k = 0$ each vertex has a zero destruction-label.

The following results are easily verified:

(1) A graph G is destruction-regular of degree 0 if and only if G is stable.

(2) The graph K_2 is destruction-regular of degree 1.

(3) All cycles are destruction-regular.

(4) Cycles of prime order are destruction-regular of degree 1.

(5) Complete graph K_n are destruction-regular of degree $\frac{(n-1)(n-2)}{2}$ for $n \geq 3$.

Conjecture 4.1 The only graphs which are destruction-regular of degree 1 are cycles of prime order and K_2.

or

Conjecture 4.2 The only graphs which are reconstructible from their collection of uniquely non-zero destruction labeled vertices are cycles of prime order and K_2.

We shall provide a partial solution to these conjectures.

Lemma 4.1 No graph on $p \geq 3$ vertices containing an end vertex can be destruction-regular.

Proof Let G have an end vertex v adjacent to a vertex w. Clearly $d(w) \geq 2$. Since every (a_i, b_i)-destruction of G which isolates w isolates v as well, we have that $dest(v) \geq dest(w)$. However v can be isolated by a (1,1)-destruction which does not isolate w so that $dest(v) > dest(w)$. Hence G cannot be destruction-regular.

Corollary 4.1 No tree on $p \geq 3$ vertices can be destruction-regular.

Lemma 4.2 The only graphs of nullity one which are destruction-regular of degree 1 are cycles of prime order.

Proof Let G be a graph of nullity one which is destruction-regular of degree 1. Clearly G must be a cycle since by lemma 4.1 we have that G cannot have an end vertex. If G is a cycle of non-prime order than $dest(v) \geq 2$ for each $v \in V$ since G is immediately annihilable and from [3] each vertex of G is $(a, 1)$-destructible for some a which divides p. Thus G must be a cycle of prime order.

Lemma 4.3 Let G be a graph on $p \geq 4$ vertices which possesses an $(a, 1)$-destruction $D_{(a,1)}$ associated with E' and $V' = \{v\}$ satisfying $d(v) < a$ and let F be the set of edges covered by v. If $dest(v) = 1$, then $D_{(a,1)}(G)$ is the disjoint union of stars $S_i, i = 1, 2..., h; h \geq 2$, such that

(i) If $p(S_j) > 2$, no edge of $E' \backslash F$ is incident with an end vertex of S_j, and

(ii) The graph G cannot contain a cycle whose edges are alternately edges are alternately edges from $E' \backslash F$ and the S_j.

Proof Let G be as above and let $H = D_{(a,1)}(G) = G - E' - v$ with $dest(v) = 1$. Note that $d(v) \geq 2$, for if $d(v) = 1$ then G will possess two distinct destructions which isolate v. Let F be the set of edges incident with v in G.

Suppose that H contains a path $v_1, v_2, \cdots, v_n, n \geq 4$. Then G is $(a,1)$- destructible with respect to E'' and V' where $E'' = (E' \backslash \{e\}) \cup \{v_2v_3\}$, where $e \in (E' \backslash F) \neq \oslash$. This means that $dest(v) \geq 2$, which is a contradiction. Thus $\ell(H) \leq 2$ so that H is the disjoint union of star graphs $S_i, i, 2, ..., h$.

(i) Let $p(S_j) > 2$ from some j. Suppose there is an edge $e \in E' \backslash F$ incident with an end vertex, w say, of S_j, and let z be the center of S_j. Then G is $(a,1)$-destructible with respect to $(E' \backslash \{e\}) \cup \{wz\}$ and V', a contradiction.

(ii) Suppose G contains a cycle C_n whose edges are alternately in $E' \backslash F$ and the S_j. Then n must be even, say $n = 2t$. Let $E(C_{2t}) = \{f_1, f_2, \cdots, f_t, e_1, e_2, \cdots e_t\}$, where $e_i \in E' \backslash F$ and $f_i \in H$. Then G is $(a,1)$-destructible with respect to $E' \backslash \{e_1, \cdots, e_t\}) \cup \{f_1, \cdots, f_t\}$, and V', which is a contradiction.

Let $h = 1$. Then $p(S_1) > 2$, for if $p(S_1) = 2$ we have that $p(G) = 3$, a contradiction. Thus we must have at least one edge of $E' \backslash F$ incident with an end vertex of S_1 which is impossible by case (i) above. Thus $h \geq 2$.

5. Regular Graphs Which Are Destruction-Regular of Degree 1

From [3] we know that every graph on a non-prime number p of vertices is $(a,1)$-destructible for some a which divides p. Thus cycles of non-prime order are destruction-regular of degree k where $k \geq 2$ and cycles of prime order are destruction-regular of degree 1. Complete graphs of order $p \geq 4$ cannot be destruction-regular of degree 1. We shall show that the only regular graph of even order which is destruction-regular of degree 1 is K_2.

Lemma 5.1 Let G be a regular graph of even order $p \geq 4$. If G has an $(a,1)$-destruction $D_{a,1}$ associated with E' and $V' = \{v\}$ with $d(v) = a$, then $dest(v) \geq 2$.

Proof Let G be a regular graph of even order $p \geq 4$ and which possess an $(a,1)$-destruction $D_{a,1}$ with respect to E' and $V' = \{v\}$.

Let $D_{a,1}(G) = H_1$ and suppose that $d(v) = a$. Clearly $a < p(G)$, so that, because a is a factor of $p, a \leq p/2$. If $a = 2$ then G is a cycle and $des(v) \geq 2$. Thus $a \geq 3$.

Case (A). Let $a = p/2$. There exists at least two nonadjacent vertices v, w of G such that $G - v - w$ contains no isolated vertices. Thus G is also $(2a, 2)$-destructible and $dest(v) \geq 2$.

Case (B). suppose $a < p/2$. Let $s = a + t$ be the next divisor of p after a.

The removal of an edge e_1 from $H_1 = G - v$ does not isolate any vertices of H_1 so that if $(a + 1)$ is a factor of p then $dest(v) \geq 2$.

We shall show that there exists a set of t edges of H_1 whose removal from H_1 does not isolate any vertices of H_1 and hence of G.

If $p = 4$, then $a < 2$ which is impossible. If $p = 6$, then $a \leq 2$ which is a contradiction. Clearly the cases $p = 8$ and $p = 10$ cannot occur. Thus $p \geq 12$. Let us remove an edge e_1 from H_1 to obtain H_2. Let

$$A_2 = \{w | d_{H2}(w) = 1\},$$
$$B_2 = \{w | d_{H2}(w, A_2) \leq 1\}.$$

The set B_2 consists of vertices of H_2 which are either end vertices of H_2 or adjacent to end vertices of H_2. Let $\overline{B_2} = V(_2) \backslash B_2$. Since $|A_2| \leq 2$ and $|B_2| \leq 4, \overline{B}_2 = p - 1 - |B_2| \geq p - 1 - 4 \geq 7$. Thus there exists a non-end-vertex y_2 of H_2 adjacent to a vertex x_2 of H_2 such that $x_2 \notin A_2$. If we let $e_2 = x_2 y_2$, then the removal of e_2 from H_2 does not isolate any vertices of H_2. Form $H_i, A_i, B_i, \overline{B_i}$ and $e_i = x_i y_i$. If $i < t$, form $H_{i+1} = H_i - e_i, A_{i+1}$ etc. and stop when $i + 1 = t$. Clearly $|A_i| \leq (i-1)2, |B_i| \leq (i-1)4$ so that

$$\begin{aligned} |\overline{B_t}| &\geq p - 1 - 4(t-1) \\ &= p - 1 - 4t + 4 \\ &= p - 4t + 3. \end{aligned}$$

If $p \geq \frac{4a-2}{3}$, then, since $a + t \leq p/2$ implies that $-t \geq \frac{2a-p}{2}$ we have that

$$|\overline{B_t}| \geq p - 4(\frac{2a-p}{2}) + 3$$
$$= 3p - 4a + 3 \geq 1.$$

If $p < \frac{4a-2}{3}$ then we must have $a + t \leq 2a$ otherwise

$$4a \leq p \leq \frac{4a-2}{3}$$

which is impossible. Hence $t \leq a$ so that $p \geq 4t$. It follows that $|\overline{B_t}| \geq p - 4t + 3 \geq 3$.

Thus there exists a t-set of edges $E_t = \{e_1, e_2, \cdots, e_t\}$ whose removal from H_1 does not isolate any vertices of H_1 so that G is $(a+t, 1)$-destructible where $t \geq 1$. Hence $dest(v) \geq 2$.

Lemma 5.2 Let G be a regular graph of even order which possesses an $(a, 1)$-destruction with respect to E' and $V' = \{v\}$ such that $d(v) < a$. Then $dest(v) \geq 2$.

Proof Let G be a regular graph of even order $p \geq 4$ as stated above with $dest(v) = 1$. Then by lemma 4.3 we have that $H = G - E' - v$ is the disjoint union of star graphs. Let F be the set of edges of G covered by the vertex v. Then there exists at least one edge e of $E' \backslash F$ whose one end u is adjacent to an end vertex w of H where uw is an edge of H. The graph G is now $(a, 2)$-destructible with respect to $(E' \backslash \{e\}) \cup \{uw\}$ and $\{v, w\}$ so that $dest(v) \geq 2$.

Corollary 5.1 The only regular graph of even order which is destruction-regular of degree 1 is K_2.

We shall now consider regular graphs of odd order.

Lemma 5.3 Let G be a regular graph of odd order. Then G cannot possess an $(a, 1)$-destruction with respect to E' and $V' = \{v\}$ such that $d(v) = a$.

Proof Let G be a regular graph of odd order which possess an $(a, 1)$-destruction with respect to E' and $\{v\}$ such that $d(v) = a$. Then, since

a must be a factor of p, a must be odd. But this means that G contains an odd number of vertices of odd degree which is a contradiction.

Lemma 5.4 Let G be a regular graph of odd order which possesses an $(a, 1)$- destruction with respect to E' and $V' = \{v\}$ such that $d(v) < a$. Then $dest(v) \geq 2$.

Proof Let G be a regular graph defined above and suppose $dest(v) = 1$. Let F be the set of edges covered by v. From lemma 4.3 we have that $H - G - E' - v$ must be the union of disjoint stars $S_1, S_2, \cdots, S_h$ where $h \geq 2$. If one of the star graphs has two edges in H then the end vertices of this star graph must be adjacent to v in G and no edge of $E' \backslash F$ can be incident in G with any of these end vertices so that G is a cycle which is a contradiction. Thus G is the disjoint union of K_2 components. If we let $H' = G - v$, then $\delta(H') \geq 2$ since $\delta(G) \geq 3$. Let Q be any component of H'. Every vertex of Q must be an end of exactly one K_2 component of H so that Q, and hence G, must contain an even cycle whose edges are alternately in $E' \backslash F$ and the K_2 components of H which contradicts lemma 4.3. Hence $dest(v) \geq 2$.

Corollary 5.2 The only regular graphs of prime order $p \geq 3$ which are destruction-regular of degree 1 are cycles of prime order.

Proof Let G be a regular graph of prime order $p \geq 3$. If G is a cycle then G is destruction-regular of degree 1. Let $\delta(G) \geq 3$. Then G is either stable, in which case G is destruction-regular of degree 0, or G can only be $(p, 1)$-destructible with respect to E' and $\{v\}$ say, in which case $dest(v) \geq 2$ by the above lemma.

Theorem 5.1 The only regular graphs which are destruction-regular of degree 1 are cycles of prime order and K_2, i.e. the only regular graphs which are reconstructible from their collection of uniquely non-zero destruction-labeled vertices are cycles of prime order and K_2.

[1] Capobianco, M. and Moluzzo, J. C. *Examples and Counterexamples in graph theory*, Elsevier North-Holland, New York (1978).

[2] Chartrand, G. and Lesniak-Foster L., *Graphs and Digraphs*, Wadsworth, Monterey (1986).

[3] Goddard W. and Winter P. A. *All graphs on a non-prime number of vertices are destructible*, Quaestiones Math. 8 (1986), 381-385.

[4] Harary, F., *On the reconstruction of a graph from a collection of subgraphs*, Theory of Graphs and its Applications (Proceedings of the Symposium held in Prague, 1964), edited by M. Fiedler, Czechoslovak Academy of Sciences, Prague, 1964, 47-52, reprinted, Academic Press. New York.

[5] Swart, Henda C. and Winter, P. A. *The reconstruction of graphs from (m,n)- matroids*, Journal of Combinatorics, Information and System Sciences, 11 (1986), 115-123.

[6] Winter P. A. and Swart, Henda C. *On (α,β)-destructible graphs*, Quaestiones Math. 7 (1984), 161-178.

[7] Winter, P. A. and Swart, Henda C. *On stable graphs*, Quaestiones Math. 7 (1984), 397-405.

[8] Winter, P. A. *On the (a_i, b_i)-chromatic number of a graph*, ARS Combinatoria, Vol. 23A (1987), 337-348.